D1065352

CORROSION INHIBITORS

CORROSION INHIBITORS

Developments Since 1980

Edited by M.J. Collie

NOYES DATA CORPORATION
Park Ridge, New Jersey, U.S.A.
1983

Library of Congress Catalog Card Number: 83-13055
ISBN: 0-8155-0957-X
ISSN: 0198-6880
Printed in the United States

Published in the United States of America by
Noyes Data Corporation
Mill Road, Park Ridge, New Jersey 07656

10 9 8 7 6 5 4 3 2 1

Library of Congress Cataloging in Publication Data
Main entry under title:

Corrosion inhibitors.

Includes indexes.
1. Corrosion and anti-corrosives--Patents.
I. Collie, M. J.
TA418.74.C66 1983 620.1'1223'0272 83-13055
ISBN 0-8155-0957-X

Foreword

The detailed, descriptive information in this book is based on U.S. patents, issued between January 1981 and March 1983, that deal with corrosion inhibitors. This title contains new developments since our previous title *Corrosion Inhibitors—Recent Developments* published in 1979.

This book is a data-based publication, providing information retrieved and made available from the U.S. patent literature. It thus serves a double purpose in that it supplies detailed technical information and can be used as a guide to the patent literature in this field. By indicating all the information that is significant, and eliminating legal jargon and juristic phraseology, this book presents an advanced commercially oriented review of recent developments in the field of corrosion inhibitors.

The U.S. patent literature is the largest and most comprehensive collection of technical information in the world. There is more practical, commercial, timely process information assembled here than is available from any other source. The technical information obtained from a patent is extremely reliable and comprehensive; sufficient information must be included to avoid rejection for "insufficient disclosure." These patents include practically all of those issued on the subject in the United States during the period under review; there has been no bias in the selection of patents for inclusion.

The patent literature covers a substantial amount of information not available in the journal literature. The patent literature is a prime source of basic commercially useful information. This information is overlooked by those who rely primarily on the periodical journal literature. It is realized that there is a lag between a patent application on a new process development and the granting of a patent, but it is felt that this may roughly parallel or even anticipate the lag in putting that development into commercial practice.

Many of these patents are being utilized commercially. Whether used or not, they offer opportunities for technological transfer. Also, a major purpose of this book is to describe the number of technical possibilities available, which may open up profitable areas of research and development. The information contained in this book will allow you to establish a sound background before launching into research in this field.

Advanced composition and production methods developed by Noyes Data are employed to bring these durably bound books to you in a minimum of time. Special techniques are used to close the gap between "manuscript" and "completed book." Industrial technology is progressing so rapidly that time-honored, conventional typesetting, binding and shipping methods are no longer suitable. We have bypassed the delays in the conventional book publishing cycle and provide the user with an effective and convenient means of reviewing up-to-date information in depth.

The table of contents is organized in such a way as to serve as a subject index. Other indexes by company, inventor and patent number help in providing easy access to the information contained in this book.

16 Reasons Why the U.S. Patent Office Literature Is Important to You

1. The U.S. patent literature is the largest and most comprehensive collection of technical information in the world. There is more practical commercial process information assembled here than is available from any other source. Most important technological advances are described in the patent literature.

2. The technical information obtained from the patent literature is extremely comprehensive; sufficient information must be included to avoid rejection for "insufficient disclosure."

3. The patent literature is a prime source of basic commercially utilizable information. This information is overlooked by those who rely primarily on the periodical journal literature.

4. An important feature of the patent literature is that it can serve to avoid duplication of research and development.

5. Patents, unlike periodical literature, are bound by definition to contain new information, data and ideas.

6. It can serve as a source of new ideas in a different but related field, and may be outside the patent protection offered the original invention.

7. Since claims are narrowly defined, much valuable information is included that may be outside the legal protection afforded by the claims.

8. Patents discuss the difficulties associated with previous research, development or production techniques, and offer a specific method of overcoming problems. This gives clues to current process information that has not been published in periodicals or books.

9. Can aid in process design by providing a selection of alternate techniques. A powerful research and engineering tool.

10. Obtain licenses—many U.S. chemical patents have not been developed commercially.

11. Patents provide an excellent starting point for the next investigator.

12. Frequently, innovations derived from research are first disclosed in the patent literature, prior to coverage in the periodical literature.

13. Patents offer a most valuable method of keeping abreast of latest technologies, serving an individual's own "current awareness" program.

14. Identifying potential new competitors.

15. It is a creative source of ideas for those with imagination.

16. Scrutiny of the patent literature has important profit-making potential.

Contents and Subject Index

Introduction

There are few industries that do not make use of corrosion inhibitors in some phase of their operations. The organization of this book, based on the recent patent literature, is mainly according to the uses for which the corrosion inhibitors described are intended.

The largest number of patents give information on compositions for the protection of various structural materials and equipment used in the production of chemicals, foods and cosmetics. Also described are materials intended for use as the equipment itself, for example, one for parts such as impellers in ore treatment equipment and another for parts used in hydrogen fluoride laser systems.

Another large group of patents describes inhibitors useful in various aqueous systems. Included are inhibitors for automobile radiator coolants as well as methods of dispensing such inhibitors. Compositions to prevent corrosion in industrial cooling water systems include one that also prevents scale formation. Oxygen scavengers for boiler systems are described. Many patents in this chapter supply inhibitors for more than one type of aqueous system. In addition to the systems mentioned above, systems addressed include reverse osmosis, metal pickling and shaping and hydraulic systems. Corrosion inhibitors to protect equipment used in the secondary recovery of petroleum by water flooding and in the disposal of wastewater and brine from oil and gas wells are also in the group.

Compositions intended for use in both aqueous and nonaqueous systems are next described. These include metal treating baths, aqueous and hydrocarbon systems of the oil industry, motor oils, heat transfer media, hydraulic fluids and greases. Because of the wide range of uses, classification of these inhibitors has been according to chemical composition.

Corrosion inhibiting coatings and films for ferrous and nonferrous metals are described in another large group of patents. Zinc-rich coatings, methods of preparing anticorrosive phosphate pigments, and low temperature curable coatings are provided. Also included are rust inhibiting pigments that are less toxic than lead- and chromium-containing pigments and coating compositions that require no rinsing, thus lowering the volume of water that must be treated for pollution control. Special applications include can coatings, one-step cleaner-primers, and high temperature corrosion protection for superalloys.

Corrosion inhibiting additives for lubricants, fuels and hydraulic fluids have been grouped together. Among the lubricant additives there may be mentioned those for the protection of silver plate areas of railway diesel engines, antioxidant and anticorrosive additives for diesel crankcase lubricants and for perfluorinated fluids. Some of the fuel additives have activity as antioxidants, detergents and antiicing agents as well as corrosion inhibitors. Fuels to which the compositions are to be added include gasoline, jet fuels, fuel oil and gasohol.

In the natural gas and oil industry applications there are described inhibitors for use in well drilling operations such as well packer solutions and aqueous drilling fluids. Two patents provide methods for preventing corrosion problems in deep gas wells where high pressure, high temperature bottomhole conditions prevail. There are compositions that prevent evolution of hydrogen sulfide while removing sulfide-containing scale from crude oil and natural gas refinery equipment. Other natural gas and oil applications will be found in the chapters on aqueous and aqueous/nonaqueous systems.

The last chapter pertains to corrosion inhibition in metal treating baths. These include surface scale and rust removers, metal working solutions, prevention of rust and discoloration in the burnishing of steel and prevention of etching in aluminum polishing solutions. Compositions for removing phenolic resins from aluminum, alloys that keep corrosion of the metal to a minimum, and one for selective stripping of gold-nickel brazing alloys used in jet engines without damage to the substrate are provided. Again, the reader is referred to the chapters on aqueous and aqueous/nonaqueous systems for other compositions useful in metal treating baths.

–1–

Structural and Manufacturing Applications

STRUCTURAL MATERIALS

Volatile Complexing Agent for Latex Paints

A method of modifying latex paints for maintenance paint applications, primarily on iron or steel industrial and transportation structures, in order to provide both short term and long term corrosion resistance, and the resulting composition, are described by *M.J. Grourke and R.W. Flynn; U.S. Patents 4,243,416; Jan. 6, 1981; and 4,243,417; January 6, 1981; both assigned to Rohm and Haas Co.* Two types of corrosion encountered in painting ferrous surfaces with latex paint are:

(1) Flash rusting which occurs as a result of contact with the water from the latex and results in rust bleed-through at the time the paint is drying, and

(2) Rust bleeding, which may occur over a period of months or years.

It has been found that the benefits of both water-soluble corrosion inhibitors, for preventing flash rusting, and water-insoluble corrosion inhibitors, for minimizing blistering and rust bleeding, are achieved by utilizing a soluble complex of a polyvalent metal, a volatile complexing agent, and a corrosion inhibiting anion. This complex may be represented by $M(Z)_xAn$ where M is the metal cation. An is the corrosion inhibiting anion, Z is the complexing volatile component, and x is the number of mols of volatile complexing agent per mol of M, being a coordination complex, x typically ranges from 2 to 6, depending upon the quantity of volatile complexing agent present, all as is well known in the art of coordination chemistry. Examples are zinc ammonium carbonate and zinc ammonium molybdate.

The mode of addition of the complex to the paint or to one of the components thereof, such as the pigmented dispersion or the latex, can be accomplished in various ways. The polymer particles of the latex are insoluble in water and in dilute aqueous acid or alkaline solutions having a pH of from about 3 to 10, and the complex can be added to the latex and dissolved therein either as a powder or as a preformed solution. Where ammonia is the volatile complexing agent, the dispersion is generally made alkaline with ammonia and usually with a pH of from about 7.5 to 9.5. While preformed complexes such as zinc ammonium carbonate and zinc ammonium molybdate can be added to the latex, they may be formed in situ. Thus, zinc oxide, ammonium carbonate, and ammonia have been added to the dispersion to form the complex zinc ammonium carbonate in situ. Similarly, ammonium molybdate, zinc oxide, and ammonia are added to form zinc ammonium molybdate in situ in the latex. Of course, where zinc oxide or the like is included as a paint pigment, all that need be added is a soluble compound of the corrosion inhibiting anion and the volatile complexing agent such as ammonia.

Preferred metals are zinc, cadmium, and zirconium. The preferred volatile complexing agents are ammonia or the volatile amines such as methyl amine, ethyl amine, dimethyl amine, diethyl amine, triethyl amine, morpholine, ethanol amine, diethanol amine, triethanol amine, etc. The complexing agent is as volatile or more volatile than water, the rate of evaporation preferably being from about the same as that of water to about twice as fast as water. Preferred anions are carbonate and molybdate.

The quantity of water-soluble of the polyvalent metal, the volatile complexing agent, and the corrosion inhibiting anion depends primarily upon the quantity of water in the latex paint. Particularly as regards flash rust inhibition, the more water that is present (and thus the longer it takes to evaporate), the larger the quantity of soluble complex is needed, assuming constant ambient conditions of temperature, wind, humidity and the like. A suitable range of soluble complex is from 0.2 to 20 mmol of complex per mol of water.

In the examples, the following paint was used:

Material	Pounds	Gallons
Water	28.0	3.36
Dispersing agent*	9.9	1.08
Wetting agent**	2.2	0.25
Defoamer	2.2	0.30
Ethylene glycol	22.0	2.37
Hydroxyethylcellulose (2% aqueous solution)	82.0	9.88
Germantown lampblack	6.0	0.41
Preservative (100%)	0.5	0.03
Rutile titanium dioxide	209.4	5.98
Water-ground mica (325 mesh)	26.0	1.11
Precipitated calcium carbonate	125.4	5.68
Zinc oxide	6.1	0.13
Basic lead silico chromate (Oncor M-50)	80.0	2.34
Acrylic latex*** (46% solids)	600.9	67.52
Coalescent	5.0	0.62
Defoamer	2.2	0.30
Ammonium hydroxide (28%)	1.0	0.13
Total	1,208.8	101.49

PVC† = 35.2%.
Volume solids = 43.7% in paint

*Sodium salt of 1:1 mol ratio diisobutylene/maleic anhydride copolymer, 25% solution.
**Benzyl ether of octyl phenol ethylene oxide adduct, 100% active.
***Copolymer containing about two-thirds ethyl acrylate, one-third methyl methacrylate, and about 1% methacrylic acid.
†Pigment volume concentration.

In the following examples, data are presented which were obtained by using the test methods listed below:

(1) Flash rust resistance

(a) Substrate–rusty cold rolled steel, mechanically wire brushed.

(b) Test paints are brush applied, 3 g/48 in^2

(c) Test paint on the panel is dried at 72°F, 90% RH for 1 hr.

(d) Panels are rated for percent rust bleeding.

(2) Maintenance testing (long term)

(a) Substrates–clean cold rolled steel, rusty cold rolled steel, mechanically wire brushed.

(b) Procedure–two brush applied coats (each 3 g/48 in^2) over clean cold rolled steel, three brush applied coats (first coat 4 g/48 in^2, second and third 3 g), over wire brushed rusty cold rolled steel. One hour air dry between coats. Seven day air dry before exposure to 5% salt spray.

(c) Rating system–panels are rated for blistering-rust bleeding (for example, 7 MD-50).

The blistering number indicates the blister size. The range is from 10 to 2 with 10 indicating no blisters. The letter indicates blister density: F-few, M-medium, MD-medium dense, and D-dense (e.g., 7 MD in above example). The rust bleeding number indicates the percent of rust bleeding (e.g., 50 in the above example).

Example 1: This demonstrates that $Zn(NH_3)_xCO_3$ gives flash rust resistance with improved maintenance properties relative to a soluble inhibitory salt, here, x is typically 2 to 4, depending on the quantity of ammonia.

(1) Flash Rust Resistance.

	Pounds of Added Salt* per 100 gal	Percent Flash Rusting**
Paint formulation	None	75
Plus zinc ammonium carbonate in complex form	18.2	2
plus sodium carbonate	15.5	10

*Salts were added on an equivalence basis, 18.2 lb zinc ammonium carbonate = 15.5 lb Na_2CO_3.

**Low numbers best.

(2) Maintenance Testing.

	Pounds of Added Salt*** per 100 gal	Blistering*-Rust Bleeding** (1 wk exposure 5% NaCl spray)	
		Clean Cold Rolled Steel	Rusty Cold Rolled Steel
Paint formulation	None	10-2 to 10	9 M–8 MD
plus zinc ammonium carbonate complex	18.2	10–2	9 M–40

(continued)

(2) Maintenance Testing.

	Pounds of Added Salt*** per 100 gal	Blistering*-Rust Bleeding** (1 wk exposure 5% NaCl spray) Clean Cold Rolled Steel	Rusty Cold Rolled Steel
plus sodium carbonate	15.5	6 M-20	8 M-90

*9 F to 10 best.
**Low numbers best.
***Salts were added on an equivalence basis, 18.2 lb zinc ammonium carbonate = 15.5 lb Na_2CO_3.

Example 2: This example shows the use of an ammonium salt in combination with ZnO in the paint to form $Zn(NH_3)_xCO_3$ in situ. An ammonium salt capable of forming the complex versus a sodium salt which cannot form the complex are used.

(1) Flash Rust Resistance.

	Pounds of Added Salt* per 100 gal	Percent Flash Rusting**
Paint (contains ZnO)	None	75
plus ammonium carbonate	14	2
plus sodium carbonate	15.5	10

*Salts were added on an equivalence basis, 14 lb ammonium carbonate = 15.5 lb Na_2CO_3.
**Low numbers best.

Maintenance Testing.

	Pounds of Added Salt*** per 100 gal	Blistering*-Rust bleeding** (1 wk exposure 5% NaCl spray) Clean Cold Rolled Steel	Rusty Cold Rolled Steel
Paint (contains ZnO)	None	10-2 to 10	9 M-8 MD-25 to 35
plus ammonium carbonate	14	10-0	8 M-40
plus sodium carbonate	15.5	6 M-20	8 M-90

*9 F to 10 best.
**Low numbers best.
***Salts were added on an equivalence basis, 14 lb ammonium carbonate = 15.5 lb $NaCO_3$.

Magnesium-Containing Complexes for Undercoats for Automotive Bodies

The process of *J.W. Forsberg; U.S. Patent 4,253,976; March 3, 1981; assigned to The Lubrizol Corporation* relates to magnesium-containing compositions of matter and methods for their preparation. In a general sense, the process comprises thixotropic noncarbonated magnesium-containing complexes which are prepared by heating, at a temperature above about 30°C, a mixture comprising:

(A) At least one of magnesium hydroxide, magnesium oxide, hydrated magnesium oxide and a magnesium alkoxide;

(B) At least one oleophilic organic reagent comprising a carboxylic acid, a mixture of a major amount thereof with a minor amount of a sulfonic acid or pentavalent phosphorus acid, or an ester or alkali metal or alkaline earth metal salt of either of these;

(C) Water, if necessary to convert a substantial proportion of component (A) to magnesium hydroxide or hydrated magnesium oxide; and

(D) At least one organic solubilizing agent for component (B);

the ratio of equivalents of magnesium to the acid portion of component (B) being at least about 5:1; and the amount of water present, if any, being sufficient to hydrate a substantial proportion of component (A) calculated as magnesium oxide.

The thixotropic magnesium complexes of this process, especially those that are viscous or solid at ambient temperatures, are useful as corrosion-resistant coatings for metal (e.g., ferrous metal, galvanized, aluminum or magnesium) surfaces, especially in the nature of undercoats for automotive bodies, coatings for structural members such as automotive frames, and the like. They may be employed as such alone or in combination with various adjuvants known to be useful in such coatings, such as other basic metal sulfonates (of the type described in U.S. Patent 3,453,124), acidic phosphate esters, and waxes and resins [component (D)].

For coating automotive frames and the like, a solid hot melt composition is suitable. Frequently, a dye or pigment is added to the hot melt composition. In the following examples, all parts are by weight.

Example 1: A mixture of 754 parts of water, 23 parts of magnesium oxide, 210 parts of mineral oil and 247 parts of Stoddard solvent is heated to about 40°C and 331 parts of a carboxylic acid having an equivalent weight of about 350 and obtained by oxidation of petrolatum, which acid has been preheated to about 50° to 60°C, is added as the temperature of the mixture is maintained at 40° to 45°C. An additional 350 parts of magnesium oxide is added, with stirring, and the temperature of the mixture is increased to 75°C. An opaque dispersion is obtained which is screened to afford the desired magnesium oxide-carboxylate complex.

Example 2: A product similar to that of Example 1 is prepared, substituting about 300 parts of sorbitan trioleate for the oxidized petrolatum.

Example 3: A mixture of 16 parts of an alkylbenzenesulfonic acid having an equivalent weight of about 430 and containing about 22% unsulfonated alkylbenzene, 305 parts of mineral oil, 180 parts of magnesium oxide and 96 parts of Hydrex 440, a mixture of hydrogenated fatty acids, is heated to 95°C and blown with steam for 2 hours. The temperature is increased to 145° to 150°C, an additional 28 parts of mineral oil is added and the mixture is blown with air as the temperature is heated to 170°C over 15 minutes. The mixture is then cooled to room temperature and an additional 44 parts of mineral oil is added to yield the desired magnesium oxide-carboxylatesulfonate complex having the consistency of a grease.

The thixotropic magnesium containing complexes in this process of *J.W. Forsberg; U.S. Patent 4,260,500; April 7, 1981; assigned to The Lubrizol Corporation* differ from those of the preceding patent in that the component (B) comprises at least one oleophilic organic reagent comprising a sulfonic acid, a pentavalent phosphorus acid, a mixture of a major amount of either of the above with a minor amount of a carboxylic acid, or an ester or alkali metal or alkaline earth metal salt of any of these. In the following examples all parts are by weight.

Example 1: Magnesium oxide, 600 parts, is added to a solution in 478 parts of Stoddard solvent and 244 parts of mineral oil, of 308 parts of an alkylbenzenesulfonic acid having an equivalent weight of about 430 and containing about 22% unsulfonated alkylbenzene. Water, 381 parts, is added and the mixture is heated under reflux for 15 minutes. It is then cooled to room temperature, yielding the desired magnesium oxide-sulfonate complex in the form of a gel.

Example 2: A mixture is prepared of 1,106 parts of water, 54 parts of magnesium oxide, 425 parts of the alkylbenzenesulfonic acid of Example 1, 495 parts of mineral oil and 856 parts of Stoddard solvent. An additional 781 parts of magnesium oxide is then added and the mixture is slowly heated to 52° to 55°C. There are then added 30 parts of tetrapropenyl succinic acid and 37 parts of a black pigment. Upon screening and cooling, the desired composition containing the magnesium oxide-sulfonate gel is obtained.

Example 3: A reaction vessel is charged with 63 parts of Epal 20+, a solid mixture consisting predominantly of C_{20-32} linear and branched aliphatic alcohols; 83 parts of Factowax R-143, a paraffin wax melting at about 62°C; and 83 parts of Bareco Polywax 655, a polyethylene synthetic wax melting at about 102°C. The mixture is melted and 21 parts of magnesium oxide is added. The mixture is agitated at 96° to 99°C, 235 parts of the alkylbenzenesulfonic acid of Example 1 is added. Following the sulfonic acid addition, an additional 185 parts of magnesium oxide is added at 96° to 99°C. Mixing is continued at that temperature for 2 hours and then 69 parts of water is added over 2½ hours at 99° to 102°C. An additional 76 parts of alkylbenzenesulfonic acid is added at 96° to 99°C and mixing is continued for 1½ hours after which the mixture is heated to 143° to 149°C for 3 hours and blown with nitrogen to remove volatiles by distillation. The residue is the desired solid (i.e., hot melt) magnesium oxide-sulfonate complex.

The products of Examples 1, 2, and 3 are exemplary of the magnesium complexes suitable for use as corrosion-resistant coatings of the undercoat type. Also useful for this purpose is the product prepared by the following example.

Example 4: A product is obtained substantially in accordance with the procedure of Example 2 by the reaction of 11.61 parts of the alkylbenzenesulfonic acid of Example 1, 9.2 parts of mineral oil, 23.71 parts of Stoddard solvent, 22.61 parts of magnesium oxide and 30.92 parts of water. To the resulting gel are added 0.97 part of a black pigment composition and 1.0 part of a vinyl acetate-ethylene copolymer comprising about 28% vinyl acetate units.

Coating for Mining Industry Bolts and Fasteners

A.L. Black; U.S. Patent 4,256,811; March 17, 1981; assigned to Placer Exploration Limited, Australia provides an improved coating composition particularly for use on fasteners and bolts used in the mining industry.

The composition comprises 5 to 20 parts by weight of zinc metal, 1 to 20 parts by weight of zinc oxide and 10 to 40 parts by weight molybdenum sulfide dispersed in a resin and solvent carrier such as methylated spirit and a phenolic resin.

Example 1:

	Percent
Methylated spirits	40.7
Union Carbide Phenolic Resin BKR 2620	21.7
Water	3.5
Union Carbide Resin XYHL	0.4
Zinc oxide powder	5.2
MoS_2 powder	28.5
	100.0

Sealing Compound
Steetly Steel Improvement 12B
Mixed 1 part to 3 of water

The following two examples are illustrative of a coating composition of this process with a conventional sealing composition and with the preferred sealing composition of this process.

Example 2:

	Percent
Methylated spirits	50.0
Union Carbide Phenolic Resin BKR 2620	18.0
Zinc dust	14.3
Zinc oxide powder	2.3
MoS_2 powder	15.4
	100.0

Sealing Compound
Steel Improvement 12B
Mixed 1 part to 3 of water

Example 3: The coating formulation of Example 2 is used in conjunction with the following sealing composition.

Sealing Compound

	Percent
Shell Solvent X 222	77.0
No. 2 Bentone Grease	21.6
Steetly Duomeen T.D.O.	1.0
Lubrizol 850	0.4
	100.0

Bentone Grease is a mixture of bentonite clay and an oil. Steetly Duomeen is a long chain aliphatic diamine oleate and acts as a water repellant. Lubrizol is a commercially available rust inhibitor.

The coating and sealing compositions of Examples 1, 2 and 3 were applied to steel panels for testing of corrosion resistance. The coating composition was applied and then cured for one hour at approximately 180°C. The sealing composition was then applied and dried.

The final coating of Example 1 was about 0.013 mm thick and that of Examples 2 and 3 was about 0.026 mm thick.

The test results for corrosion resistances are as follows:

Method of test – The salt spray testing was carried out according to Aust. Std. K173, Part III, viz.:

Sodium chloride concentration	50±5 g/ℓ
pH	6.5-7.2
Temperature	35±1°C
Fog collection rate (80 cm^2 area)	1-2 ml/hr

Results – The panels were examined at regular intervals and finally taken from the cabinet, rinsed thoroughly, dried and examined. The observations were as follows:

Two Plain Panels Coated with the Composition of Example 1

96, 168 and 240 hours–There were no rust spots on either.

312 hours–They both exhibited a few small rust spots associated with rust staining.

408 hours–One exhibited many rust spots associated with rust staining, the other exhibited rust on approximately 10% of the surface.

Two Plain Panels Coated with the Composition of Example 2

96 and 168 hours-There were no rust spots on either.

240 hours–There were a few tiny rust spots on both.

312 hours–There were many small rust spots on both.

408 hours–There were many small rust spots on both associated with rust staining.

Two Plain Panels Coated with the Composition of Example 3

96, 168, 240, 312 and 408 hours–There were no rust spots on either.

One Punched Panel Coated with the Composition of Example 3 and Containing 3 Center Punchings

48 hours–There was no visible rust at the punch marks.

96, 144, 194, 264 hours–There was rust at one punch mark, none at the other two.

360 hours–As at 264 hours, but also a few tiny rust spots on the face.

These results show that although under the testing conditions the composition of this process as illustrated in Example 2 was not superior to that of Example 1 after short periods of exposure to salt corrosion, the longer periods of exposure do illustrate the superiority of the coating composition of this process and this means that the articles so coated will remain serviceable for far greater periods than conventionally treated products.

Cyanide Solution for Inhibiting Aluminum Fatigue

The known aluminum corrosion inhibitors are often addressed in terms of reducing the uniform surface corrosion rate of aluminum. However, in addition to surface corrosion, fatigue corrosion is an important contributing factor towards shortening the life of aluminum structural members, especially in cyclical high stress conditions which are encountered in aircraft.

Aluminum, aluminum alloys and other metals are elastic and will, although to an extent much less than encountered in a highly elastic material such as a rubber band, stretch and compress in reaction to external tensile or compressive forces. In attempting to adapt to the application of such external forces which approach or exceed the yield strength of the metal, units of the metal comprised of thousands of unit cells slide along each other on slip planes.

If the external stress is continued the slip planes increase in size and cracks form which lead to the eventual fracture of the metal. In ductile metals like aluminum the fracture is transcrystalline, or across the crystal comprising the metal, at room temperature. As the temperature approaches the metal's melting point, the fracture becomes intercrystalline such that crystal are torn away from each other at their boundaries. Such intercrystalline or brittle failure is usually sudden and without significant prior deformation of the metal.

Although surface corrosion accelerates the fatigue of a metal, fatigue per se is comprised of processes which are distinct and separate from corrosion. As a result, treatments which improve the corrosion resistance of a material do not necessarily improve the fatigue resistance of the material.

Improved fatigue resistance does not inherently accompany an improvement in the corrosion resistance of a material. Each particular corrosion inhibiting process must be tested on its merits in combustion with a specific material to determine if the fatigue resistance of the material is improved or reduced.

R.L. Crouch; U.S. Patent 4,261,766; April 14, 1981; assigned to Early California Industries, Inc. has discovered a method for improving fatigue failure characteristics of aluminum and aluminum alloys comprising the steps of immersing the aluminum in an aqueous solution of a water-soluble cyanide compound at room temperature, the aqueous solution being substantially free of chromium, and continuously maintaining the aluminum in contact with the aqueous solution.

The fatigue corrosion inhibiting cyanide compound is typically incorporated into a carrying agent such as water in a minor effective amount sufficient to substantially reduce the fatigue corrosivity of aluminum or aluminum alloys. In the preferred embodiment of the process sodium ferrocyanide is used.

Example: This example illustrates the improvement in fatigue and fatigue corrosion characteristics of aluminum which results from contacting the metal with a composition of water and a cyanide component.

A test specimen of aluminum alloy (2024-T3) measuring 0.25" x 0.50" x 14" is oriented in the long transverse direction, notched at the center, degreased and inserted through slits cut in the side wall of a polyethylene bottle. The slits are sealed around the test beam with silicone caulking and the bottle is filled with the corrosion inhibiting composition of deionized water containing 0.25% by weight sodium ferrocyanide. The ends of the specimen are then attached to the vice and the crank of a Fatigue Dynamics VSP-150 plate bending machine and the loading is adjusted to 6,800 psi.

The test beam is then stressed at 100 cycles per minute at 70°F until the specimen breaks.

With only deionized water in the polyethylene bottle, the test specimen breaks at 720,000 cycles. With a solution of deionized water containing 0.25% by weight sodium chromate, the test specimen breaks at 854,000 cycles. Tests with the bottle filled with deionized water containing 0.25% by weight sodium ferrocyanide were conducted and the following data obtained.

Test Number	Cycles to Failure
1	1,010,100*
2	1,404,000*
3	1,203,100*

*The aluminum bar did not fail.

The oxide layer which normally forms on the surface of aluminum is very resistant to ordinary water. It is known that cyanide is an equally effective corrosion inhibitor for aluminum. Thus, when the aluminum bar was immersed in the aqueous solution of deionized water containing a sodium ferrocyanide, the aluminum bar was placed in a solution which would cause minimal corrosion. The failure of the aluminum bar was therefore predominantly due to the effects of fatigue.

Lanthanum Hexaboride-Carbon Composition for Hydrogen Fluoride Laser

The process of *C.E. Holcombe, L. Kovach and A.J. Taylor; U.S. Patent 4,261,753; April 14, 1981; assigned to the U.S. Department of Energy* provides a composition for use as a structural material in the fabrication of hardware contactable with high temperature fluorine-containing gases. The composition of this process consists essentially of a carbon phase uniformly dispersed in a continuous phase of lanthanum hexaboride with the carbon providing about 10 to 30 volume percent of the composition. A portion of this carbon may consist of discontinuous carbon fibers for increasing the strength of the composite.

The composition is particularly adaptable for use in gas lasing operations in which the lasing medium consists primarily of hydrogen fluoride. The composition is suitable for constructing nozzles, combustion chambers and the like using gas-laser systems in which the high temperature fluorine and hydrogen fluoride comes into contact therewith. The composition can be machined into the desired hardware configuration substantially easier than lanthanum hexaboride without the addition of the carbon. The composition is highly resistant to thermal shock and is relatively stable under the attack by the corrosive fluorine and/or hydrogen fluoride at temperatures up to a maximum use-temperature of about 1800°K.

While the term carbon is used to describe the matrix material, it is to be understood that this term is to be equally applicable to graphite in the event the matrix is graphitized during the preparation thereof. Further, while the discontinuous fibers employed in the composition are preferably graphite fibers, carbon fibers or carbon fiber precursors can be utilized and then subsequently carbonized and/or graphitized.

Example 1: A solid cylinder of the material of this process was fabricated from a homogeneous mixture of lanthanum-hexaboride particles and a thermal setting phenolic resin having a carbon yield of about 53 vol %. The cylinder was formed with a diameter of 1 inch and a height of 0.5 inch and contained approximately 74.5 vol % lanthanum hexaboride and 25.5 vol % carbon. The cylinder was prepared by coating the lanthanum-hexaboride particulates of an average particle size less than 25 micrometers with the phenolic resin. The mixture was blended and isostatically pressed at 150°C for polymerizing the resin under an argon pressure of 2,500 psi. Following this pressing operation which lasted for a duration of 15 minutes, the compact was sintered in argon at 2100°C for 1 hour during which the precursor was converted to carbon. No interaction of the lanthanum hexaboride and carbon below 2100°C was noted by metallography of the specimen.

The composite was 89% of theoretical density and had a porosity of 11%. The composite was subjected to a hydrogen-fluoride flame at 1700°K for a duration of 2.5 minutes and 1830°K for 4 minutes. During this 6.5 minute duration, the surface temperature of the composite attained a temperature of approximately 1478°K. A weight loss of 0.62% was noticed during this exposure to the flame. However, no change in diameter and height of the cylinder occurred. During this run a lanthanum-trifluoride layer was formed on the crystal. This coating of a thickness of about 6 mils was white and very adherent to the cylinder. The failure temperature (1870°C) of the lanthanum hexaboride-carbon composite is approximately the same as the failure temperature (1875°K) of theoretically dense lanthanum hexaboride, and then has approximately the same upper-use-temperature (1800°K)'

Example 2: A second composite in the form of a cylinder as in Example 1 was prepared so as to contain 15.5 vol % carbon and after sintering had an essentially 0 porosity. This composite when subjected to a hydrogen-fluoride flame for approximately the same duration and temperature as the cylinder in Example 1 showed a surface temperature of 1467°K with a weight loss of 0.37%. Again, as in Example 1, no change was detected in the diameter or height of the cylinder. The inspection of the specimen illustrated that it was gray on the top where exposed to the flame and whitish on the sides. Some reacted area was present on the bottom. The top coating was hard and approximately 6 mils thick. This coating was formed of lanthanum trifluoride and was very adherent to the composite.

Example 3: A lanthanum-hexaboride composite containing 10 vol % carbon was hot pressed at 4,060 psi and a temperature of about 2100°C for a duration of 20 minutes. This composite was tested under the hydrogen-fluoride flame to maximum temperatures from 1580° to 1610°K. Weight loss and dimensional changes were relatively negligible. The composite appeared to be black on top and white elsewhere with the total top surface coating thickness being approximately 10 mils. The coating of lanthanum trifluoride was hard and adherent.

Example 4: A composite containing 6.0 volume percent carbon formed by the thermal-setting resin and 9.8 volume percent chopped graphite fibers was formed by hot pressing the mixture at 3,200 psi and about 2100°C for a duration of twenty minutes. A first run with this composite at a surface temperature of 1545°K in a hydrogen-fluoride flame provided a hard whitish coating of approximately 10 mils thick on the top surface with negligible weight loss and dimensional changes. In a rerun of this composite in the hydrogen-fluoride flame a maximum surface temperature of 1570°K was attained. The thickness of the top coating remained at approximately 10 mils with the weight loss during the second run being 0.6%.

It will be seen that this process provides a composition which is particularly useful in the high-temperature fluorine-containing environments encountered in hydrogen-fluoride laser systems. By employing the composition of this process in such laser systems the operating range may be increased to a temperature in the range of about 1400° to 1700°K so as to considerably increase the efficiency of the hydrogen-fluoride laser. Further, while the composition is primarily directed to its use in a hydrogen-fluoride laser, it will appear clear that a torch using the hydrogen-fluorine-helium flame could have various applications such as in the cutting of metals, ceramics and the like.

Removing Aluminide Coatings from Nickel or Cobalt Base Alloys

In order to enhance their resistance to corrosion, nickel or cobalt base alloys are often provided with a thin coating of an aluminum-containing alloy usually by the technique commonly known as aluminizing. If coatings of this type are damaged, it is frequently necessary to remove the whole or a major portion of the coating from the substrate before a fresh coating may be applied. This is essential if the coated substrate is, for example, in the form of an aerofoil blade for a gas turbine propulsion engine. Such blades are manufactured to an extremely high degree of accuracy and consequently any surface discontinuities cannot be tolerated.

Aluminum containing alloy coatings are most commonly removed by chemical dissolution. However, it is difficult to achieve dissolution of the coating without the occurrence of significant intergranular substrate attack. Moreover concentrated acid mixtures frequently employed for coating removal are usually heated at temperatures in the region of 80°C in order to achieve an acceptable coating removal rate. Such temperatures seriously restrict the use of wax based masking compounds which are necessary if only partial coating removal is desired.

F. Cork; U.S. Patent 4,282,041; August 4, 1981; assigned to Rolls-Royce Limited, England provides a method of removing aluminum containing alloy coatings from nickel or cobalt base alloy substrates in which the coated substrate is immersed in an aqueous mixture of nitric and sulfamic acids until coatings dissolution is complete. A solution containing 5 to 30% v/v nitric acid and from 5 to 30% w/v sulfamic acid is preferred.

Eight nickel base alloy aerofoil blades were dry blasted in order to remove surface oxidation before being divided up into four groups of two blades. Each blade has been aluminized, i.e., each had a coating of nickel aluminide. The nickel aluminide coating varied in depth between 0.8 to 2 thousandths of an inch.

All of the blades were totally immersed in an aqueous solution containing 5% w/v sulfamic acid and 10% v/v nitric acid and maintained at a temperature of 40°C. At four hour intervals, one group of blades was removed from the solution, dry blasted and examined for intergranular attack by microsectioning. The results obtained were as follows.

Blade Group	Time in Solution (hr)	Depth of Intergranular Attack (inch)
1	4	0.0003
2	8	0.0005
3	12	0.0009
4	16	0.0011

Examination also revealed that substantially all of the nickel aluminide coating had been removed after a period of four hours immersion in the solution.

In order to compare the degree of intergranular attack resulting from mixture of this process with existing acid coating removal solutions, a further test was carried out.

Eight more aerofoil blades similar to those used in the above example but which had not been aluminized were dry blasted and then immersed in a known coating removal solution containing 1 part by volume glacial acetic acid, 1 part by volume 1.42 SG nitric acid and 2 parts by volume phosphoric acid. The solution was maintained at a temperature of 80°C.

After four hours, four of the blades were removed from the solution dry blasted and examined for intergranular attack. The remaining blades were removed from the solution after a further four hours and similarly dry blasted and examined.

The depth of intergranular attack on the blades was more severe and widespread than was the case with the nitric acid/sulfamic acid solution. Intergranular attack to a depth of 0.0016 inch was observed on the first four blades removed from the solution whilst the remaining blades had intergranular attack to a depth of 0.0035 inch. Thus intergranular attack by the nitric acid/phosphoric acid/acetic acid solution was significantly greater than was the case with the nitric acid/sulfamic acid solution in accordance with this process.

Since the nitric acid/sulfamic acid solution of this process is effective at low temperatures, it is possible to utilize wax-based masking compounds if it is desired to remove only a portion of a coating.

Protecting Titanium with Tellurium or Selenium Compounds

Titanium and titanium alloys are, largely due to their generally high corrosion-resistance properties, widely used in industry as construction material or linings for vessels, piping and the like.

According to *V.P. Gupta; U.S. Patent 4,321,231; March 23, 1982; assigned to Atlantic Richfield Company* the rate of corrosion attack upon titanium and titanium alloys by corrosive aqueous media containing dissolved acidic moieties in the substantial absence of molecular oxygen is decreased by introducing into the acidic media as corrosion inhibitor at least one soluble selenium or tellurium compound in an amount sufficient to provide at least about 1×10^{-6} gram-atoms of dissolved tellurium or selenium values per gram mol of total acidic moieties in the corrosive aqueous media.

The corrosion inhibitors of this process comprise at least one member selected from the group consisting of soluble tellurium compounds, soluble selenium compounds and mixtures thereof. Thus, suitable corrosion inhibitors comprise organic or inorganic compounds of tellurium or selenium which provide cations of these elements in any of their positive valance states, e.g., Te^{4+}, Te^{6+}, Se^{4+} and Se^{6+}, when placed in contact with the aqueous medium to be treated.

The most preferred anticorrosion agents of this process are the sources of tellurium cations, most especially the sources of Te^{4+} cation. Of the latter group of tellurium sources, inorganic tellurium (Te^{4+}) compounds are most preferred, of which TeO_2, $TeBr_4$, $TeCl_4$, $Te(SO_4)_2$, $Te(NO_3)_4$ and $Te(OH)_4$ are exemplary.

The process can be further illustrated by reference to the following examples, wherein parts are by weight unless otherwise indicated. The corrosive aqueous media in the examples possess a pH of less than 2. The gram-mols of acidic moieties in the examples is calculated based on the gram-mols thereof which are added initially in forming the test liquids.

Examples 1 through 3: In a series of experiments, 400 ml of a liquid mixture containing 9.0 wt % HBr, 20.0 wt % water and 71.0 wt % acetic acid, and in which is dissolved the selected amount of tellurium dioxide as anticorrosion agent, is charged to a 500 cc glass flask provided with a gas sparger, a reflux condenser and a stirring rod (made of titanium Ti-50A) to which is attached a 40 x 20 x 1.5 mm coupon of Ti-50A titanium metal. The flask also contains 50 g of sand in order to minimize the deposition of any solid tellurium on the coupon surface and thereby to ensure that only chemical anticorrosion properties of the added anticorrosion agent will be measured. The flask is sealed and nitrogen is sparged through the liquid at a rate of about 2 cc/min. The flask is then heated from room temperature to the desired temperature for the selected period of time.

The titanium coupon is immersed in the liquid and is rotated by means of the stirring rod at a speed of about 700 rpm throughout the heating period. The nitrogen gas sparging is continued throughout the heating period as well. The sand is substantially evenly dispersed in the liquid by the stirring. After the end of the above period of time, the stirring is discontinued, and the coupon is removed, and its weight and dimensions are measured (to an accuracy of $\pm 10^{-5}$ g; $\pm$ 0.5 mm) to determine the rate of corrosion of the coupon, thereby yielding the data set forth in the following table.

Experiment No.	Temperature (°C)	Time (hr)	Te* (ppm)	Gram-atoms Te per Gram-mol AM**	Rate of Corrosion (mpy)
Control A	108	61	0	0	412
1	104	56	384	230×10^{-6}	1
2	109	140	48	29×10^{-6}	2
3	110	94	4.8	2.9×10^{-6}	372

*Added as TeO_2.
**AM is acid moieties.

The tellurium cation therefore can effect a greatly reduced rate of corrosion of the titanium metal, and the introduction of only 2.9 x 10^{-6} gram-atoms of Te per gram-mol of acidic moieties in the liquid medium reduces the rate of corrosion by about 10%.

Protecting Freshly Produced Aluminum Structural Elements

In the manufacture of profiles, i.e., the extrusion of structural elements from aluminum (window frames, wall facings and the like), the problem of so-called precorrosion does exist due to the high reactivity of the surfaces of freshly produced aluminum elements. The term precorrosion is intended to include all damages of the surfaces affecting the aspect, for example, damages caused by hand perspiration, water or humidity, packing material and corrosive atmosphere on industrial sites prior to eloxation, which often require an expensive manual after-treatment of the finished elements. Anticorrosive agents used to avoid these damages should be stable up to 180°C (annealing temperature) and they should not cause any problem in the following eloxal process. The products hitherto used are not fully satisfactory in these respects.

The process of *R. Helwerth and H. Lorke; U.S. Patent 4,323,476; April 6, 1982; assigned to Hoechst AG, Germany* provides an improved anticorrosive agent for aluminum and the alloys thereof consisting of:

(A) 15 to 50, preferably 40 to 50%, by weight of a product obtained by reaction of sulfochlorination products of aliphatic, alkylaromatic or cycloaliphatic hydrocarbons having from 12 to 24 carbon atoms with ammonia or a C_{1-3}-alkyl or hydroxyalkylamine with subsequent reaction with a C_{2-11} halocarboxylic acid and conversion into an alkaline earth metal or the zinc salt,

(B) 40 to 90, preferably 40 to 50%, by weight of a paraffinic hydrocarbon containing from 50 to 60% of C_{13-16} paraffins, 50 to 40% of naphthenes and 0 to 1% of aromatics and having a viscosity of from 3° to 5°E/20°C.,

(C) 1 to 4, preferably 1 to 2%, by weight of the salt of a C_{8-10} alkyl amine and a C_{8-10} carboxylic acid,

(D) 1 to 4, preferably 1 to 2%, by weight of an oxethylate of 1 mol of a C_{8-12} alkylphenol and 2 to 10 mols of ethylene oxide, and

(E) 1 to 2% by weight of a C_{4-8} aliphatic alcohol.

The anticorrosive agents according to this process are simply sprayed on the metal elements to be treated. The respective amounts of components (A) through (E) are chosen in the indicated limits to obtain a total amount of 100% of anticorrosive agent.

The advantageous anticorrosive example of the agents of this process is demonstrated by the following tests. In these tests dry, freshly pickled aluminum sheets were sprayed with the respective anticorrosive agent and the sheets treated in this manner were examined under the conditions of the Kesternich Test (DIN 50,017) during a period of time of up to 4 weeks to determine their corrosion. The tendency to corrosion was also tested on aluminum sheets which had been sprayed with the anticorrosive agent and annealed for 4 hours at 180°C.

In the visual evaluation of the test sheets notes from 0 to 4 were given according to the following scheme: 0, no corrosion; 1, traces of corrosion; 2, slight corrosion; 3, distinct corrosion; and 4, very pronounced corrosion. The results obtained with anticorrosive agents 1 to 5 are summarized in the table below. The following anticorrosive agents were used.

Agent (1)

15% of a component (A) obtained by reaction of a C_{13-16} alkyl sulfochloride with ammonia with subsequent reaction with acetic acid and conversion into the barium salt, 80% of a paraffinic hydrocarbon approximately composed of 50 to 60% of C_{13-16} aliphatic hydrocarbons, at most 1% of aromatic hydrocarbons, and 50 to 40% of naphthenes and having a viscosity of about 0.8/15°C, a refractive index of about 1.4/20°C and a flash point according to Abel-Pensky of about 100°C.
2% of capryl-aminooctoate,
2% of nonyl phenol with 2 mols of ethylene oxide, and
1% of isobutanol.

Agent (2)

15% of component (A) as defined under (1), but in the form of its calcium salt,
80% of the paraffinic hydrocarbon as defined under (1),
2% of capryl-aminooctoate,
2% of nonyl phenol with 2 mols of ethylene oxide, and
1% of isobutanol.

Agent (3)

4% of fatty acid esters from C_{16-18} fatty acid having an iodine number of 27 with a fatty alcohol and low molecular weight polyhydric alcohols,
4% of fatty acid glyceride,
2% of alkanol amine salt of a sulfocarboxylic acid, and
90% of paraffin oil.

Agent (4)

a paraffinic hydrocarbon as used in component (B) in anticorrosive agents (1) and (2).

Agent (5)

a mineral oil (spindle oil 3°E/30°C).

Anticorrosive agents (1) and (2) correspond to the process, while agents (3) and (4) are prior art agents.

Anticorrosion Tests According to Kesternich (DIN 50,017)

Anticorrosive Agent	After 1 wk a	After 1 wk b	After 2 wk a	After 2 wk b	After 3 wk a	After 3 wk b	After 4 wk a	After 4 wk b
1	0	0	0	0	0	0	0	0
2	0	0	0	0	0	0	0	1
3	0	2	0	2	1	4	2	4
4	2	4	3	4	4	4	4	4
5	2	4	2	4	4	4	4	4

For the tests aluminum sheets having a magnesium content of 0.5, 1.5 and 3%, respectively, were used. The results obtained with the anticorrosive agents 1 to 5 were the same for all three types of alloy. The values listed under b are the results obtained with sheets which had been annealed for 4 hours at 180°C after spraying with the anticorrosive agent. The results a are those obtained without heat treatment.

Corrosion Resistant Reflective Sheet Material

The process of *J.E. Kropp; U.S. Patent 4,329,396; May 11, 1982; assigned to Minnesota Mining and Manufacturing Company* relates to reflective or transparent-reflective sheet material of the type comprising a self-supporting polymeric foil backing, a vapor-deposited layer of corrodible metal (especially aluminum) on the backing, and a protective polymeric barrier layer overlying the metal layer and bonded thereto. The process relates especially to energy control films of the type used in connection with windows to exclude undesired solar energy.

To some degree in any installation of such product, and particularly where conditions are both sunny and humid, there has been a tendency for the aluminum layer to corrode to a transparent oxide form. Such corrosion may take place either locally, generating pin holes which gradually increase in size, or generally, causing a gradual fading which results in an overall loss of effectiveness of the energy control sheet material. This problem has been exacerbated by the fact that it is common to include ultraviolet light absorbers in the barrier layer in order to prevent such light from entering the interior of a room, where it has a tendency to bleach and degrade any fabrics which it strikes. For reasons which have never been adequately explained, the benzophenone ultraviolet light absorbers, which are most effective in excluding UV radiation from a room interior, tend to increase the rate at which the aluminum layer corrodes.

The contribution of this process lies in incorporating, in the polymeric layer which is in direct contact with the corrodible metal, a nickel organic compound, especially one of the type commonly recognized as an ultraviolet light stabilizer. It is not known why the incorporation of nickel organic compounds is so effective in improving corrosion resistance, especially since the incorporation of various known antioxidants has been ineffective in this regard.

Examples 1 through 13: In each of the examples listed in the table, in which all parts are by weight unless otherwise noted, biaxially oriented polyethylene terephthalate foil approximately 25 micrometers thick was coated with aluminum by conventional vapor deposition to achieve a light transmission of 16 to 20% as measured with a Hunter colorimeter using a broad spectrum light source (CIE Source C, which corresponds closely to typical daylight).

	Components Included in Barrier Coat												
	UV Absorber		Anti-oxidant		Ni Organic Compound		Δ % T After Time Indicated						
Example	Type	%	Type	%	Type	%	2 wks	3 wks	4 wks	7 wks	8 wks	11 wks	12 wks
Control A	—	—	—	—	—	—	0.8						
Control B	U490	2	—	—	—	—	1.0				9.2		15.2
1 (Comparison)	"	"	BHT	8	—	—	0.6				18.1		
2 (Comparison)	"	"	A33	8	—	—	0.3				8.8		
3	—	—	—	—	N2	8	0.4				7.3		
4	"	"	—	—	N4	8	0.2				2.1		3.1
5	"	"	—	—	N5	8	0.8				4.2		5.6
Control C	"	8	—	—	—	—	3.9	14.4					
6	"	"	—	—	N2	2	1.4						
7	"	"	—	—	"	4	1.3						
8	"	"	—	—	"	8	1.1				1.3		2.5
9	"	"	—	—	"	16	0.2				2.0		6.1
10	"	"	—	—	"	32	0.0				1.9		8.4
11	"	"	—	—	N4	8	1.3			10.1		42.0	
12	"	"	—	—	N5	8	0.6			1.7		3.3	
13	"	"	—	—	N84	8	0.3		0.2				

The aluminum was then overcoated with a 25% methyl ethyl ketone solution of a soluble copolyester resin and the solvent evaporated to leave a barrier layer weighing approximately 5.4 g/m^2. (The copolyester resin was formed by reacting 12 parts sebacic-azelaic acid, 46 parts terephthalic acid and 12 parts isophthalic acid with 60 parts ethylene glycol and 40 parts neopentyl glycol.) Over the barrier layer was applied a 25% 1:3 ethyl acetate:isopropanol solution of a 96:4 isooctyl acrylate:acrylamide copolymer and the solvent evaporated to leave a pressure-sensitive adhesive layer weighing approximately 4.4 g/m^2. As is shown in the table, the only differences among the examples reside in the composition of the barrier layer, where the amounts and types of UV absorbers, antioxidants, and nickel organic compounds are varied.

When a UV absorber is included in the barrier layer, the examples in the table utilize a composition which is especially effective at wavelengths of 380 nanometers, as measured using a Beckman spectrophotometer. The wavelength of 380 nanometers was chosen because it represents the high end of the UV range, and shorter wavelengths of light are absorbed even more efficiently. When 8% of this UV absorber is incorporated in the barrier layer, less than 2% of the 380 nanometer light is transmitted.

The ability of the various constructions to resist outdoor weathering was determined by adhering a 1.3 cm x 10.2 cm sample of each construction to a glass plate and placing it in a closed chamber maintained at 52°C and 100% relative humidity, where it was then subjected to an accelerated aging cycle consisting of 12 hours of exposure to ultraviolet light followed by 12 hours of darkness. (It has been empirically found that one week of test exposure corresponds to 3 months of outdoor exposure in Florida.) Using the Hunter colorimeter, the percent light transmission (% T) was measured initially and at various intervals after commencing the test.

Corrosion of the aluminum was indicated by an increase in transmission, the percentage of increase (Δ%T) being a convenient measure. To illustrate, if the transmission were 18% initially and 24% after testing, the value of Δ%T would be 6. A 3% increase in light transmission can be visually detected, and a 10% increase renders a product commercially unacceptable.

The following abbreviations have been used to refer to various components in the barrier layer:

A33 antioxidant: Tris[2-(2-hydroxy-3-tert-butyl-5-methyl benzyl)-4-methyl-6-tert-butyl phenol] phosphate.

BHT antioxidant: Butyrated hydroxy toluene.

N2: Nickel bis[O-ethyl(3,5-di-tert-butyl-4-hydroxy benzyl)] phosphonate.

N4: Nickel dibutyldithio carbamate.

N5: Nickel bis(octylphenylsulfide).

N84: [2,2'-thiobis(4-tert-octylphenolato] n-butylamine nickel.

U490: UV absorber comprising a complex mixture of 2,2'-dihydroxy-4,4'-dimethoxy benzophenone and other tetra-substituted benzophenones.

The table indicates that the corrosion-inhibiting effectiveness of a given nickel salt depends, in part, on the amount and nature of other components incorporated in the barrier layer. Note, for instance, Examples 4 and 11, where the effectiveness of a specific corrosion inhibitor was greatly reduced when the accompanying amount of UV absorber was increased.

Comparison Examples 1 and 2 incorporate in the barrier layer antioxidants which include some of the functional groups found in nickel organic compound N2; it will be noted that neither of these examples was as satisfactory as the examples containing nickel organic salts in the barrier layer. To further illustrate this point, a sample of the N2 compound was contacted with an ion exchange resin to remove the nickel ion and replace it with hydrogen. When Example 8 was repeated, substituting this nickel-free compound for the N2 nickel compound, Δ%T was 1.6 after two weeks of testing and 7.3 after four weeks.

In selecting nickel organic compounds for use in practicing this process, consideration should also be given to the degree of color which can be tolerated. Nickel organic compounds N2 and N84, for example, are pale yellow and hence quite inconspicuous, while compounds N4 and N5 are deep shades of, respectively, purple and green. Where absence of color is important, the amount of nickel organic salt is generally the minimum needed to secure the desired corrosion inhibition. When the concentration of nickel organic salt exceeds about 15% in the polymeric layer directly contacting the corrodible metal, some reduction in adhesion may be noted.

Cable Coating Compound

The process of *P.H. Stovall and R.C. Webber; U.S. Patent 4,330,571; May 18, 1982; assigned to Lockheed Corporation* concerns the application of corrosion preventive coatings to metallic cables whereby the applied coating is substantially consistent and uniform along the cable length.

Experience has dictated that control cables formed of wires of steel, zinc coated steel, and tin coated steel should have a coating protection to avoid corrosive deterioration resulting from water (humidity) or gaseous atmospheres (carbon dioxide, sulfur dioxide, etc.). Such need for a corrosion protective coating exists notwithstanding a usual lubrication of the individual wires by cable manufacturers to keep the cable wires from fretting; such lubrication generally being applied whether the outer surface of the individual wires is steel, zinc or tin, etc.

The type of corrosion preventive coatings used on metallic cable members is typified by that defined in Military Specification MIL-C-16173 which covers solvent-dispersed corrosion preventive compounds which deposit thin films after evaporation of solvent which are in turn easily removable by subsequent use of solvent after application. This specification MIL-C-16173 further defines the compounds as composed of nonvolatile hydrocarbon material dispersed in petroleum solvent so as to form a fluid formulation that is homogeneous, free from grit, abrasives, water, chlorides or other impurities; that benzol or chlorinated hydrocarbons shall not be used in the compounds; and the compounds shall readily wet the surfaces of the material being coated with the resulting coating being continuous upon evaporation of the solvent.

The process is practiced by dipping a cable length to be coated into a corrosion preventive coating compound of the nature described above that is of specifically controlled viscosity and followed by the wiping of the cable length as it is withdrawn from the compound solution.

More specifically, this process involves establishing the viscosity of the compound, at the temperature of application, to a level of from approximately 200 to approximately 250 centistokes (equivalent to a level of about 18 to 22 seconds time drain from a No. 4 Zahn cup), followed by the immersion or dipping of the cable length therein. Such viscosity control is accomplished by either the appropriate addition of coating compound or evaporation of solvent to increase viscosity or addition of a hydrocarbon solvent such as naphtha, mineral spirits or Stoddards Solvent to reduce viscosity. Associated with this operation, a sponge or other appropriate wiping material is immersed in the compound solution, removed and with most of the compound squeezed out of it, is placed around the cable as it is withdrawn from the compound solution. The coating of the cable is then dried in any appropriate manner by evaporation of the compound solvent whereupon the coating process is complete.

It is immaterial whether the wiping contact with the coated cable length is hand or tool maintained, the primary feature being that through different wiper material-cable contact pressures, various thicknesses of coating can be obtained, and so long as the pressure is constant over the length of the cable, a substantially consistent and uniform coating thickness will result.

Two-Coat System for Structural Steel

B. van der Bergh; U.S. Patent 4,337,299; June 29, 1982; assigned to Akzo NV, Netherlands provides a process for applying a corrosion resistant system to structural steel. In the process a zinc dust paint comprising 5 to 15% by weight of a binder and 85 to 95% by weight of a metallic zinc-based pigment component is applied to the steel as a first coat, and subsequently a coat of a composition comprising an aromatic/aliphatic polyhydroxyether and/or a saturated aromatic polyester resin having a number average molecular weight of at least 4,000 and a pigment having a Mohs hardness of at least 2, the volume ratio of the binder to the pigment being in the range of 50:50 to 95:5. The thickness of the first coat is about 15 to 150 μm and that of the second coat is about 15 to 150 μm.

In the examples the degree of resistance to corrosion is designated in accordance with ISO 4628/1-1978(E), the degree Ri 0 indicating that there is no detectable corrosion and the degree Ri 5 that there is a very high degree of corrosion. The corrosion is brought about by a salt fog in accordance with ASTM B117-64 at 35°C over a period of 3,000 hr in Example 1 or for 1,000 hr in the other example.

Example 1: A blasted steel structural steel member (steel grade No. 37) was coated with a composition made up of 9 parts by weight of a binder, 90 parts by weight of zinc dust having an average particle size of 2 to 4 μm and 1 part by weight of strontium chromate. The binder was made up of 40% by weight of Bisphenol-A epoxy resin having a molecular weight of 900 (Epikote 1001), 20% by weight of a polyamino amide (Euredur 115) and 40% by weight of an aromatic/aliphatic polyhydroxyether having a molecular weight of about 30,000 (Phenoxy PKHC). Per 100 parts by weight of solid matter the composition contained 23 parts by weight of a solvent mixture consisting of 40% by weight of xylene, 30% by weight of ethylglycol acetate and 30% by weight of ethylene glycol.

After the structural member had been dried for about 24 hours at about 20°C, the thickness of the undercoat was 20 μm. Subsequently, a second coat was applied which was made up of 50 parts by weight of the aromatic/aliphatic polyhydroxyether, one of the constituents of the binder that had been used in the first coat, 49 parts by weight of micaceous iron oxide, 1 part by weight of strontium chromate and 105 parts by weight of ethylglycol acetate. In this second coat the volume ratio of the binder to the pigment was 80:20. After the second coat had been left to dry for 14 days at about 20°C and found to have a thickness of 20 μm, the structural member was subjected to a salt spray test. The degree of corrosion was Ri 0.

Example 2: The same procedure was used as in Example 1, except that use was made of an aromatic/aliphatic polyhydroxyether (Phenoxy PKHH) as binder in the two coats. In the second coat the volume ratio of the binder to the pigment was 80:20. The degree of corrosion was Ri 0.

Protecting Copper Neutral Wires of Power Conducting Cable

The process of *R.L. Martin; U.S. Patent 4,343,660; August 10, 1982; assigned to Petrolite Corporation* comprises coating copper with a barrier corrosion inhibitor, such as a water-insoluble sulfonate, for example, a petroleum or similar sulfonate as illustrated by a metal petroleum sulfonate, which contains an agent capable of complexing with copper, such as an organic sulfur, nitrogen, or sulfur-nitrogen compound, for example, an organic triazole, thiazole, or the like. This process is particularly effective in inhibiting the corrosion of copper buried in the earth such as occurs when copper is exposed to electrochemical forces when copper is employed as neutral wires wrapped around the outside of an insulated power conducting cable and buried in the earth.

The corrosion mechanism involves acceleration of the anodic polarization of the copper, probably due to leakage currents. Soluble salts that leach into the electrolyte make the problem worse but anodic polarization is the prime cause of corrosion. Therefore, the testing emphasized the importance of inhibiting anodic polarization as indicated by potentio-dynamic polarization curves.

Examples 1 through 5: The following examples illustrate the process.

Tests were conducted by dipping the copper coupons in inhibitor solution, dripping dry in quiescent laboratory air, then corrosion testing in an air saturated water solution of 5% NaCl at room temperature. Both pure copper and copper coated with tin/lead were tested. A Petrolite Corrosion Rate Instrument (also called Pair Meter) was used to follow open circuit corrosion rates while a Petrolite Potentiodyne Analyzer was used for anodic polarization.

The preferred inhibitor is a combination of benzotriazole and calcium overbased petroleum sulfonate. Since the two components are not miscible in each other, a mutual solvent or suspendant is necessary. Alcohols, glycols and surfactants are all satisfactory; the final choice can depend on the method of application.

The corrosion inhibitor is applied by any suitable means such as by painting, dipping, spraying, etc. If already buried, the metal can be treated by pumping the mixture into the spaces surrounding the metal.

The following formulations were employed in the tests described in Table 2. The ratios expressed therein are weight ratios.

Table 1

Ex. No.	
1	50/50 Calcium petroleum sulfonate/oxyalkylated Alfol 8-10
2	2/98 Benzotriazole/oxyalkylated Alfol 8-10
3	18/80/2 Calcium petroleum sulfonate/isopropanol/benzotriazole
4	18/80/2 Calcium petroleum sulfonate/oxyalkylated Alfol 8-10/benzotriazole
5	49/49/2 Calcium petroleum sulfonate/oxyalkylated Alfol 8-10/benzotriazole

The calcium petroleum sulfonate is water-insoluble and 60% active in mineral oil. It has a base number of 160, Brookfield viscosity is 45,000 cp at 77°F with a molecular weight of about 1,000.

The oxyalkylated Alfol 8-10 is a straight chained alkanol oxyethylated with 3.3 mol of EtO/Mo (or 1/1 by weight) alkanol and it is employed as a solvent/dispersant.

Table 2

	Copper Blank	Coated Copper Blank	..... Copper Treated with ..... Ex. 1	Ex. 2	Ex. 3	Ex. 4	Ex. 5	Coated Copper Treated with Ex. 3
MPY @ 1 hr	11	7	4.0	3.3	0.1	0.1	0.1	0.2
MPY @ 4 days	15	–	0.7	1.5	0.1	0.1	0.1	–
MPY @ 50 mv anodic polarization* after 4 days	75	110	1.5	3.3	0.1	0.2	0.1	0.3

*Anodic polarization of 50 mv was chosen because, at this point, blank corrosion rates approximate those seen in the field.

From Table 2 it is evident that:

(1) Copper is more protected by Ca overbased petroleum sulfonate plus benzotriazole than the additive effect of either alone;

(2) Protection is also afforded to tin/lead plated copper;

(3) The percent protection is increased at anodic polarizations of the copper;

(4) Several solvent systems and ratios work well; the optimum concentration depends on the particular solvent system.

In general, there should be sufficient barrier corrosion inhibitor to coat the metal and sufficient complexing agent in the barrier corrosion inhibitor to chelate with copper where corrosive fluids penetrate this barrier by any means such as by holidays (i.e., pinpoint entrance points) in the barrier corrosion inhibitor.

The weight ratio of barrier corrosion inhibitor to copper complexing agent may vary widely depending on the particular system, the particular copper complexing agent, etc. In general, the ratio of barrier corrosion inhibitor to copper complexing agents is from about 100/1 to 1/100, but preferably from about 25/1 to 5/1.

The amount of total corrosion inhibitor employed should be sufficient to coat copper to a thickness so as to protect it from the effects of corrosive materials. For example, a thickness of from about 0.0001" to 0.100", but preferably from about 0.005" to 0.015".

Phosphorus Oxide/Zinc Oxide/Aluminum Oxide Glasses

C.F. Drake; U.S. Patent 4,346,184; August 24, 1982; assigned to International Standard Electric Corporation has found that within the phosphorus oxide/zinc oxide/aluminum oxide glass forming region certain composition ranges are particularly effective as corrosion inhibiting materials, particularly for structural components, including, e.g., bridges, buildings and shipping containers.

According to one aspect of this process, there is provided a water-soluble glass composition comprising 54.6 to 63.3 mol zinc oxide, 35.8 to 45.3 mol % phosphorus pentoxide, the remainder comprising at least 0.1 mol % aluminum oxide. In its preferred form, the glass composition is either from 55.3 to 57.3 mol % zinc oxide, 41.4 to 43.4 mol % phosphorus pentoxide and 1 to 2 mol % aluminum oxide or, alternatively, from 61.1 to 63.3 mol % zinc oxide, 35.8 to 37.8 mol % phosphorus pentoxide and 0.1 to 2 mol % aluminum oxide. The preferred compositions are especially effective for corrosion inhibition.

According to a further aspect of the process there is provided a method of making a corrosion inhibiting paint composition, comprising providing a fine powder by fusing quantities of zinc oxide, phosphorus pentoxide and alumina or precursors thereof to form an homogeneous melt, quenching the melt to form a solid material, comminuting the solid material to the fine powder; and dispersing the powder in a paint binder medium.

For structural applications in which thick coatings are used, i.e., 50 to 100 microns thickness or even more, the glass should be comminuted to a final size in which the majority by weight of the particles are from about 10 to 60 microns, preferably 20 to 40 microns in average diameter.

Example: The compositions listed in Table 1 were individually prepared by blending together appropriate amounts of zinc oxide, ammonium dihydrogen phosphate and aluminum hydroxide, and fusing to form a melt at elevated temperature. The melts so formed were then quenched by pouring onto a cold steel plate and the glass obtained was successively crushed, granulated, pin-disc milled and finally wet milled in a nonaqueous medium in a vibratory ball mill. The wet slurry of powdered glass was then dried. These glasses were analyzed and gave the compositions listed in Table 1.

Table 1

	(mol %)		
Batch No.	ZnO	Al_2O_3	P_2O_5
1	59.8	1.0	39.2
2	59.0	2.0	39.0
3	56.9	1.3	41.8
4	62.1	1.1	36.8
5	58.8	1.4	39.8
6	54.6	2.4	43.0
7	56.3	1.3	42.4

These were then evaluated by preparing small quantities of test paints to the formulations listed in Table 2.

Table 2

Test paint formulations:		
Volume concentration of active glass	Proportion of active to total pigment (% w/w)	
pigment (% v/v)	long oil alkyd binder	Chlorinated rubber binder
3	6	6
10	23	24
25	66	67

The paints are prepared by two-stage ball-milling to a fineness of grind over 10 microns and were then applied to clean mild steel coupons by brushing and allowed to cure for several days. The coating was then crosscut and the lower half top coated with a proprietory alkyd white gloss paint.

The coupons were then subjected to accelerated and natural test schedules as defined in British Standard No. 3900 using commercially available priming paints for comparison.

The results of such testing procedures showed that the glass pigments gave effective protection against corrosion at significantly lower loadings in a paint coating than those specified for conventional pigments such as zinc orthophosphate, when evaluated either by resistance to rusting or lack of blistering of the paint film. In particular, paints containing soluble glass pigments are especially effective in preventing rusting or lack of blistering of the paint film. In particular, paints containing soluble glass pigments are especially effective in preventing rusting of steel surfaces where the dried paint film has been removed by mechanical damage.

FERROUS METALS IN CONTACT WITH CONCRETE

Mold Release Compositions

A.L. Dessaint and J. Perronin; U.S. Patent 4,295,976; October 10, 1981; assigned to Produits Chimiques Ugine Kuhlmann, France provide a process for fluorinated telomerization products. The products are produced by the reaction of:

(a) 1 mol of an ester of one or more acids of the formula

(1) $$HS{-}A(COOH)_n$$

with one or more polyols, and;

(b) 1 to 5 mols of one or more compounds possessing at least one ethylenic bond, one at least of these compounds conforming to the general formula

(2) $$Rf{-}B{-}\underset{\displaystyle R}{\underset{|}{C}}{=}CH{-}R$$

In formulae (1) and (2), A represents a hydrocarbon radical which may contain atoms of oxygen, nitrogen, phosphorus, sulfur or halogen, n is a whole number from 1 to 4, Rf is a perfluorinated chain containing 1 to 20 carbon atoms, B is a bivalent chain, one of the symbols R represents a hydrogen atom and the other a hydrogen atom or an alkyl group containing 1 to 4 carbon atoms.

In addition to their primary applications as antistaining or soil release finishes, the products of this process may also be used for other purposes, such as textile printing or the coloring of textiles by pigments, oiling or sizing of textile fibers, or to obtain special properties such as antistatic, or antipilling effects. On substrates such as paper, wood, metals or plastics one can obtain properties of interest for mold release, prevention of corrosion and problems relating to antiadhesion.

Example: A glass reactor of 1,000 cm^3 capacity, equipped with a stirrer, a condenser with separator and a heater is charged with 600 g (1 mol) of polyoxyethylene glycol of molecular weight 600 (known as Emkapol 600), 115 g (1 mol) of 80% thioglycolic acid, 1 g of paratoluenesulfonic acid and 600 g of toluene. This is heated to boiling and 41 g of water is driven off by azeotropic entrainment. After removal of the toluene by distillation, 674 g of the monoester of the thioglycolic acid and the polyoxyethylene glycol mentioned above is obtained. The crystallizing point of this liquid product, which is soluble in water is 10°C.

A glass reactor of 500 cm^3 fitted with a stirrer, a reflux condenser and a heater is charged with 43.2 g (0.1 mol) of polyfluorinated monomer of the formula:

$$C_6F_{13}{-}C_2H_4{-}O{-}CO{-}\underset{\displaystyle CH_3}{\underset{|}{C}}{=}CH_2$$

67.4 g (0.1 mol) of the thioglycolic ester prepared as described above, 260 g of dioxane and 2 g of tert-butyl perpivalate. The mixture is heated to 80°C and allowed to react for 12 hours at this temperature. After cooling, one obtains 370 g of a solution S of telomer according to this process. This solution which is autoemulsifiable in water, has a proportion of dry matter of 30% and a proportion of fluorine of 6.8%. The telomer is of low consistency; it is obtained in the form of a paste of which the liquefaction point is about 9°C.

An iron mold 305 x 405 x 40 mm is coated internally with a composition containing 6 g of solution (S) and 94 g of acetone. The coating is carried out with a brush, in two coats with an intermediate drying at ambient temperature of 30 minutes.

A concrete prepared from Portland cement CPALC-325 (1 part) gravel of granulometry 5/25 (2 parts) and river sand of granulometry 0/2 (1 part) is poured into the coated mold. The concrete is allowed to set for 48 hours at ambient temperature. After this time a slab of concrete is obtained which releases perfectly from the mold. Moreover, the mold is not rusted.

Working under the same conditions with an untreated mold, release from the mold is difficult, part of the concrete remains adhering to the mold and the latter is rusted.

In this process of *J. Perronin and A.L. Dessaint; U.S. Patent 4,302,366; Nov. 24, 1981; assigned to Produits Chimiques Ugine Kuhlmann, France* fluorinated products result from the reaction of:

(a) one molecule of one or more acids of the formula:

$$\text{(1)} \qquad HS{-}A(COOH)_n$$

and of

(b) 1 to 5 molecules of one or more compounds possessing at least one ethylenic bond, at least one of these compounds corresponding to the general formula:

$$\text{(2)} \qquad Rf{-}B{-}\underset{\displaystyle R}{\underset{|}{C}}{=}CH{-}R$$

and possibly neutralized or partially neutralized with an inorganic or organic base.

In the formula (1) and (2), A represents an aliphatic or aromatic hydrocarbon radical, n is a whole number from 1 to 4, Rf represents a straight or branched perfluorinated chain containing 1 to 20 carbon atoms, B represents bivalent, possibly branched, chaining and may comprise sulfur, oxygen or nitrogen atoms, one of the symbols R represents a hydrogen atom and the other a hydrogen atom or an alkyl group containing 1 to 4 carbon atoms.

These products, applied to substrate materials, confer upon these an oilproofing and waterproofing effect as well as a resistance to aggressive products or solvents.

On materials such as wood, metals, plastics, concrete, plaster, interesting properties for mold-stripping, and in the fight against corrosion or fouling and the problems relating to adhesion resistance may be obtained.

Calcium Nitrite for Steel Contacting Cement

The process of *A.M. Rosenberg and J.M. Gaidis; U.S. Patent 4,355,079; Oct. 19,*

1982; assigned to W.R. Grace & Co. is directed to coating metal elements with a thin film of calcium nitrite (including compositions containing same) whereby the metal is inhibited against corrosion on being placed in hydraulic (including alite) cement.

The term alite cement is defined as including neat pastes, mortars, and concretes and the mixed, dry unreacted ingredients of neat pastes, mortars, and concretes, comprising as alite cement binder, a composition containing greater than 20% tricalcium silicate based on the dry weight of the composition. The most common alite cements are portland cements, and mortars and concretes containing portland cements. Most commercially available alite cements contain binders comprising from about 20 to 75% tricalcium silicate. The alite cement binder, or concrete binder, is the component which provides the desired bonding, for example, portland cement.

Hydraulic cements (including alite cements) encounter various corrosion environments. In some, the environment is an inherent part of the cement, e.g., as by use of calcium chloride accelerator, or the use of chloride containing materials. Other environments may be extraneous, e.g., use of calcium chloride and/or salt in snow and ice removal, exposure to salt spray or brines, and the like. Such environments tend to attack and corrode metal pieces within or in contact with the alite cement.

Calcium nitrite can be used as an effective corrosion inhibiting coating on steel structures, including iron, steel, aluminum, steel-aluminum alloy grids, reinforcing rods, girders, etc., which are in contact with corrosion environments in alite cements.

There are several ways to coat the metal piece with the inhibitor. The simplest way is to make up an aqueous solution of calcium nitrite, for example, 10 to 40 weight percent, dip the work piece into this solution, pull it out, and let it air dry or oven dry. This will give an effective coating of the inhibitor on the metal piece. The metal piece can then be placed as a structural member in concrete, mortar, or the like. Spraying with a hose is also effective as is, wiping or brushing the inhibitor onto the metal.

There are various other systems of coating the metal piece with the inhibitor. The calcium nitrite can also be contained in a carrier material which is then applied to the metal piece. For example, in one system, the inhibitor is admixed into a slurry of hydraulic cement, in an amount of about 1 to 2 wt %, based on the weight of solids in the cement. The metal piece is then dipped into the slurry, withdrawn, then permitted to cure, and after curing is then available for use in alite cement matrix.

It has been found that calcium nitrite can be coated as a thin film on a metal piece. The film thickness can be from about 0.005" (5 mils) to about 0.375" and, more preferably, from about 0.005" to 0.06" thick. Such thin film coatings of calcium nitrite have been unexpectedly found to effectively inhibit corrosion of the treated metal piece when it is subsequently embedded in or in contact with cement, concrete, mortar or the like. This process has been found more effective than other known metal salts as inhibitors.

This process causes effective corrosion inhibition of metal pieces without the need for using large amounts of calcium nitrite. It is well known that large scale manufacture, such as commercial manufacture, of nitrite salts is difficult and causes the inhibitor to be expensive. To form an effective and economically efficient reinforced cement system, one should utilize the coated metal pieces in conjunction with a nitrite-free alite cement. That is to say, the cement in which the precoated metal pieces are embedded in or in contact with should be free of nitrite salt such as calcium nitrite. In this manner one forms a corrosion inhibiting system in a most effective and efficient manner.

Nitrite-Glycine-Hydrazine Hydrate Mixture

In reinforced concrete structures and the like wherein sea sands are used, there is a problem that the steel materials used in concrete, such as reinforcing bars, steel frames, lathes and the like, are rusted and corroded by the chlorides contained in sea sands.

This process of *T. Fujita and T. Kashima; U.S. Patent 4,365,999; December 28, 1982; assigned to Kiresuto Kagaku KK and Osaka Semento KK, Japan* provides a corrosion-inhibiting method for steel materials in concrete or mortar. It is characterized in that

(1) a nitrite,

(2) a compound of general formula:

$$NH(R)CH_2COOH$$

wherein R represents a H atom or an alkyl group having 1 to 4 carbon atoms, or salts thereof, and

(3) a hydrazine hydrate

are incorporated in the chlorides-containing concrete or mortar.

The nitrites used in this process are alkali metal salts or alkaline earth metal salts of nitrous acid or mixtures thereof; sodium nitrite is especially preferable.

The salts of a compound represented by the aforementioned general formula are alkali metal salts or alkaline earth metal salts of the compound or mixtures thereof; sodium salt of glycine is especially preferably.

The amounts to be used and the relative proportion of each component in the corrosion inhibitor of this process depend upon the amount of chlorides incorporated in the concrete or mortar and are therefore not critical. In general, 0.5 to 0.05 part by weight of component (1) as sodium nitrite, 0.5 to 0.005 part by weight of component (2) as glycine and 0.5 to 0.002 part by weight of component (3) as 100% hydrazine hydrate are blended based on 1 part by weight of the chlorides as sodium chloride in order that the corrosion inhibitor may display a sufficient corrosion-inhibiting effect when ordinary sea sands are used as the fine aggregate.

Methods for incorporating the corrosion inhibitor of this process into the concrete or mortar are as follows. Although each component of the corrosion inhibitor may be directly added to concrete or mortar during blending or individually at kneading time, it is preferable to dissolve the components in the water used for kneading. Alternatively, the steel materials to be protected may be immersed in an aqueous solution of the corrosion inhibitor, or an aqueous solution of the corrosion inhibitor may be sprayed or coated on the surfaces of the steel materials.

In general, although the corrosion-inhibiting effect of a corrosion inhibitor for steel materials in concrete or mortar must be excellent, the corrosion inhibitor must also not harmfully effect such properties as solidification, hardening, size-stability, durability and the like of concrete or mortar. Further, the inhibitor must not contain a component harmful to the human body and must be convenient and economical to use and so forth. The corrosion inhibitor of this process satisfies these requirements. In the following example all parts and percents are based on weight.

Example: A mortar was prepared according to JIS (Japanese Industrial Standard) R5201 (normal portland cement-Toyoura standard sand = 1:2, and water-cement ratio 0.65). The chloride concentration in the mortar was adjusted to the concentration obtained when a sand whose chloride content is 0.3% is employed by using an artificial seawater ($NaCl_2$ 12.45%, $MgCl_2 \cdot 6H_2O$ 1.11%, Na_2SO_4 0.41%, $CaCl_2$ 0.12%, and KCl 0.07%). An admixture, Pozzolith 5L (lignin sulfonate containing water-reducing agent) was added.

A test reinforcing bar was prepared by eliminating the black skin of a steel bar SR-24 for use in reinforced concrete, grinding the surface of the steel bar by means of No. 400 emery paper, washing the ground surface with benzene and acetone, and drying the steel bar; the bar's ends are hemispheres.

The mortar was used to mold a test body (10 cm x 30 cm) whose covering thickness over the bar surface was about 2 cm by immersing two such reinforcing bars (1.2 cm x 25 cm) into the mortar. After 24 hours, the test body was taken out from its mold and stored in a thermostatic humidity chamber in which the temperature was 70°±2°C and the humidity (RH) was more than 90%. After three months, the reinforcing bars were removed from the concrete and examined for the existence of rust by means of a magnifying glass of 5 magnifications. The results obtained are shown in the following table.

Chlorides in Sand (%)	. . . Corrosion Inhibitor (ppm)* . . .			Admixture (%)**	Area of Rusted Surfaces (%)
	Sodium Nitrite	Glycine (Na salt)	Hydrazine Hydrate	Pozzolith 5L	
–	–	–	–	0.25	–
0.3	–	–	–	–	~40
0.3	2,000	–	–	0.25	~10
0.3	5,000	–	–	0.25	~2
0.3	–	1,000	–	–	~20
0.3	–	–	500	–	~15
0.3	1,000	75	150	–	–
0.3	500	50	100	–	~1
0.3	500	50	100	0.25	–

*The amount of corrosion inhibitor added to the kneading water.
**The ratio of admixture to cement.

As is obvious from the table it is recognized that a corrosion of the steel is satisfactorily prevented by this process. Such a satisfactory effect would not be expected if a prior art corrosion inhibitor alone was used. Furthermore, the tests prove that corrosion-inhibiting ability is increased by using the corrosion inhibitor of this process together with a water-reducing agent.

MARINE APPLICATIONS

Pipe Coating Material

It is well known in pipeline construction of the need for providing corrosion protection to pipe for marine or overland installations. In the case of marine installations or installations over swamp lands, it is necessary to coat the joints of pipe with a relatively heavy weight coating in addition to corrosion protection. Bituminous mastic materials have long been utilized in pipe coating because of their superior corrosion resistance and also because of their ability to form a binder for particulate material to provide an antibuoyancy coating.

A conventional pipe coating composition employing a thermoplastic binder material will include from 5 to 15% by weight of the thermoplastic binder such as asphalt, 15 to 35% by weight of filler which has heretofore been lime, dust, portland cement or some combination of the two, and a particulate material such as limestone or heavy ore bearing materials present in quantities ranging from 50 to 70% and with a fibrous material such as glass fibers present in a quantity of less than 1% by weight.

If the binder material is asphalt, it is heated to a temperature within a range of 350° to 400°F and the dry components are then mixed with the heated binder so that the temperature of the final mix is 225° to 275°F. The material is normally applied using an extrusion process at a temperature of around 250°F. Other naturally occurring petroleum base binders known to those skilled in the art may be employed.

It has been found by *R.J. Harris; U.S. Patent 4,244,740; January 13, 1981; assigned to H.C. Price Co.* that, in lieu of lime dust or portland cement, a material which is obtained from the stacks of cement manufacturing plants may be substituted. The cement stack dust will normally comprise about 40 to 50% lime with the remainder comprising inorganic minerals and salts.

In general, cement stack dust may be used with a thermoplastic binder and particulate material to form a suitable coating composition. The binder material should comprise from 5 to 15% by weight, the cement stack dust filler should comprise from 15 to 35% by weight, and the particulate material which may include sand, crushed limestone, iron ore aggregate or a combination of these should comprise 50 to 70% by weight of the composition. It is to be understood that the term particulate material is intended to encompass fibrous substances. For example, glass fibers may be utilized up to approximately 1% by weight of the total coating. The fibrous material is preferably about 0.2% by weight.

The components may be varied over a range of approximately + or -10% of the weight percentages given. The preferred formulation is set forth below.

Component	Percent by Weight
Type II asphalt	14.1
Cement stack dust	16.4
Sand	69.3
Glass fiber	0.2

The pipe coating composition set forth above has been subjected to various standard tests to determine tensile strength, flexural strength, modulus of elasticity, crack time and deflection coefficient. It has been found that the formulation performs comparably with conventional formulations employing lime dust as the filler material.

Abrasive Particle for Cleaning and Rustproofing Steel

In the shipbuilding industry and in the manufacture of steel structures, especially where the steel is subjected to a rustproofing treatment prior to being subjected to further operations, there are disadvantages to the use of zinc as a corrosion-resistant metal.

An object of this process of *A. Noomen; U.S. Patent 4,244,989; January 13, 1981; assigned to Akzo NV, Netherlands* is to provide a method which obviates such drawbacks of the known methods such as the formation of white rust and weakening of the paint system as a result of a zinc-rich undersurface while at the same time maintaining the formation on the metal surface of a corrosion-resistant coat during the abrasive treatment.

This object is accomplished by providing an abrasive particle and a process for abrading an iron containing surface to be painted with abrasive particles coated with a binder and a corrosion-resisting agent which is a corrosion-resisting salt having a solubility of not more than 20 g/ℓ in water at 20°C. The corrosion-resisting agent is bonded to the abrasive-particles with a binder which is compatible with the paint to be applied to the surface. The corrosion resisting compound is deposited on the surface as it is abraded. The corrosion-resisting salt which may be metallic or nonmetallic and which is not or is only poorly soluble in water effectively prevents osmotic action following application of the paint system, as a result of which there will be no formation of blisters or other disengaging phenomena of the coating of paint. Moreover, when for instance a metal object is stored in the open the layer of the corrosion-resisting salt is prevented from being removed by rain or water of condensation before the actual paint system is applied to it.

For examples of suitable, corrosion resisting salts may be mentioned the salts of carboxylic acids containing nitro groups, and preferably the salts of aromatic nitrocarboxylic acids containing 7 to 14 carbon atoms. The heavy metal salts, such as the lead and/or the zinc salts of carboxylic acids are preferred. More particularly, use is made of the zinc salt and the lead-zinc salt of 5-nitroisophthalic acid. As examples of other suitable salts may be mentioned film-forming alkaline complex compounds of an alkaline earth metal salt of an organic sulfonic acid and an alkaline earth metal carbonate. Calcium is the preferred alkaline earth metal. A mixture of calcium carbonate and the calcium salt of an alkylphenyl sulfonic acid whose alkyl group contains 22 carbon atoms is most preferred.

As examples of suitable binders may be mentioned epoxy resins, polyamide resins, and coumarone indene resins. These binders are compatible with a paint system based on epoxy resin. If the use of a paint system based on, for example, an unsaturated polyester resin or acrylate resin is desired, then use may be made of a copolymer of styrene and an acrylate monomer such as methylmethacrylate and/or butyl acrylate as binder. Optionally, mixtures of binders may be employed.

As examples of suitable abrasives may be mentioned inorganic materials such as glass beads, copper slag, aluminum oxide granules such as corundum and sand.

Examples 1 through 7: An abrasive in the form of corundum contained in a rotating drum is coated with a corrosion-resisting composition. The corundum and the coating composition are mixed until the composition is homogeneously distributed over the surface of the abrasive particles. The epoxy resin is a diglycidyl ether of Bisphenol A and is available as Epikote 828 and has an equivalent weight of 180 to 210.

The zinc powder is commercially available as Zincomox AAA and has a minimum zinc content of 98% by weight and a particle size of 2 to 4 μm.

The compound referred to as calcium sulfonate is a mixture of calcium carbonate and the calcium salt of an alkylphenyl sulfonic acid whose alkyl group contains 22 carbon atoms. The amounts given are in parts by weight. Example 1 is a comparative example.

		Parts by Weight
Ex. 1:	Corundum	1,000
	Epoxy resin	4
	Zinc powder	200
Ex. 2:	Corundum	1,000
	Epoxy resin	4
	Zinc salt of 5-nitroisophthalic acid	50
Ex. 3:	Corundum	1,000
	Epoxy resin	4
	Lead-zinc salt of 5-nitroisophthalic acid	80
Ex. 4:	Corundum	1,000
	Epoxy resin	0.4
	Calcium sulfonate	3.6
	Zinc salt of 5-nitroisophthalic acid	50
Ex. 5:	Corundum	1,000
	Epoxy resin	0.4
	Calcium sulfonate	3.6
	Lead zinc salt of 5-nitroisophthalic acid	60
Ex. 6:	Corundum	1,000
	Epoxy resin	4
	Zinc powder	150
	Zinc salt of 5-nitroisophthalic acid	50
Ex. 7:	Corundum	1,000
	Epoxy resin	4
	Zinc powder	150
	Lead-zinc salt of 5-nitroisophthalic acid	50

Subsequently, corroded steel plates (steel 37) are abraded with corundum particles coated with the aforementioned compositions until a clean, uniformly coated steel surface is obtained. With the test panels thus obtained the following experiments are carried out.

(A) The panels are exposed to open air conditions for a period of four weeks. The panels are evaluated in accordance with ASTM Standard Method D 610. The time taken for the panels to reach rust grade 3 of the rust grade scale is measured. The panels are also examined for the time it takes to form white rust thereon, if any. For comparison a steel panel abraded with noncoated corundum is evaluated (control). The values obtained are listed in the table below.

Abrasive	Time to Value 3 (ASTM D 610) (days)	Formation of White Rust After Some Time (days)
Control	1	–
1*	28	2
2	14	None
3	28	None
4	14	None
5	28	None
6	28	None
7	28	None

*Comparison.

Coating Effective in Fresh and Salt Water

A variety of materials are known as coating compositions for preventing corrosion of metals (primarily steels), and one of the most effective materials is a composition comprising chromic (acid) anhydride, a particulate metal (mainly zinc or aluminum), a viscosity modifier, an oxohydroxy low molecular weight ether (polyglycol) and a solvent, such as described in U.S. Patent 3,940,280.

This composition exhibits superior anticorrosive effect against salt water on steel, but not the same effect against fresh water. Because it contains hexavalent chromium, some consideration may need to be given to environmental pollution.

An anticorrosive coating composition has been sought that will be effective in fresh water as well as salt that will be highly acceptable environmentally, by seeking a bonding material that would replace the chromic acid.

The process of *T. Higashiyama and T. Nishikawa; U.S. Patent 4,266,975; May 12, 1981; assigned to Diamond Shamrock Corporation* offers a substantially resin-free anticorrosive coating composition for metals comprising at least one boric acid compound and at least one water-soluble chromic acid compound and particulate metal and at least one high-boiling organic liquid and water and/or organic solvent which, when necessary, contains a pH modifier. Further, the composition may contain nonionic dispersing agent and/or viscosity modifier.

In a particular aspect, the composition comprises 10 to 40 wt % of particulate metal; 1 to 12 wt % of water-soluble acid compounds (boric acid component plus chromic acid constituent, with the boric acid component contributing from 5 to 95% of the mixture); 7 to 30 wt % of at least one high-boiling organic liquid; and with the remainder being water or water mixed with a solvent, with the composition optionally containing ingredients such as pH modifier and viscosity modifier.

The corrosion resistance tests and the evaluation of the test results employed in the examples are as follows:

(1) Salt water (fog) spray test—the neutral salt water spray test described in JIS Z-2371 was followed. The degree of corrosion of the test samples was visually observed and evaluated in accordance with the following standards.

5 points: Absolutely no formation of red rust.
4 points: Formation of 10 or less pinholes of red rust.
3 points: Rust spots are distributed and some flow of rust is observed.
2 points: The flow of rust is remarkable.
1 point: The entire surface is covered with red rust.

(2) CASS test–the test method of JIS D-0101-1971 was followed, except that a spray liquid of pH 3.5 was used. Standards for evaluation of the formation of rust are the same as above.

(3) Outdoor exposure test–the test pieces were exposed attached to exposure stands (surfaces facing the south inclined at 30°) in Yokohama, Japan. Standards for evaluation of formation of rust are the same as above.

The used test pieces were 15 x 15 cm, 0.8 mm thick soft steel plates.

Example 1: 60 parts of metallic zinc flakes (0.1 to 0.3 micron thick, about 15 microns long in average in the longest part) were dispersed in diethylene glycol containing 0.3 part of Nopco 1529 (alkylphenol polyethoxy adduct surfactant) so as to make the total amount 100 parts. (This mixture is the first component.) Separately orthoboric acid and chromic acid anhydride are dissolved in deionized water so that the orthoboric acid content was 5.17% and the chromic acid content was 1.72%, and calcium oxide was added as the pH modifier so that the content thereof would be 1.72%. The boric acid concentration, basis boric plus chromic acid, is 75 wt %. (This mixture is the second component.)

The first and second components were mixed in the weight ratio 42:58 by pouring the former into the latter while slowly stirring, and stirring was continued overnight at room temperature. In this mixture, the orthoboric acid concentration was 3%, the chromic acid anhydride concentration was 1% and the calcium oxide concentration was 1%. The mixture thus obtained was applied onto soft steel plates by means of a bar coater to form a uniform film thereon, the plates having been washed with alkali and sufficiently polished with a Scotch Bright Very Fine polishing cloth, and the plates were heated in an electrically heated hot air circulating furnace, the temperature of the soft steel plates being held at 300°C for 4 minutes after the plates reached that temperature, and they were then let stand to be cooled to room temperature. The amount of the applied composition was 1 micron in thickness and 250 mg/ft^2 (2.7 g/m^2), by weight, per area.

Example 2: Coating films were formed under the same conditions as in Example 1 except that the final mixture contained 2% orthoboric acid and 2% chromic acid anhydride, for a 50% boric acid concentration.

Example 3: Coating films were formed under the same conditions as in Example 1 except that the final mixture contained 1% orthoboric acid and 3% chromic acid anhydride, for a boric acid ratio of 25%.

Example 4: Coating films were formed under the same conditions as in Example 1 except that the final mixture contained 0.4% orthoboric acid and 3.6% chromic acid anhydride, for a 10% boric acid concentration.

Comparative Example: Coating films were formed under the same conditions as in Example 1 except that no boric acid compound was used (that is, ingredients were blended so that the chromic acid anhydride content was 4%).

The tests described above were carried out with these test pieces on which the coating films were formed as described above. The results are shown in the following table.

Sample	Boric Acid Concentration (%)	Salt Water Spray Test**	CASS Test***	Outdoor Exposure* Flat Surface	Scribe
†	0	5	1	3	1
Ex. 4	10	5	2	4	3
Ex. 3	25	5	3	5	4
Ex. 2	50	5	3	5	5
Ex. 1	75	4	3	5	5

*6 months.
**144 hours.
***20 hours.
†Comparative example.

Hydrolyzed Organotin Siloxane

Compositions have been developed for protecting marine surfaces from fouling organisms, including a linear or crosslinked polymer having a siloxane backbone and trisubstituted tin radicals such as tributyltin bonded to silicon atoms by way of an oxygen atom. The biologically active polymer can be used for forming coatings for steel, concrete, or other surfaces exposed to marine environments for inhibiting the growth of fouling organisms. In some embodiments the polymer is present as an additive in a coating composition. In other embodiments the crosslinked polymer forms a portion of the binder of the coating.

It is sometimes convenient to hydrolyze the precursor to a prepolymer before completing a coating composition since prehydrolysis can reduce the curing time of a coating. Polycondensation of the prepolymer has heretofore been inhibited by retaining the prepolymer in ethyl alcohol or similar solvent. Ethyl alcohol is a product of hydrolysis of the precursor when, for example, an ethoxysilicate is employed in the synthesis. When such a coating is applied, evaporation of ethyl alcohol results in polycondensation in the coating.

Ethyl alcohol and similar water miscible solvents may not be compatible with other ingredients in the coating composition. This is particularly true when the coating composition includes a chlorinated rubber, for example. It is therefore desirable to provide a solution containing polymeric material which is compatible with such coating compositions.

A.P. Gysegem; U.S. Patent 4,311,629; January 19, 1982; assigned to Ameron, Inc. provides a method for preparing a solution containing hydrolyzed organotin siloxane polymeric material. The method comprises the steps of:

> combining an organotin R-oxy siloxane in which the ratio of tin atoms to silicon atoms is in the range of 1:50 to 1:1, where R represents a group consisting of alkyl and alkoxyalkyl radicals containing less than 6 carbon atoms, a water

miscible solvent, water, a hydrolysis catalyst, and a water immiscible solvent; and

after hydrolysis of the organotin R-oxy siloxane, removing by distillation the water miscible solvent and R-alcohol from the hydrolysis reaction.

A prepolymer for forming a biologically active polysiloxane binder preferably has a tin to silicon atom ratio of from about 1:12 to about 1:3. In this range it has been found that a hard, clear, solvent-resistant film exhibiting effective and long-lived biological activity in preventing fouling on marine surfaces can be formed.

Example 1: A precursor is formed by reacting 533.2 g of Ethyl Silicate 40 with 623.7 g of tributyltin acetate. Ethyl acetate from the reaction is removed by distillation yielding 1,000 g of tributyltin ethoxysiloxane or tributyltin ethyl silicate. This precursor is mixed with 500 g of ethyl alcohol, 20 g of 2-ethylaminoethanol and 200 g of xylene. This solution is heated to about 50°C and stirred vigorously while 61 g of water is added dropwise to the solution over a period of about 15 minutes. The temperature of the solution is then increased to effect distillation of the ethyl alcohol. Distillation is complete when 942.5 g of distillate is collected. The product, comprising a tributyltin substituted siloxane polymeric material in xylene, is a viscous liquid.

Example 2: A marine antifouling coating composition is formulated from a solution containing polymeric material prepared in accordance with Example 1 and contains the following ingredients: 98.0 g of tributyltin silicate (from Example 1), 77.0 g of Alloprene X-10 (a chlorinated rubber binder), 9.2 g of coal tar, 36.3 g of w/w rosin, 5.6 g of pine oil, 15.8 g of Shell solvent 1693 (a petroleum distillate), 3.0 g of Chevron solvent 265 (naphtha), 39.0 g of xylene, 1.6 g of methanol, 56.0 g of zinc oxide, 42.0 g of red iron oxide, 35.0 g of talc, and 5.4 g of Bentone 34 (dimethyldioctadecyl ammonium bentonite). When this composition is coated on a substrate and the solvents evaporate, an adherent, tough, somewhat resilient marine antifouling coating results.

Example 3: A precursor is formed by reacting 745.2 g of Ethyl Silicate 40 with 453.3 g of tributyltin acetate. Ethyl acetate from the reaction is removed by distillation. A solution is formed comprising 200 g of the resultant tributyltin ethyl silicate in 100 g of ethanol. The solution is heated to 40°C and stirred vigorously during slow dropwise addition of 17.8 g of a 2% solution of sulfuric acid in water combined with an equal volume of ethanol. When the addition of this mixture is complete, 50 ml of xylene is added to the solution which is then heated to distill the ethyl alcohol. When about 150 ml of distillate has been collected, an additional 200 ml of xylene is added, distillation is considered complete when a total of 400 ml of distillate has been collected. The product is a clear liquid which becomes somewhat cloudy upon cooling. Full crosslinking and precipitation of polymer was not observed. This solution containing polymeric material is useful in a variety of marine antifouling coating compositions.

Metal Silicides or Silicon-Metal Alloys

Various compositions have been proposed in the prior art to provide varying measures of protection against corrosion of corrodible metallic surfaces. In more recent years, attention has been progressively given over to providing compositions and systems for providing cathodic-anodic corrosion protection for metallic surfaces of structural elements. Such structures as underground pipes, storage tanks, buildings and the like, as well as metallic structures continually in contact with water, such as ships, support structures for drilling rigs, docks and the like have been treated with or coated with a variety of compositions or systems to impart varying degrees of corrosion resistance thereto.

Generally, such systems or compositions as have been employed utilize either an external source of electrical current which serves to maintain as cathodic the surface to be protected or the corrosion protection composition itself forms an internal current with the metallic surface to be protected. In the latter systems, the coatings contain metallic particles which are more anodic in nature than the metallic surface to be protected and thus serve to function as sacrificial anodes.

It is known that such zinc-rich (80 to 95%) coating compositions serve to protect against harmful corrosion of corrodible metal surfaces to a greater degree than do ordinary paints, particularly when applied to iron and steel surfaces. Such zinc-rich coating compositions are effective in salt-air atmospheres and in applications where the coated metal surface contacts brine solutions. Zinc powder has for some time been recognized as having a specific use as a pigment in the antifouling and anticorrosive ship bottom paints.

It is an object of *J.P. McKaveney and V.P. Simpson; U.S. Patent 4,360,384; November 23, 1982; assigned to Occidental Chemical Corporation* to provide a corrosion protective coating composition for corrodible metal surfaces, which coating composition provides corrosion protection comparable to that provided by current zinc-rich coating compositions.

The coating composition includes a binder and a filler, with the filler present in an amount sufficient to impart corrosion-resistant characteristics to the coating composition. The filler comprises conductive metal particles and particles of a metal composition which comprises silicon and at least one active metal. The metal composition can be an alloy of silicon with an active metal which is unstable when introduced alone into water or a silicide of an active metal which is unstable when introduced alone into water.

Preferably, the active metals are selected from the group consisting of calcium, magnesium, manganese, and barium.

The conductive metal particles which are more anodic than the metal of the substrate to be protected may be particles of the various suitable metals, the specific choice in each instance depending upon the metal of the substrate. In many instances, it has been found that zinc particles produce excellent results on iron and steel and, for this reason, such particles are generally preferred. Other metal particles may be used, however, such as aluminum, magnesium, and the like, as well as particles of metal alloys, so long as these metal particles are electrically conductive and more anodic than the metal of the substrate which is to be protected.

In the following example, unless otherwise indicated, parts and percentages are by weight.

Example: Various coating compositions were formed by admixing a hydrolyzed ethyl silicate binder, comprised of approximately 20% solids, with a filler component in a weight ratio of 100 parts by weight of binder to 225 parts by weight of filler. The filler comprised 100% zinc particles as a control and varying ratios of zinc particles with active metal silicon alloys and/or silicides; varying ratios with ferrophosphorus filler; and a 100% active metal silicon alloy and/or silicide filler. The varying compositions are as presented in the table below.

Formulation	Zinc	Fe_2P	Mg—Fe—Si (5% Mg)	Mg—Fe—Si (9% Mg)	Binder
A(100% Zn)	225	—	—	—	100
B(75% Zn)	169	56	—	—	100
C(75% Zn)	169	—	56	—	100
D(0% Zn)	—	—	225	—	100
E(75% Zn)	169	—	—	56	100
F(0% Zn)	—	—	—	225	100

Standard 4" x 8" cold-rolled steel panels were sandblasted and coated with the above formulations and then air cured at room temperature at about 50% relative humidity for five days. Thereafter, the panels were scribed crosshatching and were tested for corrosion resistance by subjecting them to the known salt-fog exposure test (ASTM B-117-73) for a maximum of 840 hours. The following table sets out the results of such tests at 672 hours and at 840 hours.

	Salt-Fog Results			
	672 Hours		840 Hours	
Formulation	Scribed	Face	Scribed	Face
A	3	9+	1-2	9+
B	0	2-5	0	2-5
C	10	10	9	10
D	2-3	5	1	5
E	10	10	10	10
F	0	6	0	5

The panels were evaluated on a scale of from 0 to 10, wherein 10 is perfect and 0 is 100% rusted. As can be seen, the panel having a filler composed of 100% zinc had severe rust on the scribed portion of the panel at both the 672 and 840 hour test points, while the panels containing 25% active metal silicon alloy or silicide with 75% zinc had little, if any, rust on the scribe after the 672 and 840 hour testing points. When compared to a filler comprising 75% zinc and 25% ferrophosphorus, the activity of the active metal silicon alloy or silicide in its anticorrosion function is further dramatically accentuated.

CHEMICAL INDUSTRY

Diffusion Coating of Ferrous Metals

The process of *I.I. Zaets, I.D. Zaitsev, O.K. Yakshina, N.F. Pershina, A.K. Gorbachev, I.N. Gladky, N.M. Davydenko and G.A. Tkach; U.S. Patent 4,256,490; March 17, 1981* relates to protection of metals against corrosion, and particularly to compositions intended for diffusion coating ferrous metals. Most advantageously the process can be used for corrosion protection of parts and assemblies of the equipment used in chemical industries for the production of soda and soda products as well as magnesium chloride, barium chloride, sodium sulfates and other products of this kind.

The composition for diffusion coating ferrous metals contains a particulate mixture of titanium, chromium, alumina and ammonium halide and, further contains molybdenum and boron, and the ingredients being in the following ratio in percents by weight.

Titanium	51.5-64.0
Chromium	17.5-24.0
Alumina	15.0-21.25
Ammonium halide	1.5-2.0
Molybdenum	0.75-1.5
Boron	1.0-2.0

With such ratio of titanium and chromium in the composition it is possible to provide a high concentration of titanium on the surface of the diffusion layer and at the same time a continuous layer of titanium carbides. The introduction of molybdenum and chromium also assists in making a high-quality diffusion layer. Specifically, boron which is a more active carbide-forming element than chromium makes it possible, in combination with molybdenum, to compact the diffusion layer and to improve its plasticity. In addition, molybdenum considerably increases the passivating power of the coating. The above factors increase the corrosion resistance of the coating in highly concentrated salt solutions and chloride containing media, prevent pitting corrosion and microcracks in the coatings of both statically and dynamically loaded apparatus parts.

Example: A composition for diffusion coating ferrous metals is obtained in the following way. Titanium, chromium, alumina, crystalline ammonium halide, (ammonium chloride), molybdenum and boron, all in the powder form, are taken as initial ingredients. Particles of the metallic powders range in size from 0.8 to 1.5 mm. Alumina and ammonium chloride are taken in powder form with particles of the substances having different dispersion characteristics ranging from a dust-like fraction to particles sizing 1.5 mm. The initial ingredients are taken in the following ratio (percent by weight).

Titanium	64.0
Chromium	17.5
Alumina	15.0
Ammonium chloride	1.5
Molybdenum	0.75
Boron	1.25

At first, titanium powder and chromium powder are mixed together. Then molybdenum, boron, ammonium chloride and alumina are added into the obtained mixture, whereupon the above ingredients are stirred until a homogeneous mixture is obtained.

With the composition obtained the diffusion layer on the workpieces such as parts of pump casings and parts of housings for column apparatus intended for the production of soda and made of 2.5% carbon content cast iron is formed in the following way.

A hermetically sealed container of stainless steel is loaded with workpieces, and then filled with the composition of the process. The container is closed, placed into the furnace and heated up to a temperature of 1000°C. At this temperature the contents of the container are held for 8 hours. In the course of heating and holding, a continuous plastic diffusion layer having a total thickness of 0.16 to 0.20 mm is formed on the surface of the workpiece. After holding, the container with the treated workpieces is air-cooled.

Similarly, samples together with the workpieces were treated in the same container in order to form a diffusion layer on their surface. The samples were made of gray cast iron containing 2.5% C and were 65 x 15 x 3 mm rectangular plates. After the diffusion layer has been formed, the samples were subjected to x-ray analysis in order to determine the nature of phases constituting the diffusion layer, and hardness, continuity and corrosion resistance of the layer were determined by the methods described below. The layer hardness was determined by the Vickers method (HV kg/mm²). Continuity of the layer was determined by means of Wocker reagent (a mixture of $K_3[Fe(CN)_6]$ and NaCl). Filter paper moistened with the reagent was placed onto the surface of the samples in the pore zones. Ions of iron with ions of $[Fe(CN)_6]^{3-}$ formed the compound known as Turnbull's blue, $[Fe_3Fe(CN)_6]$. Location of pores was fixed on the filter paper by blue spots.

Corrosion resistance of the layer was determined in the following way. The samples were immersed into solutions of salts and held therein for 1200 hours, the temperature of the solution being 95°C. Corrosion resistance was evaluated as a decrease in sample weight per unit surface area, taking into account the test duration. Corrosion magnitude was determined in mm/yr, taking into account the specific weight of the material. Test results appeared to be as follows.

Vickers hardness of the diffusion layer HV, kg/mm²	950
Layer continuity, number of spots per cm²	0
Corrosion resistance, mm/yr	
in sodium chloride, 310 g/ℓ	0.001
in magnesium chloride, 250 g/ℓ	0.001
in barium chloride, 263 g/ℓ	0.001
in ammonium chloride, 271 g/ℓ	0.003
in sodium sulfate, 250 g/ℓ	0.001
in potassium sulfate, 200 g/ℓ	0.001
in sodium carbonate, 178 g/ℓ	0.001
in potassium carbonate, 100 g/ℓ	0.001
in sodium bicarbonate, 88 g/ℓ	0.001

I.I. Zaets, N.M. Davydenko, O.K. Yakshina, N.F. Pershina, V.D. Demyanenko and J.M. Pavlov; U.S. Patent 4,276,088; June 30, 1981 provide another composition for a diffusion coating of ferrous metals. It contains a mixture of ingredients taken in the following ratio (wt %): 70.0 to 82.0 titanium, 14.5 to 20.0 alumina, 2.0 to 5.0 ammonium halide and 1.0 to 2.0 graphite, and is intended for use in protecting chemical apparatus from corrosion.

The composition for a diffusion coating of ferrous metals is obtained as follows. Titanium, alumina, crystalline ammonium chloride and graphite, all in the powder form, are taken as initial ingredients. Particles of the titanium powder are ranging in size from 0.8 to 1.5 mm. Alumina and ammonium chloride are taken in the powder form with particles of the substances having different dispersion characteristics and ranging from dust-like fraction to particles sizing 1.5 mm.

Particles of the graphite powder are 0.8 mm in size. The initial ingredients are taken in the following ratio (% wt): 70.0 titanium, 27.0 alumina, 2.0 ammonium chloride and 1.0 graphite. All the above ingredients are stirred to obtain a homogeneous mixture.

Formation of the diffusion layer with the composition obtained on the workpiece such as a pump casing, valve casing, parts of housings for column apparatus made of 3.5% carbon content cast iron castings is carried out as described in the example of the preceding patent.

Similarly, samples together with the workpieces were treated in the same container in order to form a diffusion layer on their surface. The samples were made of cast iron containing 3.5% C and 65 x 15 x 3 mm rectangular plates.

After the diffusion layer has been formed, the samples were subjected to x-ray analysis in order to determine the nature of phases constituting the diffusion layer, and hardness, continuity and corrosion resistance of the layer were determined by the methods also described in the example of the preceding patent. Test results appeared to be as follows.

Vickers hardness of the diffusion layer HV, kg/mm^2	970
Layer continuity, number of spots per cm^2	0
Corrosion resistance, mm/yr	
in sodium chloride, 310 g/ℓ	0.001
in magnesium chloride, 250 g/ℓ	0.001
in barium chloride, 263 g/ℓ	0.001
in sodium sulfate, 250 g/ℓ	0.001
in the mixture of ammonium chloride and sodium chloride, respectively 180 g/ℓ and 70 g/ℓ	0.001
in the mixture of sodium carbonate and sodium bicarbonate, respectively 250 g/ℓ and 50 g/ℓ	0.002
in potassium carbonate, 100 g/ℓ	0.001

Hydrazine Compounds in Caustic Solutions

The process of *G.P. Khare; U.S. Patent 4,282,178; August 4, 1981; assigned to Vulcan Materials Company* relates to an improvement in the manufacture of concentrated caustic soda solutions wherein hydrazine or its salts or derivatives are added in an extremely small but effective amount to the caustic soda solution prior to its final dehydration such that the otherwise severe corrosion of the nickel evaporation equipment in which such dehydration is normally conducted is greatly reduced or eliminated.

Hydrazine and its salts and derivatives are used in proportions such as 1,000 ppm or less, preferably 200 ppm or less and most preferably 40 ppm or less. The effectiveness of this method of corrosion inhibition surprisingly does not seem to depend on the substantial elimination of oxidant impurities from the caustic liquor. In fact, the process is effective even when the hydrazine inhibitor is added to the caustic solution in a proportion substantially smaller than that required to reduce all of the oxidizing agent present, e.g., chlorate to chloride. For instance, the addition of as little as 1% or less of the stoichiometric amount of hydrazine offers significant protection.

As the process is of particular value in connection with the manufacture of concentrated caustic soda solutions, a representative embodiment of such a manufacture will now be described in detail for purposes of illustration of this process. However, while the diaphragm cell process is referred to in this embodiment as the source of the dilute sodium hydroxide solution to which this process is applied, it should be understood that the process is similarly applicable to sodium hydroxide solutions obtained from other sources, e.g., from membrane cells, from mercury cells and from the lime-soda process.

Only about one-half of the sodium chloride in the feed brine to a diaphragm cell is electrolytically converted. The cell liquor is a composite of the unconverted sodium chloride brine, the produced sodium hydroxide, any sodium sulfate impurity, minor amounts of decomposition products, e.g., sodium chlorate and sodium hypochlorite and water. The overall caustic system typically performs the three-fold function of (a) concentrating the caustic to a commercial 50 wt % concentration; (b) recovering the sodium chloride for recycle to the cells; and (c) purging sulfate from the overall chloralkali operation.

Concentration of the caustic has conventionally been done in three steps or effects. With greater emphasis on energy conservation newer plants are being designed featuring quadruple effect evaporation systems, as illustrated in the figure. Referring to Figure 1.1, a weak caustic solution such as the cell liquor from a diaphragm cell process (not shown) is fed from feed tank (1) to the fourth effect (40), concentrated and sent to the third effect (30), where it is concentrated further and sent to the second effect (20) and subsequently to the first effect (10), with further concentration being obtained in each effect.

Differing orders of progression between effects are sometimes employed. Two liquor flash effects (50) and (60) are incorporated as part of the basic system to partially cool and further concentrate the hot caustic liquor by flash evaporating to lower pressure and temperature prior to discharging via line (61) to a final cooling and filtering system (not shown).

Steam introduced via line (3) is used as the primary heat source in the first effect. Vapors evaporated from the first effect (10) are then withdrawn via line (11) and used as the heat source in the second effect (20). Similarly the second effect vapors are passed via line (21) to the third effect (30) where they are used as a heat source. The third effect vapors are in turn removed via line (31) and used in the fourth effect (40). A natural balance of pressure and temperatures occurs between effects, dependent upon progressive concentration of the caustic liquors in each effect.

Heaters (12), (22), (32) and (42) are used as a means where extraneous steam or the vapors produced in the process are used to preheat the caustic solutions that are fed into the effects (10), (20), (30) and (40), via caustic lines (15), (25), (35) and (45), respectively. Steam condensate is withdrawn from the process via lines (100), (101), (102), (103) and (104), while sodium chloride removal is effected in stages (26) and (16).

While the quadruple effect evaporating system is highly efficient with respect to energy, the system usually requires higher process temperatures, e.g., 160° to 175°C, in the more concentrated evaporative stage. It is primarily in this temperature range and at this point in the process that corrosion problems are most persistent and troublesome.

Figure 1.1: Quadruple Effect Evaporation Systems

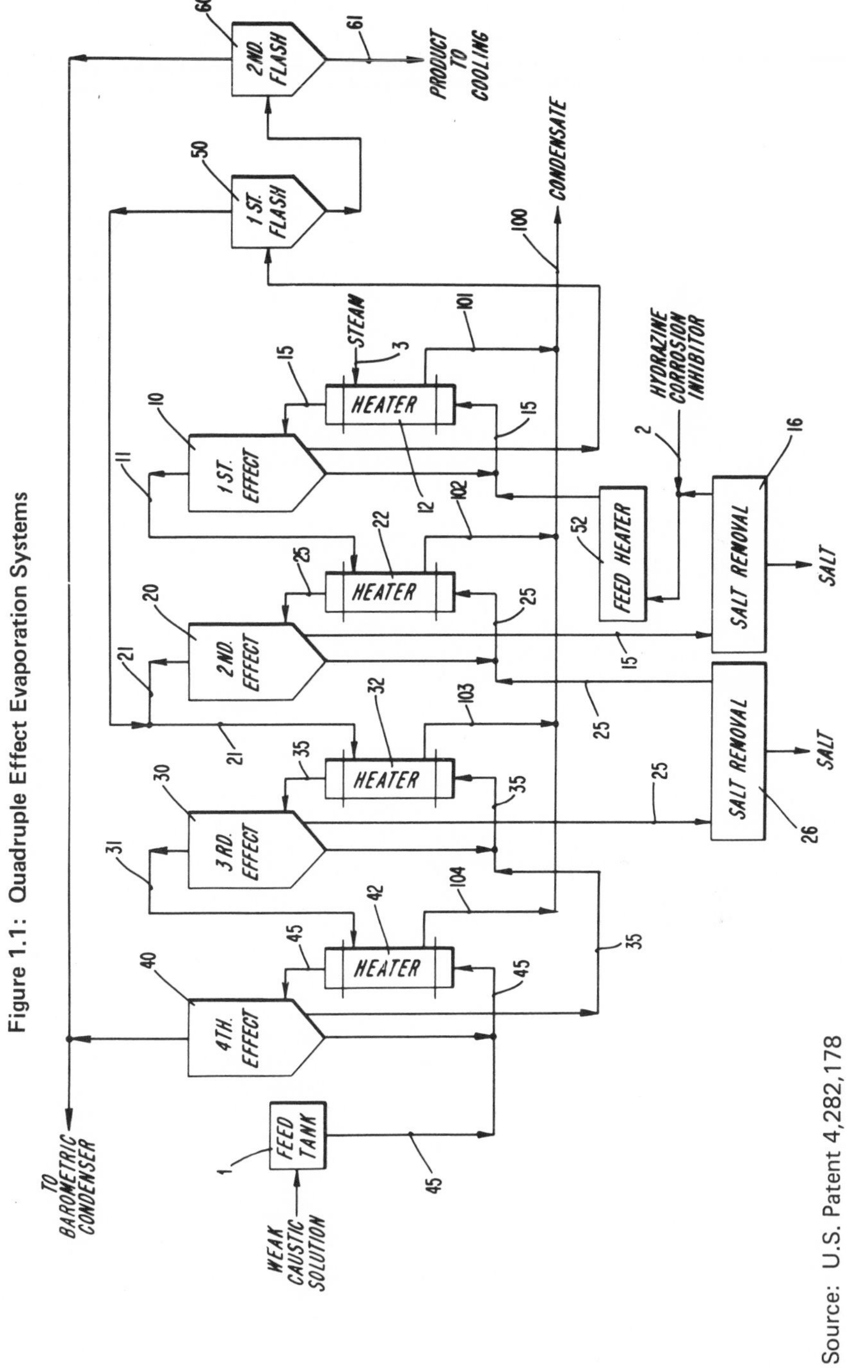

Source: U.S. Patent 4,282,178

In the system illustrated all of the hydrazine corrosion inhibitor is therefore shown as being introduced via line (2) into the caustic solution from the second effect (20) before it is further heated in heater (52) and before it is introduced into the first effect (10). However, one can introduce the inhibitor portion-wise at various stages of the process.

Example: Full-scale plant test runs illustrative of this process have been performed in a plant corresponding to the system shown in Figure 1.1. Referring to this figure, varying amounts of hydrazine were added in these test runs via line (2) to the solution being fed to the first effect evaporator (10). During each test, the plant was operated for three to six days with the continuous addition of the inhibitor in the proportion shown in the table. The chlorate concentration in the caustic solution feed was determined frequently to assure that the stoichiometric proportion between sodium chlorate and hydrazine was held substantially constant. Feed and effluent solutions were regularly analyzed for nickel using an atomic absorption technique. The results obtained are shown in the table.

The amount of nickel pickup in the effluent is a direct measure of the corrosion suffered by the first effect evaporator, i.e., the higher the nickel pickup, the higher the corrosion. The corrosion rate is known to vary considerably in commercial production over an extended period, because it is dependent on the chlorate content of the caustic feed solution and this can undergo substantial variation from day to day. Typically, for instance, referring to the series of comparative tests shown in the table, the chlorate content may vary from about 0.01 to 1% or more, more commonly from 0.03 to about 0.15%. No inhibitor was used in test 1 and the amount of nickel pickup in the effluent in this three-day test was 1.16 ppm. By comparison, in tests 2 and 3 the addition of 0.00035 wt % (3.5 ppm) and 0.0017 wt % (17 ppm) hydrazine to the caustic solution resulted in a reduction of nickel pickup to 0.28 and 0.075 ppm respectively.

	Test 1*	Test 2	Test 3
Duration of test, days	3	6	4
NaOH in feed, wt %	33.0	33.0	33.0
$NaClO_3$ in feed, wt %	0.054	0.070	0.037
Hydrazine added			
Weight percent	0	0.00035	0.0017
Parts per million	0	3.5	17.0
Percent stoichiometric	0	1.1	10.4
Nickel pickup in effluent, ppm	1.16	0.28	0.075

*Control.

The stoichiometric requirement for complete destruction of chlorate by hydrazine can be calculated from the equation:

$$2NaClO_3 + 3N_2H_4 \rightarrow 2NaCl + 3N_2 + 6H_2O$$

Thus it appears that a practical degree of corrosion protection is gained in a commercial-scale operation by as little as 1% or less of the stoichiometric amount of hydrazine as needed per the equation above, and that use of 10% of the stoichiometric amount of hydrazine can reduce the normal corrosion rate by 90% or more. Based on the weight of the caustic solution, the addition of about 2 ppm or more, preferably 3 to 40 ppm, of the hydrazine compound provides useful and effective protection. Of course, greater proportion of the hydrazine compound, e.g., up to ~1,000 ppm or more, can be used if desired or if special circumstances warrant.

Organic Amine in Aromatic Liquid Extraction

It is known in the art that a conventional process for the recovery of high purity aromatic hydrocarbons of, e.g., nitration grade from various feedstocks including catalytic reformates is liquid-liquid extraction utilizing a solvent such as diethylene glycol or sulfolane, each of which has high selectivity for the desired aromatic hydrocarbon components contained in the feedstock.

Typically, in the practice of such prior art process a hydrocarbon feed mixture is contacted in an extraction zone with an aqueous solvent composition which selectively dissolves the aromatic component of the hydrocarbon feedstock thereby forming a raffinate phase comprising one or more nonaromatic hydrocarbons and an extract phase containing dissolved aromatic components. The extract phase is then separately distilled yielding an overhead distillate containing only a portion of the extracted aromatic component, a sidecut fraction comprising aromatic hydrocarbons and a bottoms fraction comprising lean solvent suitable for reuse in the extraction zone. Frequently to prevent losses of the solvent, the raffinate phase is washed with water in a washing zone in order to remove solvent from the raffinate phase.

Also, not infrequently, the extract phase is subjected to stripping or extractive distillation in order to remove a contaminating quantity of nonaromatic hydrocarbons from the extract phase. This stripping or extractive distillation operation is normally performed in order to make possible the recovery of nitration grade aromatic hydrocarbons such as benzene and toluene. Therefore, a typical prior art process for the recovery of aromatic hydrocarbons encompasses a solvent extraction step, a stripping or extractive distillation step, and a final distillation or recovery step for recovery of high purity aromatic hydrocarbons from the solvent phase. Another prior art step is a benzene distillation step which has as its function the recovery of benzene from the other aromatics.

The solvents which are applicable to the practice of the aromatics extraction process, generally, are known to be thermally unstable. The instability is not pronounced, however, and only becomes evident upon prolonged recycling of the solvent whereupon the accumulation of the decomposition products becomes evident.

It is known that the solvent decomposition results in the production of acidic organic deterioration products as well as polymerization products of a resinous character. It is further believed that the decomposition is accelerated by the presence of oxygen. The exact nature of the final decomposition products is not fully known, but where sulfolane is the solvent, the decomposition initially produces sulfur dioxide, sulfur trioxide, and olefins.

The presence of organic solids within the aqueous solvent and of sulfurous gases within an aqueous sulfolane solvent tends to cause accelerated corrosion of the steel materials used in the construction of the process unit, particularly when water is present which is usually the case. Therefore, it is the usual prior art practice to add organic amine compounds to the solvent composition as corrosion inhibitors.

The preferred amine utilized as a corrosion inhibitor in the sulfolane solvent system is an alkanolamine, and more particularly, monoethanolamine. Because of the basic characteristics of these amine inhibitors, these materials react with the acidic solvent decomposition products to produce amine salts and amides at the temperature conditions utilized in the aromatic extraction process and thereby maintain solvent pH at a level far less conducive to corrosion.

The location in the aromatic extraction process at which the organic amine compounds are known in the art to be introduced is in the stripper receiver vessel. Addition at that location will provide inhibitor protection, of course, in the stripper overhead receiver, but also in the extractor, since hydrocarbons from the stripper are normally refluxed to the extractor. There is a problem, however, with the inhibitor addition as described above, because, due to loss in process product streams and consumption, very little inhibitor will be left in the process far downstream or upstream of the inhibitor addition point, particularly at the benzene column or overhead receiver, and, therefore, corrosion will occur at those downstream locations when solvent and solvent decomposition products are present.

Accordingly, the objective of the process of *G.R. Winter, III; U.S. Patent 4,333,823; June 8, 1982; assigned to UOP Inc.* is to modify and improve the known method of addition of organic amine corrosion inhibitor to the sulfolane type organic compound aromatic liquid extraction process and to modify the process flow itself so as to achieve corrosion inhibition throughout the process equipment.

The process comprises the steps of: (a) introducing the mixture into an extraction zone, and therein contacting the mixture with a solvent comprising a sulfolane type organic compound characteristically selected for absorbing aromatic hydrocarbons; (b) removing a nonaromatic raffinate stream from the zone, through an upper locus thereof; (c) removing an aromatic, solvent-rich extract stream from the zone, through a lower locus thereof, and introducing the extract stream into a stripper column; and (d) removing a nonaromatic concentrate from the stripper column, through an upper locus thereof, and removing a solvent-rich aromatic concentrate from the stripper column, through a lower locus thereof.

The process further comprises: (e) combining the nonaromatic concentrate with a hereafter described high water content stream containing organic amine and introducing the combination into a stripper condenser, removing nonaromatic concentrate containing organic amine from the stripper receiver vessel and passing the nonaromatic concentrate containing organic amine as reflux into the extraction zone; (f) introducing the aromatic concentrate into a recovery column, recovering a substantially solvent-free aromatic concentrate through an upper locus thereof, and removing a substantially hydrocarbon-free solvent-rich stream from a lower locus thereof; and (g) introducing the solvent-free aromatic concentrate into a benzene column, recovering a substantially pure benzene concentrate through an upper locus thereof, and removing a substantially benzene-free aromatic product stream from a lower locus thereof.

Further steps are: (h) introducing the benzene concentrate into a benzene column receiver vessel; (i) introducing an organic amine into the benzene column receiver vessel; (j) removing benzene reflux containing organic amine from the benzene column receiver vessel and introducing such reflux to the benzene column; and (k) removing a high water content stream containing organic amine from the benzene column receiver vessel, and combining the stream with the nonaromatic concentrate from the stripper column as mentioned above, thereby effecting introduction of organic amine in effective corrosion inhibiting concentrations to all process equipment in need of corrosion inhibition.

Storing Solid Chlorinating Agent

Solid chlorinating agents are usually stored and transported in closed packing containers such as paper bags, plastic containers and metallic cans. Since the solid chlorinating agents will often be stored for a period of time of as long as one or two years after manufacture before they are actually used on site, noxious gases will be generated during storage upon decomposition of the solid chlorinating agents and those gases may exert undesirable influences, causing in rare cases dangerous incidents. For example, such noxious gases may cause label information on a container to become unclear or fade away completely. The gas generation also may cause the corrosion of packing materials or the breakdown of containers themselves from an increase in internal pressure therein.

M. Ota, T. Mori and T. Taniguchi; U.S. Patent 4,334,610; June 15, 1982; assigned to Nissan Chemical Industries, Ltd., Japan provide a method of storing a solid chlorinating agent with improved storage performance. The process involves placing the solid chlorinating agent and a storage stabilizer such as calcium oxide, trisodium phosphate, magnesium oxide, ferrous oxide, melamine, ammeline or ammelide in a closed container in a manner that the stabilizer contacts the noxious gases in the same container but not with the solid chlorinating agent in the same container. An article for storing the same is composed of a closed container in which the solid chlorinating agent is placed together with the storage stabilizer in such a manner that the gases are brought into contact with the storage stabilizer.

Reference Example 1: 20 g of granular trichloroisocyanuric acid having a particle diameter of 0.25 to 1.4 mm, a water content of 0.13% and an effective chlorine content of 90.7% were charged into a three-necked glass container with one of the necks (first neck) closed with a rubber lid and with the other two (second and third necks) closed with a plug equipped with a valve.

The container was placed for given days in a room where the temperature was maintained at 35°C. The second neck was then connected to a tube through which dry nitrogen gas was passed into the container, and the third was connected to a tube through which the inner gas was led to 80 ml of 1% o-toluidine hydrochloride aqueous solution. After the completion of storage, the dry nitrogen gas was then introduced from the second neck into the container and the inside air was introduced from the third neck into the o-toluidine aqueous solution in which the decomposed gas present therein was absorbed.

The solution was measured by means of a spectrophotometer for the absorbance of yellow light at a wavelength of 400 mμ. The measurement was made after storage of 3, 10 and 30 days. The table below indicates the measured results where the reading of the spectrophotometer measured with respect to the 30 days storage was referred to at "100."

Examples 1 through 4: The procedure of Reference Example 1 was repeated with the exception that a bag of porous polyvinyl chloride film 0.12 mm thick containing a stabilizer as mentioned in the table below was hung below the glass lid closing the first neck of a container in a manner that the bag is not in contact with trichloroisocyanuric acid to be tested. The results are shown in the table.

Examples	Stabilizers	Amounts of Gases Generated		
		3 Days	10 Days	30 Days
Reference Example 1	None	32	62	100
Example 1	Calcium oxide	0	0	1
Example 2	Trisodium phosphate	3	5	10
Example 3	Ferrous oxide	5	10	10
Example 4	Magnesium oxide	below 1	below 1	2

Metal Salt Used in Olefin Hydration

It is well known that various acidic compounds, including both inorganic and organic acids, may be used to catalyze the hydration of olefins to form the corresponding alcohols. However, the use of acidic catalysts has an inherent disadvantage in that the acid catalysts are highly corrosive in nature, attacking the metallic surfaces of the various reactors in which the hydration is effected. This property of the acid catalysts is detrimental to the process inasmuch as it is not possible to use conventional types of reactors and thus necessitates the use of relatively expensive equipment which is highly resistant to acid corrosion.

In addition to using types of materials which are relatively expensive and difficult to fabricate, a relatively large amount of time is lost in producing alcohols when the corroded equipment needs to be replaced, thus adding to the final expense of the finished product, this expense including the replacement of the equipment as well as downtime in the process line.

It has been found by *T. Imai and R.J. Schmidt; U.S. Patent 4,339,617; July 13, 1982; assigned to UOP Inc.* that the corrosive nature of the various catalysts which are employed may be minimized or suppressed to a great degree by effecting the process in the presence of a salt of a metal selected from magnesium, barium, beryllium and radium. The presence of these metal salts is beneficial, not only from the standpoint of suppressing the corrosion problems relating to the equipment, but also of maintaining the high catalytic activity of the acid catalyst.

Example 1: To illustrate the corrosive nature of an acidic catalyst, a piece of stainless steel 316 tubing approximately 1" in length and weighing 2.1918 g was placed in a rotating autoclave along with 21.8 g of a sulfuric acid solution containing 5.2 wt % sulfuric acid and 400.5 g of water. The autoclave was sealed and heated to a temperature of about 200°C. The autoclave was maintained at a range of from 199° to 202°C for a period of 10 hr. At the end of this period, heating was discontinued and the autoclave was allowed to return to room temperature. The pressure which had built up in the autoclave to 16 atm was released and the autoclave was opened. The solution was recovered and it was found that the tube was completely corroded and dissolved, there being a 100% corrosion of the tube.

A repeat of the above experiment in which a similar piece of tubing weighing 2.4584 g was placed in an autoclave along with 150.2 g of water and 8.3 g of a sulfuric acid solution. The autoclave was again heated to 200°C, the steam pressure during the 10 hr period reaching 14 atm. At the end of the the 10 hr period, heating was discontinued and the pressure was discharged after the autoclave had returned to room temperature. Examination of the solution again determined that the tube was completely dissolved, there being a 100% corrosion of the tube.

Example 2: To illustrate the corrosion inhibiting properties of the metal salts of the process, a stainless steel 316 tube weighing 2.9776 g was placed in a rocking autoclave along with 150 g of water, 8.5 g of sulfuric acid solution and 10.6 g of a magnesium sulfate solution. The autoclave was sealed and heated to a temperature of 200°C, the autoclave being maintained at this temperature for a period of 10 hr during which time a steam pressure of 17 atm was reached. At the end of the 10 hr period, heating was discontinued and, after the autoclave had reached room temperature, the excess pressure was discharged. The reaction mixture containing the tube was recovered from the autoclave and the tube was weighed. The weight of the tube at the conclusion of the experiment was 2.2029 g, the weight loss corresponded to a 26% corrosion.

Example 3: To illustrate the ability of the cocatalyst system of the process to hydrate olefinic hydrocarbons with water, a catalyst system comprising 16.7 g of a sulfuric acid solution and 21 g of magnesium sulfate, along with 300 g of water, were placed in a rotating autoclave. The autoclave was sealed and 22.65 g of propylene were charged thereto. The autoclave was heated to a temperature of 202°C and maintained thereat for a period of 1 hr, the operating pressure during this period reaching 38 atm. Upon completion of the desired reaction time, heating was discontinued and, after the autoclave had returned to room temperature, the excess pressure was discharged. Analysis of the reaction mixture by gas chromatography showed that there had been a 50.7% conversion of the propylene with a 100 mol % selectivity to isopropyl alcohol.

Halogen-Containing Olefin Polymers

It is known that halogen-containing olefin resins, because of their halogen component, cause corrosion or rust on metallic component parts such as molding machines or molds during the molding of the resins, or these resins or molded products prepared therefrom undergo coloration or deterioration.

S. Miyata and M. Kuroda; U.S. Patent 4,347,353; August 31, 1982; assigned to Kyowa Chemical Industry, Co., Ltd., Japan provide an improvement in a method for inhibiting the corrosion-causing tendency and coloration of an olefin polymer or copolymer containing a halogen component by incorporating about 0.01 to 5 pbw of a hydrotalcite. The improvement comprises mixing an olefin polymer or copolymer, containing a halogen component attributed to a polymerization catalyst and/or posthalogenation, with a hydrotalcite of the formula:

$$Mg_{1-x}Al_x(OH)_2A^n{}_{x/n}\cdot mH_2O$$

wherein $0<x\leqslant 0.5$, m is a positive number and A^n represents an anion having a valence of n, or a product resulting from the surface-coating of the hydrotalcite with an anionic surface-active agent, the hydrotalcite having (1) a BET specific surface area of no more than 30 m^2/g, preferably not more than 20 m^2/g, and (2) an average secondary particle size of not more than 5 μ, preferably not more than 1.5 μ, and preferably (3) a crystallite size, in the <003> direction determined by an x-ray diffraction method, of at least 600 Å, preferably at least 1000 Å.

As mentioned above, the hydrotalcite particles having the parameters (1), (2) and (3) may be surface-treated with an anionic surface-active agent, and this frequently gives favorable results. In the surface treatment, the anionic surface-active agent is preferably used in an amount of about 1 to 10% by wt, preferably about 1 to 5% by wt, based on the weight of the hydrotalcite.

When such a surface treatment is carried out, the dispersibility of the hydrotalcite is increased, and the flowability of the resin during molding is improved, thus contributing more to the improvement of the appearance of the molded article from the resin and the inhibition of its tendency to cause corrosion.

Corrosion Resistance Test: A resin composition was prepared from 100 pbw of a polyolefin containing a halogen component, 0.4 to 0.8 pbw of hydrotalcite, 0.1 pbw of 2,2'-methylene-bis(4-methyl-5-tert-butylphenol) and 0.1 pbw of dilauryl thiodipropionate, and pelletized at 260°C. A well-polished, degreased mild steel panel, 40 x 40 mm, was embedded in the pelletized resin composition, and heated at 200°C for 30 min. After cooling, the mild steel panel was taken out of the resin composition and placed in a desiccator adjusted to a relative humidity of about 93%, and allowed to stand at 60°C for 20 hr. The degree of rust formation on the mild steel plate was then evaluated on a scale of classes 1 to 10 as follows: Class 4 and classes of lesser numbers mean that the resin composition has a practical effect of inhibiting corrosion. Class 3 and classes of lesser numbers are especially desirable.

The inhibitor was mixed with polypropylene (containing 500 ppm of Cl) obtained by using a highly active Ziegler-type catalyst without performing an ash-removing treatment. The composition was pelletized at 260°C in an extruder. The pellets were molded into a film having a thickness of about 5 μ using an extrusion molding machine. The pellets were tested for corrosion inhibiting ability, as described above, and were found to be in class 2.

ORE TREATMENT

Phosphoric Acid Circuits

The process of *G.L. Long and R.B. Humberger; U.S. Patent 4,277,454; July 7, 1981; assigned to J.R. Simplot Company* relates to the control of excessive corrosion in phosphoric acid circuits by oxidation of reduced ion species in wet process phosphoric acid and elevation of the valence state of certain dissolved reduced ion species such as ferrous iron ($Fe^{2+} \rightarrow Fe^{3+}$), uranium ($U^{4+} \rightarrow U^{6+}$), and vanadium ($V^{3+} \rightarrow V^{4+}$), among others.

In the "wet process" method, the phosphate ore is contacted with a mineral acid, such as sulfuric, to extract phosphate values. In the process many other metallic compounds present in the ore are also dissolved by the acid and remain in the phosphoric acid solution. For a reduction in acid corrosivity it is necessary to raise the valence states of these and other metals.

In accordance with the process, methods for control of excessive corrosion in phosphoric acid circuits are effected by oxidation of reduced ion species present in the acid and maintenance of EMF values above about 190 mV through the digestion process.

Supplementary red-ox characterization and monitoring of the acid can be achieved by ceric bisulfate titration of an analyzed sample of the acid in 10% sulfuric acid.

In one embodied form, the process is utilized to control corrosion in a phosphoric acid plant digestion system operating by dihydrate (gypsum) process even when such process employs a calcined western U.S. phosphate ore feed derived from ore with relatively high amounts of carbonaceous materials. The process is also applicable for treating wet process phosphoric acid in phosphoric acid plants having a digester system operating by a hemihydrate process.

In more detail, when corrosion control is to be effected in a phosphoric acid plant digestion system operating by a dihydrate process, a temperature range of 165° to 185°F is maintained through the digesters, a temperature range of about 165° to 174°F is maintained through the filter and a temperature of about 160° to 185°F is maintained in the bulk phase of acid through evaporators. In a phosphoric acid plant digestion system operating by a hemihydrate process, a temperature of about 160° to 205°F is maintained through the hemihydrate digesters and a temperature of below about 180°F is maintained through the filter of the hemihydrate digestion system. The hemihydrate system may also include evaporators which should be maintained at below about 185°F. In a preferred embodiment, additions of manganese dioxide to a primary digester in a sufficent amount effective to achieve an EMF value over about 190 mV will significantly control the occurrence of excessive corrosion in phosphoric acid circuits.

In yet another preferred embodiment an oxygen autoclave oxidation method is used to oxidize the reduced ion species present in the acid for inhibiting corrosion. Accordingly, the use of deleterious chemical oxidants is avoided, especially such oxidants that can themselves cause corrosion and supply deleterious amounts of impurities.

Example: In accordance with the process a full scale plant application was employed over a period of two months. Native, minus 325 mesh 68% MnO_2 (available oxygen analysis basis) was added to the primary digester of a phosphoric acid (30% P_2O_5) dihydrate circuit. The EMF of the acid in the slurry was controlled above 190 mV the majority of the time by addition of a sufficient amount of MnO_2 each operating hour. EMF was the basis of control, but the control procedure was checked in the initial stages of the addition by the ceric titration method.

At the end of the application period, the primary digester was partly drained and the metal components of agitators, baffles and pumps examined by operating personnel. These metal parts included 316 L stainless steel, alloy 20Cb-3 and 317 stainless steel. There appeared to be a significant reduction in combined corrosion-erosion in the circuit relative to previous periods of operation. Further, there appeared to be no serious problems of corrosion in later stages of the plant, including the evaporators.

In addition, over this period, there appeared to be somewhat less scaling in the phosphoric acid plant. The scale in the plant is typically alkali fluorosilicates or gypsum and alkali fluorosilicates. The mechanism of any such scale inhibition by this process is not known but may be due to alteration of the scaling surfaces or a steadier state of complexation equilibria due to a more constant EMF.

Corrosion and Abrasion Resistant Material

The process of *A. Heidingher and I. Sinko; U.S. Patent 4,336,178; June 22, 1982; assigned to Centrala Minereurilor si Metalurgiei, Neferoase Baia Mare, Romania* relates to a material resistant to abrasion and chemical corrosion which may be utilized in the manufacture of machinery such as engines.

This material, owing to its remarkable anticorrosive properties and its mechanical durability, may be utilized in machine construction and for manufacturing wearing parts of various equipment especially in the ore treatment industry, such as impellers and stators of centrifugal pumps; impellers, stators and liners for flotation cells; hydrocyclones; liners for gangue carrying pipes; crocks, etc.

On the basis of results industrially obtained, the anticorrosive properties of the material are as follows: when used in the form of impellers of centrifugal pumps which pump the ore suspension in nonferrous ore treating plants, that were run under the same conditions as other commonly used pumps, the impellers made with the material of the process were usable for a duration 15 to 20 times as long as those manufactured from manganese steel and 3 to 4 times as long as rubber-coated impellers.

The material of the process is resistant against corrosion by HCl, H_2SO_4, HF media and acetic acid, in any concentration and up to 80°C temperature, as well as by dilute HNO_3 and H_3PO_4 media. It is also durable in a concentrated alkaline medium at ambient temperatures and in a dilute alkaline medium up to 60°C.

The material is formed from 80 to 85% by wt of an erosion resistant component consisting essentially of a mixture of Al_2O_3 and SiC in a weight ratio of 1:2, each having a particle size below 5 mm; 0.5 to 3% by wt of a filler material consisting essentially of spun glass having a fiber length of between 3 and 5 mm; and from 15 to 20% by wt of a binder formed from two epoxy resins, each of the resins being formed from epichlorohydrin and 2,2-bis-p-hydroxyphenylpropane, and having a molecular weight between 350 and 1,400, but each having a different molecular weight, in a ratio of 1:2 by weight, and a phthalic anhydride hardener.

Example: 333 g Al_2O_3 having a particle size between 0.3 and 0.6 mm, 660 g SiC having a particle size between 0.6 and 1 mm and 5 g spun glass with a fiber length of 2 to 5 mm are mixed. The mixture is prewet with 200 cc of a solution of 7% by wt of the below-described binder in acetone, by continuous mixing, after which it is allowed to remain until the acetone solvent is completely evaporated. If desired, solvent removal can also be accomplished by heating.

The binder is formed by mixing two epoxy resins, in a weight ratio of 1:2, each having a different degree of polycondensation and having a molecular weight between 350 and 1,400, both prepared by epichlorohydrin and 2,2-bis-p-hydroxyphenylpropane, and adding an effective amount of a phthalic anhydride hardener. The two resins are mixed at 120°C and the phthalic anhydride is added with mixing until complete dissolution is achieved.

For preparation of the abrasion and corrosion resistant material, the prewet filler is preheated, with mixing, at approximately 100°C and it is added to 170 g of binder heated to 120°C, while continuing slow mixing until the product is homogenized. In the homogenized state, the resulting paste is cast in preheated molds at 130°C, and is then heated at a temperature of 160°C for 1 to 5 hr, after which the cast piece is relieved of stress through slow cooling (3 to 5 hr, as a function of size).

HEAT TRANSFER SYSTEMS

Preventing Cold End Corrosion of Boiler Systems

A composition is provided by *P.T. Colombo; U.S. Patent 4,298,497; November 3, 1981; assigned to Nalco Chemical Company* for neutralizing acid corrosive agents in boiler flue gases and like vapors. In large industrial boilers and like systems, flue gases carrying sulfur trioxide contact metal surfaces of heat recovery equipment. Under some circumstances, the sulfur trioxide combines with moisture in the boiler flue gases to produce sulfuric acid that condenses on the "cold end" of the boiler. This acid can corrode the surface of this equipment. The composition offered here comprises a particulate carrier having a relatively large surface area, and a particulate, active neutralizing agent having a relatively small surface area surface-coated on the carrier. Here, the particulate carrier is bentonite clay and the active neutralizing agent is sodium aluminate.

Magnesium oxide or other neutralizing agents having relatively small surface areas can be substituted for sodium aluminate.

A preferred source of the carrier or suspending agent is the product Syler Gel, a purified bentonite clay offered by the Southern Clay Company of Texas. As is well-known, bentonite clay is a colloidal hydrated aluminum silicate which consists principally of montmorillonite ($Al_2O_3 \cdot 4SiO_2 \cdot H_2O$) and has the property of forming highly viscous suspensions or gels. Bentonite has a three-layer molecular structure which consists of a sheet of Al_2O_3 and MgO molecules sandwiched between two sheets of silica molecules.

The sodium aluminate is deposited on the bentonite clay in a preferred ratio of from 1 to 50% by wt. This may be accomplished by slurrying or mixing in water with subsequent drying in a manner known in the art.

The resultant powder has provided a relative surface basicity of three to four times that of the sodium aluminate alone. This increase in surface basicity apparently results from the sodium aluminate migrating to the surface of the bentonite clay, thus greatly increasing the effective surface-to-volume ratio and the consequent neutralization power of the sodium aluminate.

The resultant powder is advantageously compatible with commonly used dry chemical injection systems or feeders. The powder flows easily, and effects a substantial cost savings over such prior additives as magnesium oxide. The powder may be fed with less difficulty than magnesium oxide based products because fluidizing or aspirating air is not required to ensure a constant feed rate.

The increase in neutralization efficiency of the process is shown by the technique of surface titration. The quantity of basic sites on a finely divided solid can be measured by forming a suspension of the solid in benzene with an indicator adsorbed in its conjungate base form. This suspension is then titrated with benzoic acid dissolved in benzene.

The benzoic acid titres are a measure of the amount of basic sites and have a basic strength corresponding to the pK_a value of the indicator used.

A 1.00 g sample of the powder to be analyzed was suspended in 200 ml of benzene. A small amount (approximately 0.01 g) of bromothymol blue (pK_a = + 7.2) was added to the suspension and mixed for 4 to 5 min. The suspension was then titrated with 0.01 N benzoic acid in benzene until a color change was no longer observed. Once the titration was started it was brought to completion within 5 min. The results are reported as μmol/g (1×10^{-6} mols/g).

	Titrant, ml	μmol/g
Bentonite	<0.01	12
Sodium aluminate	0.1	122
Bentonite:Sodium aluminate 1:1 by wt	0.3-0.4	366-488

As is shown by this laboratory titration, the surface basicity of the neutralizing agent was greatly improved through its mixture with the colloidal carrier.

In a further test, a powder was made by spraying a concentrated (25 to 50% by wt) sodium aluminate solution into a Littleford powder mixer containing bentonite clay. Upon drying, the result was a free-flowing powder. When dispersed in a flue gas stream, the acid dew point was reduced by 5° to 19°F and the SO_3 in the gas stream was reduced by 30 to 60% by wt. In addition, the powder also coats the metal in the gas stream and prevents corrosion from condensing sulfuric acid. Heretofore, such coating with prior art additives was to be avoided as it produced a thermal barrier to the proper operation of the air heater. However, the clay suspended sodium aluminate washes substantially easier than prior art powders and allows faster air heater cleaning without any significant additive build-up.

Nitrate and Triazole in Refrigerator Using Lithium Bromide Solution

The process of *M. Itoh, H. Midorikawa and A. Minato; U.S. Patent 4,311,024; January 19, 1982; assigned to Hitachi, Ltd., Japan* relates to a hermetically circulating, absorption type refrigerator, where refrigeration is produced by repetitions of concentration, refrigerant dilution, and heat exchange of an aqueous lithium bromide solution as an absorbing solution.

The absorbing solution contains a sufficient amount of the following compounds (a) and (b) for functioning as an inhibitor: (a) a nitrate; and (b) a triazole compound selected from benzotriazole and tolyltriazole.

The concentration of nitrate is practically in a range of 0.005 to 0.1% by wt, preferably 0.01 to 0.05% by wt. On the other hand, the triazole compounds have a low solubility in water, particularly a very low solubility in an aqueous concentrated lithium bromide solution. Thus, the concentration of the triazole compound is practically in a range of 0.001 to 0.12% by wt, preferably 0.005 to 0.1% by wt.

Examples 1 through 9: 0.2% by wt of lithium hydroxide was added to an aqueous 22 wt % lithium bromide solution, and then admixed with an aqueous solution containing lithium nitrate and benzotriazole separately prepared, and an absorbing solution was prepared thereby.

The resulting solution was sealed into a refrigerator using structural carbon steel, oxygen-free copper (ASTM 102) and 9:1 cupronickel (ASTM 703) as the materials of construction, and subjected to corrosion test at 160°C in vacuum for 200 hr. The amount of corrosion at that time is shown in the following table. The structure of the refrigerator used is a steam-heated, double effect, hermetically circulating, absorption type refrigerator. In the table, test results of the so far used absorbing solution containing an inorganic inhibitor is also given for comparison as conventional example.

	Lithium nitrate concentration (wt. %)	Benzotriazole concentration (wt. %)	Corrosion (mg/dm^2)		
			Carbon steel	Copper	Cupronickel
Example 1	0.035	0.005	320	45	40
Example 2	"	0.01	280	23	15
Example 3	"	0.05	250	20	12
Example 4	"	0.1	252	20	13
Example 5	0.003	0.05	350	18	10
Example 6	0.01	"	310	19	11
Example 7	0.05	"	340	50	18
Example 8	0.1	"	360	80	19
Example 9	0.15	"	430	130	23
Conventional Example	Lithium chromate (0.3 wt. %)		500	300	180

In the conventional example, sharp and deep pitting corrosion was generated, and the amount of corrosion fluctuated, and thus an average of 10 test pieces were employed. On the other hand, in the examples of the process, test pieces were covered with a uniform black thin film after the test.

As is evident from the table, the corrosion in the examples of the process is considerably reduced, as compared with the conventional example using the inorganic inhibitor, and particularly the anticorrosion effect upon the copper material is remarkable. The effect upon the iron material is also remarkable, and the anticorrosion effect upon both the materials can be said to be balanced in view of the structure of the apparatus.

Example 10: An absorbing solution was prepared in the same manner as in the foregoing examples except that 0.08% by wt of tolyltriazole was used in place of benzotriazole.

The resulting absorbing solution was applied to the same refrigerator as used in the foregoing examples, and subjected to a corrosion test. It was found that the amount of corrosion of carbon steel was 280 mg/dm^2, that of copper 18 cm/g^2, and that of cupronickel 14 mg/dm^2.

Freon-Stable Heat Transfer Oil

As one of the processes for sparing resources, there is a tendency to positively utilize exhaust heat and subterranean heat. One of the examples is generation of electric power by Freon turbine. In such a case, the heat is absorbed in a heat transfer oil and this oil is made to contact Freon directly for producing Freon steam for rotating a turbine, thus providing the merit of very high efficiency enabling the apparatus to be minimized. However, during such process, the oil contacts with Freon at high temperature resulting in the production of HCl. This HCl causes corrosion of the metal of the apparatus and also deteriorates the property of the insulating material in the apparatus. Therefore, it was necessary to apply an indirect heat exchange system through metal wall, etc., instead of a more effective direct heat exchange system.

S. Komatsuzaki and M. Sato; U.S. Patent 4,313,840; February 2, 1982; assigned to Agency of Industrial Science and Technology, Japan have succeeded in obtaining an excellent Freon-stable heat transfer oil by means of using a specific ester and adding jointly two kinds of additives thereto.

The Freon-stable heat transfer oil comprises a polyol ester containing tris(alkylphenyl) phosphite and benzotriazole

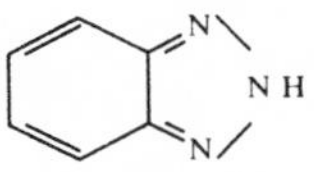

or benzotriazole derivative

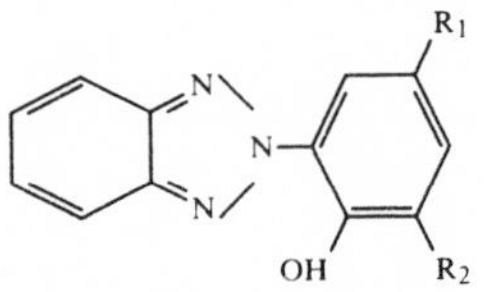

in which R_1 and R_2 each represents a hydrogen atom or an alkyl group. It is desirable to add tris(alkylphenyl) phosphite or tris(phenyl) phosphite in an amount in the range between 0.02 to 5% by wt and to add benzotriazole or its derivative in an amount in the range between 0.02 to 1% by wt to the heat transfer oil.

Polyol esters useful for the process are esters of polyhydric alcohol such as neopentyl glycol, trimethylolethane, trimethylolpropane, pentaerythritol, and saturated monocarboxylic acid having 7 to 16 carbons in the molecule.

Alkyl Cyanide to Prevent Denickelification by Halogenated Hydrocarbons

Dealloying, a particularly deleterious corrosion process, is the selective removal of one component of a metal alloy by a chemical medium. The process generally occurs over a relatively long time period and may often be unobservable since the physical dimensions and other qualities of the affected material may be unaltered, but the structure is changed and, therefore, the strength of the metal alloy is often severely decreased.

The dealloying or more specifically, denickelification of nickel-containing alloys exposed to aqueous systems has been previously reported, but this process has not been known to occur in an organic nonpolar, nonaqueous medium.

It has now been found that when halogenated hydrocarbons are exposed to nickel alloys and especially alloys of copper and nickel, particularly at elevated temperatures and pressure over long time periods, a dangerous and heretofore unobserved condition may result from the denickelification of such alloys under the action of the halogenated hydrocarbon. That this newly found condition is not the result of known corrosion processes involving halogenated hydrocarbons is demonstrated by the fact that the previously mentioned known corrosion and decomposition inhibitors useful in the halogenated hydrocarbon solvent systems are ineffective in protecting against this denickelification process.

Particular operating conditions wherein the abovedescribed denickelification process may occur are in, e.g., heat recovery and transfer systems employing the aforementioned halogenated hydrocarbons in contact with nickel alloys at elevated temperatures and pressures. Included are such systems employing halogenated hydrocarbons as the working fluid whether cooling is occasioned through using the specific heat of the fluid to transport heat to a recovery or exchanger means or through using the latent heat of vaporization of the fluid, such as in systems where the vapor is condensed in a cooler or exchanger and the liquid recirculated to be revaporized.

The process of *J.B. Ivy and T.S. Boozalis; U.S. Patent 4,324,757; April 13, 1982; assigned to The Dow Chemical Company* comprises the use of a lower alkylcyanide for the inhibition of denickelification by halogenated hydrocarbons.

The halogenated hydrocarbons for use in this process are the halogenated aliphatic hydrocarbons having from 1 to 10 carbon atoms and substituted with from 1 to 10 halogen atoms. Also included are those compounds substituted with a mixture of halogen atoms and mixtures of such halogenated aliphatic hydrocarbons.

The inhibitors that effectively prevent the denickelification of nickel alloys when exposed to the abovedescribed halogenated hydrocarbons under the conditions hereinafter described are lower alkylcyanide compounds containing from 1 to 6 carbon atoms in the alkyl group or mixtures thereof. A preferred inhibitor is methylcyanide. The amount of inhibitor required to effectively prevent the denickelification process may vary depending on the particular nickel alloy, the halogenated hydrocarbon, and temperature employed. Generally, amounts expressed in wt % from 0.01 to 10% may be employed, preferably from 0.5 to 5%.

Preferred temperatures at which the inhibitors of this process are effective are from about 40° to 200°C. Most preferred temperatures are from about 80° to 140°C.

The denickelification inhibitor of this process also may be effectively used in combination with known decomposition inhibitors of the prior art such as dioxane, nitromethane, cyclohexane, the lower aliphatic alcohols having 1 to 10 carbon atoms and propylene oxide. It has been found that lower alkylcyanide derivatives do not effectively prevent the crevice corrosion of some metals and alloys in contact with halogenated hydrocarbons, whereas nitromethane does prevent such corrosion. If inhibition of crevice corrosion in addition to denickelification is desired, an effective amount of a known corrosion inhibitor such as nitromethane may advantageously be used in combination with the denickelification inhibitor of this process. Effective amounts of nitromethane range from about 0.01 to 10% by wt.

Example 1: The behavior of halogenated hydrocarbons exposed to metal alloys at elevated temperatures is investigated under the following conditions. Test coupons of mild steel, 70:30 copper/nickel and 90:10 copper/nickel were installed in a pilot plant scale reflux unit containing dry, uninhibited methylene chloride and in sealed metal bombs filled with dry, uninhibited methylene chloride maintained at 140°C in constant temperature ovens. Methylene chloride samples were collected periodically and analyzed for evidence of decomposition. After approximately one year the metal test coupons were removed and examined for corrosion.

Careful examination indicated the overall corrosion rates were quite low for all metals tested; generally less than 0.0004 cm/yr. However, there was evidence of the occurrence of dealloying in the copper/nickel coupons. In particular the coupon of 90:10 copper/nickel suffered from general denickelification and the coupon of 70:30 copper/nickel evidenced localized or plug-type denickelification.

Example 2: A screening program was instituted to test various inhibitors for use in halogenated hydrocarbon systems. These experiments were carried out in sealed metal bombs containing coupons of mild steel, 70:30 copper/nickel and 90:10 copper/nickel. The bombs were charged with dry methylene chloride containing about 1% by wt of an inhibitor, and mounted in constant temperature ovens maintained at about 140°C. The inhibitors tested included: nitromethane, cyclohexane, propylene oxide, dioxane, methylbutynol, 2,4-pentanedione, and methylcyanide.

Samples of methylene chloride solution were collected periodically and analyzed for decomposition. After about four months the metal coupons were removed and examined for crevice corrosion and denickelification. Of all the inhibitor compounds tested only the bombs containing methylene chloride inhibited with methylcyanide showed no evidence of denickelification.

The coupons exposed to methylene chloride, inhibited with methylcyanide showed evidence of crevice corrosion. Also, the rate of decomposition of methylene chloride inhibited by methylcyanide was significantly accelerated resulting in fouling of the test coupon by deposition of decomposition products.

Example 3: The reaction conditions of Example 2 were repeated using methylene chloride inhibited with approximately 1% by wt of a mixture of 50% methylcyanide and 50% nitromethane. Again after about four months exposure at 140°C the coupons were removed for examination. The results showed no evidence of denickelification or of significant crevice corrosion. Decomposition of methylene chloride as determined by analysis of the methylene chloride solution for the presence of decomposition products was found to be extremely low.

Heterocyclic Arenes

A wide variety of lubricants and heat transfer fluids are known in the art. These materials can be paraffinic/naphthenic hydrocarbons, aromatic hydrocarbons containing one or more rings, silicones, fluorinated compounds, polyolefins, esters and the like. However, these fluids, in general, suffer from one or more deficiencies. Thus, for example, the paraffinic/naphthenic hydrocarbons are not thermally stable at or above 300°C. Lubricants and fluids used in Rankine Cycle and heat transfer systems should be stable in the range of about 325° to 375°C. Although silicones are thermally stable, they are relatively expensive and tend to decompose in contact with aqueous alkaline media. Fluorinated compounds are also often quite stable but are also quite expensive. Esters hydrolyze in aqueous media thus severely limiting their use.

Aromatic hydrocarbons having two or three rings have been found to be thermally stable, but are often exposed to conditions where the need for additives having antiwear and anticorrosion properties are needed. It is therefore an object of this process of *G.S. Somekh and R.A. Cupper; U.S. Patent 4,346,015; August 24, 1982; assigned to Union Carbide Corporation* to provide such additives, which themselves, are adequately thermally stable in the lubricants or heat transfer liquids for which they are used, i.e., at temperatures in the range of about 325° to 375°C.

A method for improving the antiwear and corrosion properties of high temperature fluid compositions has been devised which comprises adding about 0.001 to 5% by wt of the total composition of at least one arene having 2 or more aromatic rings selected from the group consisting of acridine, quinoxaline, phthalazine and mixtures thereof.

Iron Oxide Film for Condenser Tube Surface

According to *T. Fukutsuka, K. Shimogori, E. Yamamoto and K. Miki; U.S. Patent 4,369,073; January 18, 1983; assigned to Kabushiki Kaisha Kobe Seiko Sho, Japan* a thin corrosion protective film of an iron oxide can be formed on the inner surface of a condenser tube made of a copper alloy by first applying on the inner surface a thin layer of an acidic suspension containing iron powder and then exposing the thus-coated inner surface to an atmosphere of an oxidizing gas. The above method can be carried out in a relatively short period of time and also in a relatively simple fashion. The resultant protective film is uniform and hardly reduces the thermal conductivity of the condenser tube, thus ideal as a protective film for condenser tubes of a heat exchanger.

Iron powder to be employed in the process is not accompanied by any special limitation with respect to its production method and particle size. Thus, gas-atomized and water-atomized iron powders may be used extensively. In this process, the liquid layer can be formed particularly thin. It is accordingly desired to make the particle diameter of iron powder as small as possible.

It is preferred to use iron powder which pass in its entirety through a sieve of 400 mesh (not greater than about 37 μm, or preferably iron powder not greater than 10 μ in mean diameter. Among iron powder of a small particle size, there is carbonyl iron powder having a mean particle diameter of about 5 μ. There is no fear that such iron powder would protrude from an applied liquid layer, providing an extremely advantageous effect to obtain a uniform film thickness.

The acidic solution adapted to suspend therein the abovedescribed iron powder may be selected from those capable of dissolving iron powder and allowing smooth oxidation and hydration reactions thereof to take place. Hydrochloric, sulfuric, or nitric acid may generally be used. Its preferred concentration normally ranges from 0.1 to 2.0 N. Although the amount of iron powder to be incorporated therein is not specifically limited, it is usually preferred to contain about 10 to 500 g of iron powder/every 100 ml an iron powder-contaning acidic solution. A portion of such iron powder is dissolved in the acidic solution, but most of the iron powder is contained in a suspended state.

After the completion of an application of the acidic suspension on the inner surface of a condenser tube, the thus-applied liquid layer is exposed to an oxidizing gas atmosphere to form a film through the oxidation reaction of the iron powder.

As the oxidizing gas, oxygen or any oxygen-bearing gas may be employed, but air is most advantageous from the economical viewpoint. Air having a high humidity is further meritorious. Such an oxidizing gas may be applied onto the liquid layer by allowing a condenser tube coated with the liquid layer to stand in an atmosphere of an oxidizing gas, or positively blowing an oxidizing gas against the liquid layer. The latter method is preferred. Here, the oxidizing gas may preferably be fed at a rate of 20 ml to 50 ℓ/min when the inner surface of a brass tube having a diameter of 1" is treated. If the former method is followed, it is highly recommended to move the oxidizing gas in the atmosphere by means of a fan or the like. Even at normal temperature, the film-forming speed is high enough. However, the oxidizing gas may be slightly heated to further increase the film-forming speed.

A further study revealed that, by using as the acidic suspension containing iron powder an iron powder-containing suspension to which has been added a water-miscible organic solvent having a surface tension smaller than water at normal temperature, the surface tension of the acidic suspension can be lowered, thereby making it possible to apply the suspension in the form of a thin, uniform layer. It was also revealed that, by blowing an oxidizing gas in advance into the acidic suspension, the iron powder in the acidic suspension is made finer, thereby improving the uniformity of a resulting film and also the properties of the same.

Among representative examples of such organic solvents, are included methanol, ethanol and acetone. They may be used solely or in combination. The amount of such an organic solvent to be incorporated is selected from a range of 5 to 70% by wt based on the total weight of an acidic suspension to be applied.

Reference Example: A suspension containing 500 ml of 0.2 N aqueous hydrochloric acid solution and 50 g of iron powder of 350 mesh was charged into a 1 ℓ three-necked flask equipped with an air-blowing tube and thermometer. An aluminum brass plate (25 x 40 x 1 mm) was hung down at a position 20 mm above the surface of the suspension. The suspension was heated to 60°C and air was blown thereinto through the air-blowing tube at a speed of 1 ℓ/min. Thus, film formation was performed in a splash-zone. 5 hr later, a reddish-brown film 0.8 mg/cm^2) was formed. An x-ray diffraction analysis of the film revealed 70% of γ-FeOOH and 30% of α-FeOOH as their components. Treated aluminum brass plate was immersed in seawater for 12 months, developing extremely slight corrosion of the base metal. (Corrosion depth 0.02 mm.)

On the other hand, when the aluminum brass plate was hung down at a position 2 mm above the surface of the suspension, a film principally made of Fe_3O_4 was formed.

Example 1: An aluminum brass tube (25 mm in diameter and 100 mm in length) was held upright and a suspension similar to that employed in Reference Example was passed intermittently from the top of the tube and downwardly therethrough at a flow rate of 60 ml/min (equivalent to about 2 mm of liquid layer thickness) for 1 min at every 15 min. At the same time, air was continuously charged into the tube from the lower end thereof at a flow rate of 5 ℓ/min. This operation lasted 8 hr, forming 2.5 mg/cm^2 of a reddish-brown film on the entire surface of the tube. An x-ray diffraction analysis of the film determined that it consists of 80% γ-FeOOH and 20% α-FeOOH. A seawater passing test (flow rate: 2 m/sec) was carried out using this tube for 6 months. The corrosion depth of its inner surface was as little as 0.015 mm.

The above procedures were followed except the suspension was continually passed down at a flow rate of 30 ml/min (equivalent to about 1 mm of liquid layer thickness). A black film with a thickness of 0.4 mg/cm^2 was formed in 3 hr. Its components were confirmed to be 90% Fe_3O_4 and 10% γ-FeOOH. Similar to the above, a seawater passing test was carried out in the thus-treated tube. The corrosion depth of its inner surface was extremely little, namely, 0.02 mm.

Example 2: An aluminum brass tube (25 mm in diameter and 1,200 mm in length) was arranged aslant (elevation angle: 5°). A suspension similar to that employed in the Reference Example was intermittently flowed from the top end thereof and downwardly therethrough at a flow rate of 50 ml/mm (equivalent to about 1 to 3 mm of liquid layer thickness) for 1 min at every 20 min while rotating the tube in the circumferential direction thereof at a speed of 1 rpm. On the other hand, from the lower end opening thereof was supplied continually wet air of 100% relative humidity at a flow rate of 2 ℓ/min.

This treatment was carried out for 6 hr, forming 10 mg/cm^2 of a reddish-brown film on the entire inner surface of the tube. As a result of an x-ray diffraction analysis of the film, it was determined that the film consists of 70% γ-FeOOH, 20% of α-FeOOH and 10% of Fe_3O_4. A seawater passing test (flow rate: 2 m/sec) was conducted for 6 months using this tube. The corrosion depth of its inner surface was as little as 0.01 mm.

NUCLEAR POWER PLANT AND WASTE DISPOSAL

Oxidizer for Protection of Tantalum Equipment

As methods for decomposing various kinds of radioactive sludges and used ion-exchange resins after their use in treating radioactive materials, an acid digestion process has been employed for decomposing them in hot concentrated sulfuric acid at 200°C or higher. For the construction material of equipment in which the decomposing treatment is effected, ordinary metals are not applicable for lack of their corrosion resistance to hot concentrated sulfuric acid and only tantalum has been known to be practically unsable.

N. Kagawa, K. Yamamoto, R. Sasano, T. Kusakabe and Y. Moriya; U.S. Patent 4,356,148; October 26, 1982; assigned to JGC Corporation, Japan have provided a method for preventing the corrosion and hydrogen embrittlement of equipment made of tantalum handling therein hot concentrated sulfuric acid at 200°C or higher, characterized in that at least one oxidizer selected from the group consisting of nitric acid, nitrogen oxides, ferric ion, cupric ion, stannic ion or plumbic ion is incorporated in the sulfuric acid.

Example 1: Tantalum to be tested here was annealed at 1200°C in a 10^{-4} torr atmosphere and had a chemical composition meeting JIS specifications. The specimen thus obtained was 80 mm long, 10 mm wide and 2 mm thick in size and was introduced into 1.5 ℓ of hot concentrated sulfuric acid boiling at a temperature of 260°C, after which the whole was continuously incorporated with concentrated nitric acid (98 wt % conc.) at a feeding rate of 30 ml/hr to investigate how general corrosion and hydrogen absorption of the tantalum progressed.

For comparison, the procedure of the above Example 1 was followed except that the concentrated nitric acid was not incorporated.

The results are as shown in Figures 1.2a and 1.2b . It is seen from Figure 1.2a that the corrosion rate of tantalum in concentrated sulfuric acid incorporated with concentrated nitric acid is about one-half that of tantalum in concentrated sulfuric acid alone.

It is also seen from Figure 1.2b that the amount of hydrogen absorbed on the tantalum in concentrated sulfuric acid incorporated with concentrated nitric acid is remarkably small as compared with that in concentrated sulfuric acid alone.

The specimens of tantalum already subjected to the corrosion test were further subjected to a bend test. The results show that the tantalum specimen treated in the concentrated sulfuric acid alone produced cracks caused by the hydrogen embrittlement only 168 hr after the start of the corrosion test, while those treated in the concentrated sulfuric acid incorporated with the concentrated nitric acid produced no cracks at all even 500 hr after the start of the corrosion test.

Figure 1.2: Effect of HNO_3 on Corrosion and Hydrogen Absorption of Tantalum

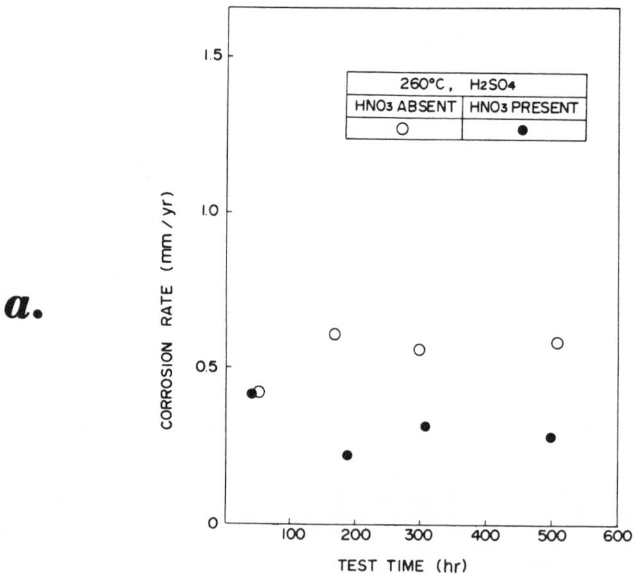

(continued)

Figure 1.2: (continued)

b.

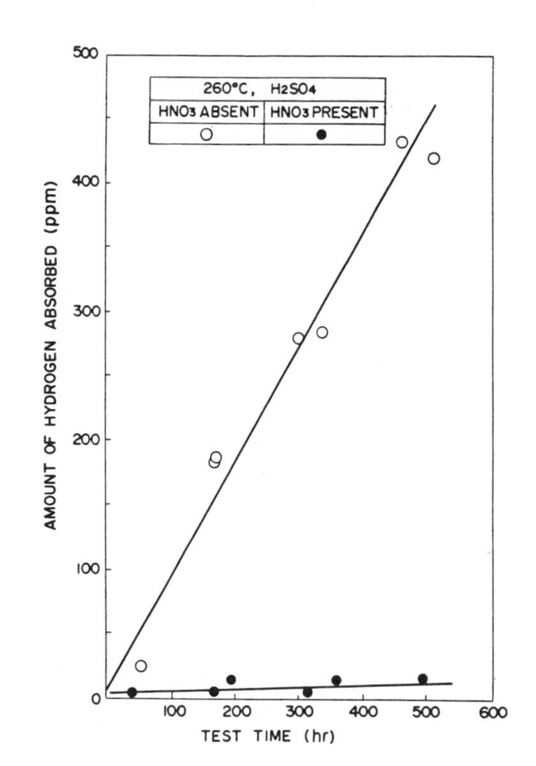

Source: U.S. Patent 4,356,148

Example 2: The procedure of Example 1 was followed except that HNO_3 was added in an amount by weight of 3% at the time of start of the test without subsequent further addition thereof.

The result is that the one-time addition of 3 wt % of HNO_3 at the time of start of the test unlike the cases of Examples 1 and 2 was found to be still effective in preventing hydrogen absorption even when the HNO_3 decreased to about 0.1 wt % in concentration during the test.

Cathodic Corrosion Protection System

Many nuclear power plants in use today are of the pressurized variety. These power plants generally include a steam generator which utilizes high temperature, pressurized water from the reactor on the primary side to heat pure water on the secondary side. A number of these power plants which have been in operation for various periods of time have begun to encounter operating problems due to certain corrosion phenomena occurring in the secondary plant and the steam generator. One of the more serious of these problems involves denting of the primary heat transfer tubes which can adversely affect the flow of pressurized water through the tubes and may ultimately cause a tube wall integrity failure and force a shutdown of the steam generator for repairs.

The process of *R.G. Lucas; U.S. Patent 4,376,753; March 15, 1983; assigned to Electric Power Research Institute* features a cathodic corrosion protection system for use in a power plant adapted to use fresh water as the working fluid and having a steam generator and a feed system including a main feed pipe for supplying fresh water to the steam generator.

A cathodic corrosion protection system in accordance with this process generally comprises a source of electrically conductive fluid such as hydrazine; an iron tank located adjacent the main feed pipe and having a first inlet communicating with the main feed pipe to draw fresh water into the ion tank, an outlet communicating with the main feed pipe to return treated water thereto, and a second inlet communicating with the source of electrically conductive fluid, a valve means associated with the second inlet to control the introduction of metered amounts of the conductive fluid into the ion tank to render the water flowing therethrough substantially conductive.

Also included in this cathodic corrosion protection system is at least one ionizable metal anode mounted in the ion tank; and means supplying a positive direct current voltage to the anode and a negative direct current voltage to a wall of the outlet to produce an electric current through the conductive liquid in the tank, resulting in treated water being returned to the main feed pipe to form a protective coating on interior walls of the feed pipe and the interior surfaces in the steam generator for reducing the rate of formation of corrosion products on these surfaces.

In a preferred embodiment, the outlet of the ion tank comprises a pipe having a diameter substantially smaller than the diameter of the main feed pipe and a discharge section of the pipe is located centrally in the main feed pipe to distribute the treated water more uniformly throughout the stream of fresh water flowing through the main feed pipe. Further, in accordance with one feature of the process, the main feed pipe includes an orifice of a prearranged size between the inlet and outlet of the ion tank to produce a fluid pressure differential of prearranged magnitude therebetween to cause fresh water from the main feed pipe to flow through the ion tank and past the anode at a prearranged flow rate. In some instances, the use of two or more anodes may be preferred.

When utilized in a pressurized water nuclear power plant, this process features the use of two separate cathodic corrosion protection systems, one located immediately prior to the steam generator to provide principal corrosion protection for the surface of the steam generator, and a second system located immediately after the condensate pump which returns fresh water to the main feed pipe to provide corrosion protection for surfaces of structures involved in the liquid feed train.

While the actual mechanism of corrosion protection is not completely understood, it is believed that corrosion protection is provided by the system of this process by the formation of a protective alkaline coating on the surfaces of the main feed pipe and the steam generator. This alkaline coating is apparently deposited as free metal ions are released from the anode in the ion tank as electric current passes between the anode and the cathode system. The thin alkaline coating built up on the surfaces of the steam generator will not substantially inhibit heat transfer between the heat exchanger tubes and the water flowing through the secondary heat exchanger compartment. However, the coating will inhibit formation of corrosion products on steam generator surfaces which contain copper and nickel and thus inhibit the formation of magnetite deposits which cause denting of the heat exchanger tubes. When a corrosion protection system is also employed at the outlet of the condensate pump, the protective coating formed on surfaces of pipes and other apparatus in the fresh water feed train will reduce corrosion formation on these surfaces and reduce sludge buildup in the steam generator.

PACKING MATERIALS

Open Cell Carriers for Volatile Inhibitors

B.A. Miksic and R.H. Miller; U.S. Patent 4,275,835; June 30, 1981 provide a volatile corrosion inhibitor carrier comprising an open cell isocyanate-derived polymer known as polyurethane. Typically, the carrier fastens to a wall or other surface, normally, on or near the top in a closed area to provide a continuous source of volatile corrosion inhibitor from an appropriately selected composition of the corrosion inhibiting systems suitably matched to the specific hostile corrosive problem.

The combination of the open cell isocyanate-derived polymers foams and the volatile corrosion inhibitors have been found to provide both a high storage of volatile corrosion inhibitor and a far more effective dispersion of the corrosion inhibitor than prior art cellulosic materials. It was found that the open cell reticulated structure provides more sites for the deposition and/or crystallization of the corrosion inhibitor and far greater surface area for the more efficient dispersion of the volatile corrosion inhibitors into the desired atmosphere than either the conventional closed cell or those foams which are often called open cell foams.

The following three groups of vapor pressure inhibitors provide rapid protection. Group 1 comprises the low vapor pressure inhibitors. These inhibitors are characterized by a vapor pressure of less than 10^{-4} mm Hg at ambient conditions and 20°C. Group 2 comprises the intermediate vapor pressure inhibitors. These inhibitors are characterized by a vapor pressure ranging from 10^{-3} to 10^{-4} mm Hg at ambient conditions and 20°C. Group 3 comprises the high vapor which are characterized by vapor pressure above 10^{-3} mm Hg at ambient conditions and 20°C.

The first group consists of the following corrosion inhibitors: cyclohexylamine chromate, cyclohexylamine m-mononitro benzoate, dicyclohexylamine chromate and dicyclohexylamine nitrite. The second group of vapor phase corrosion inhibitors consists of cyclohexylamine benzoate, diethanolamine benzoate, and benzotriazole. The third group consists of monoethanolamine benzoate and tolyltriazole. The total vapor phase corrosion inhibitors located in the carrier includes a minimum of 5% by wt of vapor phase corrosion inhibitors selected from at least two of the first group, the second group or the third group, and comprises a minimum density of vapor phase corrosion inhibitor of 0.05 g/cm^3.

The volatile corrosion inhibitors can be transported and solvent dispersed through the open cell structure of the isocyanate-derived polymer foam by immersion and deaeration. After the corrosion inhibitor is dispersed through the structure, the liquid solvent system is selectively removed through controlled evaporation leaving volatile corrosion inhibitors located at and within the cell sites throughout the large, comparatively rigid structure of this isocyanate-derived polymer foam. Maximum loading per cycle is achieved by using a near saturated solution of the applicable chemicals. After solvent removal via slow evaporation normally at room temperature in the presence of moving air and a slight negative pressure, the previously treated isocyanate-derived polymer foam may be impregnated a second, (or more) time with a like saturated solution of the same chemicals to substantially increase the loading (0.7 to 0.8 more by weight, based on the first load considered as 1). Drying follows each impregnation.

A completely loaded carrier can be covered on one side with a protected, peelable pressure sensitive adhesive for later attachment of the volatile corrosion inhibitor carrier system to a storage container well.

Nitrite, Trisubstituted Phenol and Fumed Silica in Polyolefin

Volatile or vapor phase corrosion inhibitors have been used in a variety of applications where visible coatings such as oil or grease or other permanent coatings such as paints are unacceptable for various reasons. Sodium nitrite, amine nitrite salts, organic amines, carboxylic acids and organic amine carboxylic acid salts have all been described singularly and in combination in volatile corrosion inhibiting compositions. Sheet materials useful for packaging metal parts containing volatile corrosion inhibiting compositions have also been described. Sodium nitrite is known as an effective volatile corrosion inhibitor when carbon dioxide and water are present in the air. However, in the absence of carbon dioxide, its effectiveness as a volatile corrosion inhibitor drops off abruptly.

F.P. Boerwinkle and D.A. Kubik; U.S. Patent 4,290,912; September 22, 1981; assigned to Northern Instruments Corporation have found that a mixture of an inorganic nitrite, a trisubstituted phenol and fumed silica, when mixed with polyolefin polymer, yields a plastic article having highly effective volatile corrosion inhibiting properties regardless of carbon dioxide concentration in the environment.

The process is further illustrated by means of the following examples wherein the term part refers to parts by weight unless otherwise indicated.

Examples 1 and 2 describe the preparation of volatile corrosion inhibiting articles by means of plastisol formation. The plastisols were prepared by mixing all of the following ingredients together to form a uniform paste and then heating.

Example 1: The ingredients used were: 3 parts sodium nitrite; 2 parts Ionol (2,6 ditert-butyl-4-methyl phenol); 0.1 part Cab-O-Sil (fumed silica); 3 parts oleyl alcohol; and 8 parts Microthene FE-532 organic polymer (ethylene-vinyl acetate copolymer).

The plastisol formed by heating the mixture at 250°C for 30 min showed little degradation and demonstrated excellent volatile corrosion inhibiting properties using test method FTM-101B Method 4031.

Example 2: A plastisol was formed as in Example 1 with the following composition similar to one described in U.S. Patent 2,829,080: 10.7 parts urea; 21.5 parts sodium nitrite; 21.5 parts oleyl alcohol; and 35.5 parts Microthene FE-532 organic polymer.

The resulting plastisol exhibited considerable darkening and decomposition after being heated at 250°F for 30 min.

Example 3: This example describes a volatile corrosion inhibiting article in the form of an extruded thermoplastic film. The following ingredients were mixed uniformly: 1.485 parts sodium nitrite; 1.485 parts Ionol; 0.03 part fumed silica; and 97 parts low density polyethylene.

This mixture was extruded into film at a temperature of 300°F. The resultant film showed no discoloration or gas formation and possessed excellent volatile corrossion inhibiting properties when tested against mild steel.

Flexible Graphite Sheet Containing Metal Molybdate

The manufacture of flexible graphite layer material is known and according to German Patent 66 804 this is accomplished by dipping flaky, natural graphite particles into a liquid oxidant and heating them to about 1000°C within a few seconds. The individual graphite particles expand in this treatment by a considerable amount and can be densified without the addition of a special binder into flexible planar structures such as thin foils or thicker laminar shapes, as for instance, by rolling, pounding or hammering. The layer material, which is impervious to liquids and gases, and also temperature-resistant and self-lubricating, is used among other things as a seal or packing, where the layer material is in contact with other materials such as metals.

E. Freundlinger and A. Hirschvogel; U.S. Patent 4,296,177; October 20, 1981; assigned to Sigri Elektrographit GmbH, Germany provide flexible graphite layer material containing at least one metal molybdate with a solubility in water of 10^{-2} to 10^{-4} g/100 ml to inhibit corrosion. The corrosion inhibitor may be incorporated by first immersing the graphite layer material in a solution of an alkali metal molybdate, drying, and then immersing in a solution of very soluble metal salt in which the metal when combined with molybdate to form metal molybdate has the defined solubility. The graphite layer material is then dried and contains the desired molybdate corrosion inhibitor.

The graphite layer preferably contains one or more molybdates from the group silver, calcium, barium molybdate. A particularly favorable effect is obtained with a graphite layer material having a molybdate content of 0.5 to 1.0% by wt of the graphite layer material.

Example: Graphite foil with a raw or bulk density of 0.7 g/cm^3 and a foil thickness of 0.5 mm was placed in the form of a roll in an impregnating tank and the tank was evacuated to a pressure of about 10^{-3} bar. After a degassing time of 15 min, the valve to the vacuum pump was shut. An aqueous solution containing

0.8% sodium molybdate was drawn into the tank and the pressure was then increased to about 10 bar. After a holding time of about 15 min, the foil was dried in a vacuum dryer at a temperature of about 110°C and a pressure of about 10^{-2} bar for 2 hr and then replaced in the impregnating tank. The cycle began over again; and aqueous barium nitrate solution with about 1% salt was used as the second impregnation medium. At the end of the cycle, the dried foil contained about 1% barium molybdate.

To test the inhibition effect, discs were stamped from the graphite foil and each disc of graphite foil clamped between two chrome-nickel steel discs. The surface layers of the chrome-nickel steel discs had first been removed by grinding because of their possible passivating effect. This arrangement was hung in a vessel filled with completely demineralized water (conductivity less than 1 μS/cm). The water temperature was 100°C and the evaporation losses were continuously made up by fresh water. After a testing time of 60 days, the steel plates in contact with the graphite layer material prepared as described in the preceding paragraph to contain 1% barium molybdate according to this process showed no corrosion. Plates in contact with graphite layer material containing sodium molybdate as the inhibitor exhibited some pitted corrosion spots.

Oxygen Scavenger

T. Yamaji and E. Yoshisato; U.S. Patent 4,317,742; March 2, 1982; assigned to Teijin Limited, Japan provide an oxygen scavenger composition. The composition consists of (1) as essential components, an alkali metal sulfide and at least one oxidation promoter selected from the group consisting of silica, alumina, silica-alumina, silica-magnesia and inorganic materials containing one of them as a main ingredient; and (2) as optional components, at least one oxidation promoter aid, a water-insoluble or sparingly water-soluble, inert filler, and/or water in a hydrous or hydrated condition.

The oxygen scavenger composition of this process is characterized by capturing oxygen by a chemical reaction upon contact with it. Hence, in a closed vessel, it is suitable for capturing oxygen within the vessel and making the atmosphere within the vessel oxygen-free or markedly reducing the oxygen content of the atmosphere.

Accordingly, the composition finds a wide range of applications. Among these is the rust-proofing of metallic products such as iron, zinc, copper and aluminum products, and electrical appliances, component parts, household appliances, musical instruments and precision machines using these metals.

FOOD AND COSMETIC INDUSTRIES

Quaternary Ammonium Phosphate for Hair Spray

The process of *A. Nandagiri; U.S. Patent 4,263,275; April 21, 1981; assigned to American Cyanamid Company* provides a hair spray composition containing a phosphate salt of quaternary ammonium compound, which greatly reduces the corrosion potential of hydroalcoholic formulations, especially acidic formulations. The general formula of the quaternary ammonium compound is represented as follows:

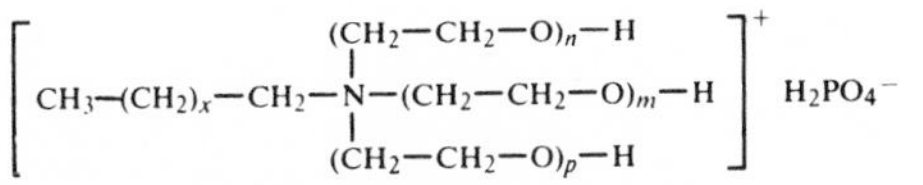

$$\left[CH_3-(CH_2)_x-CH_2-\overset{\displaystyle (CH_2-CH_2-O)_n-H}{\underset{\displaystyle (CH_2-CH_2-O)_p-H}{N}}-(CH_2-CH_2-O)_m-H \right]^+ H_2PO_4^-$$

wherein x is an integer from about 12 to 18 and m, n, and p are integers, the sum of which total from about 3 to 12. Quaternary ammonium compounds, representative of the above class, are commercially available, and one such compound is Dehyquart SP.

The addition of from about 0.1 to 3% of the quaternary compound to hydroalcoholic formulations greatly reduces the corrosion of tin plate aerosol cans. The concentration is critical, since below 0.1% the corrosion inhibition is not readily observed and above 3% the formulation would have negative tactile properties on hair. In addition to having corrosion inhibiting activity, the quaternary compounds act as a plasticizer and give the hair a soft feel. Since they are quaternary ammonium compounds, they have excellent antistatic and emulsifying properties and have been recommended for use in cosmetics, especially in hair conditioning preparations such as home permanents, hair rinses, hair-setting sprays and hair lotions.

Example 1: An aerosol hair spray is prepared by mixing the following ingredients:

Ingredients	W/W %
Ethanol anhydrous	44.18
2-Amino-2-methyl-1-propanol	0.17
Gantrez ES 225* (as is)	5.00
Water	30.00
Fragrance	0.15
Dehyquart SP (50% solution)	0.50
Hydrocarbon A-31 (Isobutane)	20.00

*Copolymer of monoethyl ester of maleic acid and methyl vinyl ether

Example 2:

Ingredients	W/W %
Ethanol anhydrous	44.68
2-Amino-2-methyl-1-propanol	0.17
Gantrez ES-225* (as is)	5.00
Water	30.00
Fragrance	0.15
Hydrocarbon A-31 (Isobutane)	20.00

*Copolymer of monoethyl ester of maleic acid and methyl vinyl ether

Example 3:

Ingredients	W/W %
Ethanol anhydrous	57.68
2-Amino-2-methyl-1-propanol	0.17
Gantrez ES 225* (as is)	5.00
Methylene chloride	12.00
Fragrance	0.15
Hydrocarbon A-31 (Isobutane)	25.00

*Copolymer of monoethyl ester of maleic acid and methyl vinyl ether

Accelerated storage testing was done at 45°C with tin plate aerosol cans with all three compositions shown above. It was found that the composition shown in Example 2 (no quaternary compound) deteriorated rapidly with can perforations being observed within 12 to 16 weeks. In contrast, the composition shown in Example 1, which contained 0.5% Dehyquart SP, and composition shown in Example 3, which did not contain any water, did not show any changes to the product or the container.

Phosphonic and Polyhydroxy Acids for Sterilization Equipment

According to *R.R. Waegerle; U.S. Patent 4,303,546; December 1, 1981; assigned to Benckiser-Knapsack GmbH, Germany* scale formation as well as corrosion are suppressed by adding amino methylene phosphonic acids, hydroxy alkane diphosphonic acids, amino alkane diphosphonic acids, polyhydroxy acids, their alkali metal salts, or mixtures thereof to the aqueous heating medium of heating systems used for heat treating, such as sterilizing and pasteurizing, goods enclosed in glass, metal, and the like containers.

The compositions of the process comprise (a) a phosphonic acid compound selected from the group consisting of diethylene triamino penta(methylene phosphonic acid), nitrilo tris(methylene phosphonic acid), propylene diamino tetra(methylene phosphonic acid), ethylene diamino tetra(methylene phosphonic acid), and alkali metal salts thereof, and (b) a polyhydroxy acid component selected from the group consisting of gluconic acid and its salts.

The proportion of phosphonic acid compound to polyhydroxy acid compound in the composition is between about 3:1 and 1:3, and the amounts added to the heating system are between about 0.1 and 100 g for 100 ℓ of aqueous heating medium.

Example: The following tests were carried out in upright autoclaves of a capacity of 10 ℓ of water. The autoclaves were operated at about 4 atm gauge and at 140°C. They were charged with conventional tin plate cans.

Tap water of the following composition was used for sterilization: total hardness, 17.3° dH; hardness due to carbonates, 17.3° dH; pH value, 7.2; chlorides, 164.2 mg/ℓ; and sulfates, 36.0 mg/ℓ.

(a) Autoclave 1 was charged with tap water of the above given composition without additive. On sterilizing the cans at 140°C for 45 min, scale formation on the cans and in the autoclave was observed.

(b) Tap water of the above given composition was filled in autoclave 2. 5 cc of a mixture of a 50% aqueous solution of diethylene triamino penta(methylene phosphonic acid) and a 16% aqueous solution of gluconic acid in the proportion of 1:3, corresponding to 625 mg of the phosphonic acid and 600 mg of gluconic acid in 10 ℓ of tap water, were added. The cans were sterilized at 140°C for 45 min.

(c) Tap water of the following compositions was filled in autoclave 3: total hardness, 29.2° dH; hardness due to carbonates, 17.3° dH; pH value, 7.1; chlorides, 164.2 mg/ℓ; and sulfates, 36.0 mg/ℓ.

5 cc of a mixture of a 50% aqueous solution of diethylene triamino penta(methylene phosphonic acid) and a 16% aqueous solution of gluconic acid in the proportion of 1:1 were added to the water. The amounts of phosphonic acid and gluconic acid added to 10 ℓ of water thus were, respectively, 1,250 mg and 400 mg.

Autoclaves 2 and 3 did not show any scale formation and the sterilized cans were free of incrustations and were glossy and shiny.

Citric Acid, Metal Chloride and Film Former as Rust Remover

J.G.M. Frew; U.S. Patent 4,326,888; April 27, 1982 provides an antirust composition comprising citric acid, a metal chloride and film-forming material which when applied to a rusty substrate will convert the rust into a harmless reaction product and provide a coating for receiving a finish paint.

The composition of the process can contain, based on the dry weight of the film-forming material, citric acid, and metal chloride, about 50 to 75% film-forming material, about 20 to 30% citric acid and 10 to 20% of the metal chloride.

The mechanism of the rust removing or converting action of this composition is not entirely clear, but it appears that the metal chloride acts as a catalyst or promoter for the action of the citric acid, thus speeding up reduction or complexation of the rust by the citric acid. The citric acid may reduce the ferric oxide of the rust to ferric and ferrous citrate or some other action may occur.

Thus, the chloride ions may solubilize the rust forming iron chloride which can react with the citric acid to form iron citrate, releasing chloride ions for further reaction. In any case, the effect of citric acid on rust is distinctly improved in association with the metal chloride compared to citric acid alone.

The composition is generally in the form of an aqueous emulsion with the film-forming polymer being the disperse phase, and as the composition dries on the rusty surface the film-forming material forms a coating over the reacting citric acid and metal chloride, and the reaction product appears as a brown coloration on or within the coating.

If the dried composition coating is left uncovered for some time moisture in the air reacts with the brown reaction products to darken them, but this does not affect the properties of the composition coating on the surface, which continues to act as a rust preventative.

A cooperation or coaction appears to take place in a way not at all understood, between the film-forming material and the complex reaction product of the rust, citric acid and metal chloride in the sense that the process compositions form a tough durable coating more effectively upon rusted surfaces than upon rust-free iron surfaces.

In the latter case, the desired coating does eventually result but by that time the surface has had an opportunity to undergo rusting so as to produce in situ a stratum of rust for participation in the reaction.

Example: An antirust composition was prepared from the following three groups of ingredients:

	Wt. lbs.	% Dry Wt. Basis	As % of resin, acid & chloride
Group 1			
Cellulose gum CMC 7ML	3.375	1.04	
Cold water	32.5	—	
Group 2			
Titanium dioxide RTC4	44	13.53	
Vegetable black	1.187	0.365	
Vinacryl 7175 (48% resin solids)	417.62	61.63	73.1
Defoamer 1512M	1.25	0.384	
Grease dispersant and actuator (Surfynol 104H)	1.25	0.384	
Group 3			
RAC (citric acid crystals)	49	15.1	17.9
RSA (zinc chloride)	24.75	7.61	9.03
Boiling water	35	—	
Total	609.932	100.04	100.03
(Total dry weight basis = 325.27)			

The components of group 1 are stirred together thoroughly and allowed to stand from 12 to 24 hr. The components of group 2 are mixed together and high-speed dispersed until an even mixture is obtained. The components of group 3 are stirred together until a solution is obtained. The group 3 solution is then added to the group 2 mixture and high-speed dispersed, and the group 1 components which have been stirred together and allowed to stand are added and high-speed dispersed. The resulting emulsion can then be fed into containers through a shaker, for storage and transportation.

When the composition of this example has been prepared, it is ready for application to a rusted substrate. The substrate should first be freed from oil and grease, and loosely adhering rust flakes should be removed. The composition can then be applied by, e.g., brush or roller, or, after the addition of 10% of cold water as a thinner, by spraying. In each case the entire rusted portion of the substrate should be covered by the composition with care being taken to ensure that air bubbles and pinholes are precluded. The composition can be removed from the brushes, rollers, spray guns, etc., by washing with water.

The time within which the composition dries is controlled by the amount of rust on the substrate as well as by environmental conditions, and the drying time is shorter on a rusty surface than on clean metal or painted surfaces, which denotes the cooperative action noted above. In warm dry conditions, the composition dries on a rusted surface in about 2 hr, but it takes about 7 to 10 days for the full reaction of the composition with the rust to be completed, after which time the maximum rust resistance of the composition is attained.

Once the composition has dried it can be overpainted with almost any air drying paint with the exception of water-based emulsions and water-soluble paints. The adhesion of paints such as synthetic enamels, chlorinated rubber paints, epoxies, acrylic and hammer finishes and decorative and maintenance finishes to the acrylic film which forms on the treated surface is excellent.

The composition can also be used as a primer for stoving allowing rusty ferrous components to be stoved without the need for expensive pretreatments such as shot blasting. In stoving, the final stoving cycle should be at a temperature of below 120°C.

The composition of this example is noninflammable and can be used to avoid the expensive step of halting continuous production processes in industry by eliminating the need for shot blasting or completely cleaning rusty equipment and structures in a factory. This is particularly important in factories where food and drink are prepared and packaged.

This composition also enables a smart attractive appearance to be regained even where rust has broken through previously-applied finishes.

Adjusting pH of Toothpaste with Alkali Hydroxide or Carbonate

According to *C.J. Taylor; U.S. Patent 4,328,205; May 4, 1982; assigned to Beecham Group Limited, England* the tendency of gel toothpastes comprising silica thickeners, such as silica aerogel or precipitated silica, and a fluoride source, such as sodium monofluorophosphate or fluoride, to corrode aluminum tubes can be reduced by incorporating at least 0.01% by wt sodium or potassium hydroxide or carbonate, so that the pH of the toothpaste is at least 8.5.

Example: Collapsible aluminum tubes were filled with certain translucent gel formulations, described below, and stored at room temperature or at 50°C. The tubes were then examined for swelling, gas production, dye fading and corrosion of the tube surface. The results are shown in the table below.

Formulations 1A and 1B –

	Percent by Weight
70% Sorbitol	60
Synthetic magnesium lithium silicate	3.9
Sodium carboxymethylcellulose	1.3
Polyethylene glycol	4.0
Silica aerogel	2.0
Sodium lauryl sulfate	1.7
Sodium monofluorophosphate	0.8
Color, flavor, preservative	1.5
Sodium hydroxide	0 (A) or 0.2 (B)
Water	to 100

The formulations are stored at room temperature for 2 years.

Formulations 2A and 2B – Same as above, except that they contained no synthetic clay and a total of 8% silica aerogel. Stored at 50°C for 23 days.

Formulations 3A and 3B – Same as Formulations 1A and 1B, except that they contained 0 and 0.5% of potassium hydroxide, respectively, in place of the sodium hydroxide. Stored at 50°C for 31 days.

Formulations 4A and 4B – Same as Formulations 1A and 1B, except that they contained 4% precipitated silica in place of the silica aerogel. Stored at 50°C for 27 days.

Formulations 5A and 5B – Same as Formulations 1A and 1B, except that they contained 3.9% of synthetic magnesium aluminum silicate in place of the magnesium lithium silicate. Stored at 50°C for 25 days.

Formulations 6A and 6B – Same as Formulations 1A and 1B, except that they contained 0 and 1.05% sodium carbonate respectively, in place of the sodium hydroxide. Stored at 50°C for 5 days.

Formulations 7A and 7B – Same as Formulation 1B, except that they contained 0.02 and 0.05% of sodium hydroxide, respectively. Stored at 50°C for 6 days.

In the above formulations, the customary dental polishing agent has been omitted to make standardization of the test easier and to facilitate measurement of, for example, dye fading. Practical toothpastes based on these formulations would normally contain a finely-divided polishing agent such as calcium carbonate (perhaps applied as stripes on a given gel formulation) or silica aerogel (dispersed throughout the formulations to produce a translucent product).

Formulation	pH	Swelling*	Gas	Fading*	Corrosion
1A	8.2	+3 to +5	Severe	−4	Slight tarnish
1B	9.4	0	None	0	None
2A	6.7	0 to +3	Slight	−3½	Severe tarnish
2B	9.1	0	None	0	None
3A	8.1	+3 to +5	Severe	−3	Slight tarnish
3B	9.9	0	None	0	None
4A	8.1	+3 to +5	Severe	−3	Slight tarnish
4B	9.1		None	0	None
5A	7.9	+3 to +5	Moderate to severe	−3	Severe tarnish
5B	9.7	0	None	0	None
6A	8.4	+2 to +4	Moderate	−2	Slight tarnish
6B	9.8	0	None	0	None
7A	8.6	0 to +1	None	0	None
7B	8.9	0	None	0	None

*Swelling scores are 0 (no detectable swelling) to 5 (tube crimp about to open)
Fading scores are 0 (no fading) to −5 (completely colourless)

Substituted Succinic Acid and Imidazoline Derivatives as Methylene Chloride Stabilizers

According to *R.W. Simmons; U.S. Patent 4,347,154; August 31, 1982; assigned to The Dow Chemical Company* certain substituted succinic acid, e.g., sodium alkenyl succinate, and substituted imidazoline derivatives, e.g., (ethoxypropionic acid) imidazoline, provide corrosion resistance for tin-plated steel and solvent stability to aerosol formulations containing alcohol, methylene chloride and water, which employ a hydrocarbon as the propellant.

The following experiments will serve to illustrate the use of various substituted succinic acid and imidazoline derivatives. The compounds contemplated as useful inhibitors in the aerosol formulations of the process are designated by succinic acid derivatives having the following formulas wherein X is sodium, potassium, ammonium or an amino group and R is an alkyl or alkenyl radical having from 3 to 16 carbon atoms.

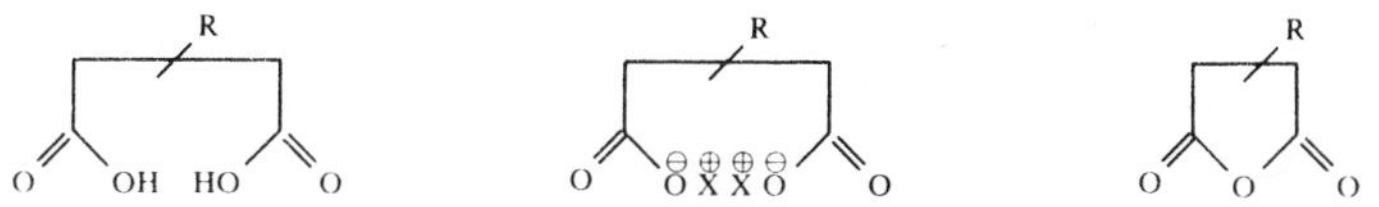

Also useful are imidazoline derivatives having the formulas:

where R_1 is an aliphatic hydrocarbon radical having from 1 to 12 carbons and R_2 is a divalent hydrocarbon chain having from 2 to 6 carbon atoms. The sodium, potassium, ammonium and amine salts of the carboxylic acid derivatives are also useful.

A representative example was conducted as follows. A solvent blend of 12% water, 50.4% ethanol, 20% isobutane, or isobutane-propane mix, and 17.6% Aerothene MM (an inhibited methylene chloride) was used in the following tests. Either 0.25 or 0.50 lb tin plate coupons were placed into a Wheaton glass aerosol bottle and charged with the above formulation. Corrosion inhibitors were introduced into this system at 100 ppm and up. Corrosion testing was conducted at 100°F. All percentage compositions are given in weight percents based on total weight of the composition.

The table below shows the concentration of inhibitor employed, the number of days observed and the degree of corrosion at the end of the observation period. If corrosion occurred in a short period of time, the observation was discontinued. The degree of corrosion is indicated on a scale of 0 to 5.

Example No.	Inhibitor**	Conc. (ppm)	Degree*	Time (days)
Comparative	None	—	3	5
1 A	SAS	5000	0	40
B		"	1	97
2 A	TAS	100	3	156
B		500	2	"
C		1000	1	"
D		2500	1	"
E		5000	0	"
3 A	MAC	100	4	156
B		500	4	"
C		1000	3	"
D		2500	3	"
E		5000	2	"
4 A	MAT	40	2	140
B		100	0	156
C		5000	0	"
5 A	ASA	100	4	140
B		500	3	"
C		2500	3	"
D		5000	2	"

*0 = none, 1 = very slight, 2 = spotty, 3 = moderate, 4 = heavy
**SAS = sodium alkenyl succinate (alkenyl = C_6-C_9).
TAS = triethanolamine alkenyl succinate (alkenyl = C_6-C_9).
ASA = alkenyl succinic anhydride (alkenyl = C_6-C_9).
MAC = Monazoline C 2-coco-(1-ethoxypropionic acid) imidazoline
MAT = Monateric 1000 2-heptyl-1-(ethoxypropionic acid) imidazoline, sodium salt

VIDEO TAPE AND SEMICONDUCTOR APPLICATIONS

Amines, Amides or Imides for Ferromagnetic Powder

M. Aonuma and Y. Tamai; U.S. Patent 4,253,886; March 3, 1981; assigned to Fuji Photo Film Co., Ltd., Japan provide a method of preparing a corrosion-resistant ferromagnetic metal powder after forming the ferromagnetic metal powder.

The method comprises washing the formed ferromagnetic metal powder with a solution containing (a) at least one volatile corrosion inhibitor, the volatile corrosion inhibitor being an organic amine, amide or imide, or an organic or inorganic salt thereof, containing 1 to 4 nitrogen atoms and 1 to 12 carbon atoms and having a molecular weight of 50 to 200 and a vapor pressure of about 0.1×10^{-5} to 1×10^{2} mm Hg at a temperature of 15° to 25°C; (b) water, an organic solvent miscible with water, or a mixture of water and at least one organic solvent miscible with water; and (c) an anionic surface active agent selected from the group consisting of a carboxylic acid or salt thereof, a salt of a sulfuric acid ester, a salt of a phosphoric acid ester, a salt of a dithiophosphoric acid ester and a salt of a sulfonic acid.

Unless otherwise indicated, in the following examples all parts and percentages were by weight, and all processings were at atmospheric pressure and at room temperature.

Example 1: An M_1 solution was prepared which consisted of 0.36 mol/ℓ of ferrous chloride and 0.04 mol/ℓ of cobalt chloride. An R_1 solution was also prepared which consisted of 1.6 mol/ℓ of $NaBH_4$ (0.01 N NaOH solution).

80 parts of the above M_1 solution was placed in a nonmagnetic container. A dc magnetic field (max. 1,000 Oe) was applied to the container, and 20 parts of the above R_1 solution was added to the M_1 solution over a 10 sec period while slowly stirring to effect oxidation-reduction. The temperature of the solution was 20°C at the beginning of the reaction, but due to the exothermic reaction, it tended to rise.

By cooling externally, the peak temperature was controlled at 35°C. The reaction rapidly occurred with the generation of H_2 gas and was stopped after 2 min. There were obtained black ferromagnetic powders. The powders were separated from the reaction mother liquid, added to 100 parts of distilled water containing 0.01% dicyclohexylamine nitrite/1 part of powder and dispersed therein for 1 min by applying ultrasonic waves of 29 kHz and 150 W. The supernatant liquid was then removed, and the residue was washed three times with 50 parts of distilled water containing 0.01% dicyclohexylamine nitrite (24°C, 30 min/wash).

Thereafter, the residue was washed with 50 parts of acetone three times to remove water and then dried using a hot-air dryer at 40°C to obtain ferromagnetic powders (referred to as Sample P-1). The reaction yield was 60%. The ferromagnetic powders obtained were ferromagnetic Fe-Co-B alloy powders having an average particle size of 300 Å in a needlelike shape in which the spherical particles formed a chain. The powders were added to a small amount of water. When m-phenylenediamine reagent was added thereto, a yellow coloration resulted, which indicated the presence of nitrite ion.

From the result of an analysis of the powder, it was found that it contained 2 wt % B and 11 wt % of Co/100 wt % of Fe. 1,200 parts of butyl acetate was added to 300 parts of the powder, and then stirring at a high shearing force was effected at ambient temperature for 2 hr. The following composition was then added thereto: 30 parts polyester polyurethane (MW about 30,000; addition polymerization product of polyester and m-xylylenediisocyanate, the polyester being produced by the condensation of ethylene glycol and adipic acid (Desmocoll 400); 35 parts nondrying oil-modified alkyl resin (reaction product of glycerin, terephthalic acid and a synthetic nondrying oil of an oil length of 30%; hydroxyl value of about 130; Burnock DE-180-70); and 2 parts silicone oil (dimethylpolysiloxane).

The resulting mixture was ball-milled for 10 hr. Then, 22 parts of a triisocyanate compound (Desmodur L-75, a 75% by wt ethyl acetate solution of an adduct of 3 mols of toluene diisocyanate and 1 mol of trimethylolpropane, molecular weight about 560) was added thereto, and the system dispersed with high shearing force for 1 hr to form a magnetic coating composition.

The coating composition was coated by doctoring on one side of a polyethylene terephthalate film (25 μ thick) so as to provide a film thickness of 5 μ (dry basis) while applying a 2,500 gauss dc magnetic field for 0.02 sec, and then dried while heating at 100°C for 2 min with a 3 kl/m^2 of air flow for 2 min. The broad magnetic web thus obtained was super-calendered at 60°C, 60 kg/cm of pressure, and a 40 m/min rate and slit into a ½" width video tape. The tape obtained had excellent surface properties and was referred to as Sample T-1.

Example 2: An M_2 solution was prepared consisting of 0.695 mol/ℓ ferrous chloride, 0.285 mol/ℓ of cobalt chloride and 0.02 mol/ℓ chromium chloride. An R_2 solution which consisted of 3.5 mol/ℓ $NaBH_4$ (0.01 N NaOH solution) was also prepared.

The procedure of Example 1 was repeated using the above reaction solutions to obtain black ferromagnetic powders which were subjected to aftertreatments in the same manner as in Example 1 to obtain powders and then a tape. The average particle diameter of these powders was 350 Å. The analysis of the powders showed 43 wt % of Co and 6 wt % of Cr/100 wt % of Fe. The powders obtained were referred to as Sample P-2, and the tape obtained referred to as Sample T-2.

Comparative Example 1: The procedure of Example 1 was repeated except for omitting dicyclohexylamine nitrite to prepared powders and then a tape. The powders obtained were referred to as Sample PC-1, and the tape obtained was referred to as Sample TC-1.

Powder Sample PC-1 was added to a small amount of water, m-phenylenediamine reagent added thereto, and the coloration observed. No nitrite ions were detected.

Comparative Example 2: Reaction was effected in the same manner as in Example 2 except for adding 0.001 mol/ℓ or 0.01 mol/ℓ of dicyclohexylamine nitrite to the M_2 solution. The aftertreatment step was effected in the same manner as in Example 2 except for using distilled water free of dicyclohexylamine nitrite, thus preparing powders and then tapes. The powders obtained were referred to as Samples PC-2A and PC-2B, respectively. The tapes obtained were referred to as Samples TC-2A and TC-2B, respectively. Various characteristics of the Samples obtained in Examples 1 and 2 and Comparative Examples 1 and 2 were measured, and the results obtained shown in Tables 1 and 2. In Table 1 the values for magnetic field on measurement: Hm = 3,000 Oe, and in Table 2, Hm = 2,000 Oe.

Table 1

Sample No.	Saturation Magnetization, σ	Magnetization After 7 Days, σ*
	. . . . (emu/g)	. . . .
Process:		
P-1	105.0	90.3
P-2	107.6	98.5
Comparison:		
PC-1	90.4	46.5
PC-2A	88.7	75.8
PC-2B	79.9	69.6

*60°C, 90% RH.

Table 2

			. . . . After 7 Days	. . . . .
Sample No.	Hc (Oe)	Br/Bm	Hc (Oe)	Decrease in Br (%)
Process:				
T-1	1,100	0.84	1,120	10
T-2	980	0.83	–	–
Comparison:				
TC-1	1,100	0.81	1,180	34
TC-2A	870	0.80	–	–
TC-2B	630	0.76	–	–

*60°C, 90% RH.

As is apparent from the above results, the ferromagnetic powders can be improved in σ by treating them with a solution containing a volatile corrosion inhibitor according to this process. Moreover, the lowering of the σ can be improved even when the powders are allowed to stand in an atmosphere of 60°C and 90% RH.

Regarding the tapes, an improvement in the squareness ratio was observed in all cases. Furthermore, on comparing Example 1 with Comparative Example 1, the percentage of decrease in the Br and the increase in the Hc after standing at 60°C and 90% RH were lower when the ferromagnetic metal powders of this process were used; thus, a stable tape can be obtained using such powders.

The above results show that the ferromagnetic powders of this process are stable when stored for a long period of time. The initial Br was 2,550 gauss and 2,150 gauss for T-1 and TC-1, respectively. It can be seen from Example 2 and Comparative Example 2 that it is ineffective to add a volatile corrosion inhibitor to the reaction bath and then form ferromagnetic powders therein.

Bromine-Containing Plasma for Aluminum Films

In the process of *K.J. Radigan; U.S. Patent 4,351,696; September 28, 1982; assigned to Fairchild Camera and Instrument Corp.* bromine-containing plasma is generated whereby bromine atoms bombard the surface of a chlorine-etched metallization pattern and displace more reactive chlorine atoms. The lower chemical reactivity of bromine with respect to chlorine prevents damage to the substrate during this process. With the loss of retained surface chlorine, the patterned aluminum or aluminum alloy film is much more resistant to corrosion induced by hygroscopic pickup from the ambient.

A typical step in the fabrication of semiconductor devices is the formation of a conductive metallization pattern by etching a thin film of aluminum or aluminum alloy, such as aluminum/silicon or aluminum/silicon/copper, which covers an underlying layer of a semiconductor device. The underlying layer may be an insulating material such as silicon dioxide or silicon nitride. The metallization pattern to be etched from the film is defined by a patterned photoresist mask formed over the film such that only regions unmasked by photoresist will be etched. For best results, it is required that etching proceed to the interface between the metallization layer and the underlying layer without unduly undercutting the metallization. Chlorinated plasma etching provides this result.

According to one conventional aluminum or aluminum alloy plasma etching process, a suitable chlorinated etch gas, such as carbon tetrachloride is introduced to a reaction chamber, such as a conventional parallel plate reactor, containing a metallized wafer at about 60 to 70 cc/min utilizing nitrogen gas at about 6 psia as a carrier gas. In this initial power-up stage of the etching process, the reactor power is about 3.5 A and the reaction chamber pressure is maintained at about 250 mμ. After maintaining these conditions for about 3.5 min, the power is reduced to about 2.5 A and the pressure is reduced to about 150 mμ. The reactor is then maintained at these conditions for about 20 min while plasma etching of the aluminum or aluminum alloy film proceeds.

Following the chlorine plasma etch, the reactor is purged, utilizing a combination of oxygen at a regulated flow of about 20 psia and nitrogen at a regulated flow of about 6 psia. During this clean-up portion of the process, the reactor power is maintained at about 3.0 A and the reactor pressure is maintained at about 400 mμ for about 5 min.

Following the purging step, power to the reactor is turned off and a bromine-containing gas, preferably an organic such as methyl bromine or ethyl bromide, is introduced to the reactor at a regulated flow of about 15 psia along with oxygen at about 9 to 20 psia. The pressure of the reactor is held at about 300 mμ. The reactor is maintained at these conditions, i.e., with just gas in the reactor and no power applied, for a waiting period of about 2 min. After the 2-min waiting period, power of about 3.0 A is applied to the reactor for about 15 min to generate a bromine-containing plasma which acts to passivate the patterned aluminum or aluminum alloy film.

Plasma generated from methyl bromide or ethyl bromide sources gases will retain these gases since only a percentage of the source gas actually becomes an ionized plasma. Both methyl and ethyl bromide will react with aluminum and aluminum chloride to form alkyl aluminum compounds which subsequently react with water to form the alkane and aluminum oxide. This reaction aids in clearing the etched areas and passivating the patterned film.

OTHER

Silicone Penetrating Agent

The process of *D.H. MacIntosh; U.S. Patent 4,248,724; February 3, 1981* relates to penetrating agents which are applied to seized, rusted, or corroded metal parts to free them by seeping into the minute cracks separating the parts and dissolving any corrosion present.

The penetrating and lubricating composition consists of a silicone lubricant dissolved in a glycol ether carrier having a very low surface tension. The glycol ether gives the composition the ability to penetrate into the cracks between two seized or corroded elements and dissolve any rust or corrosion present. The glycol ether then volatilizes leaving a coating of silicone on the parts to serve as a lubricant and prevent further corrosion.

Example 1: 1 fluid ounce of Dow Corning 200 fluid and 100 cs viscosity (at 25°C) was mixed with 1 gal of Dowanol EB (ethylene glycol butyl ether). This formulation was found to have superior penetrating lubricating properties when applied to corroded metallic parts.

Example 2: A mixture of 1 fluid ounce of Dow Corning 200 fluid of 100 cs viscosity and 1 gal of Dowanol DE (diethylene glycol ethyl ether) was made and tested with good results.

Example 3: A mixture of 1 fluid ounce of Dow Corning 200 fluid of 100 cs viscosity and 1 gal of Dowanol PM (propylene glycol methyl ether) was made and tested with good results.

Example 4: A mixture of 1 fluid ounce of Dow Corning 200 fluid of 100 cs viscosity and 1 gal of Dowanol DPM (dipropylene glycol methyl ether) was made and tested with good results.

It has been found that the upper limit on the volume of dimethyl siloxane polymer that can be dissolved in 1 gal of glycol ether is approximately 2 fluid ounces. Above this limit it is difficult to insure that the polymer will remain in solution. The lower limit, below which the mixture fails to exhibit adequate lubricating properties, is on the order of 1 fluid ounce/gallon of glycol ether. These functional limits will vary somewhat depending on the viscosity of the polymer used.

Aqueous Sulfur Dispersion as Fertilizer

In view of limitations on the natural supply of sulfur from both air and soil, a sulfur deficiency is not uncommon in agricultural soils. Such a deficiency frequently occurs in well-leached soils containing little organic matter which are located in areas far from sources of sulfur dioxide atmospheric pollution. Crops which appear to be particularly sensitive to a sulfur deficiency include corn, sugar cane, wheat, sugar beets, and legumes such as alfalfa and peanuts.

One of the most satisfactory sources of sulfur for fertilizer purposes is elemental sulfur. Although elemental sulfur cannot be directly utilized by plants, it is slowly oxidized by microorganisms in the soil to sulfate which can be metabolized by plants. In view of the water-insolubility of elemental sulfur, it serves as a highly effective slow release source of sulfur for plant nutrition which is not susceptible to leaching by rain water or irrigation. In addition to its suitability as a fertilizer, elemental sulfur is also an excellent fungicide and is widely used in agriculture for this purpose.

Elemental sulfur can be applied either to soil or plants as a dry solid. It is frequently preferable, however, to apply elemental sulfur in the form of an aqueous dispersion since a dispersion avoids the hazards associated with dust formation. In addition, an aqueous sulfur dispersion can be applied with conventional spray equipment and can be diluted to any desired concentration with water. Further, an aqueous sulfur dispersion can be readily blended with other liquid fertilizers.

The use of aqueous dispersions of elemental sulfur has been hampered, however, as a consequence of their highly corrosive action with respect to ferrous metal, such as carbon steel. In the past, the use of these dispersions has required the use of storage and application equipment which is constructed of stainless steel, high chromium steel, or certain plastics.

It has been found by *L.E. Ott; U.S. Patents 4,256,691; March 17, 1981; and 4,321,079; March 23, 1982; both assigned to Standard Oil Company (Indiana)* that the corrosion of ferrous metal by aqueous dispersions of elemental sulfur can be reduced if the aqueous phase of the dispersion contains dissolved therein an effective amount of a combination of ammonia and at least one metal compound selected from the group consisting of soluble zinc and magnesium compounds.

The amount of elemental sulfur is from about 1 to 70 wt % based on the total dispersion. The mols of ammonia/g-atom of metal from the metal compound is from about 0.1 to 200, and the atomic ratio of metal to sulfur is from about 0.0002 to 0.1.

The concentration of metal compound in the inhibitor solution is desirably in excess of about 0.01 molar and preferably in excess of about 0.1 molar.

Example 1: To a solution of 27.0 parts of anhydrous citric acid in 60.3 parts of water was slowly added with stirring 8.5 parts of magnesium oxide while keeping the temperature below 52°C. After the addition was completed, stirring was continued until the mixture cooled to 38°C. Anhydrous ammonia (4.2 parts) was then added to the mixture at a rate such that the temperature remained below 49°C during addition of the last 50% of the ammonia. The resulting magnesium citrate solution contained 1.2 mols of ammonia/g-atom of magnesium, had a pH of 9.7, and had a specific gravity of 1.233 at 20°C.

Example 2: To a solution of 22.0 parts of anhydrous citric acid in 55.6 parts of water was slowly added with stirring 9.8 parts of anhydrous ammonia while keeping the temperature below 32°C. The resulting mixture was cooled to 10°C, and 12.6 parts of zinc oxide was then slowly added while keeping the temperature below 32°C. Stirring was continued until a clear solution was obtained. The resulting zinc citrate solution contained 3.7 mols of ammonia/g-atom of zinc, had a pH of 9.75, and had a specific gravity of 1.226 at 21°C.

Example 3: To a mixture of 20.6 parts water and 26.8 parts of ammonia hydroxide solution (28% NH_3) was added with stirring 12.6 parts of zinc oxide followed by 40.0 parts of 58 to 60% aqueous ammonium thiosulfate solution. The resulting zinc thiosulfate solution contained 4.9 mols of ammonia/g-atom of zinc, had a pH of 11.2, and had a specific gravity of 1.226 at 22°C.

Example 4: A series of test dispersions were prepared from Super-Six Sulfur, a commercially available aqueous sulfur dispersion containing 52 to 54 wt % sulfur and having a sulfur particle size of from 1 to 5 μ, by mixing with sufficient amounts of the magnesium citrate solution of Example 1 to give the magnesium to sulfur ratios which are set forth in the following table.

A carbon steel test coupon was then weighed and suspended in each test dispersion at 21° to 24°C for 336 hr. At the end of this time, the coupons were removed, cleaned, dried, and reweighed. From the weight loss of each coupon, a corrosion rate, expressed in millimeters per year, was calculated. The results are as follows.

Weight of Sulfur Dispersion, g.	Weight of Magnesium Citrate Solution	Atomic Ratio of Magnesium to Sulfur	pH	Corrosion Rate, millimeters/year
100	0	0	8.08	2.583
99	1	0.0013	9.31	0.846
98	2	0.0026	9.45	0.0023
95	5	0.0067	9.52	0.0010
90	10	0.0142	9.61	0.0008

The results shown above demonstrate that the corrosive activity toward carbon steel of an aqueous dispersion of elemental sulfur is substantially reduced at a magnesium to sulfur ratio in excess of about 0.001.

Example 5: A series of test dispersions were prepared from Flo-Sul Sulfur, a commercially available aqueous dispersion of elemental sulfur containing 52 to 54 wt % sulfur and having a sulfur particle size of from 1 to 5 μ, by mixing with sufficient amounts of the zinc citrate solution of Example 2 to give the zinc to sulfur ratios which are set forth in the following table. A carbon steel test coupon was then weighed and suspended in each test dispersion at 43°C for 168 hr. At the end of this time, the coupons were removed, cleaned, dried, and reweighed. From the weight loss of each coupon, a corrosion rate, expressed in millimeters per year, was calculated. The results are as follows.

Weight of Sulfur Dispersion, g.	Weight of Zinc Citrate Solution, g.	Atomic Ratio of Zinc to Sulfur	Corrosion Rate, millimeters/year
100	0	0	22.46
99.9	0.1	0.00009	25.47
99.8	0.2	0.00019	20.71
99.6	0.4	0.00038	17.70
99.4	0.6	0.00057	6.464
99.2	0.8	0.00076	0.013
99.0	1.0	0.00095	0.005
98.0	2.0	0.00191	0.001

The results shown above demonstrate that the corrosive activity toward carbon steel of an aqueous dispersion of elemental sulfur is substantially reduced at a zinc to sulfur atomic ratio in excess of about 0.006.

Example 6: A series of test dispersions were prepared from Flo-Sul Sulfur by mixing with sufficient amounts of the zinc thiosulfate solution of Example 3 to give the zinc to sulfur ratios which are set forth in the following table. A carbon steel test coupon was then weighed and suspended in each test dispersion at 43°C for 312 hr. At the end of this time, the coupons were removed, cleaned, dried, and reweighed. From the weight loss of each coupon, a corrosion rate, expressed in millimeters per year, was calculated. The results are as follows.

Weight of Sulfur Dispersion, g.	Weight of Zinc Thio-Sulfate Solution, g.	Atomic Ratio of Zinc to Sulfur	Corrosion Rate, millimeters/ year
100	0	0	15.99
99.8	0.2	0.00019	16.72
99.6	0.4	0.00038	11.12
99.4	0.6	0.00057	1.458
99.2	0.8	0.00076	0.0008
99.0	1.0	0.00095	0.0000

The results shown above demonstrate that the corrosive activity toward carbon steel of an aqueous dispersion of elemental sulfur is substantially reduced at a zinc to sulfur atomic ratio in excess of about 0.0006.

Denaturation of Water-Based Paints

The "denaturation" of a paint denotes the treatment by which the physico-chemical characteristics of a paint are modified so that the latter cannot form a strongly adherent deposit on the articles with which it happens to be in contact.

It is essentially when the paint is applied by spraying that the problem arises. In fact, a nonnegligible portion of the paint used is not applied to the articles to be coated. So it is necessary to avoid these unused paint-forming troublesome deposits in the equipment.

To treat the excess paint and to avoid its decomposition on the walls, the spray booths may be arranged in various ways. A customary arrangement consists of producing a flow or circulation of air in the booth, the air loaded with paint particles then being brought into contact with running water in order to be freed from the paint aerosol. In the booth the air generally passes from the ceiling to the floor, but may also be removed laterally, the principle remaining the same. The water loaded with particles is collected, then sent to the treatment installation to remove the paint.

Whatever the device adopted, in all cases, the presence in the running water, of denaturating compounds must ensure treatment of the particles in the water thereby reducing or eliminating their adhesive power although they may still be brought into contact with the walls of all of the installation.

It has appeared that salts with a monovalent cation and a majority of the salts with a trivalent cation do not permit one to achieve satisfactory denaturation. In the same way, it has appeared that chlorides, although effective for certain ones, should be avoided by reason of their corrosive properties with respect to metals. The presence of additional calcium cation in a denaturation solution may also have a troublesome influence by facilitating the bacterial proliferation which may be a considerable handicap, taking into account the necessary recycling of the composition used.

According to the process of *P. Schlicklin, A.-M. Mertzweiller and J. Ploussard; U.S. Patent 4,294,617; October 13, 1981; assigned to Air Industrie, France* it has been shown that denaturation compositions for water-base paints, responding well to the exigencies of practice, are constituted by aqueous solutions containing effective amounts of bivalent metal sulfates of the group comprising iron, magnesium and manganese. The denaturation solutions contain between 1 and 10 g/ℓ of at least one of the sulfates. The pH of the solution is kept advantageously in the range of neutrality.

In the finely-divided state, the denaturation of the paint in contact with the treatment solution is rapid. The rapidity of denaturation is advantageous in that the fine particles in contact with the treatment solution lose their adhesive power very quickly and there is no longer a risk of the formation of strongly adherent deposits. This rapid denaturation is accompanied by an equally rapid precipitation which is not always desirable from the point of view of industrial use. This is the case in particular when the spraying installation has not been designed exclusively for the settling of a water-base paint. In this case, if the previously indicated denaturation solutions are used as such, the precipitation can occur outside of the areas provided for this purpose, which substantially complicates the maintenance of the installation.

It is therefore desirable, while preserving the excellent denaturation properties of these solutions, to arrange that it is possible to control the conditions under which the precipitation takes place, and in particular the speed of precipitation.

To reach this objective, denaturating solutions are used such as abovedescribed to which is added an effective amount of a surface active agent of the type of polyether of the form $R{-}O(CH_2{-}CH_2O)_n{-}[CH_2{-}CH(CH_3){-}O]_p{-}H$, in which R is an alkyl, aryl, aryl-alkyl or alkyl-aryl group, containing at least 10 carbon atoms, and n and p are whole numbers of which one may be zero, n + p being comprised between 12 and 100.

It is also possible to add to the denaturation solution agents traditionally used to prevent corrosion such as, e.g., thiourea or its derivatives, notably diphenylthiourea. Corrosion phenomena having a tendency to manifest themselves principally at the beginning of operations, it is particularly at this moment that the inhibitor must be present in the bath. These agents may be used in the amounts customarily relied on for this type of application. A concentration of about 1 to 2 g/ℓ of diphenylthiourea is preferred at the start. This concentration may then be reduced substantially and it is advantageously kept in the range of 0.5 g/ℓ.

Example 1: A solution containing 3 g/ℓ $MnSO_4{\cdot}H_2O$, 1.5 g/ℓ of thiourea, and 1 cm^3/ℓ of formol was tested to determine its effectiveness in a paint booth. The experimental booth was equipped with a ventilator, a washer and a flow channel. In operation, the flow rate of the solution running over the walls and into the washer was 4 m^3/hr. The flow rate of air was 1,500 m^3/hr. The paint treated was based on acrylic resin modified with melamine.

Good denaturation was observed and almost total decantation. Slight flotation of the treated paint in the low channel, in the form of a fine discontinuous film was eliminated by using a stirrer.

Example 2: Several tests were carried out with the same solution as in Example 1 in a larger experimental booth of the type used in the automobile industry, having a ground surface of 6 m^2, and equipped with a washer under the grating. A portion of the recycled solution was sent to a decanter.

A mixture of paints containing alkyd resins and acrylic resins modified with melamine were treated in this way.

In general, it was observed that the paint was well denatured and that the paint agglomerates on the surface were friable and easily dispersed. They were constituted of denatured paint with trapped air bubbles. The paint particles trapped by the bath water form a solution of which the major portion had a high sedimentation rate (about 1 mm/sec) and a very small portion had a tendency to float.

The paint in suspension decanted to at least 60%. If the surface of the bath was stirred, there was almost no supernatant liquid; the bath was almost clear and there was almost no paint in suspension.

In continuous operation, the amounts of manganese sulfates necessary to maintain the concentration of the bath at about 3 g/ℓ were evaluated. This amount was about 15 g of manganese sulfate/kg of paint treated.

A similar test in the same equipment but using a solution containing 1.5 g/ℓ instead of 3 g/ℓ of manganese sulfate leads also to satisfactory results.

Triazole and Olefinic Fatty Acid for Detergent

The process of *J. McGrady; U.S. Patent 4,321,166; March 23, 1982; assigned to The Procter and Gamble Company* relates to unbuilt heavy-duty liquid detergent compositions containing, as a corrosion inhibiting system, a mixture of an aromatic triazole and an oligomeric, olefinic fatty acid. The combination of these individual ingredients results in improved corrosion inhibition in washing machines exposed to dilute aqueous solutions of the heavy-duty liquid detergents.

The unbuilt liquid detergent composition comprises from about 20 to 75% of a detergent surfactant; from about 1.0 to 75% water; and from about 0.85 to 2.0% of a corrosion inhibiting system consisting essentially of a mixture of an oligomeric C_{14} to C_{22} olefinic fatty acid and an aromatic triazole, wherein the weight ratio of the fatty acid to the aromatic triazole is from about 40:1 to 1:1.

Example: The following unbuilt storage-stable liquid detergent composition was produced:

Component	Percent by Weight
$C_{14\text{-}15}(EO)_7$*	15.0
Magnesium C_{14} linear alkyl benzene sulfonate	30.0
Ethanol	6.5

(continued)

Component	Percent by Weight
Triethanolamine	3.0
Monoethanolamine	0.45
Trimeric C_{18} olefinic fatty acid	1.0
Benzotriazole	0.05
Citric acid	0.1
Perfume, brightener, dye	1.22
Water	42.68
(pH = 8.4)	

*Condensation product of C_{14-15} alcohol with 7 mols of ethylene oxide, commercially available as Neodol 45-7.

The above composition provided excellent fabric cleaning when used either full strength as a pretreatment or for through-the-wash detergency at a level of about ½ cup usage/17 gal of wash water. Moreover, the composition was stable and provided a corrosion inhibition effect to the steel surface of a washing machine wherein the composition was used. The corrosion inhibition provided by the combination of the benzotriazole and the trimeric C_{18} olefinic fatty acid was greater than that provided by each individual corrosion inhibiting element.

Substantially similar corrosion inhibiting benefits are obtained when the benzotriazole is replaced with other substituted aromatic triazoles, especially when replaced with tolyltriazole, acylated benzotriazoles, naphthotriazole, and mixtures thereof.

Substantially similar corrosion inhibiting benefits are obtained when the trimeric C_{18} olefinic fatty acid is replaced with other oligomerized C_{14-22} olefinic fatty acids (such as dimerized C_{18} olefinic fatty acid, dimerized C_{16} olefinic fatty acid, dimerized C_{20} olefinic fatty acid, trimerized C_{16} olefinic fatty acid and trimerized C_{20} olefinic fatty acid) and mixtures thereof.

Substantially similar corrosion inhibiting benefits are obtained when the anionic/nonionic surfactant mixture described above is replaced with other detergent surfactants selected from the group consisting of nonionic surfactants, anionic surfactants and mixtures thereof.

This is especially true when the above composition is replaced with any mixture of a nonionic surfactant produced by the condensation of from about 5 to 11 mols of ethylene oxide with 1 mol of a C_{13} to C_{16} alcohol, the nonionic surfactant being characterized by an HLB of from about 9.5 to 15; and an ionic surfactant which is a mixture of an alkanolamine and an alkali metal salt of an alkylbenzene sulfonic acid where the alkyl group contains from about 9 to 15 carbon atoms and wherein the alkanolamine is selected from the group consisting of mono-, di-, and triethanolamines and the alkali metal is selected from the group consisting of sodium, potassium, magnesium, and calcium; at a weight ratio of nonionic surfactant to anionic surfactant of from about 1.8:1 to 3.5:1 based on the free acid form of the anionic surfactant.

Protection of Titanium Equipment by Calcium, Strontium or Barium Ions

Titanium and its alloys are materials frequently used in the manufacture of equipment for industrial installations for bleaching of cellulose materials. These installations often include multipurpose apparatus in which various kinds of widely differing reagents can be used. This is the case, e.g., in textile bleaching installations, in dynamic paper pulp bleaching installations, and in certain conventional kraft pulp bleaching installations, which include an all-purpose final stage. Equipment in such installations is made at least in part of titanium or one of its alloys.

Since titanium and its alloys can be corroded by certain aqueous solutions commonly used in bleaching, such as alkaline aqueous solutions of peroxy compounds, the choice of reagents or the concentration at which they should be used in installations having surfaces which are in contact with such solutions, and which are made of titanium or one of its alloys, is limited.

L. Clerbois and L. Plumet; U.S. Patent 4,372,813; February 8, 1983; assigned to Interox (Société Anonyme), Belgium provide a process for inhibiting the corrosion of equipment made of titanium or of alloys containing titanium by a solution containing a peroxy compound. To effect the inhibition, a solution containing calcium, strontium or barium ions is used.

The alkaline earth metal ions can be supplied to the solution in the form of various types of compounds. In general, compounds are used which are soluble in the solution at the concentrations used; soluble organic or inorganic compounds can be used. Preferably, the acetates, nitrates, hydroxides, sulfates, chlorates, hypochlorites or halides such as chlorides are used. The best results have been obtained with the acetates, carbonates and bicarbonates, nitrates, sulfates and chlorides. Mixtures of these compounds as well as mixtures of ions can also be used. Where calcium ions are being used, hard water can advantageously be used to constitute the solution by adjusting the level of ions where necessary to the value required by addition of ions.

The quantity of ions used in the solution is, in general, between 0.0001 and 0.5 g-atom/ℓ of solution.

–2–

Aqueous System Applications

AUTOMOBILE COOLANT SYSTEMS

Device for Dispensing Corrosion Inhibitor

R.H. Krueger; U.S. Patent 4,273,744; June 16, 1981; assigned to Borg-Warner Corporation provides a device designed to automatically add a suitable corrosion inhibitor to the coolant in an engine cooling system in the event that the operator replaces lost ethylene glycol solution with water. The device is designed as a hydrometer with a solid or liquid inhibitor in the hollow upper tube or stem above the weighted end. The hydrometer could be placed in the coolant overflow tank or, if small enough, in the radiator tank. As the specific gravity of the coolant decreases due to the addition of water, the hydrometer tube will gradually drop until the inhibitor is contacted by the coolant solution.

The hydrometer has an upper tube with small openings therein at a level opposite the corrosion inhibitor. To provide for gradual additions, separators may be placed between levels of corrosion inhibitor so that the device will be operative over an extended period of time.

Figure 2.1 shows a hydrometer (10) of conventional shape with an elongated tubular stem (11) terminating in an enlarged bulbous lower end (12) wherein a suitable weight (13), such as steel shot, is located.

The tubular stem has a graduated scale (14) thereon to indicate specific gravity of the coolant and a plurality of small openings (15) in the stem generally opposite a quantity of a suitable corrosion inhibitor located within the stem and resting on a glass separator (17). Assuming that the hydrometer (10) is placed in the overflow tank for the radiator (not shown) where temperatures are unlikely to exceed 200°F, at the proper specific gravity for the ethylene glycol-water mix, the stem (11) containing the corrosion inhibitor will not contact the coolant. As water is added to the coolant system to replace the loss of coolant, the hydrometer will gradually drop as the specific gravity decreases until the coolant can pass through the openings (15) to contact the inhibitor (20) and a portion of it or all of it will dissolve, depending on the contact area, which in turn depends on the water-ethylene glycol concentration.

Figure 2.1: Hydrometer Containing Corrosion Inhibitor

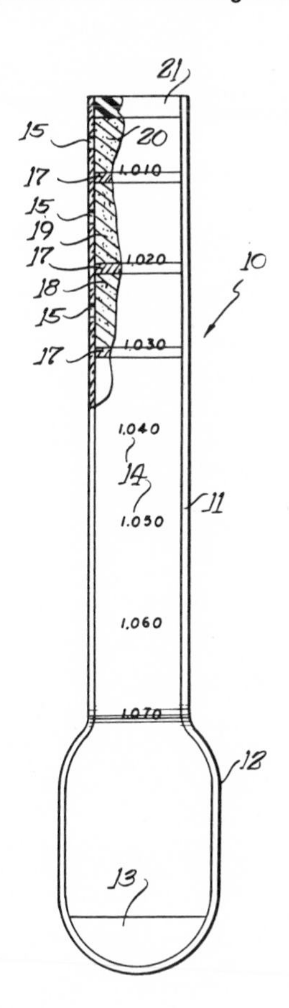

Source: U.S. Patent 4,273,744

As more specifically shown in the table, the specific gravity of the coolant will vary with concentration and temperature:

Volume Percent Solution			Specific Gravity t°/60° F. in air		
Ethylene Glycol	Water	Freezing Pt. °F.	60° F.	150° F.	200° F.
50	50	−33.5	1.080	1.050	1.030
40	60	−11.0	1.064	1.037	1.018
30	70	4.3	1.050	1.024	1.005
20	80	15.7	1.034	1.010	0.992
10	90	24.6	1.020	0.997	0.980
0	100	32.0	1.004	0.982	0.966

Considering the data in this table, assuming the hydrometer (10) is located in the overflow tank at approximately 200°F, a 50-50 mixture will have a specific gravity of 1.030 and no inhibitor will contact the coolant. However, if the specific gravity decreases to 1.020 through the addition of water, the hydrometer will drop allowing part of the solid inhibitor to dissolve. As the coolant dissolves the solid material, the weight of the hydrometer may initially increase and then decrease as the solid dissolves and is replaced by water. To overcome this, a separator is located between each level (18),(19),(20) of inhibitor.

The weight of the inhibitor is dependent on its solid density. Mixed powdered inhibitors can be compacted to increase their density by compression molding. For an average coolant volume of about 15 ℓ, requiring 0.1% inhibitor for metal protection, the weight of inhibitor needed will be 15 g. To increase the inhibitor weight, it is necessary to increase the cross sectional area of the hydrometer. In the hydrometer a rubber stopper (21) closes the open end of the stem (11).

Tertiary, Cyclic, Bridgehead Amines

Tertiary, cyclic, bridgehead amines are described by *P. Davis, J.F. Maxwell and J.C. Wilson; U.S. Patent 4,292,190; September 29, 1981; assigned to BASF Wyandotte Corporation* for use as corrosion inhibitors under basic or alkaline conditions to inhibit the corrosion of metals below sodium in the electromotive displacement series and especially the corrosion of iron, steel and aluminum. It has been unexpectedly found that not only are the amines useful alone as corrosion inhibitors for metals but the use of the amines in combination with conventional corrosion inhibitors provides unexpectedly improved corrosion resistance when the compositions are used as components of an aqueous alkaline solution with which the metals are in contact.

The corrosion inhibitors are especially useful in protecting metals against corrosion by aqueous antifreeze compositions containing an alcohol. The corrosion inhibitors are effective under alkaline conditions to inhibit the corrosion of all the metals and alloy components commonly found in present day internal combustion engines in contact with the coolant system of the engines.

The tertiary, cyclic, bridgehead amine is selected from the group consisting of tertiary amines having the general formulas:

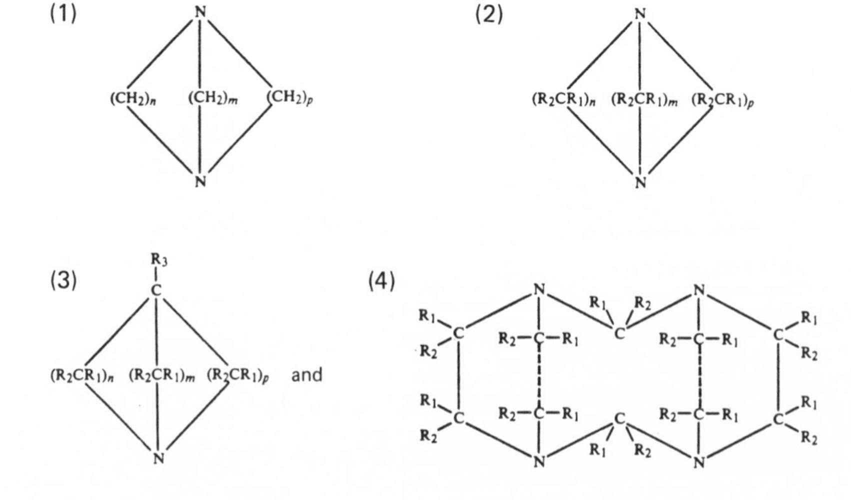

wherein n is 1 to 3, m is 1 to 2, and p is 1 to 2; R_1, R_2 and R_3 are each independently selected from the group consisting of:

> hydrogen and (a) a monovalent straight or branched chain alkyl or alkylene group having 1 to 6 carbons; and which can contain one or more of the following functional groups: –OH, $-CO_2H$ (or the alkali metal salt), $-CO_2R$, halogen, $-NH_2$, $-SO_3H$ (or the alkali metal salt), –CN, or –OR (where R is a monovalent alkyl or alkylene as defined above) or alkylene oxide derived groups;
>
> hydrogen and (b) a monovalent monocyclic aromatic hydrocarbon group which can contain one or more ring substituents selected from the following: –H, –R [where R is a monovalent alkyl or alkylene as defined in (a) above], –OH, $-NH_2$, $-CO_2H$ (or the alkali metal salt), OR (R as defined above), $-NO_2$, $-SO_3H$ (or the alkali metal salt), or halogen;
>
> hydrogen and (c) a monovalent monocyclic alkyl or alkylene hydrocarbon group of from 3 to 6 carbons which can contain one or more ring substituents selected from the following: monovalent alkyl or alkylene hydrocarbon groups, –OH, –OR (R as defined previously), $-CO_2H$ (or the alkali metal salt), $-NH_2$, $-SO_3H$ (or the alkali metal salt), $-PO_3H_2$ (or the alkali metal salt), halogen, or alkylene oxide derived units, and provided only one or R_1 and R_2 in Formula (2) can be hydrogen and no more than two of R_1, R_2 and R_3 can be hydrogen.

The cyclic, tertiary, bridgehead amine is used in a corrosion inhibiting amount, generally about 0.01 part by weight to about 5 parts by weight per 100 parts total weight of the antifreeze concentrate.

The corrosive additive composition of the process comprises about 0.3 part to about 50 parts, preferably about 1 to about 25 parts by weight of a conventional corrosion inhibitor and/or alkaline pH buffer selected from the conventional corrosion inhibitors and/or alkaline pH buffers, about 0.3 part to about 25 parts, preferably about 1 to about 20 parts by weight of the bicyclic, tertiary, bridgehead amine compound. The composition can contain additionally about 5 parts to about 150 parts, preferably about 20 to about 50 parts by weight of water or an alcohol, as defined above.

The pH of the aqueous liquid compositions of the process is alkaline and generally adjusted to a pH of above 7 to about 12.

Fluoroaliphatic Radical-Containing Phosphonic Acid

In accordance with the process of *R.G. Newell and D.C. Perry; U.S. Patent 4,293,441; October 6, 1981; assigned to Minnesota Mining and Manufacturing Company* there is provided a corrosion inhibiting liquid composition, useful as or in a heat transfer medium for a heat exchanger, such as the cooling system of an internal combustion engine. The composition comprises (1) ethylene glycol, propylene glycol or mixtures thereof, including aqueous solutions thereof, and (2) as a corrosion inhibitor, fluoroaliphatic radical-containing phosphonic acid or salt or hydrolyzable ester whereof, which imparts corrosion resistance to aluminum articles in contact with the composition, particularly aluminum metal defining at least in part the flow passageways of the heat exchanger.

The fluoroaliphatic radical-containing phosphonic acids, salts and esters used as corrosion inhibitors are preferably those of the general formula:

$$R_f{-}Q{-}\overset{\overset{\displaystyle O}{\|}}{\underset{\underset{\displaystyle OM^2}{|}}{P}}{-}OM^1$$

in which

R_f is a fluoroaliphatic radical,

Q is a divalent linkage through which R_f and the phosphorus atom, P, are covalently bonded together, the skeletal chain of such linkage being free of moieties, such as ester groups, which are more readily hydrolyzable than amido moieties; and M^1 and M^2 are preferably hydrogen but can be independently alkyl (e.g., with 1 to 8 carbon atoms) or any salt-forming cation, preferably a Group I or Group II metal cation, ammonium, or an aliphatic primary, secondary, tertiary, or quaternary ammonium cation, e.g., RNH_3^+, $R_2NH_2^+$, R_3NH^+ and R_4N^+, where R is alkyl or alkaryl or substituted alkyl or alkaryl (such as hydroxyalkyl) with 1 to 18 carbon atoms.

Example 1: A mixture of N-butyltrifluoromethanesulfonamide (62.0 g, 0.302 mol), potassium carbonate (45.54 g, 0.330 mol) and methanol (250 ml) was refluxed with stirring for 2 hr. Allyl bromide (39.93 g, 0.330 mol) was then added and this mixture stirred under reflux for 24 hr, cooled, filtered, and the solvent evaporated in vacuo. The residue was fractionally distilled yielding 37.0 g (BP of 53° to 58°C at 0.07 torr) of the desired compound, N-allyl-N-butyltrifluoromethanesulfonamide.

Over a 60 minute period, a solution of the above-prepared N-allylsulfonamide (24.60 g, 0.10 mol) and diethylphosphite (15 g) was added dropwise concurrently with a solution of di-tert-butyl peroxide (0.90 g) in diethylphosphite (5 g) to diethylphosphite (80 g) being stirred at 150°C with an argon purge. The resulting mixture was stirred for an additional 1 hr under these conditions. The excess diethylphosphite was distilled at 10 to 20 torr and the residue fractionally distilled to give 19.64 g (BP 150° to 153°C at 0.20 torr) of the desired compound, diethyl 3-(N-butyltrifluoromethanesulfonamido)propanephosphonate. A mixture of the phosphonate prepared above (15.36 g, 0.040 mol) and bromotrimethylsilane (13.46 g, 0.088 mol) was stirred under a reflux condensor and calcium chloride drying tube for 96 hr. Then water (75 ml) was added and the resulting mixture stirred for 15 minutes followed by extraction with three 50 ml portions of diethylether. The ether portions were combined, dried with $MgSO_4$ and evaporated in vacuo to give 12.25 g of the desired compound, 3-(N-butyltrifluoromethanesulfonamido)propanephosphonic acid.

Example 2: The procedure of Example 1 was repeated with the exception that N-ethylperfluorobutanesulfonamide was used in place of N-butyltrifluoromethanesulfonamide to make N-allyl-N-ethylperfluorobutanesulfonamide (BP

50° to 55°C at 0.12 to 0.10 torr). The latter was reacted with diethylphosphite to produce the desired compound, diethyl 3-(N-ethylperfluorobutanesulfonamido)-propanephosphonate (BP 148° to 152°C at 0.4 to 0.5 torr). A solution of the phosphonate prepared above (30.0 g, 0.062 mol) and bromotrimethylsilane (20.0 g, 0.130 mol) was stirred under a reflux condenser and calcium chloride drying tube for 24 hr, water (75 ml) was added, and the mixture stirred for 15 minutes. The resulting white precipitate was filtered out and washed with water (50 ml) and diethylether (100 ml), then dried in an oven at 100°C for 24 hr at 10 to 20 torr, to yield the desired compound, 3-(N-ethylperfluorobutanesulfonamido)propanephosphonic acid (MP 138.5° to 140°C).

Example 3: Following the procedure of Example 1, starting with N-ethylperfluorooctanesulfonamide, diethyl 3-(N-ethylperfluorooctanesulfonamide)propanephosphonate (BP 170° to 173°C at 0.05 torr) was made. The latter was hydrolyzed according to the procedures of Example 2 to produce the corresponding phosphonic acid (MP 146° to 151°C).

Example 4: The ability of the fluoroaliphaticphosphonic acid to impart corrosion resistance to aluminum was determined by measuring the corrosion current density, j_{corr}, in microamperes per square centimeter, a/cm^2, which is directly proportional to the corrosion rate. This means the larger the j_{corr} value, the more corrosion that is taking place. The j_{corr} was measured by a well-known electrochemical technique called linear polarization, as explained by F. Mansfeld in *Adv. Corr. Sci. Technol.*, 6, 163 (1976).

A freshly cleaned bare aluminum panel (Al 7072, typical of that used in the fabrication of aluminum radiators for automobiles) was immersed at 25°C in a bath of 50 volume percent aqueous ethylene glycol solution containing 0.1 weight percent $C_8F_{17}SO_2N(C_2H_5)(CH_2)_3P(O)(OH_2)$ dissolved therein. The bath contained "corrosive" water, viz, one hundred times the chloride, sulfate, and bicarbonate concentration specified in ASTM D 1384. During the immersion, the j_{corr} values were measured after 1 hr to get an initial value and after 1 to 2 days by the linear polarization method described in Example 4. The exposure was run under nitrogen at 25°C. For purposes of comparison, j_{corr} was obtained under the same conditions for an uninhibited solution of the ethylene glycol and for a solution of a commercially available inhibited ethylene glycol antifreeze. Results are set forth in the table below.

		j_{corr}, μa/cm^2	
Run	Liquid Composition	Initial	1-2 days
1	Aqueous ethylene glycol containing no inhibitor	3.6	2.6
2	Aqueous ethylene glycol solution of $C_8F_{17}SO_2N(C_2H_5)(CH_2)_3P(O)(OH)_2$	0.32	0.12
3	Aqueous solution of commercially available corrosion inhibited ethylene glycol anti-freeze	0.22	0.45

Nucleus-Substituted Benzoic Acids

The process of *K. Barthold and G. Liebold; U.S. Patent 4,324,675; April 13, 1982; assigned to BASF AG, Germany* concerns anticorrosive additives for aqueous liquids and coolants or antifreeze compositions for use in contact with metals present in the cooling system of water-cooled internal combustion engines. The coolants are based on an aqueous glycol solution, for instance, ethylene glycol or propylene glycol and which are free of nitrites, amines and phosphates. They preferably contain about 0.05 to about 5% by weight, relative to the glycol, of at least one nucleus-substituted benzoic acid having a pKa value of less than 3. These include, for instance, nitro and halogen substituents on the ring of benzoic acid. Preferred among these are the nitrobenzoic acids and particularly the ortho- and p-nitrobenzoic acids. These acids are most preferably added in quantities of about 0.5 to about 2.5% by weight relative to the glycol to provide excellent corrosion protection. It has been determined that the aromatically bonded nitro group is chemically stable. When mixed with coolants containing alkanolaminophosphate, this material does not result in nitrosamines.

The coolant concentrates of the process contain an effective amount of at least one nucleus-substituted benzoic acid together with an effective corrosion inhibiting amount of at least one conventional metal corrosion inhibitor in an ethylene glycol or propylene glycol base. Useful conventional metal corrosion inhibitors include alkali metal benzoates, alkali metal silicates, borax, alkali metal benzotriazole and alkali metal benzothiazole.

The corrosion behavior of the new formulations was tested according to ASTM D-1384-70. The water used for diluting purposes contained 100 ppm chloride, sulfate, and bicarbonate.

In the examples where not otherwise specified, temperatures are given in degrees centigrade and parts, percentages and proportions are by weight. The concentration of the coolant concentrate in water was 20 volume percent.

Example 1: *(Control)* – A commercially available coolant based upon benzoate/nitrite was obtained. Proportions are in percent by weight; 5.0 sodium benzoate, 0.5 sodium nitrite and 94.5 ethylene glycol.

Example 2: A coolant was prepared according to this process as follows: (The proportions are in parts by weight); 1.0 p-nitrobenzoic acid, 2.5 sodium benzoate, 1.4 borax·$10H_2O$, 0.05 sodium silicate·$5H_2O$, 0.1 sodium nitrate and 94.9 ethylene glycol.

Example 3: A coolant of the process was prepared. (The proportions are in parts by weight); 2.0 p-nitrobenzoic acid, 3.0 sodium benzoate, 0.7 borax·$10H_2O$, 0.05 sodium silicate·$5H_2O$, 0.05 sodium nitrate, 0.07 benzotriazole and 94.15 ethylene glycol.

Example 4: A coolant of the process was prepared. (The proportions are in percent by weight); 0.05 o/p-nitrobenzoic acid (1:1), 2.5 sodium benzoate, 2.0 borax·$10H_2O$, 0.05 potassium silicate, 0.1 $NaNO_3$, 2.0 water and 93.70 ethylene glycol.

Loss (-) or Gain (+) in Weight

	.... mg/cm²					
Example	Copper	Solder	Brass	Steel	Cast	Aluminum
1	-0.01	-0.04	-0.01	-0.01	-0.78	-0.75
2	±0.00	-0.01	±0.00	-0.05	-0.21	+0.03
3	±0.00	-0.07	-0.17	±0.00	+0.03	-0.09
4	-0.13	±0.00	-0.28	-0.14	-0.32	-0.29

Beta-Dicarbonyl Compounds

It has been found by *H.S. Gilbert; U.S. Patent 4,324,676; April 13, 1982; assigned to The Dow Chemical Company* that the corrosion performance of single phase glycol-water compositions containing the usual inhibitors can be improved by the addition of an effective amount of one or more soluble β-dicarbonyl compounds. The compounds have the general formula

$$R_1-C(O)-C(R_2)(R_3)-C(O)-R_4$$

where R_1 and R_4 are independently alkyl groups of 1 to 4 carbons, and alkoxy groups, of 1 to 4 carbons,

R_2 and R_3 are hydrogen, hydroxyl, alkyl groups of 1 to 4 carbons, or acetyl groups, with the proviso that R_1 and R_2 can be joined to form a 5 or 6 membered ring when R_3 is hydrogen and R_4 is an alkyl group of 1 to 4 carbons,

with the proviso that R_1 and R_4 can be joined to form a 5 membered ring when R_2 is hydrogen and R_3 is an alkyl group of 1 to 4 carbons,

with the proviso that R_1 and R_4 can be joined to form the 1,2-phenylene group when R_2 and R_3 are hydrogen, and with the proviso that R_1 and R_4 can be joined to form the 1,2-phenylene group when R_2 is hydroxyl and R_3 is the 2-hydroxy 1,3-dioneindanyl-2 group.

The corrosion inhibitors of this process are effective in the presence of the other well-known corrosion inhibitors generally present in such compositions such as alkali metal silicates, borates, mercaptobenzotriazoles, nitrates, nitrites, phosphates, benzoates and the like.

An effective amount of the β-dicarbonyl compound which will give an improved composition is an amount ranging from about 0.01 to about 1.0% by weight based on the total weight of the composition and preferably in the range from 0.05 to 0.3% by weight.

Example 1: A sample of a commercial antifreeze concentrate (hereinafter Control A) was tested by the hot aluminum test. In this test 70 ml of the concentrate was mixed with 140 ml of water having a standard hardness, (i.e. ASTM water which has 100 ppm each of chloride, sulfate, and bicarbonate ions). An aluminum specimen 1½" dia. x ¾" was exposed to this solution and heated under a nitrogen pad at 125°C for 168 hr (one week). Under these conditions, it was found that the specimen lost 17 mg of weight.

Under identical conditions, the test was repeated using Control A plus 1000 ppm acetylacetone. The specimen lost 6 mg of weight and visual inspection indicated that the surface was significantly improved, i.e. less pitting and less discoloration as compared to the original formulation without the acetylacetone.

Example 2: Ethyl acetoacetate (886 ppm) was added to the Control A solution. One part of the solution was diluted with 2 parts of ASTM water and was tested on the hot aluminum test equipment. The aluminum weight loss was reduced from 17 mg (Control A) to 1.7 mg.

Example 3: Specimens of metals shown in the table below typical of those present in automotive cooling systems were totally immersed in the test coolant solution which consisting of a 33⅓% solution of Control A in ASTM water with aeration for 336 hr at 190°F. The corrosion-inhibitive properties of the test solutions with and without acetylacetone are evaluated on the basis of the weight changes incurred by the specimens.

ASTM Glassware Corrosion Data (D-1384)

	Weight Loss (mg/specimen) (Average of 3)					
Antifreeze	Copper	Solder	Brass	Steel	Cast Iron	Aluminum
Control A Solution	2.4	3.4	3.1	0.7	3.1	+1.1
Control A Solution + 1000 ppm acetylacetone	0.9	2.3	1.4	0.3	0.4	0.0
Average weight loss:	for all coupons			all coupons except Aluminum		
Control A	2.0 mg			2.6 mg		
Control A + 1000 ppm acetylacetone	0.7 mg			1.7 mg		

Control A with acetylacetone gave 65% less weight loss than Control A when the combined weight losses of all six coupons were considered.

Control A with acetylacetone gave 35% less weight loss than Control A when the combined weight losses of all coupons except aluminum were considered.

Automatic Addition of Inhibitor Using Filled Polymer

The process of *R.H. Krueger; U.S. Patent 4,333,850; June 8, 1982; assigned to Borg-Warner Corporation* relates to a composition and device to automatically and continuously add a suitable corrosion inhibitor to a coolant solution in a cooling system for an automotive vehicle engine or other system utilizing a coolant circulating through a heat exchanger. The composition is a filled polymer containing one or more corrosion inhibitors for the metals commonly found in the coolant system. The composition in a suitable form is inserted into or incorporated as an integral part of the coolant system to be exposed to the circulating coolant so that the corrosion inhibitor would be leached out of the polymer into the coolant.

One suitable insoluble polymer for the filled polymeric composition is polypropylene, available as a fine powder, which was mixed with a slightly water soluble or water penetrable polymer, such as polyvinyl alcohol, and a corrosion inhibitor; ball milled for 30 minutes; and then compression molded at a temperature in the range of 400° to 450°F for 4 minutes to form a plaque. Lithium phosphate (Li_3PO_4) was chosen as the corrosion inhibitor because of its slow solubility in water and because phosphate has good corrosion inhibitor characteristics. For the slightly water soluble or water penetrable polymer, polyoxyethylene or a cellulose acetate could be substituted for the polyvinyl alcohol; the function of the slightly water soluble polymer being to slow the solubility of the inhibitor and to prevent encapsulation of the inorganic salt, lithium phosphate, by the insoluble polypropylene.

After the filled polymer plaques were formed, they were cut into strips and exposed to water or ethylene glycol solution to determine the amount of the corrosion inhibitor extracted as a function of time and temperature. A test of filled polymer strips in water at room temperature showed the lithium phosphate gradually dissolved reaching a maximum solubility of approximately 400 ppm after 48 days. After 105 days the water was replaced with new deionized water and the test continued. Analysis over the period of 112 to 150 days indicated that a filled polymer strip composed of 62.5% polypropylene, 25% lithium phosphate and 12.5% polyoxyethylene gave the greatest lithium phosphate solubility.

At 160 days, the water was again changed, and the test continued for 362 days before replacing the water. At 369 days, the specimen was weighed showing a 17.8% weight loss. Assuming lithium phosphate and polyoxyethylene dissolve at rates equivalent to their initial concentration, then approximately one-half of the inhibitor is still available. After 537 days, the specimen showed an 18.7% weight loss and, after 675 days, the specimen showed a 19.1% weight loss.

A similar extraction test was run in a 50-50 solution of Prestone II and water at room temperature. During this test, the specimens were taken out after 42 days, placed in 100% water and the test continued for 253 days. Then the specimens were dried and weighed. Again, the specimen composed of 62.5% of polypropylene, 25% lithium phosphate and 12.5% polyoxyethylene showed the greatest solubility by weight loss.

From test results it can be concluded that at room temperature lithium phosphate gradually leaches from filled polypropylene plastics containing water soluble or penetrable polymers; this continuing over more than a one year period with water replaced three times. Lithium phosphate is also extracted from a 50-50 Prestone II-water mixture, but at a slightly slower rate. Also, approximately 50% of the filler of lithium phosphate-polyoxyethylene is extracted in approximately one year's time.

Four Component Composition for Year-Round Use

I. Manabe and A. Inubushi; U.S. Patent 4,338,209; July 6, 1982; assigned to Otsuka Chemical Co., Ltd., Japan provide known metal corrosion inhibitors having the disadvantage that those employed in combination with anti-freezing agents in winter cannot be employed in summer and those employed in summer cannot be employed in combination with anti-freezing agents in winter. For instance, borax which is a metal corrosion inhibitor employed in combination with anti-freezing agents in winter has difficulty in preparation, because of low solubility in water, and also cannot be employed in summer as a metal corrosion inhibitor, because of insufficient anti-corrosive property in water.

Also, chromates employed in summer as a metal corrosion inhibitor cannot be employed in combination with anti-freezing agent in winter, because they accelerate the oxidation of anti-freezing agents.

The process provides a metal corrosion inhibitor which is applicable to various metals employed in cooling systems of internal-combustion engines and is usable for a long term without lowering the excellent anti-corrosive property.

The process provides a metal corrosion inhibitor comprising per 100 parts by weight of (a) benzoic acid and/or a benzoate (calculated as benzoic acid), 1.3 to 20 parts by weight of (b) nitrous acid and/or a nitrite (calculated as nitrous acid), 3.8 to 120 parts by weight of (c) phosphoric acid and/or a phosphate (calculated as phosphoric acid), and 1 to 20 parts by weight of (d) at least one member selected from mercaptobenzothiazole, its salts, benzotriazole and tolyltriazole (the amount of the salts of mercaptobenzothiazole being calculated as mercaptobenzothiazole).

The metal corrosion inhibitor of the process may be employed in the solid form. In that case, the inhibitor is added to a cooling water as it is. The corrosion inhibitor of the process may also be prepared as a liquid corrosion inhibitor to provide a commercially available product. In that case, the inhibitor is dissolved in an appropriate amount of water or an antifreezing agent such as ethylene glycol. Although the concentration of the inhibitor at the time of the preparation is not particularly limited, in the case of dissolving in water alone or water containing a small amount of an antifreezing agent, the concentration is usually from 30 to 50% by weight, and in the case of dissolving in an antifreezing agent, the concentration is usually from 2 to 15% by weight. The thus prepared liquid corrosion inhibitor is employed by adding to a cooling water.

It is desirable that pH of a cooling water to which the inhibitor of the process is added falls within the range of 6.5 to 9.5

The amount of the corrosion inhibitor added to a cooling water varies depending on the kind of the metals. In general, the corrosion inhibitor is employed so that the concentration of the inhibitor in a cooling water falls within the range of 6,000 to 70,000 ppm.

The metal corrosion inhibitor of the process is applicable to various metals employed in cooling systems such as aluminum, cast aluminum, cast iron, steel, brass, copper and solder, and can exhibit the excellent anticorrosive effect for a long term. Also, the corrosion inhibitor of the process can be employed in combination with antifreezing agents, such as ethylene glycol irrespective of season. Therefore, the corrosion inhibitor of the process can be employed not only to prevent the corrosion of the cooling systems of internal-combustion engines, but also to prevent the corrosion of the cooling systems in chemical factories and thermoelectric power plants. It is also possible to prevent rust by immersing a metal in an aqueous solution of the corrosion inhibitor of the process.

COOLING WATER SYSTEMS

Copolymer of Acrylic Acid and Hydroxy Alkyl Acrylate

The polymers used in the four patents that follow contain moieties derived from an acrylic acid compound (AA), i.e.,

(I)

$$\left[-CH_2-\underset{\underset{\underset{R_1}{|}}{C=O}}{\overset{\overset{R}{|}}{C}}- \right]$$

where R is hydrogen or a lower alkyl of from 1 to 3 carbon atoms and R_1 = OH, NH_2 or OM, where M is a water-soluble cation, e.g., NH_4, alkali metal (K, Na), etc.; and moieties of an hydroxylated lower alkyl (C_{2-6}) acrylate (HAA) as represented, for example by the formula:

$$\left[-CH_2-\underset{\underset{\underset{\underset{\displaystyle R_2-OH}{|}}{\underset{\displaystyle O}{|}}}{\underset{\displaystyle C=O}{|}}}{\overset{\displaystyle R_3}{\overset{|}{C}}}- \right]$$

where R_3 is H or lower alkyl of from 1 to 3 carbon atoms, and R_2 is a lower alkyl having from about 2 to 6 carbon atoms.

In terms of mol ratios, the polymers are considered, most broadly, to have a mol ratio of AA:HAA of from about 1:4 to 36:1. This mol ratio is preferably about 1:1 to 11:1, and most preferably about 1:1 to 5:1. The only criteria that is considered to be of importance with respect to mol ratios is that it is desirable to have a copolymer which is water-soluble. As the proportion of hydroxylated alkyl acrylate moieties increases, the solubility of the copolymer decreases. It is noted that, from an efficacy point of view, the polymers having a mol ratio of AA:HAA of 1:1 to 5:1 are considered the best.

The polymers could have a molecular weight of from about 1,000 to about 50,000 with from about 2,000 to about 6,000 being preferred.

F.G. Vogt and P.T. Sparrell; U.S. Patents 4,297,317; October 27, 1981; and 4,329,250; May 11, 1982; both assigned to Betz Laboratories, Inc. provide a composition for treating corroding aqueous media to reduce the corrosion of metal surfaces exposed thereto. The treatment comprises water-soluble chromate and the copolymer of acrylic acid and hydroxy alkyl acrylate.

From an environmental acceptability point of view, chromate levels of less than 5 ppm of active chromate are most desirable. However, at such low levels pitting is extremely severe. It was unexpectedly discovered that a corrosion inhibitor treatment comprising less than 5 ppm water-soluble chromate in combination with AA/HAA significantly reduced such pitting.

These chromate compounds would include alkali metal or any water-soluble compound that contains hexavalent chromium and provides chromate radical in water solutions.

Although this process is considered to have applicability to any aqueous system, it is particulary useful in cooling water systems.

Example 1: To demonstrate the corrosion inhibition efficacy of the combination of chromate and AA/HAA polymer, various mixtures were prepared by dissolving varying ratios of the components in water. The combinations were tested using a spinner testing technique.

The tests were each conducted with two non-pretreated low carbon steel coupons which were immersed and rotated in aerated synthetic cooling water for a 3- or 4-day period. The water was adjusted to the desired pH and readjusted after one day if necessary; no further adjustments were made. Water temperature was 120°F, and rotational speed was maintained to give a water velocity of 1.3 feet per second past the coupons. The total volume of water was 17 ℓ. Cooling water was manufactured to give the following conditions:

Ca as $CaCO_3$ = 400 ppm
Mg as $CaCO_3$ = 200 ppm
Chloride ion = 281 ppm
Sulfate ion = 192 ppm

Corrosion rate was determined by weight loss measurement. Prior to immersion, coupons were scrubbed with a mixture of trisodium phosphate-pumice, rinsed with water, rinsed with isopropyl alcohol and then air dried. Weight measurement to the nearest milligram was made. At the end of one day, a weighed coupon was removed and cleaned. Cleaning consisted of immersion into a 50% solution of HCl for approximately 20 seconds, rinsing with tap water, scrubbing with a mixture of trisodium phosphate-pumice until clean and then rinsing with tap water and isopropyl alcohol. When dry, a second weight measurement to the nearest milligram was made. At the termination of the tests, the remaining coupon was removed, cleaned and weighed.

Corrosion rates were calculated by differential weight loss according to the following equation:

$$\text{Corrosion Rate} = \frac{N^{th}\text{ Day Weight Loss} - 1^{st}\text{ Day Weight Loss}}{N - 1}$$

where N = 3 or 4.

The results of these tests are reported in the following table in terms of percent (%) corrosion inhibition of various treatments as compared to an untreated control test. The polymer tested was an acrylic acid/hydroxypropyl acrylate copolymer (AA/HPA) having a mol ratio of AA to HPA of 3:1 and a nominal molecular weight of 6000. The chromate compound used was sodium dichromate dihydrate. The amounts of chromate reported are active chromate dosages in ppm, and the pH of the test water was 8. The corrosion rates are reported in mils per year (mpy).

Test	Polymer (ppm)	$CrO_4^{=}$ (ppm)	Corrosion Rate (mpy)	% Corrosion Inhibition
A	—	—	143-147	0
B	—	1	120-139	10
C	1	—	145	0
D	1	1	86.5	40
E	10	—	83	43
F	10	1	43.5	70

It can be seen from the above table that the chromate-polymer combinations were quite effective in inhibiting corrosion, even at very low levels of active chromate.

Example 2: As already noted above, the use of insufficient amounts of chromate as a corrosion inhibitor can lead to rather severe pitting of metal surfaces. The ability of chromate-polymer combinations to reduce this pitting is demonstrated in the results of additional tests which were obtained in accordance with ASTM Pit Rating Methods. According to ASTM Pit Rating Methods, the density, size and depth of such localized corrosion is measured as described in ASTM Method STP-576 (1976). Testing procedures and conditions were substantially the same as those described in Example 1, with the exception that the coupons were additionally microscopically examined for pit rating.

The results of these tests are reported in the table below in terms of ASTM Pit Ratings. The polymer tested was AA/HPA having a mol ratio of AA to HPA of 3:1 and a nominal molecular weight of 6,000. The chromate was sodium dichromate dihydrate. The amounts of chromate reported are active chromate dosages. The ASTM pit ratings are in terms of density, in number of pits/square decimeter (dm^2); pit sizes, in square millimeters (mm^2), and pit depths, in millimeters (mm).

		$CrO_4^=$	Polymer	ASTM Pit Rating		
Test	pH	(ppm)	(ppm)	Density	Size	Depth
A	7	5	—	100-300/ dm^2	0.5-2.0 mm^2	≦0.4 mm
B	7	2.5	2.5	25-100/ dm^2	≦0.5 mm^2	≦0.4 mm

R.C. May and G.E. Geiger; U.S. Patent 4,303,568; December 1, 1981; assigned to Betz Laboratories, Inc. provide a method of inhibiting the corrosion of ferrous metal parts in contact with an aqueous solution with little or no attendant deposition of scale on the ferrous parts. Corrosion inhibition is obtained by providing for the formation of a protective passive oxide film on the metal surface in contact with the aqueous medium, the method comprises

(1) assuring that the pH of the aqueous medium is 5.5 or above;

(2) assuring that the aqueous medium contains a calcium or other appropriate ion concentration selected from the group consisting of zinc, nickel and chromium and mixtures thereof; and

(3) adding to the aqueous medium

(a) The water-soluble polymer comprising moieties derived from an acrylic acid or water-soluble salt thereof and moieties of an hydroxylated lower alkyl acrylate.

(b) a water-soluble orthophosphate compound; the polymer being such that together with the orthophosphate compound is effective to promote the metal oxide film with no attendant deposition, and the amount of (a) and (b) being added being sufficient to provide a substantially scale-free protective passive oxide film on the metallic surface.

Optionally but desirably, polyphosphates, organophosphonates, and copper corrosion inhibitors may be included. For additional corrosion protection zinc may be included in the treatment.

Specific Embodiments: Since compositions containing the polymer (AA/HAA) and the orthophosphate are not particularly stable when mixed as a highly active product for delivery to a user, it is desirable to treat the systems using what the industry considers a two-barrel approach, i.e., feed the ingredients separately. Since the polyphosphates, the phosphonates, and the copper corrosion inhibitors mix quite well with the orthophosphate, these are included in one drum when utilized as the treatment. It is possible, however, to blend all of the ingredients dry as powders or crystals and make the liquid form at the use site. Stability is not a significant problem when the make is used within a short time.

The preferred rate of application of this treatment to cooling water systems and the ratios of various components depends on the calcium concentration of the cooling water. The treatment is preferably applied in waters having between 15 and 1,000 ppm calcium. Within this range the weight ratio of calcium to orthophosphate is varied from 1:1 to 83.3:1, the weight ratio of AA/HAA to orthophosphate is varied from 1:3 to 1.5:1.

Field Studies: Actual testing of the concept was conducted at various industrial sites under production operating conditions. These case studies are detailed below.

Case Study No. 1 – A midwestern refinery was successfully using a zinc/phosphate/dispersant treatment program to control corrosion in an open recirculating cooling system, but was experiencing fouling in critical heat exchange equipment which impeded heat transfer and necessitated periodic shut-downs. Prior to using the zinc/phosphate/dispersant treatment, conventional phosphate/dispersant treatments were unsuccessful in controlling both fouling and tuberculation corrosion in mild steel heat exchangers.

Comparative studies were run with the zinc/phosphate/dispersant treatment (applied at a concentration that would give 2 ppm zinc and 3 to 5 ppm orthophosphate in the recirculating water) and a composition of this process (applied at a concentration that would give 5 ppm AA/HPA polymer and 10 to 15 ppm orthophosphate in the recirculating water) for a 6 week period in an open recirculating cooling system having a total recirculation rate of 12,000 gpm and a 15°F temperature drop across the cooling tower. Cold lime-softened well water was used as make-up to maintain the recirculating water at 3 cycles of concentration.

Analysis of the recirculating water typically gave 384 ppm total hardness (as ppm $CaCO_3$), 283 ppm calcium hardness (as ppm $CaCO_3$), 10 ppm methyl orange alkalinity (as ppm $CaCO_3$), 60 ppm silica (as ppm SiO_2), and 1139 μmhos conductivity. The recirculating water was controlled at pH = 7 with sulfuric acid. A critical mild steel process heat exchanger was cleaned prior to each trial and inspected after the 6 week trial period to assess the performance of each of the treatments. During the trials, mild steel (AISI-1010) and admiralty brass corrosion rates were measured.

At the end of the zinc/phosphate/dispersant trial, a uniform deposit coated the entire heat transfer surface of the process exchanger and tuberculation corrosion was noticeably evident. Corrosion rates during the trial period were 5.6 mpy for mild steel and 0.9 mpy for admiralty brass.

At the end of the AA/HPA-phosphate treatment program trial, the heat transfer surface of the process exchanger was deposit-free and did not show any signs of active corrosion. Additionally, inspection of other heat exchangers which were not cleaned prior to the AA/HPA-phosphate treatment trial revealed they were significantly cleaner than before the trial, indicating the treatment facilitated the removal of existing deposits. Corrosion rates during this trial were 1.0 mpy for mild steel and 0.0 mpy for admiralty brass.

In the above case it should be noted that the actual treatment was: 10.2 ppm $O{-}PO_4^{3-}$; 5.1 ppm pyrophosphate; 2 ppm hydroxyethylidene diphosphonic acid; and 2.6 ppm tolyltriazole; 5 ppm AA/HPA (3:1; Molecular weight about 6,000).

Case Study No. 2 – A midwestern petrochemical plant was using a conventional phosphate/dispersant program in their open recirculating cooling system to control corrosion. With the phosphate/dispersant treatment (applied at a concentration that would give 3 ppm orthophosphate in the recirculating water), mild steel (AISI-1010) corrosion rates averaged 15 mpy with severe pitting corrosion present, and fouling of heat transfer surfaces in process equipment was a chronic problem. The cooling system had a history of pH upsets and control problems which made corrosion and deposition control difficult. The cooling system operated at a recirculation rate of 15,000 gpm with an 18°F temperature drop across the cooling tower. Untreated well water was used as make-up to maintain the recirculating water at 2.5 cycles of concentration. Analysis of the recirculating water typically gave: 1230 ppm total hardness (as ppm $CaCO_3$), 740 ppm calcium hardness (as ppm $CaCO_3$), 40 ppm methyl orange alkalinity (as ppm $CaCO_3$), 34 ppm silica (as ppm, SiO_2) and 2360 μmhos conductivity.

Application of a composition of this process applied at a concentration that would give 15 ppm AA/HPA polymer and 10 to 15 ppm orthophosphate was made to the cooling system, with the pH being controlled at 7. Over an 8-month period, mild steel corrosion rates were 1.9 mpy without any significant pitting corrosion present. Additionally no fouling problems were incurred with process equipment. Monitoring of heat transfer coefficients (U) of a process exchanger indicated fouling had subsided after the application of the AA/HPA-phosphate treatment.

In this case study the actual treatment concentrations were: 10.2 ppm $O{-}PO_4^{3-}$, 5.1 pyrophosphate; 2 ppm hydroxyethylidene diphosphonic acid; and 1.7 ppm benzotriazole (BZT); 15 ppm AA/HPA (3:1 molecular weight about 6,000).

According to *G.E. Geiger and R.C. May; U.S. Patent 4,324,684; April 13, 1982; assigned to Betz Laboratories, Inc.* a corrosion inhibitor treatment for metal surfaces exposed to an aqueous medium comprises (1) water-soluble zinc compound (2) water-soluble chromate compound and (3) the water-soluble polymer composed essentially of moieties derived from acrylic acid or derivatives thereof and hydroxylated lower alkyl acrylate moieties. The treatment could additionally comprise (4) orthophosphate.

Although this process is considered to have general applicability to any aqueous system where zinc precipitation is a problem, it is particularly useful in cooling water systems.

Based on experience, compositions according to the process could vary widely and would comprise on a weight basis:

(1) about 1 to about 95% of water-soluble zinc compound,

(2) about 1 to about 98% of water-soluble chromate compound, and

(3) about 1 to 95% AA/HAA polymer of the total amount of zinc compound, chromate compound and polymer.

Similarly, in those instances where orthophosphate is also present, compositions according to the process could comprise on a weight basis:

(1) about 1 to 95% water-soluble zinc compound,

(2) about 1 to 97% AA/HAA polymer,

(3) about 1 to 97% water-soluble chromate compound, and

(4) about 1 to 95% orthophosphate (or precursor thereof) of the total amount of zinc compound, chromate compound, polymer and orthophosphate.

The cooling water preferably will have a pH of about 6.5 to about 9.5. Since zinc precipitation problems most commonly occur at pHs above about 7.5, the most preferred pH range is from about 7.5 to about 9.5.

Example: Tests were conducted to demonstrate the efficacy of AA/HPA copolymer in combination with a cooling water treatment containing zinc, orthophosphate and water-soluble chromate. This cooling water treatment is described in U.S. Patent 2,900,222. The copolymer was also tested in combination with zinc and water-soluble chromate, the latter two components individually being well-known cooling water treatment compounds.

The test procedures were generally the same as those described in Example 1 of U.S. Patent 4,297,317 above; spinner tests were used, and coupon weight loss measurements provided the basis for calculating corrosion rates. Test-conditions were as follows:

spinner volume = 17 ℓ

ppm calcium as $CaCO_3$ = 170

ppm magnesium as $CaCO_3$ = 110

ppm chloride = 121

ppm sulfate = 107

ppm silica = 15

M-alkalinity = 80 ppm

bulk water temperature = 120°F

pH = 8.0

The results of these tests are reported below in the table in terms of corrosion rates in mils per year (mpy). The copolymer had a mol ratio of AA:HPA of 3:1 and a molecular weight of 6,000. The chromate used was sodium dichromate, $Na_2Cr_2O_7 \cdot 2H_2O$.

Copolymer (ppm)	CORROSION INHIBITION o-PO_4 (ppm)	Zinc (ppm)	Chromate (ppm)	Corrosion Rate[1]
16	—	—	—	25[2]
—	7.3	7.3	1.4	0.44
5	5	5	1	0.33
17	—	—	—	33
—	7.1	7.1	2.8	0.56
5	5	5	2	0
22	—	—	—	44
—	9.2	9.2	3.6	0.66
10	5	5	2	0
20	—	—	—	39
—	6.67	6.67	6.67	0.33[2]
5	5	5	5	0
—	8.33	8.33	8.33	0.44
5	—	5	5	0
5	—	5	2	0.25

[1] Unless indicated otherwise, these values each represent an average for 3 runs.
[2] Average for 2 runs.

As can be seen from the results of the above table, the AA/HPA in combination with zinc/chromate and zinc/chromate/orthophosphate proved to be efficacious for corrosion inhibition.

Three-Component System

E.J. Levi; U.S. Patent 4,317,744; March 2, 1982; assigned to Drew Chemical Corp. provides a corrosion inhibitor comprised of corrosion inhibiting amounts of a water-soluble phosphonic acid or salt thereof; polymer of acrylic, methacrylic, maleic acid or its anhydride; and tolyltriazole. The corrosion inhibitor is particularly employed for inhibiting corrosion of ferrous containing metal; e.g., mild steel employed in cooling water systems.

As representative examples of phosphonic acids which are preferably employed, there may be mentioned: ethane-1-hydroxy-1,1-diphosphonic acid, aminotri(methylene phosphonic acid), ethylenediaminetetra(methylene phosphonic acid), hexamethylenediaminetetra(methylene phosphonic acid); and the water-soluble salts thereof.

The preferred polymer is a homopolymer of maleic acid or its anhydride. The composition includes from about 65 to about 80% of the phosphonate, from about 5 to about 20% of the polymer and from about 15 to about 25% of the triazole, based on the three components, all by weight.

Unless otherwise specified, all parts and percentages in the examples are by weight.

Examples: The following compositions were tested in a standard hard water (SHW)(Ca^{2+} 120 ppm; Mg^{2+} 24 ppm; HCO_3^- 24 ppm; SO_4^{2-} 500 ppm; Cl^- 500 ppm) to test corrosion inhibition with mild steel specimens.

	A	B
Ethane-1-hydroxy-1,1-diphosphonic Acid	11.0%	11.0%
Poly Maleic Anhydride	1.7%	1.7%
Benzotriazole	1.75%	—
Tolyltriazole	—	1.75%
KOH (45%)	21.7%	21.7%
KELIG 32 (40%) (a lignosulfonate)	10.7%	10.7%
Water	53.15%	53.15%

Composition A, which includes benzotriazole, is a prior art composition, whereas Composition B, which includes tolyltriazole as a replecement for benzotriazole is in accordance with the process.

	Treatment Level (ppm)				Avg. Corrosion Rate
	24 hrs.	48 hrs.	Water	pH	Mild Steel
1. Composition A	300	150	SHW	7.0-7.5	13.6
2. Composition A	300	150	SHW 180 ppm $NaHCO_3$	8.0-8.5	7.1
3. Composition B	300	150	SHW	7.0-7.5	9.6
4. Composition B	300	150	SHW 180 ppm $NaHCO_3$	8.0-8.5	3.6

The above examples show the superiority of the composition of the process (Examples 3 and 4) as compared to the prior art composition (Examples 1 and 2).

Simultaneous Control of Scale and Corrosion

The process of *G.D. Hansen; U.S. Patent 4,328,180; May 4, 1982; assigned to Atlantic Richfield Company* relates to the inhibition and prevention of corrosion of metal in contact with cooling water, which corrosion can be adequately controlled only if there is simultaneous prevention of calciferous deposits.

In accordance with the process, cooling water is inhibited so as to minimize corrosion by a method of treating such cooling water by controlling injection of additives, so that after such injection of additives, a million parts of cooling water contain: sufficient hydrogen ion to provide a pH within the range from about 6.5 to 8.2; from about 10 to 600 parts of hardness (Ca plus Mg cations); plus from about 3 to about 40 parts of a phosphate; plus from about 0.3 to about 3 parts of a polyacrylic acid having a molecular weight of about 1,000, the polyacrylic acid being a thioglycolate terminated polymer; plus from about 0.1 to about 1 part of a mixture of a plurality of organic phosphonates of the group comprising diphosphonates, triphosphonates, tetraphosphonates, and polyphosphonates; plus about 0.6 to about 5 parts of dispersants comprising salts of both sulfonic acids and phosphate ester acids, the ratio of sulfonic to phosphate ester salts being within a range from about 1.5 to 1 to about 3 to 1; plus from about 0.1 to about 10 parts of aromatic triazole; plus about 0.1 to about 10 parts of tartaric acid.

Example: Acrylic acid is polymerized by providing a mixture of acrylic acid, a solvent, and a peroxide catalyst, and heating the mixture to a polymerization temperature, maintaining polymerization conditions for only a controlled period of time, and then injecting into the mixture a solution of thioglycolic acid to halt the polymerization at a controlled degree of polymerization. The thioglycolic acid terminates the polymerization and leaves a thioglycolate group at the end of the polymer.

The range of molecular weights is relatively narrow so that it is possible to prepare a product from which there can be separated a fraction having a molecular weight of approximately 1,000 in which most polymer molecules have 9 acrylic acid units and in which substantially all of the molecules in the polymer are in the range from about 7 to 11 units. An important advantage of this polymer is that a 50% solution of the 1,000 molecular weight material has a relatively low viscosity so that it can be pumped and promptly mixed throughout the cooling water. Thus, the problems inherent in the utilization of high viscosity polymers are avoided. Moreover the 1000 MW polymer is not readily degraded when recycled through the cooling water system, thus differing from some higher molecular weight polymers.

Dispersants which are alkali metal salts of sulfonic acids can be effective in dispersing precursors for scale and in inhibiting corrosion in the presence of appropriate corrosion inhibitors. Dioctyl (2 ethylhexyl) maleate can be coverted to the sulfonic acid derivative of the succinic ester. The sodium salt of such succinic ester derivative having a sulfonic acid group on the 2 carbon atom has been used as a surfactant for many years. Such salt can be very effective dispersant. Such a dispersant is available as Aerosol OT75. However, such dispersant is not readily dissolved and mixed with an aqueous solution. Accordingly, it is advantageous to dissolve the salt in an appropriate alcohol such as isopropyl alcohol, and to add such solution to the aqueous system.

Tartaric acid can enhance the action of both the sulfonate dispersant and the polyacrylic acid dispersant. Moreover, the tartaric acid is useful in solubilizing compounds of iron, nickel, etc. which might otherwise participate in scale formation.

A composition was prepared by mixing approximately 6% of the polyacrylic acid polymer having a molecular weight of about 1,000 and about 4% of the sodium sulfonate derivative of dioctyl ester of succinic acid. The sodium sulfonate derivative was dissolved in 3 parts of isopropyl alcohol so that the composition included 16% of such solution for providing the 4% of the sodium sulfonate derivative of dioctyl ester of succinic acid. The composition contained 3% tartaric acid and 75% water. The composition consisted of:

	Percent
Water	75
Polymer of acrylic acid, 1000 MW	6
Tartaric acid	3
Sulfonate dispersant	4
Isopropyl alcohol	12

The composition was injected into cooling water at the rate of 20 ppm so that it could combine with the phosphate composition also injected at a rate of about 20 ppm. Such dilution corresponds to about 1 part of the composition which is injected per 50,000 parts of water.

A phosphate composition for use with the polyacrylate was prepared by mixing:

	Parts per Million
Polyacrylic acid molecular weight 1000 MW	1
Tartaric acid	0.4
Sodium sulfonate derivative of dioctyl ester of succinic acid	1.2

(continued)

	Parts per Million
Sodium dihydrogen phosphate	5
Sodium tripolyphosphate	1
Sodium amino tris(methylene phosphonic acid)	0.1
Ethylenediaminetetra(methylene phosphonic acid)	0.2
Sodium salt of phosphate ester of tetra ethylene glycol	1.0
Tolyl triazole	0.4

The ethylene glycol is employed to permit the preparation of a solution of the tolyl triazole, which can then be dispersed in the remaining aqueous system.

This composition is injected at the rate of about 20 ppm together with 20 ppm of the polyacrylate solution so that the cooling water contains the following active ingredients:

	Percent
Sodium hydrogen phosphate	24
Sodium tripolyphosphate	4
Sodium sulfonate derivative of dioctyl ester of succinic acid	2
Sodium amino tris(methylene phosphonic acid)	1
Ethylenediaminetetra(methylene phosphonic acid)	0.5
Sodium salt of phosphate ester of tetra ethylene glycol	12
Tolyl triazole	3
Sodium hydroxide	4
Ethylene glycol	9
Water	40.5

The cooling water has a pH within the range from 6.5 to 8.2, and the concentration of calcium type ions is within a range from about 50 to about 600 ppm. The cooling water, after injection of the two solutions, is effectively protected so that there is not troublesome deposition of calciferous deposits and/or corrosion of the metal in the areas through which the cooling water flows. The method is advantageous by reason of its flexibility in adapting to the variations in the composition of the cooling water. The corrosion of mild steel in the aerated hot water containing the inhibitor combination is about 1 mpy, and well below the about 40 mpy which was measured for untreated water.

BOILER SYSTEMS

Carbohydrazide as Oxygen Scavenger

Efficient operation of boilers and other steam-run equipment requires chemical treatment of feedwater to control corrosion. Corrosion in such systems generally arises as a result of oxygen attack of steel in water supply equipment, preboiler systems, boilers and condensate return lines. Unfortunately, oxygen attack of steel is accelerated by the unavoidably high temperatures found in boiler equipment. Since acid pHs also accelerate corrosion, most boiler systems are run at alkaline pHs.

In most modern boiler systems, dissolved oxygen is handled by first mechanically removing most of the dissolved oxygen and then chemically scavenging the remainder.

Chemical scavenging of the remaining dissolved oxygen is widely accomplished by treating the water with hydrazine. Unfortunately, however, it has become widely recognized that hydrazine is an extremely toxic chemical.

M. Slovinsky; U.S. Patent 4,269,717; May 26, 1981; assigned to Nalco Chemical Company provides a method of removing dissolved oxygen from boiler water by adding to the water an oxygen scavenging amount of carbohydrazine. Since carbohydrazide is a high melting solid (MP 157° to 158°C) which is freely soluble in water, the carbohydrazide may be used in either dry powdered form or in solution form.

Although the carbohydrazide may be added to the boiler system at any point, it is most efficient to treat the boiler feedwater, preferably as it comes from the degasifier. Residence times prior to steam formation should be maximized to obtain maximum corrosion protection. While the carbohydrazide will control corrosion even if residence times are as low as 2 to 3 minutes, residence times of 15 to 20 minutes or more are preferred.

It is generally desirable that at least 0.5 mol of carbohydrazide be used per mol of oxygen. These minimum levels of carbohydrazide will have the added benefit of effectively passivating metal surfaces. Of course, levels of carbohydrazide considerably in excess of 0.5 mol per mol of oxygen may be required, particularly for treating boiler water under static storage conditions. Under such static conditions, for example, treatment levels of 160 mols or more of carbohydrazide per mol of oxygen have proven effective in controlling corrosion.

Carbohydrazide is an effective oxygen scavenger and metal passivator over the entire range of temperatures to which boiler water is generally subjected. Typically, these temperatures will lie in the range of 190° to 350°F.

While it is well known that each molecule of carbohydrazide is capable of being hydrolyzed to 2 molecules of hydrazine, the extent of hydrolysis under typical boiler conditions is very minor.

While carbohydrazide may be used alone in the application, it is preferred that it be catalyzed. For this purpose, it is desirable to use catalysts which undergo oxidation-reduction reactions. For example, hydroquinone and other quinones can be used to catalyze the carbohydrazide since they are capable of undergoing oxidation-reduction reactions. When a quinone catalyst is used, the amount of quinone added in relation to the carbohydrazide should be in the range of 0.2 up to about 20% by weight of the carbohydrazide.

Another oxidation-reduction catalyst useful with carbohydrazide is cobalt, preferably in a stabilized form. The amount of cobalt used in relation to the carbohydrazide should be in the range of 0.2 to about 20% by weight.

Example 1: The oxygen scavenging efficiency of carbohydrazide was compared to hydrazine in a series of tests, with results reported in the table below. A once-through experimental apparatus for simulating boiler water transport was utilized in this example. Water temperature in this example was maintained at about 227° to 239°F and pH at about 10.5

The data generated in this example shows that oxygen removal levels achieved with a given concentration of hydrazine can generally be accomplished with half as much carbohydrazide. In addition, this data showed little difference between catalyzed and uncatalyzed scavengers.

Carbohydrazide vs Hydrazine in Oxygen Removal

Scavenger	ppm/ ppm O_2	Initial O_2 (ppm)	Final O_2 (ppm)	% O_2 Removal
Catalyzed	9.9/1	7.4	0.06	99.2
Hydrazine	1.0/1	7.4	2.02	72.7
(1% hydroquinone)	0.1/1	7.4	7.12	3.8
Hydrazine	10.0/1	7.4	0.06	99.2
	0.9/1	7.4	2.18	70.5
	0.1/1	7.4	6.26	15.4
Catalyzed	5.0/1	7.7	0.10	98.7
Carbohydrazide	0.55/1	7.7	3.50	54.5
(1% hydroquinone)	0.05/1	7.7	6.70	13.0
Carbohydrazide	5.0/1	7.7	0.10	98.7
	0.5/1	7.7	3.28	57.4
	0.05/1	7.7	6.60	14.3

Example 2: Commercial scale comparisons were made between carbohydrazide and hydrazine. The tests were run on a system including a deaerator which treats a mixture of feedwater and condensate at 60 psi and 300° to 310°F before it is fed to a first stage heater at 160 psi and 360° to 366°F and a second stage heater at 400 psi and 446° to 450°F. The second stage heater than feeds into three boilers: a combination boiler at 1,275 psi, a first power boiler at 1,275 psi and a second power boiler at 400 psi. The treatment chemicals were fed alternatively at the deaerator and just down line from it.

Sampling at various points in this system indicated that the carbohydrazide was comparable or superior to the hydrazine, at carbohydrazide levels half those of the hydrazine. In particular, sampling at 300°F (just past the deaerator) showed results comparable with the hydrazine whereas sampling at 360°F (just past first stage heater) and at 446°F (just past second stage heater) indicated that the carbohydrazide scavenged the oxygen more efficiently than the hydrazine (carbohydrazide still at 50% of hydrazine level).

Hydroquinone and Related Compounds

H. Kerst; U.S. Patent 4,278,635; July 14, 1981; assigned to Chemed Corporation provides a method for boiler water treatment to retard corrosion due to dissolved oxygen. Hydroquinone and certain related compounds are highly effective oxygen scavengers for use in boiler water and thereby effect reduction of corrosion resulting from dissolved oxygen.

The oxygen scavengers for use in this process are the o- or p-dihydroxy, diamino, and aminohydroxy benzenes, and their lower alkyl substituted derivatives:

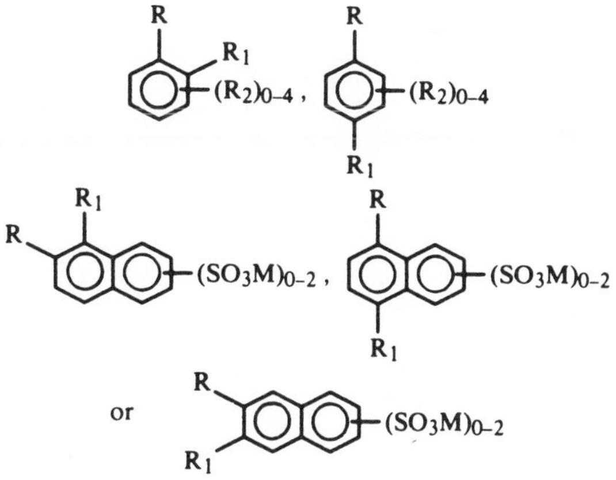

in which R and R_1 are independently selected from $-OH$ or $-NH_2$, R_2 (when present) is one or more of low molecular weight alkyl groups, in which the alkyl group has 1 to 8 carbons, M is H, Na, or K or permutations thereof.

These additives may be added to feedwater in an effective amount depending on the amount of oxygen present, so as to maintain a small residual of the additive at the point where it enters the boiler. The amount fed should be from 0.1 to 20, preferably 1 to 5 times the oxygen concentration, on a weight basis, and residuals of 0.1 to 1 ppm are generally adequate.

Example: The following experimental laboratory work shows that above a pH of 8.5, hydroquinone is an effective deoxygenation agent for water at room temperature (20°C). The rate of the reaction is increased by higher pH and higher temperature. No catalyst is needed.

Tests were run in an oxygen bottle of 2 ℓ volume, with stirrer and oxygen electrode, at room temperature. Results are compared with those given by sodium sulfite, and hydrazine. Oxygen residuals as a function of time are shown in the table. The results with hydroquinone are very favorable when compared with the results given by other oxygen scavengers. In addition, there is none of the build-up of dissolved salts given by sulfite and hydroquinone is much cheaper than hydrazine.

Dissolved Oxygen Concentration, ppm

Time, Minutes	Na_2SO_3	Hydroquinone			N_2H_4
	76 ppm	5 ppm	10 ppm	20 ppm	10 ppm
0	8.75	8.85	8.85	8.80	8.40
1	6.40	4.15	0.50	0.20	—
2	4.70	3.90	0.12	0.02	—
3	3.50	3.78	0.11	0.00	—
4	2.60	3.70	0.10	0.00	—
5	1.95	3.60	0.10	0.00	—
10	0.48	3.15	0.05	0.00	—
15	0.15	2.82	0.00	0.00	—
30	0.00	2.20	0.00	0.00	8.00
24 hrs.	0.00	1.00	0.00	0.00	3.05

Hydroquinone plus Mu-Amines

An improved oxygen scavenger for aqueous mediums described by *J.A. Muccitelli; U.S. Patents 4,279,767; July 21, 1981; and 4,289,645; September 15, 1981; both assigned to Betz Laboratories, Inc.* comprises hydrazine-free solution of hydroquinone and mu-amine.

To provide convenient single-drum treatments, it was decided to test-formulate various dioxo aromatic compounds with neutralizing amines. In combining the two most commonly used neutralizing amines, morpholine and cyclohexylamine, with hydroquinone, they precipitated out of solution. It was discovered that while some additional amines precipitated out, others did not.

Those neutralizing amines which were found to be compatible with hydroquinone did not fall within any readily discernible chemical class. Accordingly, those amines which are compatible with hydroquinone for purposes of oxygen scavenging were classified "mu-amines."

A simple test has been developed for identification of those amines which are suitable for stable oxygen scavenger formulations with hydroquinone, that is, mu-amines. The steps of this test are as follows.

Step 1: To a 1 pint glass jar which can be equipped with a cover lid add 80.0 g of demineralized water and 10.0 g of the amine to be tested.

Step 2: Stir the contents of the jar for one minute, preferably via a magnetic stirrer. If the amine is a solid, stir until complete dissolution into the water is attained.

Step 3: To the contents of the jar, after the amine has been thoroughly mixed with the water, add 10.0 g of hydroquinone.

Step 4: Stir the contents of the jar for 3 minutes. If precipitation within the jar occurs, then the amine is not a mu-amine. If no precipitate formation within the jar is evident, the amine may possibly be a mu-amine, so proceed to Step 5.

Step 5: Seal the jar and stir vigorously for seven minutes.

Step 6: After stirring, filter the contents of the jar through a 5 micron filter paper via suction. If a nonfilterable and/or insoluble mass remains on the filter paper (not merely discoloration of the paper) after filtration, or if an insoluble mass clings to the interior of the jar, then the amine is not a mu-amine. If the contents of the jar pass unhindered through the filter paper, the amine may possibly be a mu-amine and Steps 7 and 8 should be performed.

Step 7: Remove a 1.0 ml aliquot from the filtration flask or receiver (i.e., an aliquot of the former jar contents) with a syringe.

Step 8: Inject the contents of the syringe into the room temperature oxygen scavenger apparatus described below which contains air-saturated, demineralized water and sufficient sodium hydroxide to result in a pH 9 to 10 range in the reaction flask. If at least 70% of the dissolved oxygen initially present in the air-saturated water is removed within 1 minute after injection, the amine is definitely a mu-amine. If at least 70% of the dissolved oxygen is not removed within 1 minute, then the amine is not a mu-amine.

When making or using a two-component treatment in accordance with the process, that is, hydroquinone in combination with mu-amine, the ratio of hydroquinone to mu-amine should be about 1:1.3 or less (weight basis). When making or using an aqueous solution of hydroquinone and mu-amine, the ratio of hydroquinone to mu-amine will depend on the total amount of hydroquinone (weight basis) in solution. If the hydroquinone concentration is less than 7%, below the solubility limit for hydroquinone in water, any ratio of hydroquinone to amine would be suitable. In such instances, even though the mu-amine wouldn't be necessary to overcome hydroquinone-solubility problems, it would still, nonetheless, increase the oxygen scavenging efficacy of the hydroquinone. When the hydroquinone concentration in aqueous solution exceeds 7%, the ratio of hydroquinone to mu-amine should be about 7:1 or less.

According to the results of experiments, compositions in which the ratio of hydroquinone to mu-amine ranged from about 7:1 to 1:99 proved to be effective for oxygen scavenging. The preferred range for this ratio is about 5:1 to 1:10.

The amount of treatment added could vary over a wide range and would depend on such known factors as the nature and severity of the problem being treated. Based on experimental data, it is believed that the minimum amount of treatment composition could be about 0.01 total parts of active hydroquinone and mu-amine per million parts of aqueous medium being treated. The preferred minimum is about 0.1 ppm. Also based on experimental data, it is believed that the maximum amount of active treatment composition could be about 10,000 ppm. The preferred maximum is about 100 ppm.

Because it was the best hydroquinone solubilizer and the most thermally stable of the amines tested, methoxypropylamine is the most preferred mu-amine. The preferred ratio of hydroquinone to methoxypropylamine is about 1:1.

In keeping with standard practices for treating boiler feedwater, an excess amount of hydroquinone should be used to provide a residual amount thereof in the boiler water for the uptake of oxygen from other sources.

Example: Tests were performed using an apparatus that consisted of a 1 ℓ, three-necked flask fitted with a dissolved oxygen probe and a pH electrode. The intent of this apparatus was to show reaction kinetics differences rather than precise measurements (within seconds) of reaction times.

Oxygen scavenging experiments were performed by injecting 1 ml of stock solution into a glass vessel containing 1070 ml of air-saturated, demineralized water adjusted to the appropriate pH with 7 N sodium hydroxide. The results of these tests are given in the table below in terms of % oxygen removal at 60 seconds. The tests were performed at ambient temperature (23° to 25°C), and the feedwater dosage was 103 parts of active hydroquinone per million parts of water and 103 parts of active mu-amine per million parts of water. The aqueous stock feed solution, on a weight basis, was 11% hydroquinone/11% amine (actives). The final pH values indicated are for the test water.

mu-amine Combined With Hydroquinone	Final pH	Initial O_2 (ppm)	% O_2 Removed
Aminomethylpropanol	9.8	8.4	100
Triethylenetetramine	9.6	6.8	89
Diisopropanolamine	9.3	6.5	86
Ethylenediamine	9.9	7.0	85
Diethylaminoethanol	9.7	7.9	77
Dimethylaminopropylamine	10.1	7.0	76
Monoethanolamine	10.0	7.2	75
sec-Butylamine	10.5	6.4	74
Dimethyl(iso)propanolamine	9.8	7.6	73

As can be seen from the results, stable solutions of hydroquinone and mu-amines demonstrated efficacy as oxygen scavengers.

Hydroquinone

A method is disclosed by *S.J. Ciuba; U.S. Patent 4,282,111; August 4, 1981; assigned to Betz Laboratories, Inc.* for reducing the oxygen content of water using a hydrazine-free aqueous solution of hydroquinone. The process is considered to be particularly useful for treating boiler feedwater.

The minimum amount of hydroquinone compound could be about 0.05 part of active compound per million parts of aqueous medium being treated. The preferred minimum is about 0.2 ppm. It is believed that the amount of hydroquinone compound used could be as high as about 200 ppm, with about 35 ppm being the preferred maximum.

Example: In order to compare the performance of hydroquinone with that of hydrazine under field-type conditions, a series of experiments were conducted on the feedwater of a working boiler. The test materials were fed to the deaerator storage tank, and the resultant change, if any, in a dissolved oxygen level was measured on a sample flowing from the feedwater line through a membrane-type dissolved oxygen probe.

The results of these tests are reported below in Table 1 in terms of % oxygen removed from the boiler feedwater. During the tests, a relatively wide range of experimental conditions was encountered among the test parameters, reflecting the difficulty of attempting to obtain precise data during experiments performed on a working boiler.

In order to provide some point of reference for judging oxygen scavenging efficacies while taking into account the diverse conditions experienced on a daily basis, hydrazine control runs were performed, whenever possible, under feedwater and boiler conditions similar to those encountered during testing of other test materials. The hydrazine control runs are reported below in Table 2 also in terms of % oxygen removal.

Table 1

Material	Feedwater Concentration (ppm Actives)	Feedwater pH	Steam Load (1000 lbs./hr.)	Feedwater[1] Initial O_2 (ppb)	% O_2 Removed
hydroquinone	1.9	8.4	15	10.0	90
hydroquinone	1.9	8.4	15	8.0	88

[1] Dissolved oxygen concentration of feedwater exiting the deaerator with no chemical feed.

Table 2

Hydrazine Control Run #	Feedwater[1] Concentration (ppm Actives)	Feedwater pH	Steam Load (1000 lbs./hr.)	Feedwater Initial O_2 (ppb)	% O_2 Removed
1	1.6	8.8	11	13.0	69
2	1.1	9.4	15	4.0	50
3	1.2	8.6	15	6.0	60
4	1.0	8.8	15	5.8	55
5	0.8	9.4	11	3.5	31
6	1.6	9.6	14	4.0	30

[1] All hydrazine concentrations.

Based on the results reported in Tables 1 and 2, hydroquinone was considered to be efficacious in treating boiler feedwater and compared favorably with hydrazine.

Hydroxylamine-Neutralizing Amine Combination

The process of *D.G. Cuisia and C.M. Hwa; U.S. Patent 4,350,606; September 21, 1982; assigned to Dearborn Chemical Company* is directed to the use of a hydroxylamine compound in combination with one or more volatile, neutralizing amines such as cyclohexylamine, morpholine, diethylaminoethanol, dimethylpropanolamine, and 2-amino-2-methyl-1-propanol. The hydroxylamine compound has the following general formula:

$$\begin{matrix} R_1 \\ & \searrow \\ & & N{-}O{-}R_3 \\ & \nearrow \\ R_2 \end{matrix}$$

wherein R_1, R_2, and R_3 are either the same or different and selected from the group consisting of hydrogen, lower alkyl having between 1 to about 8 carbon atoms, and aryl such as phenyl, benzyl and tolyl. Specific examples of hydroxylamine compounds usefully employed herein include hydroxylamine, oxygen-substituted and nitrogen-substituted derivatives.

It was found that the combinations of a hydroxylamine compound and one or more neutralizing amines will reduce both the carbon dioxide and oxygen gases that may be present in the steam condensate. Furthermore, the presence of neutralizing amines provides a catalytic effect in the reaction of a hydroxylamine compound and oxygen, making the removal of oxygen fast enough even at relatively low temperature for immediate corrosion protection in the steam condensate systems.

The oxygen scavenging activity of N,N-diethylhydroxylamine (DEHA) in combination with neutralizing amines was compared to the activity of N,N-diethylhydroxylamine alone. The effect of neutralizing amines by itself to the dissolved oxygen was also determined.

The tests were performed in the laboratory using a 4.5 ℓ reaction vessel containing distilled water saturated with dissolved oxygen and 10 ppm CO_2. A 5-gallon batch of distilled water was saturated with oxygen by bubbling air through a fritted dispersion tube. The carbon dioxide was naturally present in the distilled water.

The 4.5 ℓ container was filled up with the oxygen-saturated water containing 10 ppm CO_2. The water temperature was adjusted at 70°±2°F. The dissolved oxygen was determined by means of a commercially available oxygen meter equipped with selective membrane electrode. The oxygen meter probe after calibration was inserted into the top of the container. The first test was conducted by injecting 36 ppm, N,N-diethylhydroxylamine. The subsequent decrease in oxygen concentration was measured as a function of time.

Similar experiments were performed by using the same amount of DEHA and adding neutralizing amines to pH 8 to 8.5. Other tests with neutralizing amines but without DEHA were conducted to determine the effect of the amines by itself. The table illustrates the catalytic activity of the neutralizing amines in promoting the reaction of DEHA and oxygen in a low temperature water containing both dissolved oxygen and carbon dioxide.

Removal of Oxygen

		Dissolved Oxygen, ppm O_2					
Ex.	Time, Minutes	0	15	30	60	90	120
1.	N,N-Diethylhydroxylamine (DEHA)	9.70	8.76	8.08	6.50	5.60	5.40
2.	Morpholine (I)	9.43	9.26	8.85	8.70	8.61	8.60
3.	Cyclohexylamine (II)	9.50	9.03	8.88	8.76	8.66	8.60
4.	Diethylaminoethanol (III)	9.86	9.60	9.57	9.50	9.50	9.50
5.	Dimethylpropanolamine (IV)	9.65	9.04	8.63	8.43	8.39	8.36
6.	2-Amino-2-methyl-1-propanol (V)	8.63	8.52	8.45	8.25	8.12	8.12
7.	DEHA + I	8.22	5.54	3.90	1.97	1.23	0.87
8.	DEHA + II	8.60	4.70	2.63	1.05	0.54	0.33
9.	DEHA + III	9.48	4.53	2.21	0.80	0.42	0.32
10.	DEHA + IV	8.36	5.30	3.31	1.66	0.92	0.66
11.	DEHA + V	8.10	5.45	3.81	2.07	1.36	1.05
12.	DEHA + (I & II)	9.52	4.70	2.33	0.71	0.31	0.21
13.	DEHA + (I, II, III, IV & V)	9.80	3.50	1.40	0.34	0.18	0.13

It is evident from the table that the combinations of DEHA and one or more neutralizing amines were more effective than the DEHA alone when the water contained both carbon dioxide and oxygen. As expected, the neutralizing amines alone did not significantly reduce the oxygen content. With the DEHA alone the oxygen was reduced by 44.3% as compared to 89.4% with a combination of DEHA and morpholine and 98.7% with a combination of DEHA and a mixture of five amines.

In the table, in Example 12, the weight ratio of I:II was 1:1, and in Example 13, the ratio of I:II:III:IV:V was 1:1:1:0.5:0.5.

The following hydroxylamine compounds according to this process show similar unexpected oxygen scavenging activities when tested in combination with one or more neutralizing amines.

Example No.	
14	N,N-dimethylhydroxylamine
15	N-butylhydroxylamine
16	O-pentylhydroxylamine

(continued)

Example No.	
17	N,N-dipropylhydroxylamine
18	N-heptylhydroxylamine
19	O-ethyl N,N-dimethylhydroxylamine
20	N-benzylhydroxylamine (β-benzylhydroxylamine)
21	O-benzylhydroxylamine (α-benzylhydroxylamine)
22	O-methyl N-propylhydroxylamine
23	N-octylhydroxylamine
24	N-methyl N-propylhydroxylamine
25	N-hexylhydroxylamine

At equilibrium operating conditions it is preferred to maintain the level of the hydroxylamine compound in the condensate at 0.001 to 100 ppm (more preferably, about 5 ppm); and the second amine (or amine mix) at 1 to 1,500 ppm (more preferably, about 100 ppm).

The components can be added separately or in admixture, and can be added to the boiler feedwater and/or directly to the condensate lines. When added as a mix, the weight ratio of the hydroxylamine compound:amine is about 0.001 to 1:1, or more preferably about 0.05:1.

One good way to add the composition is first to add the preselected amount of the hydroxylamine compound and after that, add the second amine or amine mix until the pH of the condensate or the like is 8 to 8.5. This method was used in the runs for the table.

MULTIPLE USE COMPOSITIONS

Amino Methylene Phosphonic Acid and Hydroxyl-Containing Carboxylic Acid

It has been found by *R.J. Lipinski; U.S. Patent 4,246,030; January 20, 1981; assigned to The Mogul Corporation* that certain amino alkylene phosphonic acids and their derivatives in combination with water-soluble carboxylic acids having at least one hydroxyl group per molecule and/or the alkali metal salts thereof either alone or with a metal molybdate when used in effective amounts, is capable of protecting various metals and its alloys such as copper, brass, steel, aluminum and iron. This corrosion inhibiting composition, which also helps to minimize mineral deposits generally formed on the surface of metal, may be used in various water systems, including, for example, air conditioning systems, steam generating plants, refrigeration systems, heat-exchange apparatus, engine jackets, pipes and the like.

The composition comprises, in parts based on a million parts by weight of water, from about: (a) 0 to 50 parts by weight of an azole; (b) up to 100 parts by weight of at least one water-soluble carboxylic acid having at least one hydroxyl group per molecule and/or the alkali metal salt of the carboxylic acid; (c) 0 to 100 parts by weight of a metal molybdate; and (d) 2.0 to 50 parts by weight of an amino alkylene phosphonic acid and/or derivatives or salts thereof.

It is important that either the carboxylic acid, the alkali metal salt of the carboxylic acid or a combination of the carboxylic acid and its alkali metal salt either alone or further in combination with a metal molybdate be present in the water in an amount of at least 3.0 ppm. In addition to the carboxylic acid or its alkali metal salt, it is essential to have a corrosion inhibiting amount of at least one amino alkylene phosphonic acid and/or its derivative having the formula

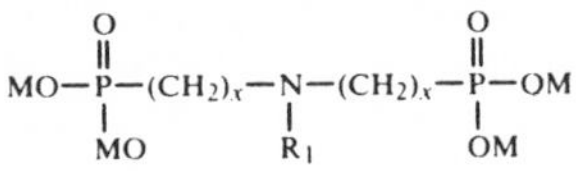

$$MO-\overset{\overset{\displaystyle O}{\|}}{\underset{\underset{\displaystyle MO}{|}}{P}}-(CH_2)_x-\underset{\underset{\displaystyle R_1}{|}}{N}-(CH_2)_x-\overset{\overset{\displaystyle O}{\|}}{\underset{\underset{\displaystyle OM}{|}}{P}}-OM$$

wherein R_1 is a monovalent radical selected from the class consisting of the formulas:

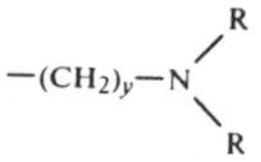

(1) 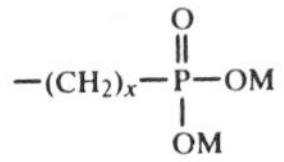 $-(CH_2)_x-\overset{\overset{\displaystyle O}{\|}}{\underset{\underset{\displaystyle OM}{|}}{P}}-OM$ (2) $-(CH_2)_y-N\begin{matrix} R \\ R \end{matrix}$ and (3) 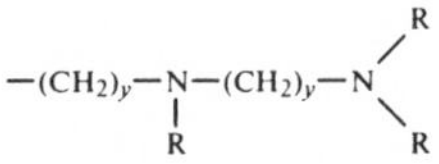 $-(CH_2)_y-\underset{\underset{\displaystyle R}{|}}{N}-(CH_2)_y-N\begin{matrix} R \\ R \end{matrix}$

wherein R is:

$$-(CH_2)_x-\overset{\overset{\displaystyle O}{\|}}{\underset{\underset{\displaystyle OM}{|}}{P}}-OM$$

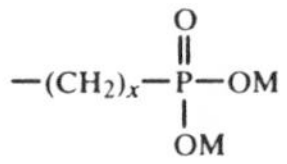

and y has a value of 1 to 8, x has a value of 1 to 4, and M is a radical selected from the class consisting of hydrogen, an alkali or alkaline earth metal, ammonium, an amino radical, and an alkyl or substituted alkyl radical having 1 to 4 carbon atoms.

The derivatives of the phosphonic acids, as defined herein, e.g. the salts and esters, etc. may be one or the other or a combination thereof provided the derivative is substantially soluble in water.

It is of particular importance to recognize that as the molecular weight of the amino methylene phosphonic acid increases, i.e. by increasing the number of methylene groups in the molecule, the effectiveness of the phosphonate as a corrosion inhibitor likewise increases. There is a relationship between the chemical structure of the amino methylene phosphonates and their effect on corrosion inhibition of metal. It was found that the corrosion rate of metal decreases as the chain length of the methylene group increases between the phosphonate groups.

Aminoalkylenephosphonic Acids

The process of *J.G.E. Fenyes and J.D. Pera; U.S. Patent 4,243,524; Jan. 6, 1981; assigned to Buckman Laboratories, Inc.* relates to phosphonoalkylene derivatives of tetrahydrothiophenamine 1,1-dioxides and the use of the same for inhibiting the deposition of scale and sludge on heat transfer surfaces of cooling water systems and boilers. The compositions of matter may be also defined as aminoalkylenephosphonic acids and their alkali metal salts.

The aminoalkylenephosphonic acids and their alkali metal salts prepared by a reaction utilizing a primary or secondary tetrahydrothiophenamine 1,1-dioxide or the hydrochloride thereof, an aldehyde or a ketone, and orthophosphorous acid are useful in the control of scale and sludge deposition in aqueous systems. When used in combination with known corrosion inhibitors they show synergistic results in inhibiting corrosion of metal surfaces in contact with an aqueous system that is normally corrosive to such metals.

Example 1: *Tetrahydro-3-thiophenamine 1,1-dioxide hydrochloride* – A solution of 2,5-dihydrothiophene 1,1-dioxide (50.0 parts) in 29% NH_4OH (180 ml) was heated in a 1 ℓ stainless steel autoclave at 80° to 86°C for 7 hr. The mixture was concentrated under reduced pressure to a yellow oil which was filtered, dissolved in ethanol (150 ml) and treated with concentrated HCl (100 ml). Addition of ethyl ether (100 ml) to the resultant mixture precipitated the crystalline hydrochloride, which was collected, washed with ether and dried in vacuo over P_2O_5, MP 220°C; Yield 54.9 g (75.5% of the theory).

Example 2: *Tetrahydro-N,N-bis(phosphonomethyl)-3-thiophenamine 1,1-dioxide* – Method 1: A mixture of tetrahydro-3-thiophenamine 1,1-dioxide hydrochloride (171.6 parts, 1.0 mol); 258 parts of 70% aqueous orthophosphorous acid (2.2 mols); and 73.4 parts of 90% paraformaldehyde (2.2 mols) was refluxed for 3 hr. After cooling the mixture was stirred into 750 ml of ethanol. The white, sticky precipitate obtained was triturated under ethanol to give a filterable white hygroscopic solid that upon drying over P_2O_5 in a vacuum dessicator weighed 163.7 g (50.6% of the theory).

Method 2: To tetrahydro-3-thiophenamine 1,1-dioxide (270.4 parts, 2.0 mols); and concentrated hydrochloric acid (197.2 parts, 2.0 mols) was added at such a rate as to keep the temperature below 50°C. This was followed by the addition of 70% aqueous orthophosphorous acid (468.6 parts, 4.0 mols). The resulting mixture was heated to 60° to 65°C at which temperature 37% aqueous formaldehyde (356.8 parts, 4.4 mols) was introduced during a period of 30 to 40 minutes. Heating was continued and the temperature rose to 108°C and the mixture was refluxed for 1 hr after the addition of formaldehyde was completed. A 50% aqueous solution of the title compound was obtained.

Example 3: *Corrosion inhibiting properties of tetrahydro-N,N-bis(phosphonomethyl)-3-thiophenamine 1,1-dioxide in combination with poly(acrylic acid) and water-soluble zinc compounds* – This example illustrates the corrosion-inhibiting properties of compositions containing tetrahydro-N,N-bis(phosphonomethyl)-3-thiophenamine 1,1-dioxide, poly(acrylic acid), (molecular weight 4000), and water-soluble zinc compounds.

The test apparatus included a sump, a flow circuit, a circulating pump, and a heater. The test fluid was tap water from the City of Memphis water system. The water did not come in contact with any metal except for test coupons placed within the circuit in a manner simulating flow, impingement, and sump conditions. In addition to the normal sump coupon, an additional steel coupon was coupled with a copper coupon and placed in the sump. The test coupons were 1010 mild steel, and the circulating water had a calcium hardness as $CaCO_3$ of 25 ppm, a magnesium hardness as $CaCO_3$ of 18 ppm, chloride as Cl of 10 ppm, and sulfate as SO_4 of 2.5 ppm.

The temperature during the test was maintained about 50°C and the pH was adjusted to 6.5 at the beginning of the test. The water was circulated continuously through the system containing the coupons for a period of 72 hr. The steel coupons were then removed and examined for scale. No scale was observed on any of the coupons protected by the compositions of this process.

Propane-1,3-Diphosphonic Acids

In a process by *H.-D. Block, H. Kallfass and R. Kleinstück; U.S. Patent 4,246,103; January 20, 1981; assigned to Bayer AG, Germany* water is conditioned by adding thereto a corrosion-inhibiting, concretion inhibiting, precipitation retarding, deflocculating, dispersing or fluidifying amount of an acid of the formula below

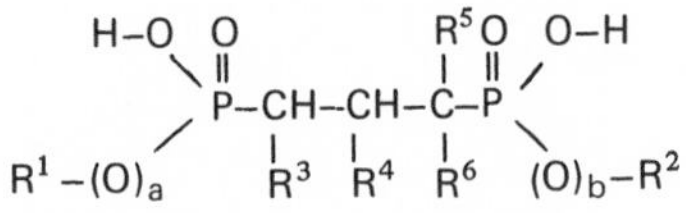

$$\begin{array}{c} H{-}O \quad O \qquad\qquad R^5 O \quad O{-}H \\ \diagdown \; \| \qquad\qquad\quad | \;\; \| \;\diagup \\ P{-}CH{-}CH{-}C{-}P \\ \diagup \qquad | \qquad | \qquad | \qquad \diagdown \\ R^1{-}(O)_a \qquad R^3 \quad R^4 \quad R^6 \qquad (O)_b{-}R^2 \end{array}$$

In the formula, or a salt thereof

R^1 and R^2, independently of one another, represent hydrogen where a or b = 1, or where a or b = 0 represent an optionally substituted alkyl radical with 1 to 8 carbon atoms or an optionally substituted phenyl radical,

R^3 represents hydrogen, an optionally substituted alkyl radical with 1 to 4 carbon atoms, an optionally substituted phenyl radical, a halogen atom from the group comprising fluorine, chlorine and bromine, a hydroxyl group, an alkoxy group containing 1 to 8 carbon atoms, a carboxyl group or a group $-P(O)-(O)_a-R^1OH$,

R^4 represents hydrogen, an alkyl radical with 1 to 4 carbon atoms, an alkenyl radical with up to 3 carbon atoms, an optionally substituted phenyl radical, a carboxyl group, a fluorine or chlorine atom or a group $-P(O)-(O)_a-R^1OH$, or

R^3 and R^4 together form a ring bridge with 3 to 5 methylene groups,

R^5 represents hydrogen, an optionally substituted alkyl radical with 1 to 4 carbons, a carboxymethyl or carboxyethyl group optionally substituted by methyl, succinyl, or a group having the structure $-CHR^4-CHR^3-P(O)-(O)_a-R^1OH$ or the structure $-CH_2-P(O)-(O)_a-R^1OH$,

R^6 represents an optionally substituted phenyl, a carboxyl, or a group having the structure $-P(O)-(O)_b-R^2OH$, and

a and b independently of one another represent the numbers 0 or 1.

The corrosion-inhibiting effect may be used in a plurality of aqueous systems which are used, for example, for dissipating or supplying heat or in which water of high purity are obtained from salt-containing water which may optionally contain organic constituents. Examples of systems such as these are water-cooled installations with fresh water cooling, with effluent cooling, with open or closed back cooling, and dry cooling towers which may be operated with prepurified or even with unpurified surface water from seas, lakes and rivers or with spring water. Installations for the production of substantially salt-free water may use salt-containing seawater, brackish water or river water as starting material and may function, for example, in accordance with the principles of evaporation, electrodialysis or reverse osmosis.

To develop the favorable effect of the compounds, it is advisable to add from about 0.05 to 5000 ppm to the aqueous medium to be treated, the concentrations preferred for concretion and encrustation prevention amounting to from about 0.05 to 20 ppm, for corrosion prevention to from about 5 to 500 ppm and for the fluidification of slips to from about 500 to 5000 ppm, depending upon the principal effect required and upon the main application involved.

Example: *Corrosion inhibition of St 35 carbon steel* – 4 HCl-pickled steel tube rings of St 35 were each attached to a plastic stirrer and stirred in Leverkusen tapwater at ambient temperature at a speed of 0.6 m/s. The concentration of active substance amounted to 50, 100 and 300 ppm, respectively.

Analytical data of the water: total hardness, 15°day; carbonate hardness: 10°d; chloride: 190 mg/kg; sulfate (SO_4): 100 mg/kg and total ions: 770 mg/kg.

After 4 days under test, the tube rings were again pickled and the weight was determined. During the test, the pH-value was adjusted to about 7 with H_2SO_4.

	Input ppm	Corrosion rate $g/m^2 \cdot d$
	no addition	21.06
	50	0.59
A	100	0.57
	300	1.82
	50	2.26
C	100	1.60
	300	5.84
	50	2.37
F	100	1.94
	300	6.24

A = 1-carboxypropane-1,3-diphosphonic acid
C = 2-phosphonobutane=1,2,4-tricarboxylic acid (comparison)
F = polycarboxylic acid (Belgard Ev(R), a product of Ciba-Geigy)

Quaternary Alkynoxymethyl Amines and Their Halogen Derivatives

The process of *P.M. Quinlan; U.S. Patents 4,248,796; February 3, 1981 and 4,343,720; August 10, 1982; both assigned to Petrolite Corporation* relates to quaternaries of alkynoxymethyl amines and uses thereof. These may be summarized by the following formulas:

(1) $$R-\overset{\oplus}{\underset{\underset{R'}{|}}{N}}(CH_2OR''')_2\ X^{\ominus}$$

where R and R' are substituted groups such as alkyl, aryl, etc.; R''' is an acetylenic group; and X is an anion; and

(2) $$\begin{array}{l} R-\overset{\oplus}{N}(CH_2OR''')_2 \\ \quad\ \ | \\ \quad\ \ Z \qquad\qquad 2X^{\ominus} \\ \quad\ \ | \\ R-N^{\oplus}(CH_2OR''')_2 \end{array}$$

where R and R''' having the same meaning as in (1) and Z is a bridging group, preferably hydrocarbon such as alkylene, alkynylene, alkenylene, arylene, etc.

Example 1: Into a 500 ml three-necked flask provided with a reflux condenser and stirrer were placed 84 g (0.4 mol) of N,N-di(propynoxymethyl)butyl amine, 57 g (0.4 mol) of methyl iodide, and 100 ml of ethanol. The reaction mixture was heated to reflux and held there for 24 hr.

The solvent was removed under reduced pressure on a rotary evaporator. The product was a water-soluble viscous liquid. It had the following structure:

$$\left[C_4H_9\overset{\oplus}{N}(CH_2-O-CH_2C\equiv CH)_2 \atop \quad | \atop \quad CH_3 \right] I^{\ominus}$$

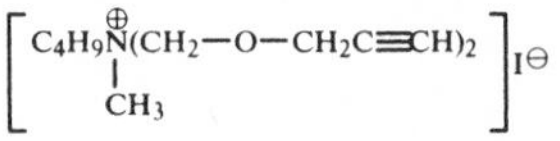

Analysis – % I^- calculated 36.18; %I^- found 35.92.

Example 2: In a similar manner 164 g (0.5 mol) of N,N-di(propynoxymethyl)-tert-dodecyl amine, 71 g (0.5 mol) of methyl iodide and 200 ml of 2-propanol were refluxed together for 60 hr. The isolated product had the following structure:

$$\left[t\text{-}C_{12}H_{25}\overset{\oplus}{N}(CH_2OCH_2C\equiv CH)_2 \atop \quad | \atop \quad CH_3 \right] I^{\ominus}$$

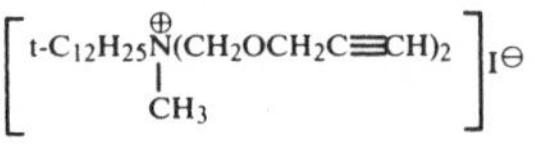

Example 3: In a similar manner 164 g (0.5 mol) of N,N-di(propynoxymethyl)-tert-dodecyl amine, 64 g (0.5 mol) of benzyl chloride and 200 ml of 2-propanol were refluxed together for 60 hr. The isolated product had the following structure:

$$\left[t\text{-}C_{12}H_{25}\overset{\oplus}{N}(CH_2OCH_2C\equiv CH)_2 \atop \quad | \atop \quad CH_2-C_6H_5 \right] Cl^{\ominus}$$

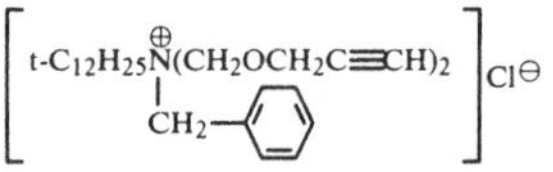

Analysis – % Cl^- calculated 7.8; % Cl^- found 7.1.

Example 4: In a similar manner 160 g (0.5 mol) of N,N-di(propynoxymethyl)-n-dodecyl amine, 71 g (0.5 mol) of methyl iodide, and 200 ml of 2-propanol were refluxed together for 60 hr. The isolated product was a viscous water soluble liquid that foamed in water. It had the following structure:

$$\left[C_{12}H_{25}\overset{\oplus}{N}(CH_2OCH_2C\equiv CH)_2 \atop \quad | \atop \quad CH_3 \right] I^{\ominus}$$

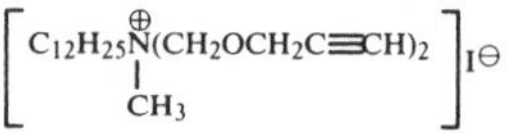

Analysis – % I^- calculated 27.4; % I^- found 25.4.

Example 5: Into a 500 ml three-necked flask equipped with a stirrer and reflux condenser were introduced 164 g (0.5 mol) of N,N-di(propynoxymethyl)-n-dodecyl amine, 54 g (0.25 mol) of 1,4-dibromobutane and 150 ml of n-propanol. This mixture was heated at reflux for 72 hr. The isolated product was a sticky solid that foamed greatly in water. It had the following structure:

$$\left[\begin{array}{c} C_{12}H_{25}-\overset{\oplus}{N}(CH_2OCH_2C{\equiv}CH)_2 \\ | \\ (CH_2)_4 \\ | \\ C_{12}H_{25}-\underset{\oplus}{N}(CH_2OCH_2C{\equiv}CH)_2 \end{array}\right] 2\ Br^{\ominus}$$

The process avoids problems in pickling ferrous metal articles and provides a pickling composition which minimizes corrosion, overpickling and hydrogen embrittlement. Thus the pickling inhibitors described herein not only prevent excessive dissolution of the ferrous base metal but effectively limit the amount of hydrogen absorption thereby during pickling. A pickling composition for ferrous metal is provided which comprises a pickling acid such as sulfuric or hydrochloric acid and a small but effective amount of the compounds of this process, for example at least about 5 ppm, but preferably from about 3,000 to 10,000 ppm.

The compositions of this process can also be used as corrosion inhibitors in acidizing media employed in the treatment of deep wells to reverse the production of petroleum or gas therefrom and more particularly to an improved method of acidizing a calcareous or magnesium oil-bearing formation.

Corrosion Test Procedure: In these tests the acid solutions were mixed by diluting concentrated hydrochloric acid with water to the desired concentrations.

Corrosion coupons of N-80 steel (ASTM) were pickled in an uninhibited 10% HCl solution for 10 minutes, neutralized in a 10% solution of $NaHCO_3$, dipped in acetone to remove water and allowed to dry. They were then weighed to the nearest milligram and stored in a dessicator.

In most of the tests, a 25 cc/in^2 acid volume to coupon surface area ratio was used. After the desired amount of acid was poured into glass bottles, the inhibitor was added. The inhibited acid solution was then placed in a water bath which had been set at a predetermined temperature and allowed to preheat for 20 minutes. After which time, the coupons were placed in the preheated inhibited acid solutions. The coupons were left in the acid solutions for the specified test time, then removed, neutralized, recleaned, rinsed, dipped in acetone, allowed to dry, then reweighed.

The loss in weight in grams was multiplied times a calculated factor to convert the loss in weight to lb/ft^2/24 hr. The results of these tests are included below.

Inhibitor	Conc. in ppm	Test Temp. °F	Test Time hr	Acid	Metal Type	Corrosion Rate (lb/ft^2/day)
Example 1	6000	200	4	15% HCl	N-80	0.385
Example 2	6000	200	4	15% HCl	N-80	0.033
Example 3	6000	200	4	15% HCl	N-80	0.045
Example 4	6000	200	4	15% HCl	N-80	0.024
Example 5	6000	200	4	15% HCl	N-80	0.025
Blank	6000	200	4	15% HCl	N-80	2.240

Compositions of this process are also useful in prevention of corrosion in apparatus used in the secondary recovery of petroleum by water flooding and in the disposal of wastewater and brine from oil and gas wells.

Static Weight Loss Tests: The test procedure involves the measurement of the corrosive action of the fluids inhibited by the compositions herein described upon sandblasted SAE 1020 steel coupons measuring ⅞ by 3¼ inches under conditions approximating those found in an actual producing well, and the comparison thereof with results obtained by subjecting identical test coupons to the corrosive action of identical fluids containing no inhibitor.

Clear pint bottles were charged with 200 ml of 10% sodium chloride solution saturated with hydrogen sulfide and 200 ml of mineral spirits and a predetermined amount of inhibitor was then added. In all cases the inhibitor concentration was based on the total volume of the fluid. Weighed coupons were then added, the bottles tightly sealed and allowed to remain at room temperature for 72 hr. The coupons were then removed, cleaned by immersion in inhibited 10% HCl, dried and weighed.

The changes in the weight of the coupons during the corrosion test were taken as a measurement of the effectiveness of the inhibitor compositions. Protection percentage was calculated for each test coupon taken from the inhibited fluids in accordance with the following formula:

$$\frac{L_1 - L_2}{L_1} \times 100 = \% \text{ Protection}$$

in which L_1 is the loss in weight of the coupons taken from the uninhibited fluids and L_2 is the loss in weight of coupons which were subjected to the inhibited fluids.

Static Weight Loss Test

Example	Concentration ppm	% Protection
2	100	97.5
3	100	98.6
4	100	96.4
5	100	97.4

This process of *P.M. Quinlan; U.S. Patent 4,252,743; Feb. 24, 1981; assigned to Petrolite Corp.* relates to quaternaries of halogen derivatives of alkynoxymethyl amines and uses thereof. These may be summarized by the following formulae:

(1)
$$\begin{array}{l} R\text{–}N^+(CH_2OR''C{\equiv}CX)_2A^- \\ \;\;\;| \\ \;\;\;R' \end{array}$$

where R and R' are substituted groups such as alkyl, aryl, etc.; R'' is an alkylidene group; and X is halogen and A is an anion; and

(2)
$$\begin{array}{l} R\text{–}N^+(CH_2OR''C{\equiv}CX)_2 \\ \;\;\;| \\ \;\;\;Z \qquad\qquad\qquad 2A^- \\ \;\;\;| \\ R'\text{–}N^+(CH_2OR''C{\equiv}CX)_2 \end{array}$$

where R, R' and R'' have the same meaning as in (1) and Z is a bridging group, preferably hydrocarbon such as alkylene, alkynylene, alkenylene, arylene, etc.

Example 1: Into a 500 ml three-necked flask provided with a reflux condenser and stirrer were placed 18.4 g (0.04 mol) of $C_4H_9N(CH_2OCH_2C{\equiv}CI)_2$, 5.7 g (0.04

mol) of methyl iodide, and 20 ml of ethanol. The reaction mixture was heated to reflux and held there for 24 hours. The solvent was removed under reduced pressure on a rotary evaporator. The product had the following structure:

$$\left[C_4H_9N^+(CH_3)(CH_2{-}O{-}CH_2C{\equiv}CI)_2 \right] I^-$$

Example 2: In a similar manner 28.7 g (0.05 mol) of $C_{12}H_{25}N(CH_2OCH_2C{\equiv}CI)_2$, 7.1 g (0.05 mol) of methyl iodide, and 40 ml of 2-propanol were refluxed together for 60 hours. The product had the following structure:

$$\left[C_{12}H_{25}N^+(CH_3)(CH_2OCH_2C{\equiv}CI)_2 \right] I^-$$

Example 3: In a similar manner 28.7 g (0.05 mol) of $C_{12}H_{25}N(CH_2OCH_2C{\equiv}CI)_2$, 6.4 g (0.05 mol) of benzyl chloride and 40 ml of n-propanol were refluxed together for 48 hours. The product had the following structure:

$$\left[C_{12}H_{25}N^+(CH_2C_6H_5)(CH_2OCH_2C{\equiv}CI)_2 \right] Cl^-$$

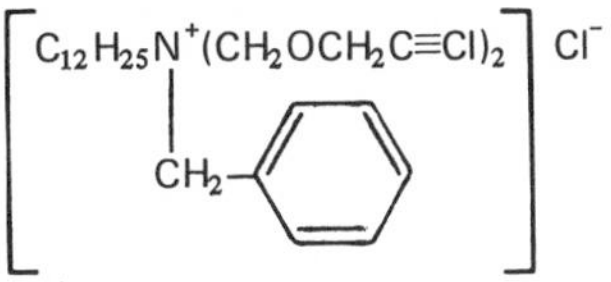

Example 4: Into a 500 ml three-necked flask equipped with a stirrer and reflux condenser were introduced 28.7 g (0.05 mol) of $C_{12}H_{25}N(CH_2OCH_2C{\equiv}CI)_2$, 5.4 g (0.25 mol) of 1,4-dibromobutane and 35.0 ml of n-propanol. The reaction mixture was heated at reflux for 72 hr. The isolated product had the following structure:

$$\left[\begin{array}{c} C_{12}H_{25}N^+(CH_2OCH_2C{\equiv}CI)_2 \\ | \\ (CH_2)_4 \\ | \\ C_{12}H_{25}N^+(CH_2OCH_2C{\equiv}CI)_2 \end{array} \right] 2Br^-$$

The compounds of this process have the same uses as the compounds of the preceding patent and were tested as described above.

Oil-Free Industrial Fluids

H.N. Singer; U.S. Patent 4,257,902; March 24, 1981; assigned to Singer & Hersch Industrial Development (Pty.) Ltd., South Africa provides oil-free aqueous compositions of matter useful as the lubricants or functional liquids. They comprise (A) a major amount of water including up to as much as 99.9% by weight of water (based on the total weight of the composition), (B) a minor amount of at least one substantially water-insoluble functional additive stably dispersed therein, and (C) a minor amount of at least one substantially water-soluble, liquid organic dispersing agent, the dispersing agent being capable of stably dispersing the functional additive in the aqueous composition.

Optionally, but preferably, these compositions can also contain (D) at least one water-soluble polymeric thickener for the aqueous composition and (E) at least one inhibitor of corrosion of metal. As a further option, they can also contain (F) at least one shear stabilizing agent, especially when the thickener (D) is present. In addition, these compositions can also contain (G) at least one glycol of inverse solubility, (H) at least one bactericide, (J) at least one transparent dye, (K) at least one water softener, (L) at least one odor masking agent and (M) at least one antifoamant, including one of these or mixtures of two or more.

The compositions are useful in the shaping of solid materials and hydraulic systems. The aqueous compositions can also be used to inhibit corrosion of ferrous metal and as mold release agents.

Example 1: A liquid concentrate useful in preparing an aqueous composition of matter according to this process is made as follows. To make 1 ℓ of concentrate, the following ingredients are assembled in the indicated amounts:

(a)	A first portion of hydroxyethylcellulose (Natrosol 250 GR)*	10 g
(b)	A first portion of sodium carboxymethylcellulose (Hercules 7M8S)*	10 g
(c)	Molyvan L	1 g
(d)	A first portion of polypropylene glycol (Pluriol P900)	1 g
(e)	Anglamol 32	1 g
(f)	Lubrizol 5315	1 g
(g)	Tributyl tin oxide	1 g
(h)	A second portion of polypropylene glycol (Pluriol P900)	2 g
(i)	A first portion of diethanolamine	5 g
(j)	Emulan SH**	10 g
(k)	A second portion of diethanolamine	5 g
(l)	Emcol TS 230***	10 g
(m)	Para tertiary butyl benzoic acid previously neutralized with triethanolamine as 50% solution in water	10 g
(n)	A second portion of hydroxyethylcellulose (Natrosol 250 GR)	10 g
(o)	A second portion of sodium carboxymethylcellulose (Hercules 7M8S)	8 g
(p)	Green metal acid dye	1 g

*Hercules Incorporated
**A nitrogenous fatty acid condensation product in the form of the free carboxylic acid, anionic; functions as a corrosion inhibitor; BASF Corporation, Germany.
***A corrosion inhibitor, Witco Chemical Corp.

The above ingredients are combined as follows: (a) and (b) are mixed as dry powders and then dispersed into 600 ml of water and allowed to hydrate. The thickened water mixture is cooled to about 5°C with ice and a portion thereof is mixed with ingredients (c) and (d) and the mixture again well dispersed. Ingredients (e), (f), (g) and (h) are mixed and dispersed well into the balance of the thickened water. The two portions of thickened water containing the various other ingredients are then recombined and mixed. Ingredients (i), (j), (k), (l) and (m) are individually added to the thickened mixture which is thoroughly agitated after each addition to form a homogeneous mixture. Ingredients (n), (o) and (p) are then individually dispersed into the thickened mixture. The volume of mixture is then brought up to 1 ℓ total volume and stored for about 24 hr with occasional agitation.

Example 2: One liter of an aqueous composition according to this process useful as a hydraulic fluid is made by assembling the following ingredients in the indicated amount.

(a)	Hydroxyethylcellulose (Natrosol LR)	40 g
(b)	Molyvan L	1 g
(c)	A first portion of polypropylene glycol (Pluriol P900)	1 g
(d)	Anglamol 32	1 g
(e)	Lubrizol 5315	1 g
(f)	Tributyl tin oxide	1 g
(g)	A second portion of polypropylene glycol (Pluriol P900)	2 g
(h)	Diethanolamine	5 g
(i)	Emulan SH	10 g
(j)	Para tertiary butyl benzoic acid previously neutralized with triethanolamine as a 50% solution	10 g
(k)	Pluracol V10	20 g
(l)	Ethylene glycol	50 g
(m)	Dye	1 g

Ingredient (a) is dispersed in 600 ml of water and allowed to hydrate. The thickened mixture is then cooled to about 5°C with ice. Ingredients (b) and (c) are mixed and then dispersed well into a portion of the thickened water. Ingredients (d), (e), (f) and (g) are dispersed well into the remainder of the thickened water. The two portions of thickened water are recombined and agitated to form a homogeneous dispersion. The remaining ingredients are added individually to the thickened mixture which is agitated after each addition. The mixture is then brought up to a total volume of 1 ℓ with water.

Example 3: A concentrate useful in making an aqueous composition for use as a heavy-duty machining fluid is made as follows. The following ingredients are assembled in the indicated amounts:

(a)	Hydroxyethylcellulose (Natrosol HHR)	6 g
(b)	Sodium nitrite	30 g
(c)	Sodium tripolyphosphate	30 g
(d)	Dye	1 g
(e)	Anglamol 32	3 g
(f)	Lubrizol 5315	3 g
(g)	Molyvan L	3 g
(h)	Tributyl tin oxide	2 g
(i)	A first portion of Hostacor KS1	10 g
(j)	A second portion of Hostacor KS1	140 g

The dry solids (a), (b), (c) and (d) are thoroughly mixed. The liquid ingredients (e), (f), (g), (h) and (i) are mixed; then the liquid mixture is slowly added to the solid mixture with good agitation so as to evenly distribute it. The resulting mixture is stored for approximately 16 hr and then any lumps are broken by screening through a 10 mesh screen.

The resulting powder is mixed with 800 ml of water and ingredient (j) is added. The resulting mixture is thoroughly dispersed. Water is added to bring the total mixture up to a volume of 1 ℓ. This concentrate can be diluted with 3 to 5 parts of water to provide a machining fluid.

Additives for Poly(Alkylene Oxide) Compositions

Poly(alkylene oxide) polymers have many and diverse industrial applications, providing not only compatibility with water, but good lubricating qualities and stability as well. Thus, these materials have found wide use both as substantially 100% active compositions and as aqueous solutions in applications such as hydraulic fluids, metal working lubricants, metal treating formulations, and the like. As with all materials that may contain some water, these must protect metals with which they come in contact, and ferrous metals are of particular interest. Corrosion-inhibitive additives are therefore commonly included in the poly(alkylene oxide) containing compositions for this purpose, and success or failure in a particular application may well depend upon the quality of protection for metals realized in service.

A potential hindrance to the protection accorded is that, like most organic polymers, poly(alkylene oxides) are oxidized by oxygen at elevated temperatures. Even in an aqueous medium, oxidation of the polymer can have a significant effect on the lubricating quality of the fluid. Moreover, the products of oxidation are usually acids that foster corrosion of most metals and further limit the useful life of the polymer-containing compositions.

Heretofore, alkali metal nitrites such as sodium nitrite have been widely used as additives for poly(alkylene oxide) solutions to inhibit corrosion of metals, their powerful passivating effect on ferrous metals being well known. Recently, however, carcinogenic properties of N-nitrosamines, which are the reaction products of secondary amines with nitrites, have caused serious concern about the advisability of using nitrites; and the replacement of nitrites with additives that do not present a health hazard, yet are effective in corrosion inhibition, is of considerable interest.

B.F. Mago; U.S. Patent 4,263,167; April 21, 1981; assigned to Union Carbide Corporation provides poly(alkylene oxide) compositions which exhibit excellent resistance to oxidative degradation and inhibit the corrosion of ferrous metals. The compositions have incorporated therein a small effective amount of a bridged dimer of a hydroxy-substituted aromatic carboxylic acid and salts thereof of the general formula:

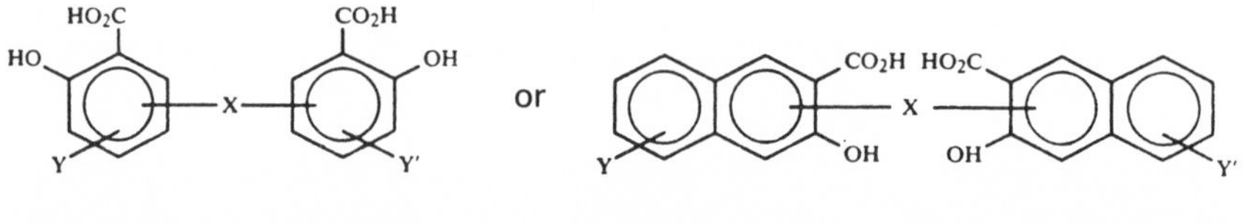

HO, HO_2C, X, CO_2H, OH, Y, Y'

or

Y, CO_2H, OH, X, HO_2C, OH, Y'

wherein X is a chemically stable group selected from lower alkylene, sulfonyl, and amino groups, and a sulfur atom, and Y and Y' may be the same or different and are a hydrogen atom, hydroxyl group, amino group, alkyl group, or sulfonyl group.

Preferably, aqueous poly(alkylene oxide) compositions contain at least about 5 millimols of the additive per liter of composition.

Suitable acids are, e.g., methylene or sulfur bridged, hydroxyl-substituted aromatic carboxylic acids such as 5,5'-methylenedisalicylic acid, pamoic acid, and thiodisalicylic acid. The compound may be present in the composition as a salt such as an alkali metal or ammonium salt.

The compositions of this process as well as the controls which demonstrate the prior art were evaluated for oxidation resistance and corrosion inhibition as follows. The inhibition of oxidation of aqueous solutions of poly(alkylene oxide) were studied using a 10% solution of the polymer in distilled water heated to 70°C, above which temperature some polymer tends to separate from water due to its known inverse solubility. All glass reaction cells 75 mm in diameter and 200 mm high were used. The cells were equipped via 71/60 ST joints with heads that contained glass joints for inserting a thermometer, an aeration tube, and a reflux condenser. Test solutions of 400 g charged to each cell were heated to 70°C in a thermostatically controlled liquid bath for a period of 8 days while sparging continuously with filtered air at a flow rate of 50 to 100 cc/min. The extent of oxidation that occurred during this period was determined by titrating a sample with standard base and by noting the change in its viscosity at 40°C.

Studies of corrosion inhibition of aqueous solutions of poly(alkylene oxide) involved immersion tests of panels of steel for 8 days in a 10% polymer solution at 70°C while sparging lightly with air and then measuring weight loss and noting the appearance of the metal and appearance of the solutions. Inasmuch as the corrosion of steel by aqueous poly(alkylene oxide) solutions is more severe under film disruptive conditions as found in heat exchangers, the corrosion evaluation studies were also made with steel as a heat-transfer surface for a boiling solution using the test apparatus and test procedure described in U.S. Patent 3,951,844. The corrosion inhibiting ability of the various polymer solutions was determined for solutions prepared with distilled water to which, typically, 500 ppm each of sodium chloride, sodium sulfate and sodium formate were added to stimulate typical impurities.

The samples used in this study were weighed, cold-rolled 4" x 4" x 1/16" mild steel plates which had been polished and scrubbed with a wet bristle brush and commercial kitchen powder cleaner, rinsed, and dried. The degree of corrosion was determined from the weight change of the steel panel sample after cleaning, with weight loss being recorded in units of mils per year (mpy).

In the following example which illustrates the process all parts and percentages are by weight unless otherwise indicated.

Example: The apparatus and procedures outlined above were used to evaluate the oxidation resistance and corrosion inhibition of steel for 10% solutions of poly(alkylene oxide). In the oxidation and immersion corrosion tests, the poly(alkylene oxide) used was a liquid, water-soluble commercial product available as Ucon 75-H-90000 (1). In the heat-transfer corrosion tests, the poly(alkylene oxide) used was a liquid, water-soluble commercial product available as Ucon 75-H-1400 (2). The oxidation and immersion corrosion rates were measured after 8 days exposure and the corrosion rates using the heat-transfer apparatus were measured after 3 days.

In these tests compositions containing disodium salt of 5,5'-methylenedisalicylic acid in various proportions were compared with compositions prepared without any inhibitor and those containing sodium nitrite inhibitor. The results obtained during the oxidation tests are summarized in Table 1 and the corrosion tests are summarized in Table 2. The data thus presented show that where no inhibitor was used in the aqueous poly(alkyene oxide) composition, significant acidity developed in the solutions during the oxidation test period and that there was more than a 50% decline in viscosity at 40°C. Moreover, the weight loss of the steel test panels during the corrosion test periods was very high. All of this was almost completely prevented by including as little as 0.1 weight percent (14.5 millimols per liter) of sodium nitrite in the solution. The sodium methylenedisalicylate additive offered considerable oxidation resistance to the poly(alkylene oxide) solution as well as being efficient in its corrosion inhibition for steel, when suitable amounts of the additive were employed.

Table 1: Oxidation Inhibition of Poly(Alkylene Oxide) in 10% Aqueous Solutions

Additive	Additive Concentration Weight %	Additive Concentration M Moles/liter	pH Initial	pH Final	Acidity Formed (Millinormality)	Viscosity Change at 40° C. (%)	Solution Appearance
None	—	—	10.2	2.8	4.6	−60	clear colorless
Sodium Nitrite	0.1	14.5	8.5	5.8	0.2	+2	clear colorless
	0.5	72.	9.1	7.0		0	clear colorless
Methylene-disalicylic Acid, sodium salt	0.3	9.0	9.7	7.9	0.3	+2	clear colorless
	0.6	18.0	9.4	8.8	0.0	+2	clear yellow
	1.2	36.0	7.3	6.3	0.7	+2	clear amber

Table 2: Corrosion Inhibition of Poly(Alkylene Oxide) in 10% Aqueous Solutions

Poly(alkylene-oxide)	Additive	Additive Concentration Weight %	Additive Concentration M Moles/liter	Corrosion Test	Corrosion Rate Mils per year (mpy)	Solution Appearance
I	none	—	—	Immersion (8 days)	14.8	rusty
II	none	—	—	Heat-transfer (3 days)	81	rusty
I	Sodium nitrite	0.2	29	Immersion (8 days)	<0.04	clear
II	Sodium nitrite	0.05	7.2	Heat-transfer (3 days)	5.6	turbid-orange
	Sodium nitrite	0.1	14.5	Heat-transfer (3 days)	0.4	clear colorless
	Sodium nitrite	0.25	36	Heat-transfer (3 days)	0.5	clear colorless
I	Methylenedisalicylic	0.4	12.0	Immersion (8 days)	0.7	clear amber
II	acid, sodium salt	0.3	9.0	Heat-transfer (3 days)	0.9	clear red
II	acid, sodium salt	0.6	18.0	Heat-transfer (3 days)	13.0	very turbid
II	acid, sodium salt	1.19	36.0	Heat-transfer (3 days)	0.3	clear amber

In this process of *B.F. Mago; U.S. Patent 4,277,366; July 7, 1981; assigned to Union Carbide Corporation* the additive in the poly(alkylene oxide) compositions is a mononuclear aromatic compound having at least one substituent nitro group. Aqueous poly(alkylene oxide) compositions should contain at least 15 millimols, and preferably, at least 25 millimols of the additive per liter of composition.

Suitable nitroaromatic compounds are the nitro-substituted aromatic acids and compounds such as nitroaromatic salts, esters, and the like that, in situ, effect the formation of the acid anion. Exemplary suitable nitroaromatic compounds are 3-nitrobenzoic acid, 4-nitrophthalic acid, 5-nitroisophthalic acid, 3,5-dinitrobenzoic acid, p-nitrocinnamic acid, and the like, and alkali metal or ammonium salts thereof. Tests were conducted as described in the preceding patent.

Phosphonoadipic Acids

J.G. Dingwall, B. Cook and A. Marshall; U.S. Patent 4,265,769; May 5, 1981; assigned to Ciba-Geigy Corporation provide a compound or mixture of compounds of the general formula:

$$X^1OOC-\left(\begin{array}{c}H\\ |\\ C\\ |\\ R^1\end{array}\right)_m-\begin{array}{c}R^2\\ |\\ C\\ |\\ R^3\end{array}-\begin{array}{c}H\\ |\\ C\\ |\\ R^4\end{array}-\begin{array}{c}R^5\\ |\\ C\\ |\\ P{=}O\\ \diagup\ \diagdown\\ O\quad O\\ Z^1\quad Z^2\end{array}-\left(\begin{array}{c}H\\ |\\ C\\ |\\ R^6\end{array}\right)_n-COOX^2$$

In the formula m and n may be 0 or 1 but both cannot be 1, R^1, R^2, R^3, R^4, R^5 and R^6 are independently H or CH_3, X^1, X^2, Z^1 and Z^2 are independently hydrogen or straight or branched chain C_{1-4} alkyl; and the water-soluble inorganic or organic salts thereof, with the proviso that at least one of R^1, R^2, R^3, R^4, R^5 must be CH_3, and when m and n are both 0, R^2 and R^3 are each methyl and R^4 and R^5 have their previous significance.

When added to an aqueous system the compound imparts one or more of the following beneficial effects to the treated system: (a) the corrosion of ferrous metals in contact with the system is inhibited; (b) the precipitation of scale-forming salts of calcium, magnesium, barium and strontium from the treated aqueous system is inhibited; and (c) inorganic materials present in the treated aqueous system are dispersed.

The compounds of the above formula could find use in cooling water systems; steam generating plant; sea-water evaporators; and hydrostatic cookers. The following examples further illustrate the process. Parts and percentages, shown therein are by weight unless otherwise stated.

Example 1: Diethyl 1,3,3-trimethyl-5-oxocyclohexanephosphonate (BP 128° to 130°C/0.6 mm) was prepared by base catalyzed addition of diethyl phosphite to isophorone. Hydrolysis of this ester with concentrated HCl gave 1,3,3-trimethyl-5-oxocyclohexanephosphonic acid (MP 167° to 168°C).

22 parts of 1,3,3-trimethyl-5-oxocyclohexanephosphonic acid were added portion wise, over 6 hours, to a stirred solution of 0.05 part of ammonium metavanadate in 32 parts of 70% nitric acid at 55° to 60°C; when the addition was complete the resulting solution was heated at 55° to 60°C for a further 5 hours, cooled to room temperature, and diluted by the addition of 50 parts of water. This solution was evaporated to dryness, the solid residue redissolved in 100 parts of water and again evaporated to dryness to give 21.7 parts of a hygroscopic solid which was substantially a 1:1 mixture of 2,4,4-trimethyl-2-phosphonoadipic acid and 3,5,5-trimethyl-3-phosphonoadipic acid which had ^{31}P chemical shifts of -25 and -32 ppm respectively and a minor proportion of 2,4,4-trimethyl-2-phosphono glutaric acid having a ^{31}P chemical shift of -24 ppm.

Example 2: Diethyl 1-methyl-5-oxocyclohexanephosphonate (BP 130° to 132°C/0.8 mm) was prepared by the base catalyzed addition of diethylphosphite to 3-methyl-2-cyclohexen-1-one. Hydrolysis of this ester with concentrated HCl gave 1-methyl-5-oxocyclohexanephosphonic acid as a viscous oil.

Oxidation of 19.2 parts of the phosphonic acid with 70% nitric acid, as in Example 1, gave 20.7 parts of an hygroscopic solid which was substantially a 1:1 mixture of 2-methyl-2-phosphonoadipic acid and 3-methyl-3-phosphonoadipic acid which had ^{31}P chemical shifts of -24 and -32 ppm respectively.

Example 3: A concentrated aqueous solution of the mixed product from Example 1 was allowed to stand at room temperature for several days during which time a white solid precipitated. This was collected by filtration and dried to give 3,5,5-trimethyl-3-phosphonoadipic acid (MP 167° to 168°C, decomposing, ^{31}P chemical shift of -32 ppm) which had the following elemental analysis by weight. Required for $C_9H_{17}O_7P \cdot 1\frac{1}{2}H_2O$: C, 36.61%, H, 6.78%; P, 10.50%. Found: C, 36.71%; H, 6.66%; P, 10.43%.

Example 4: *Demonstration of corrosion inhibitor activity* – Corrosion inhibitor activity of the products was demonstrated in the following way by the Aerated Solution Bottle Test and using a standard corrosive water made up as follows: 20 g $CaSO_4 \cdot 2H_2O$; 15 g $MgSO_4 \cdot 7H_2O$; 4.6 g $NaHCO_3$; 7.7 g $CaCl_2 \cdot 6H_2O$; and 45 gallons distilled water. Mild steel coupons, 5 x 2.5 cm are scrubbed with pumice immersed for 1 minute in hydrochloric acid and then rinsed, dried and weighed.

The desired proportion of additive combination is dissolved in 100 ml of standard corrosive water. A steel coupon is suspended in the solution, and the whole is stored in a bottle in a thermostat at 40°C. During the storage period, air is passed into the solution at 500 ml/min, the passage of the air being screened from the steel coupon; any water losses by evaporation are replaced as they occur with distilled water from a constant head apparatus.

After 48 hours, the steel coupon is removed, scrubbed with pumice, immersed for 1 minute in hydrochloric acid inhibited with 1% by weight of hexamine and then rinsed, dried and reweighed. A certain loss in weight will have occurred. A blank test, i.e., immersion of a mild steel specimen in the test water in the absence of any potential corrosion inhibitor, is carried out with each series of tests. The corrosion rates are calculated in milligrams of weight loss/square decimeter/day (mdd) but for convenience the results are shown as percentage protection, which is defined as follows:

$$\%\ \text{Protection} = \frac{\text{Corrosion rate for blank (in mdd) - corrosion rate for sample (in mdd)}}{\text{Corrosion rate for blank (in mdd)}} \times 100$$

The results obtained using 100 ppm of the product of Examples 1, 2 and 4 are given in the table below.

Product (100 ppm)	Protection (%)
Product of Example 1	92
Product of Example 2	93
Product of Example 4	86

Stable Solutions of Vinyl Copolymers

Anionic polyelectrolytes such as sodium polyacrylate are added as hardness-stabilizers in the preparation of cooling water and feedwater. Zinc salts are often used as corrosion inhibitors and, to be sure, in amounts which are stoichiometrically far above the equivalent amount of carboxyl groups of the anionic polyelectrolytes. Further, effective cationic surface active agents are used in cooling water systems as biocides to avoid the growth of algae and bacteria. The zinc salts as well as the cationic surface active agents are as a rule not compatible with the hardness-stabilizers and cannot be used together.

In phosphatizing baths a disturbing crust formation occurs as a result of the separation of heavy metal sulfates. In order to avoid this, special stabilizing agents which are difficult to prepare are added to the phosphatizing baths, for example, polymers of α-oxyacrylic acid. Polyacrylic acid and polymethacrylic acid are as a rule not suitable for this purpose because they are precipitated by the heavy metal ions.

The process of *H. Trabitzsch, J. Frieser, A. Koschik and H. Plainer; U.S. Patent 4,271,058; June 2, 1981; assigned to Röhm GmbH, Germany* solves the problem of preparing stable aqueous solutions of vinyl copolymers having anionic groups and of such water-soluble ionic or ionizable compounds which have a precipitating action on aqueous solutions of polyacrylic acid or other anionic polyelectrolytes.

The aqueous solutions comprise: (1) from 20 to 99% by weight of water; (2) 0.5 to 79% by weight of a water-soluble vinyl copolymer; and (3) 0.5 to 79% by weight of at least one water-soluble ionic or ionizable compound which has a precipitating effect on an aqueous solution of polyacrylic acid. The vinyl copolymer in turn comprises: (A) from 10 to 90 mol percent, preferably from 50 to 80 mol percent, of an α,β-unsaturated carboxylic acid; (B) from 10 to 90 mol percent, preferably from 20 to 50 mol percent, of a vinyl monomer containing ammonium groups, and, as an optional component, from 0 to 80 mol percent of units of at least one further nonionic comonomer copolymerizable with monomers (A) and (B).

Example 1: For the preparation of water-treating agents according to this process, the following predominantly anionic vinyl copolymers were used (all parts are molar parts): (A) 75 parts acrylic acid, 25 parts methacryloxyethyltrimethylammonium chloride; (B) 73 parts acrylic acid, 27 parts acryloxyethyltrimethylammonium chloride; (C) 73 parts of acrylic acid, 27 parts of dimethylaminoethylmethacrylate hydrochloride; and (D) 47 parts acrylic acid, 26 parts methacrylic acid, 27 parts methacryloxyethyltrimethylammonium chloride. For comparison, the following pure anionic polymer was used: (F) 100 parts acrylic acid.

Example 2: The feedwater of a steam generator contains a water-soluble vinyl copolymer having anionic character as a hardness-stabilizing agent and a polyfunctional amine as a corrosion inhibiting component. In addition to the vinyl polymers A, B, C, D, and F (a comparision substance) from Example 1, the following polymers were used (parts are molar parts): (G) 25 parts acrylic acid, 60 parts methacrylic acid, 15 parts methacryloxyethyltrimethylammonium chloride; (H) 44 parts acrylic acid, 36 parts methacrylic acid, 20 parts methacryloxyethyltrimethylammonium chloride; and (I) 50 parts acrylic acid, 20 parts maleic acid, 30 parts methacryloxyethyltrimethylammonium chloride.

For testing compatibility, 10 weight percent aqueous solutions of the polymers are mixed at room temperature with 10 weight percent solutions of ethylenediamine, diethylenetriamine, or triethylenetetramine in such ratios that one amino group is present for each carboxyl group of the copolymer. The stable mixtures remain clear. Incompatibility can be recognized by a cloudiness.

	Compatibility with		
Polymer	**Ethylenediamine**	**Diethylenetriamine**	**Triethylenetetramine**
A	++	++	++
B	++	++	++
C	++	++	++
D	++	++	++
G	+	+	+
H	++	++	++
I	++	++	++
F*	-	-	-

Note: ++ = clear, compatible; + = very slight clouding, limit of compatibility; - = cloudy, incompatible.

*Comparison test.

Polyamine plus Alkylenephosphonic Acid Derivative

The anticorrosion composition according to the process of *F. Moran; U.S. Patent 4,276,089; June 30, 1981; assigned to Union Chimique et Industrielle de l'Ouest SA, France* is a water-insoluble composition. It contains (a) at least a polyamine with a molecular weight greater than or equal to 320 of the general formula:

(1) $$R{-}[NH{-}(CH_2)_3]_{n1}{-}NH_2$$

wherein R is a saturated or unsaturated aliphatic $C_{12\text{-}22}$ hydrocarbon radical; and n1 is an integer being such that the molecular weight of the polyamine is greater than or equal to 320; and (b) at least one alkylenephosphonic acid derivative. The composition is useful to inhibit the corrosion of metallic surfaces caused by water, in liquid or steam form.

Among the suitable aminoalkylenephosphonic acid derivatives according to this process, the following can be mentioned and their alkyl esters of $C_{1\text{-}4}$ (fatty acids being the preferred derivatives), namely, acids of the formula:

(2) $$N\left({-}Alk{-}P(=O)(OH)_2\right)_3$$

wherein Alk is an alkylene group of $C_{1\text{-}6}$ with a straight or branched hydrocarbon chain, acids of the formula:

(3) $$(H_2O_3P{-}CH_2)_2N{-}(CH_2)_{n2}{-}N(CH_2{-}PO_3H_2)_2$$

wherein n2 is an integer varying between 1 and 6 inclusive; di(hydroxyethyl)-aminomethylphosphonic acid of the formula:

(4) $$(HOCH_2CH_2)_2N{-}CH_2{-}PO_3H_2$$

acids of the formula:

(5) $$Z^1{-}O{-}(Z^2{-}O)_{n3}{-}Z^3N[(CH_2)_{n4}{-}PO_3H_2]_2$$

wherein Z^1 is H or an alkyl group of $C_{1\text{-}5}$; Z^2 is an alkylene group of $C_{2\text{-}5}$; Z^3 is an alkylene group of $C_{3\text{-}5}$; n3 is an integer varying between 1 and 20 inclusive, and n4 is an integer varying between 1 and 4.

Among the suitable alkylenepolyphosphonic acid derivatives are, for example, the acids, esters and salts represented by the formula:

(6) $$(M_4O)(M_3O)P(=O){-}A{-}P(=O)(OM_1)(OM_2)$$

A is a bivalent alkylene group comprising a straight and saturated $C_{1\text{-}10}$ hydrocarbon chain, each carbon atom of which chain can be, if necessary, substituted by at least a group selected from the OH, $C_{1\text{-}4}$ alkyl and phosphonic groups:

$$P(O)OM_5OM_6$$

and M_1, M_2, M_3, M_4, M_5 and M_6 whether identical or different are each H, an alkyl group of $C_{1\text{-}4}$, NH_4^+ or a metal cation.

Advantageously, the composition according to the process will contain (a) 5 to 80 parts by weight of polyamine 1 and (b) 20 to 95 parts by weight of the alkylenephosphonic acid derivative and, preferably (a) 15 to 70 parts by weight of polyamine 1 and (b) 30 to 85 parts by weight of derivative of the alkylenephosphonic acid of formulas 2 through 6.

The method for preparing the anticorrosion composition is performed in a known manner which consists in mixing one or more polyamines 1 with one or more derivatives of alkylenephosphonic acid.

According to the best mode which is recommended, the method used is characterized in that the polyamine or polyamines selected is/are brought to a liquid state by adequate heating, and then introduced progressively while slightly or strongly stirred, depending on the case, into an aqueous solution of the alkylenephosphonic acid(s) selected, heated beforehand to a temperature less than that of the polyamine.

Depending on the nature of the means (a) or (b) which are used, the resulting mixture may be in gel, or paste or wax form.

In practice, the polyamine(s) will be melted at a temperature varying between 30° and 85°C approximately, and then poured into the alkylenephosphonic acid(s) brought to a temperature varying between 15° and 60 °C.

Acrylic Acid Polymer and Phosphorous Acid

R.J. Lipinski; U.S. Patent 4,277,359; July 7, 1981; assigned to Mogul Corporation provides a process for protecting metals in the presence of water by adding as essential to the water an effective amount of phosphorous acid with a water-soluble acrylic acid polymer of relatively low molecular weight, for example, between 500 and 2500, with the principal monomeric therein being derived from acrylic acid.

Other (amounts less than 10% by weight) monomers interpolymerizable therewith include acrylamide monomers, vinyl phosphonic acid monomers and vinyl sulfonic acid monomers, as well as prepolymerized acrylic acid polymers which have been modified with PCl_3. Upon steam distillation, the products obtained contain from traces to about 5% of combined phosphonate along with traces of up to about 20% by weight of phosphorous acid after hydrolysis of any residual PCl_3.

The surprising discovery was that small amounts of H_3PO_3, in combination, contributed most remarkably to the increase in the corrosion inhibition in conjunction with the acrylic polymers which had been known to be useful as corrosion inhibitors in industrial waters or nonpotable waters.

The corrosion inhibiting compositions of this process minimize mineral deposits generally formed on metal and may be used in various water systems including, for example, air conditioning, steam generating plants, refrigeration systems, heat exchange apparatus, engine jackets, pipes, etc.

These polymers of acrylic acid are identified for use in the examples in accordance with the information shown below.

Polymer Identification	Known Identifying Information
Polymer A Class (BFG K-752)	An unmodified acrylic acid polymer of ~1800 MW with an excess of 90% acrylic acid monomer.
Polymer B (BFG CS5517)	Polymer A Class. Treated with PCl_3 and hydrolyzed. No bound P found. Analysis: 9% total P.
Polymer C (BFG CS5543)	Polymer A Class. PCl_3 treated and hydrolyzed 15.6% total P; 3% + 0.2% bound P (phosphonic acid group). 13% P as H_3PO_3.
Polymer D (BFG K-752)	Polymer A Class + 11% P as H_3PO_3 added without chemical processing.
Polymer E (BFG K-752)	Polymer A Class + 14.3% P as H_3PO_3 added without chemical processing.
Polymer F (BFG 5543-B)	Polymer A Class. Treated with PCl_3 and hydrolyzed. Total P = 6.6% by analysis. 3% P present in polymer bound. 3.5% free P as in H_3PO_3.
Polymer G (BFG 5543-C)	Polymer A Class. Treated with PCl_3 and hydrolyzed. 15.6% total P by analysis. 2.6% of total P bound to polymer phosphonate 13% P as H_3PO_3.
Polymer H (Cyanamer P-70 BFG)	Copolymer of acrylic acid and acrylamide MW ~1000.

The PCl_3 treatment of the substantially acrylic acid polymers were produced by treating one part of the polymer by weight with from about 28% of the polymer weight as PCl_3 to about 130% of the polymer weight as PCl_3 in tetrahydrofurfural as the solvent-diluent at reflux temperatures for 2 to 2½ hours with subsequent hydrolysis of unreacted PCl_3 with varying quantities of water. The reaction product was then steam distilled to about 22% by weight of water. Note: H_3PO_3 results from hydrolysis of PCl_3.

The foregoing polymers were formulated into a series of water treatment products and tested for inhibition of corrosion and scale deposition by electrochemical test methods to yield comparative data, reported primarily as percent corrosion inhibition in two test waters.

The first, identified as OCW (Open Cell Water) is distilled water containing 50 ppm of chloride ion. The second, identified as FCPW (Filtered Chagrin Plant Water) has the following analysis:

($CaCO_3$), ppm	162
Ca ($CaCO_3$), ppm	108
Mg ($CaCO_3$), ppm	54
Cl (Cl^-), ppm	74
PHT, Alk ($CaCO_3$), ppm	n
Mo, Alk ($CaCO_3$), ppm	218
pH	7.7
Specific conductance	780

Example 1: Polymer A, an unmodified homopolymer of acrylic acid containing no phosphorous was used in a direct test against Polymer C containing about 11% of free phosphorous acid resulting from PCl_3 treatment and hydrolysis of an acrylic acid homopolymer as 10, 15, 20, 25 and 30 parts polymer per million parts of FCPW. After 19 hours at 100°±2°F using electrochemical test methods and pH of 7.5 (adjusted), the following results were obtained:

Test Polymer	Percent H_3PO_3	Percent Corrosion Inhibition				
		10	15	10	25	30
		(ppm)				
A	0.0	29.1	22.2	58.2	75	90
C	11.0	33.5	41.7	70.8	90	93.3

Example 2: Electrochemical corrosion tests were run in OCW water at 5, 10 and 15 ppm levels wherein Polymer D contains merely a physical admixture of phosphorous acid. Polymer A is the straight acrylic acid homopolymer.

Test Polymer	% P as H_3PO_3	Percent Corrosion Inhibition		
		5	10	15
		(ppm)		
A	0	33.3	33.3	92.5
B	9	33.3	93.3	92.5
D	11	33.3	83.5	93.3

Polymer B varies over Polymer D in that the PCl_3 treatment in this instance failed to produce measurable phosphonic acid bound in the polymer. At 10 ppm, there is strong indication of value in the initial chemical treatment of the acrylic polymer. With the OCW, the corrosion inhibition at 15 ppm fails to show appreciable difference in the method of inclusion of H_3PO_3 in the admixture.

A further test run, similar to the above, but replacing Polymer B with Polymer C (Polymer C having some 3±0.2%) bound phosphonate in conjunction with the free H_3PO_3 from hydrolysis of PCl_3, in situ, shows practically no appreciable increase in percent corrosion inhibition. There was about 1.7% increase for Polymer C at 10 and 15 ppm over the mere physical admixture.

Example 3: Using the electrochemical corrosion test method, the following table sets out data on 19 hour test runs made and therein summarized.

Test Polymer	% P as H_3PO_3	Percent Corrosion Inhibition				
		OCW		FCWP		
		5	10	15	25	30
		(ppm)				
A	0	25	40	58	75	90
F	3.6	56.7	90	62	–	90
B	9.0	77.0	90	82	–	90
C	13.0	77.0	87	82	90	90
G	13.0	80.0	88	82	–	90

It is observed that a general increase above 5 ppm of polymer of acrylic acid increased corrosion inhibition in both test waters. Also one observes that at low ppm there is a marked increase in corrosion inhibition, showing the effect of phosphorous acid H_3PO_3 inclusions. Thus, at 3.6% P as H_3PO_3 (Polymer F), 10 ppm of Polymer F were equivalent or better than 30 ppm of Polymer A without H_3PO_3.

Polymers C and G with the highest percent (13%) of P as H_3PO_3 were not materially superior at 10 ppm to Polymer F at 3.6% P as H_3PO_3. This is an indication that about 25% of H_3PO_3 is optimally effective. Large amounts of phosphorous (as H_3PO_3) are not essential to optimum corrosive inhibition.

Boric Acid-Diethanolamine Reaction Products and Sulfonamidocarboxylic Acids

H. Diery, R. Helwerth, H. Fröhlich and H. Lorke; U.S. Patent 4,297,236; October 27, 1981; assigned to Hoechst AG, Germany provide water-miscible corrosion inhibitors. They consist of: (A) reaction products of boric acid and diethanolamine; and (B) arylsulfonamidocarboxylic acids of the formula:

$$(R_1)(R_2)-Ar\left(-\overset{\overset{\displaystyle O}{\|}}{\underset{\underset{\displaystyle O}{\|}}{S}}-\underset{\underset{\displaystyle R_3}{|}}{N}-R_4-CO_2H\right)_n$$

in which R_1 and R_2 each represent hydrogen, fluorine, chlorine, bromine, an alkyl or alkoxy radical having from 1 to 4 carbon atoms, with the proviso that the sum of the carbon atoms of R_1 and R_2 does not exceed 7; Ar is a benzene, naphthalene or anthracene radical; R_3 is hydrogen, an aryl radical having up to 4 carbon atoms, a β-cyanoethyl or hydroxyalkyl radical having from 2 to 4 carbon atoms; R_4 is an alkylene radical having more than 3 carbon atoms, optionally substituted by one or more methyl or ethyl radicals; and n is 1 or 2; or alkyl- and/or cycloalkylsulfonamidocarboxylic acids obtained by sulfochlorination of a saturated aliphatic and/or cycloaliphatic hydrocarbon having from 12 to 22 carbon atoms and a boiling temperature range of from about 200° to 350°C, subsequent reaction with ammonia and final condensation with chloroacetic acid.

The corrosion inhibitors of the process are prepared by simply mixing the components at room temperature or slightly elevated temperatures of up to about 100°C. In general, they consist preponderantly of the reaction products of boric acid and diethanolamine, while the amount of component (B), that is, the aryl- or alkylsulfonamidocarboxylic acids, in the corrosion inhibitors is normally from about 10 to 50, preferably 10 to 30 percent by weight. These indications relate to the pure acids, even when using the alkyl- or cycloalkylsulfonamidocarboxylic acids, because the amount of unreacted hydrocarbons or chloroparaffin accompanying these sulfonamidocarboxylic acids is eliminated by phase separation after the mixture with the reaction products of boric acid and diethanolamine is complete. In order to accelerate the phase separation, the mixture is allowed to settle advantageously at slightly elevated temperature, preferably at 50° to 70°C.

The corrosion inhibitors of the process are transparent water-soluble or easily emulsifiable products which are generally present in the form of viscous liquids. They can be applied with special advantage as corrosion-inhibiting component of aqueous cooling formulations, especially drilling, cutting or laminating liquids, furthermore of circulation cooling systems and power water. For preparing the aqueous cooling formulations, the corresponding inhibitors are stirred into the required amount of water. The concentration of application is generally 0.5 to 10, preferably 2 to 5 percent by weight.

Example 1: 315 g (3 mols) diethanolamine and 61.8 g (1 mol) pulverulent boric acid are mixed at room temperature and stirred at this temperature until a transparent yellow viscous liquid has formed, that is, for about 8 hours.

40 g ϵ-[benzenesulfonyl-methylamino]-n-capronic acid are added to 160 g of the product so obtained, and the mixture is stirred until a transparent yellow viscous liquid has formed which can be used as corrosion inhibitor for aqueous liquids.

Example 2: 315 g (3 mols) diethanolamine are heated to 100°C and 61.8 g (1 mol) boric acid are added; after 10 to 20 minutes, a clear yellow liquid is obtained. 40 g ϵ-[benzenesulfonyl-methylamino]-n-capronic acid are added at 60°C with agitation to 160 g of the above liquid. The transparent viscous liquid so obtained can be used as corrosion inhibitor.

Polyphosphate and Polymaleic Anhydride

The method of *B.P. Boffardi; U.S. Patent 4,297,237; October 27, 1981; assigned to Calgon Corporation* for inhibiting corrosion in an aqueous system comprises the step of treating the system with 1.0 to 300 ppm by weight of the total aqueous content of the system, of a composition comprising polyphosphate and polymaleic anhydride or amine adducts thereof in a weight ratio of from 10:1 to 1:10. The corrosion inhibiting composition may optionally contain zinc.

The polyphosphate material employed in the compositions is a "molecularly dehydrated phosphate," by which is meant any phosphate which can be considered as derived from a monobasic or dibasic orthophosphate, or from orthophosphoric acid, or from a mixture of any two of these by elimination of water of constitution therefrom. There may be employed alkali metal tripolyphosphates, or pyrophosphates, or the metaphosphate which is often designated as hexametaphosphate. Any molecularly dehydrated phosphate may be employed, but it is preferred to use those which have a molar ratio of alkali metal to phosphorus pentoxide of from about 0.9:1 to about 2:1, the latter being the alkali metal pyrophosphate.

The polymaleic anhydride material employed in the compositions may be prepared by a number of different polymerization methods well known in the art. Such polymaleic anhydride may be hydrolyzed very readily, for example, by heating with water, to form a polymer which contains free carboxylic acid groups, and possibly some residual anhydride groups, on a carbon backbone. The polymaleic anhydride should have a weight average molecular weight of from about 200 to about 10,000, and preferably not more than about 3,000.

Since polymerized maleic anhydride is so readily hydrolyzed, treatment of water or an aqueous system with polymerized maleic anhydride is the same as treatment with hydrolyzed polymaleic anhydride. Consequently, this process includes the use of such proportion of polymerized maleic anhydride as will yield the desired amount of hydrolyzed polymaleic anhydride on hydrolysis.

In addition to, or instead of, the polymaleic anhydride there may be utilized amine adducts of polymaleic anhydride selected from the group consisting of: (a) polymers having recurring units of the formula:

$$\left[\begin{array}{cc} -CH- & -CH- \\ | & | \\ C=O & C=O \\ | & | \\ N(R_1)(R_2) & O^{\ominus}\ M^{\oplus} \end{array} \right]_n$$

wherein M^+ may be H^+, alkali metal cation, or quaternary ammonium cation of the formula:

$$R_3R_4N^+R_5R_6$$

wherein for all of the above formulas, R_1, R_2, R_3, R_4, R_5 and R_6 are each independently selected from the group consisting of hydrogen, alkyl of from 1 to 10 carbon atoms, and substituted alkyl of from 1 to 10 carbon atoms, where the substituent is hydroxyl; carbonyl; and carboxylic acid groups, and alkali metal ion and ammonium salts thereof; and wherein n is an integer of from 2 to 100; and (b) polymers having recurring units of the formula:

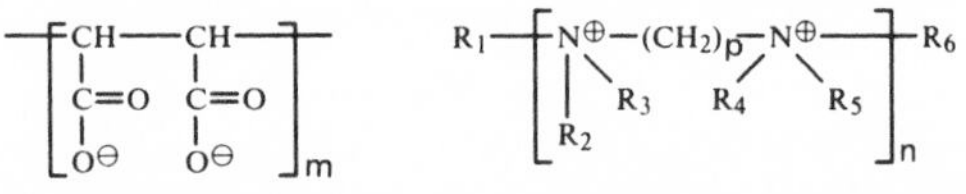

$$\left[\begin{array}{cc} -CH- & -CH- \\ | & | \\ C{=}O & C{=}O \\ | & | \\ O^{\ominus} & O^{\ominus} \end{array}\right]_m \qquad R_1 - \left[\begin{array}{c} N^{\oplus}(R_2)(R_3) - (CH_2)_p - N^{\oplus}(R_4)(R_5) \end{array}\right]_n - R_6$$

wherein R_1, R_2, R_3, R_4, R_5 and R_6 are each independently selected from the group consisting of hydrogen, alkyl of from 1 to 10 carbon atoms, and substituted alkyl of from 1 to 10 carbon atoms, where the substituent is hydroxyl, carbonyl, and carboxylic acid groups, and alkali metal ion and ammonium salts thereof; wherein p is an integer of from 1 to 6; m is an integer of from 2 to 100; and n is an integer of from 2 to about 100, provided that, n not equal to m, the lesser of m or n is multiplied by a factor such that n = m.

3-Amino-5-Alkyl-1,2,4-Triazole for Nonferrous Metals

The process of *W. Gerhardt, V. Wehle, A. Syldatk, G. Rogall, J. Reiffert and J. Conrad; U.S. Patent 4,298,568; November 3, 1981; assigned to Henkel Kommanditgesellschaft auf Aktien, Germany* relates to the use of 3-amino-5-alkyl-1,2,4-triazoles, hereafter called AAT, to prevent corrosion of nonferrous metals in aqueous, particularly industrial water, systems.

Because of their corrosion resistance, nonferrous metals, such as copper, brass, bronze, etc., are preferred materials in the construction of water-conveying plants, for example, steam generating plants, heating systems, cooling water circulating systems and the like. These materials are of particular importance for condenser tubes in steam power plants. Despite their relatively good resistance to corrosion, it is unavoidable, however, that analytically determinable amounts of copper will be given off to the surrounding water in normal use. These copper traces become cemented onto the following cooling water pipes of steel or other base materials, and cause pitting corrosions which are sometimes disastrous. For this reason an additional treatment of the water coming in contact with the nonferrous metal is technically important to reduce this copper transfer.

In practice, there are very few inhibitors which are suitable for this purpose. Essentially, these are mercaptobenzothiazole, benzotriazole and tolyltriazole. These compounds are relatively effective as inhibitors of copper corrosion, but they have the great disadvantage that they are chemically relatively difficult to produce and thus can find only limited application for economic reasons. Another disadvantage of the abovementioned compounds is their very poor solubility at acid pH values, so that a practical manufacture of these products is difficult.

An object of the process is the development of a method for inhibiting corrosion of nonferrous metals in contact with circulating water comprising the steps of adding to circulating water in contact with nonferrous metals from 0.05 to 10 g/m^3 of at least one 3-amino-5-alkyl-1,2,4-triazole wherein the alkyl has from 2 to 8 carbon atoms and adjusting the water to a pH of from 6 to 10.

A corrosion inhibitory composition for use in water-conveying systems in contact with nonferrous metals consists of from 0.1 to 90% by weight of at least one 3-amino-5-alkyl-1,2,4-triazole wherein the alkyl has from 2 to 8 carbon atoms, from 5 to 50% by weight of at least one concretion prevention agent and/or dispersion agent, from 0 to 50% by weight of a water-soluble zinc salt, from 0 to 10% by weight of a water-soluble complexing phosphonic acid, and from 0 to 94.5% by weight of water.

Example 1: The corrosive behavior was determined according to the following method. Three carefully cleaned copper test plates (75 x 12 x 1.5 mm) were dipped at room temperature for 24 hours into a 1 liter beaker containing 1 liter of water and the indicated amount of the substances being tested. During the test period, the aqueous solutions were stirred at 100 rpm in a series arrangement of 10 beakers per test. Subsequently, the Cu content in the water was determined by means of atom absorption.

The test water used as a corrosive medium had the following analytical data: 8° dH (calcium hardness); 2° dH (magnesium hardness); 1° dH (carbonate hardness); 1,000 ppm (Cl^-); 8.2 pH. The test results are given in the table.

Substance	Dosage ppm	Material of Test Plate	μ gm Cu
—	—	Brass	2,010
AAT*	0.3	Brass	62
Tolyltriazole	0.3	Brass	117
Mercaptobenzo-thiazole	0.5	Brass	126
—	—	Copper	892
AAT*	0.3	Copper	46
Benzotriazole	0.3	Copper	69

*3-amino-5-heptyl-1,2,4-triazole.

The results demonstrate the very great corrosion inhibiting effect of AAT.

Example 2: A technical cooling system with a volume of 1.2 m^3 and a circulation of 8 m^3/hr was operated with Dusseldorf, Germany, city water. The evaporation losses were compensated by the addition of fresh water to such an extent that the salt content does not exceed twice the original value. The system contained a heat exchanger of brass. Without any anticorrosion treatment of the circulating water, the copper content in the system was 240 μg/ℓ.

After the addition of the corrosion inhibitor according to the process (3-amino-5-pentyl-1,2,4-triazole) in amounts of 0.5 g/m^3, based on the circulating water, the copper content was reduced to 40 μg/ℓ. This value must be considered excellent.

Aminoalkanols

The process of *V. Wehle, W. Rupilius, J. Reiffert and G. Rogall; U.S. Patent 4,299,725; November 10, 1981; assigned to Henkel Kommanditgesellschaft auf Aktien and Deutsche Gold- und Silber-Scheideanstalt vormals Roessler, both of Germany* relates to aqueous media of decreased corrosiveness towards steel and other corrodible metals. The process includes industrial water systems, coolant water systems, steam generating systems and heating systems of decreased corrosiveness, and includes methods for the preparation of such media.

The aqueous media has a very small content (in the range of 0.1 to 100 ppm by weight) of an aminoalkanol material selected from the group consisting of (1) mixtures of vicinal aminoalkanols having the formula shown below

$$
\begin{array}{l}
R_1-\underset{|}{C}H-\underset{|}{C}H-R_2 \qquad\qquad R_3 \\
\;\;\;\; OH \quad [NH-(CH_2)_x]_yN \diagup \diagdown R_4
\end{array}
$$

in which formula R_1 and R_2 each represent a substituent selected from the group consisting of H and unbranched alkyl having 1 to 18 carbon atoms and the sum of the carbon atoms in R_1 and R_2 is from 9 to 18 inclusive, R_3 and R_4 each represent a substituent selected from the group consisting of H, C_{1-4} alkyl and C_{2-4} hydroxyalkyl; and x represents a value from 2 to 6 and y represents a value from 0 to 1 inclusive; the

$$R_1-\underset{|}{C}H-\underset{|}{C}H-R_2$$

units in the aminoalkanols of the mixture being of at least two different chain lengths in the range from 11 to 20 carbon atoms, and (2) salts of the aminoalkanols.

A synergistic increase in the corrosion-inhibiting action of the aminoalkanol mixtures occurs when other corrosion-inhibiting polyvalent ions like zinc ions are present. Amounts in the range of 0.1 to 10 ppm give good results.

Beyond that it was found that sequestering phosphonic acids and/or their salts are of advantage. Effective phosphonic acids, for example, are hydroxyethanediphosphonic acid, aminotrimethylene-phosphonic acid and 2-phosphonobutane-1,2,4-tricarboxylic acid as well as mixtures thereof. Amounts of these agents in the range of 0.3 to 30 ppm give good results.

Best results are obtained when the aqueous medium contains both zinc ions and a phosphonic acid sequestering agent.

Testing Method: Aqueous solutions of aminoalkanol mixtures described above were tested for their corrosiveness to iron-containing metal as follows. In each instance a carefully cleaned steel test plate (75 x 12 x 1.5 mm) is immersed at room temperature for 24 hours in a 1-liter glass beaker filled with 1 liter of Düsseldorf (Germany) city water and the substance to be tested is added thereto. The tests are run in groups of 10, and are all stirred at 100 rpm. Subsequently the plates are cleared of corrosion products, and the weights lost by the plates are determined. The corrosion inhibiting action of the products are determined from the mean values of three tests each, as a percentage of the weight loss by the blank.

Aminoalkanol Designation	Solidification Point (°C)	Starting Olefin Mixture	Amine Used
N–T 58 A	16	C_{15-18}	Ammonia
N–T 58 AE	18	C_{15-18}	Ethylenediamine
N–T 58 AP	-28	C_{15-18}	Propylenediamine
N–T 58 AT	–	C_{15-18}	Tetramethylene-diamine
N–T 58 HE	-17	C_{15-18}	Monoethanolamine
N–T 58 DHE	-21	C_{15-18}	Diethanolamine
N–T 58 HE–AE	2	C_{15-18}	N-ethanolethylene-diamine

Example 1: The aminoalkanols were tested for their corrosion-inhibiting action by the method shown above. The aminoalkanols were added as 10% by weight aqueous solutions in each instance in amount sufficient to provide 10 ppm of the aminoalkanol. No other material was added. The results are compiled in the table in terms of percents of the corrosion of the blank.

Aminoalkanol Used	Corrosion, % of Blank
Blank	100
N–T 58 A	68
N–T 58 AE	41
N–T 58 AP	26
N–T 58 AT	54
N–T 58 HE	74
N–T 58 DHE	75
N–T 14 HE–AE	55

All aminoalkanols showed a good corrosion inhibiting action. A particularly good corrosion-inhibiting action, where the corrosion was only 26% of the blank, was displayed by product N–T 58 AP, which was prepared by reacting a mixture of epoxidized C_{15-18} monoolefins with propylenediamine.

Example 2: The synergistic increase in the protective action afforded by an aminoalkanol corrosion inhibitor by zinc ions and hydroxyethane diphosphonic acid (HEDP) in combination is shown by the following test data.

Composition of Inhibitor Solution Parts by Weight	Name	Dose, ppm	Corrosion, % of Blank
6	Zinc(II) ions	30	44
4	HEDP		
90	Water		
6	Zinc(II) ions	30	12
4	HEDP		
3	N–T 58 AP		
87	Water		

The table shows that the dissolved zinc ions synergistically assist or fortify the corrosion inhibitory effect of a mixture of aminoalkanols.

Synergistic Thickening of Hydraulic Fluid or Metalworking Lubricant

In the process of *A. Nassry, J.F. Maxwell, J.W. Compton, E.J. Panek and P. Davis; U.S. Patent 4,312,768; January 26, 1982; assigned to BASF Wyandotte Corporation* water-based hydraulic fluids and metalworking lubricants are disclosed which are thickened with a polyether polyol having a MW of about 1000 to 75,000 modified with an alpha-olefin epoxide having about 12 to 18 carbon atoms. Unexpectedly, synergistic thickening results from a combination of the polyether polyol with the components of a water-based hydraulic fluid or metalworking lubricant. The particularly effective components of the hydraulic fluid or metalworking lubricant are the phosphate ester and water-soluble amine corrosion inhibitor components. The hydraulic fluid and metalworking fluids of the process also contain a water-soluble polyoxyethylated ester of an aliphatic acid and a monohydric or polyhydric aliphatic alcohol, either one or both the acid and the alcohol being polyoxyethylated, a sulfurized molybdenum or antimony compound and a metal deactivator as well as other adjuvants conventional in this art.

Representative amine type corrosion inhibitors are morpholine, N-methylmorpholine, N-ethylmorpholine, ethylenediamine, dimethylaminopropylamine, dimethylethanolamine, alpha- and gamma-picoline, piperazine and isopropylaminoethanol.

Particularly preferred vapor phase corrosion inhibiting compounds are morpholine and isopropylaminoethanol. As corrosion inhibitors, a proportion of from about 0.05 to 2% by weight is used based upon the total weight of the hydraulic fluid or metalworking composition of the process. Preferably, about 0.5 to 2% by weight of these amines are used.

A typical high water-base hydraulic fluid or metalworking additive of the process will contain the components shown below.

Components	Parts by Weight
Water (distilled or deionized)	2.5-32.5
Polymeric thickener	80-50
Water-soluble ethoxylated ester	3-10
Molybdenum or antimony compound at 40% solids emulsion	1-5
Water-soluble alkyl phosphate ester	0.1-1.0
Metal deactivator	0.1-0.5
Corrosion inhibitor	0.5-1.0

The hydraulic fluid and metalworking compositions of the process, when formulated as above, are transparent liquids having a viscosity of up to 400 SUS at 100°F, which are stable over long periods of storage at ambient temperature. In addition, the hydraulic fluids and metalworking additives are oil-free and will not support combustion in contrast to those flame-resistant fluids of the prior art based upon a glycol and water or petroleum oils. The hydraulic fluids and metalworking additives are ecologically clean and nonpolluting compositions when compared to existing petroleum-based hydraulic fluids. Since the hydraulic fluids and metalworking additives of the process are largely based upon synthetic materials which are not derived from petroleum, the production of such fluids is relatively independent of shortages of petroleum oil and not materially influenced by the economic impact of such shortages.

The hydraulic fluids of the process can be used in various applications requiring hydraulic pressures in the range of 200 to 2,000 psi since they have all the essential properties required such as lubricity, viscosity and corrosion protection.

Tertiary Amino-Substituted Thiazines and Their Quaternary Salts

P.M. Quinlan; U.S. Patent 4,312,830; January 26, 1982; assigned to Petrolite Corporation provides a process of inhibiting corrosion which comprises treating a system with a tertiary amino-substituted thiazine. The thiazine is selected from the group consisting of:

```
          H  H                                  H  H
          |  |                                  |  |
          C--C                                  C--C
         /|  |\         R'                     /|  |\
        / H  H \        |                     / H  H \
(1)    Z        N--A--N          (2)         Z        N--A--N⊃
        \ H  H /        |                     \ H  H /
         \|  |/         R'                     \|  |/
          C--C                                  C--C
          |  |                                  |  |
          H  H                                  H  H
```

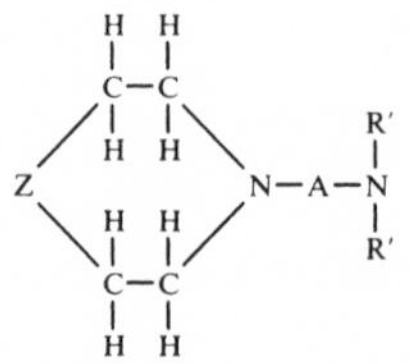

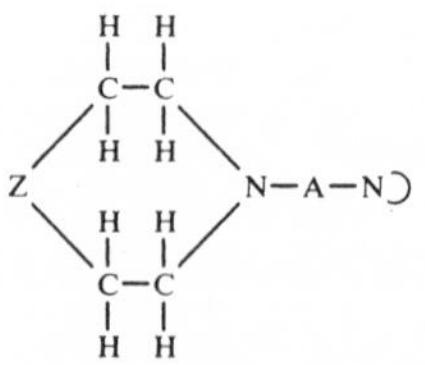

(3) $\left[Z\begin{matrix}CH_2-CH_2\\CH_2-CH_2\end{matrix}N-A- \right]_2 N-R'$

(4) $Z\begin{matrix}CH_2-CH_2\\CH_2-CH_2\end{matrix}N-A-N\frown N-A-N\begin{matrix}CH_2-CH_2\\CH_2-CH_2\end{matrix}Z$

where Z is SO or SO_2, A is alkylene having 2 to 10 carbon atoms, R' is alkyl or hydroxyalkyl, N) represents a cyclic amine group and

represents a cyclic diamine group.

The compounds are useful in the acidizing or treating of earth formations and wells traversed by a bore hole. It may also be used in metal cleaning and pickling baths which generally comprise aqueous solutions of inorganic acids. They are also useful in the prevention of corrosion in the secondary recovery of petroleum by water flooding and in the disposal of wastewater and brine from oil and gas wells and as a microbiocide.

The amount of composition employed as corrosion inhibitor or microbiocide can vary widely depending on many variables such as the particular composition employed, the particular system, the particular corrodant or microorganism, etc. In general, about 1 to 10,000 ppm or more can be employed.

Example 1:

$SO_2\begin{matrix}CH_2-CH_2\\CH_2-CH_2\end{matrix}N-CH_2CH_2CH_2-\overset{CH_3}{N}-CH_2CH_2CH_2-N\begin{matrix}CH_2-CH_2\\CH_2-CH_2\end{matrix}SO_2$

To 236.05 g (2 mols) of divinylsulfone dissolved in 350 ml of 2-propanol was slowly added with cooling 145.0 g (1 mol) of methyliminobispropylamine. After the addition had been completed, the reaction mixture was heated at reflux for 1 hour. The 2-propanol was removed on a rotary evaporator and the crude liquid product crystallized upon standing. It was recrystallized from ethanol. Analysis – Calculated for $C_{15}H_{31}N_3O_4S_2$: N, 9.07%; S, 16.80%. Found: N, 9.00%; S, 16.72%. The above structure was characterized by NMR and IR spectra.

Example 2:

$SO_2\begin{matrix}CH_2-CH_2\\CH_2-CH_2\end{matrix}N-CH_2CH_2CH_2N\begin{matrix}CH_2-CH_2\\CH_2-CH_2\end{matrix}N-CH_2CH_2CH_2N\begin{matrix}CH_2CH_2\\CH_2CH_2\end{matrix}SO_2$

To 100 g (0.5 mol) of N,N'-bis(3-aminopropyl)piperazine dissolved in 200 ml of 2-propanol was slowly added with cooling 118 g (1 mol) of divinylsulfone. After the addition had been completed, the reaction mixture was heated at reflux for 1 hour. The 2-propanol was removed using a rotary evaporator leaving a white powder. The solid was purified by washing with acetone and recrystallized from a mixture of benzene and petroleum ether. Analysis – Calculated for $C_{18}H_{36}N_4O_4S_2$: N, 12.84%; S, 14.68%. Found: N, 12.74%; S, 14.73%.

When tested in the corrosion test procedure of U.S. Patent 4,248,796 the compounds gave the following results.

Inhibitor	Concentration (ppm)	Test Temperature (°F)	Duration of Test (hours)	Corrodent	Coupon Metal Type	Corrosion Rate ($lb/ft^2/day$)
Ex. 1	2,000	150	6	15% HCl	N-80	0.0499
Ex. 2	2,000	150	6	15% HCl	N-80	0.0474
Blank	–	150	6	15% HCl	N-80	0.1956

In the static weight loss test of the same patent the following results were obtained.

Example	Parts per Million	% Protection
1	1,000	74.1
2	1,000	76.7

The process of *P.M. Quinlan; U.S. Patent 4,312,831; January 26, 1982; assigned to Petrolite Corporation* relates to quaternaries of tertiary amino-substituted thiazines; the method of preparing such quaternary thiazines; and uses thereof, for example, as corrosion inhibitors, microbiocides, etc.

The quaternaries have the formulas:

(1) 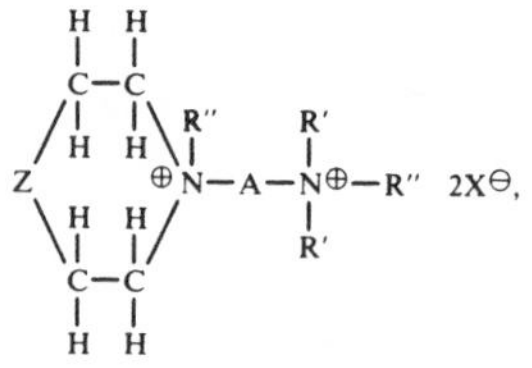 ... $2X^{\ominus}$,

(2) 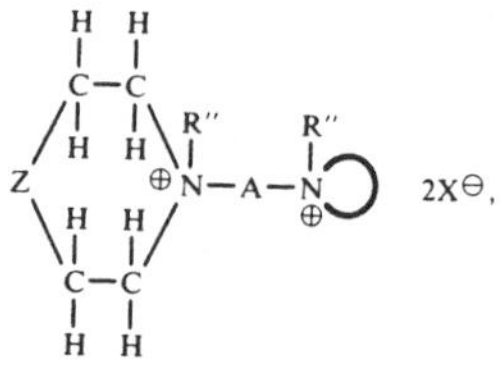 ... $2X^{\ominus}$,

(3) 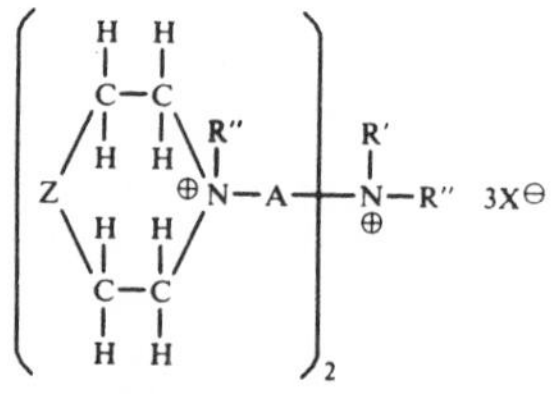 ... $3X^{\ominus}$

(4) 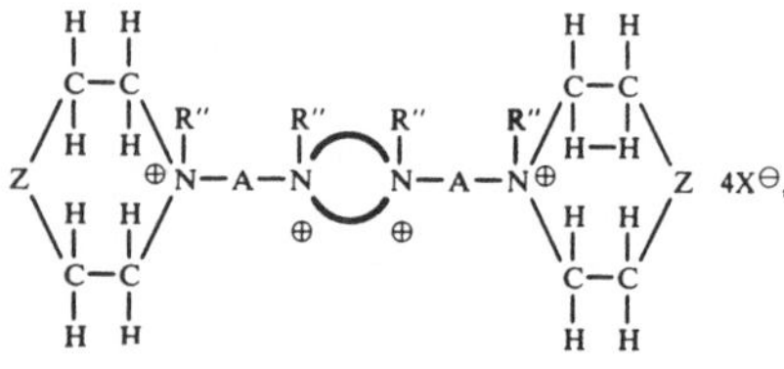 ... $4X^{\ominus}$,

where Z is SO or SO_2, A is alkylene having 2 to 10 carbon atoms, R' is alkyl or hydroxyalkyl, R" is alkyl, alkenyl or alkynyl,

N⌒

represents a cyclic amine group,

N⌒N

represents a cyclic diamine group and X is an anion selected from the group consisting of chloride, iodide, bromide, sulfate, sulfonate and carboxylate.

Example 1:

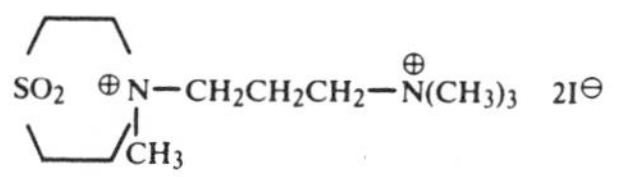

$$SO_2 \; {}^{\oplus}N(CH_3)-CH_2CH_2CH_2-\overset{\oplus}{N}(CH_3)_3 \; 2I^{\ominus}$$

A solution of 28.4 g (0.2 mol) methyl iodide and 22.0 g (0.1 mol) of

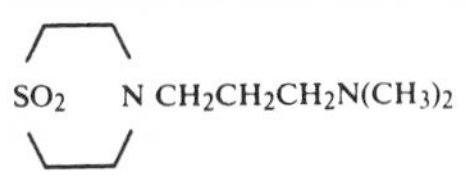

$$SO_2 \; N\,CH_2CH_2CH_2N(CH_3)_2$$

in 40 g of methanol was stirred and warmed at 40°C. After about 2 hours a solid precipitated from solution. The mixture was heated for an additional 2 hours. It was then cooled in an ice bath and the solid product was filtered and washed several times with cold methanol. It was recrystallized from ethanol. The yield was 49.5 g (98.2%). Analysis – Calculated for $C_{11}H_{26}N_2I_2SO_2$: I, 50.4%; N, 5.55%; S, 6.35%. Found: I, 49.8%; N, 5.52%; S, 6.46%.

Example 2:

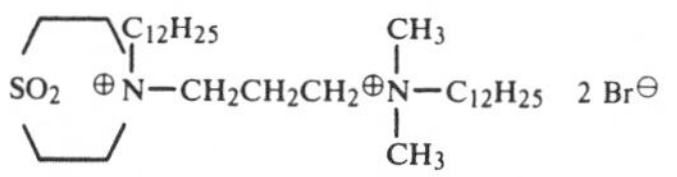

$$SO_2 \; {}^{\oplus}N(C_{12}H_{25})-CH_2CH_2CH_2{}^{\oplus}N(CH_3)_2-C_{12}H_{25} \; 2\,Br^{\ominus}$$

A solution of 25 g (0.1 mol) 1-bromododecane and 11.0 g (0.05 mol) of the thiazine used in Example 1 in 40 ml of ethanol was heated with stirring to reflux and held there for 18 hours. The solid product 34.7 g (96.5%) was recovered by adding the reaction mixture to 600 ml of ice water. The solid was filtered, washed several times with cold ethanol, and recrystallized from an ethanol-water mixture. Analysis – Calculated for $C_{33}H_{70}Br_2N_2O_2S$: Br, 22.26%. Found: Br, 22.17%.

Example 3:

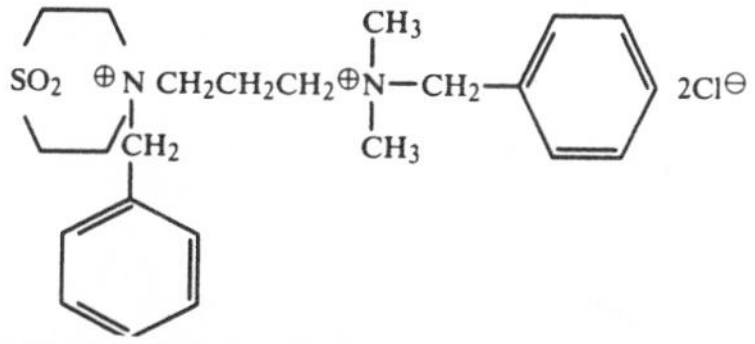

$$SO_2 \; {}^{\oplus}N(CH_2C_6H_5)\,CH_2CH_2CH_2{}^{\oplus}N(CH_3)_2-CH_2-C_6H_5 \; 2Cl^{\ominus}$$

A solution of 24.4 g (0.2 mol) benzyl chloride and 22.0 g (0.1 mol) of the thiazine used in Example 1 in 50 ml of 2-propanol was stirred and heated at reflux for 4 hours. The 2-propanol was removed on a rotary evaporator. The viscous liquid, 46.8 g (99%), was washed several times with acetone and dried. It crystallized upon standing. It was recrystallized from ethanol. Analysis – Calculated for $C_{23}H_{34}Cl_2N_2O_2S$: Cl, 15.01%. Found: Cl, 14.98%.

In the corrosion test procedure of U.S. Patent 4,248,796 at a test temperature of 150°F for 6 hours in 15% HCl as corrodent employing a coupon of metal type N-80, the results shown below were obtained.

Inhibitor (conc. 2,000 ppm)	Corrosion Rate ($lb/ft^2/day$)
Ex. 2	0.0135
Ex. 3	0.0120
Blank	0.1872

In the static weight loss test of the same patent using 1,000 ppm of compound, results were as shown on the following page.

Example	% Protection
1	87.2
2	95.4
3	90.3

Polymerized Aromatic Nitrogen Heterocyclics

P.M. Quinlan; U.S. Patent 4,312,832; January 26, 1982; assigned to Petrolite Corporation describes a process of inhibiting corrosion of metals which comprises treating a system with the composition obtained by polymerizing aromatic nitrogen heterocyclic compounds. The compounds are selected from the group consisting of quinoline, acridine, benzoquinoline, 2-methyl quinoline, pyridine, 2-methyl pyridine, 4-methyl pyridine and 2-butyl pyridine. They are prepared by heating the compounds with alkyl halides at temperatures of from about 250° to 400°C in the pressure reaction vessel for about 8 to 30 hours.

The process is especially useful in the acidizing or treating of earth formations and wells traversed by a bore hole. It may also be used in metal cleaning and pickling baths.

Example 1: Quinoline (pure) was dried over KOH pellets and distilled in vacuo. The fraction with a BP of 112° to 113°C/14 mm was retained. The CH_3I (AR grade) was dried with anhydrous calcium chloride and distilled. The fraction with a boiling point of 42°C was retained.

129 g (1.0 mol) of quinoline and 5.7 g (0.04 mol) of methyl iodide were placed in a small pressure reaction vessel and heated in a bath to 229°±2°C. It was held there for 8 hours. A change in color from red to dark brown to black of the reaction mass was noted during the heating period. An increase in viscosity was also noted. The isolated product was a black, brittle, amorphous solid. Analysis – Theoretical: C, 83.72%; N, 10.85%; H, 5.42%. Found: C, 82.74%; N, 10.62%; H, 5.20%.

It had an average molecular weight, as determined by an osmometer in dimethylformamide, of 850. From its IR spectra, the product was presumed to have the following structures:

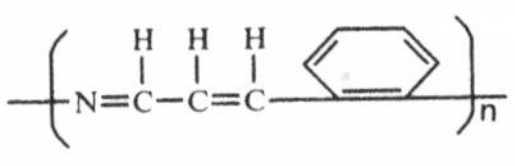

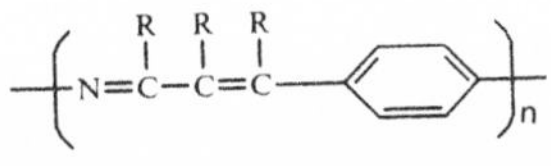

$$\left[N{=}\overset{H}{\overset{|}{C}}{-}\overset{H}{\overset{|}{C}}{=}\overset{H}{\overset{|}{C}}{-}C_6H_4 \right]_n \qquad \left[N{=}\overset{R}{\overset{|}{C}}{-}\overset{R}{\overset{|}{C}}{=}\overset{R}{\overset{|}{C}}{-}C_6H_4{-} \right]_n$$

Example 2: Pyridine (pure) was dried over barium oxide and distilled in a current of N_2 over KOH pellets. A fraction was collected with a boiling point of 115°C. The methyl iodide (AR grade) was dried with anhydrous calcium chloride and distilled. The fraction with a boiling point of 42°C was retained.

Pyridine, 79 g (1.0 mol) and methyl iodide, 7.1 g (0.05 mol) were placed in a small pressure reaction vessel and heated in a bath, at 350°±3°C for 24 hours. The recovered product was a dark brittle solid. Analysis – Theoretical: C, 75.9%; H, 6.3%; N, 17.7%. Found: C, 74.8%; H, 5.8%; N, 17.0%.

It had an average molecular weight, as determined by an osmometer in dimethylformamide, of 500. The product, from its IR spectra, was presumed to have the structure shown on the following page.

$$\left[\overset{H}{\underset{|}{C}}\!=\!\overset{H}{\underset{|}{C}}-\overset{H}{\underset{|}{C}}\!=\!\overset{H}{\underset{|}{C}}-\overset{H}{\underset{|}{C}}\!=\!N \right]_n$$

Using 1020 steel (AISI) coupons in the corrosion test procedure of U.S. Patent 4,248,796 the results were as follows:

Composition	Concentration (ppm)	Test Temperature (°F)	Test Time (hours)	Acid	Corrosion Rate ($lb/ft^2/day$)
Ex. 1	2,000	120	4	15% HCl	0.072
Ex. 2	2,000	120	4	15% HCl	0.083
Blank	–	120	4	15% HCl	1.243

Metal Soap Compositions

The process of *H. Suzuki; U.S. Patent 4,324,797; April 13, 1982; assigned to the Agency of Industrial Science and Technology, Japan* relates to metal soap compositions comprising essentially a metal soap and a chelating agent, which can easily be made up into a transparent aqueous solution.

The compositions comprise essentially a metal soap expressed by the general formula $(RCOO)_xM$ (wherein M represents lithium or a nonalkali metal atom, x represents its valency and R represents a hydrocarbon radical having 4 to 20 carbon atoms) and a chelating agent such as polybasic carboxylic acids, polyaminocarboxylic acid salts, etc., the metal soap being contained therein in an amount in the range of 10 to 1/1,000 part by weight based on one part of the chelating agent. The aqueous solutions of the metal soap compositions exhibit such a behavior as if they were a single surfactant, and also exhibit excellent surface active properties which are similar or superior to those of conventional surfactants. Further, addition of conventional surfactants to the compositions can notably enhance the performances of the chelating agent such as chelate effect and metal soap-solubilizing effect.

Such aqueous solution exhibits not only general surface active properties such as wetting power, emulsifying power, dispersing power, etc., but also particular effect such as antimicrobial action, rust-preventive power, etc.

The test for rust-preventive power is as follows: A sample aqueous solution (about 4 ml) is introduced into a test tube equipped with a stopper, a test metal piece (a wire of about 20 mm or a metal plate of 4 x 20 x 0.8-1.6 mm) is immersed in the solution; and after a lapse of 24 hours on a hot bath at 90°C, the change of the metal piece is observed. Evaluation standards are as follows:

Evaluation	Surface Condition
5	No change
4	Luster is slightly reduced; rust formed at only a small part; solution is slightly turbid
3	Changes described above in evaluation 4 are somewhat enlarged
2	Considerably changed; rust is formed on about half of the surface area
1	Greatly changed; total surface is coated by rust; solution is notably discolored or precipitate is formed

Metal pieces employed for the above corrosion test are shown below.

Aluminum — High-strength aluminum alloy
Steel (A) — Mill shape for general structure
Steel (B) — Cold-rolled steel
Copper
Brass
Phosphor bronze — Wire
Zinc plate
Tin plate
Solder

Examples 1 through 9: With aqueous solution of six kinds of compositions consisting of metal soaps and chelating agents shown in Table 1, rust-preventing properties were measured. At the same time, three representative surfactants were compared. The name of sample, blending ratio, etc., of the respective examples are shown in Table 1, and the measured results are shown in Table 2.

Table 1

Example	Metal Soap or Surfactant	Chelating Agent	Amount
1	Mg salt castor oil fatty acids	Na citrate	0.5:0.8
2	Mg salt castor oil fatty acids	Na tartrate	0.5:6.5
3	Trialuminum octoate (2-ethylhexoate)	Na pyrophosphate	0.5:1.25
4	Co laurate	Na ethylenediamine-tetraacetate	0.2:2.0
5	Li laurate	Trisodium imidobis-sulfate	0.2:1.6
6	Trialuminum octoate (2-ethylhexoate)	Trisodium imidobis-sulfate	0.1:3.0
7*	Na dodecylsulfate	—	0.5
8*	Na straight-chain alkylbenzene sulfonate	—	0.5
9*	Polyoxyethylene (20 mols) nonylphenyl ether	—	0.5

Note: Example 6 was slightly opaque.
*Comparative examples.

Table 2

	Examples								
	1	2	3	4	5	6	7	8	9
	(Rust Preventive Power)								
Aluminum	2	3	3	2	5	1	3	3	4
Steel (A)	5	4	5	5	5	1	4	3	1
Steel (B)	3	3	5	5	5	2	3	3	2
Brass	4	3	5	4	4	2	3	5	4
Copper	3	3	5	5	3	2	3	4	5
	(3.4)	(3.3)	(4.2)	(3.3)	(4.1)	(1.6)	(3.3)	(3.3)	(3.3)
Phosphor bronze	3	5	4	5	5	1	5	4	4
Zinc plate	3	3	2	3	3	2	3	2	2
Tin plate	4	3	4	4	4	2	3	3	4
Solder	4	3	5	3	3	1	3	3	4

Stabilization of Silicates

Aqueous silicates are known as metal corrosion inhibitors for aqueous systems. One of the major disadvantages of such silicates, however, has been the fact that they are unstable and after prolonged use at elevated temperatures they tend to gel and eventually precipitate out of solution. There have been many efforts, therefore, to stabilize silicates so that they could be more persistent in their corrosion inhibiting properties.

E.P. Plueddemann; U.S. Patent 4,344,860; August 17, 1982; assigned to Dow Corning Corporation has found that salts of substituted nitrogen or sulfur containing siliconates are effective stabilizers for aqueous silicates in such applications as treating boiler water, geothermal water and other aqueous silicates. They are also useful in antifreeze and coolant solutions.

Example 1: A mixture of 26 grams of methyl acrylate (0.3 mol) and 36.4 grams of $(CH_3O)_3SiCH_2CH_2SH$ was catalyzed by adding 1 ml of N/2 alcoholic KOH into a 250 ml, round bottomed glass flask, with stirring. An exothermic reaction raised the temperature to 60°C. The mixture was refluxed for 30 minutes and then distilled under vacuum to recover 46 grams of water-white product with a boiling point of 115° to 125°C at 0.7 mm Hg pressure for an 87% yield of $(CH_3O)_3Si(CH_2)_2S(CH_2)_2COOCH_3$; d_4^{20} = 1.115, N_D^{25} = 1.4546. One-tenth gram mol (26.8 g) of this product was saponified by refluxing for one hour with 4 grams of NaOH (0.1 mol) in 80 grams of H_2O. Methanol and other volatiles were taken off until the temperature, with auxiliary heating, reached 100°C. The residue was diluted to 100 grams with water to give a 1 molal solution of $(Na)O_{1.5}Si(CH_2)_2S(CH_2)_2COONa$.

Example 2: 24 grams (0.15 mol) of dimethyl itaconate, 18.2 grams (0.1 mol) of $(CH_3O)_3Si(CH_2)_2SH$ and 0.23 gram Na° in 10 ml of methanol were warmed to 100°C for one hour after initially exotherming. The mixture was distilled under vacuum to give 21 grams (65% yield) of

$$(CH_3O)_3Si(CH_2)_2SCH_2\underset{\displaystyle CH_2COOCH_3}{\underset{|}{C}}HCOOCH_3$$

with a boiling point of 145° to 155°C at 0.3 mm Hg pressure; d_4^{21} = 1.55 and N_D^{25} = 1.4572.

The product, 17 grams (0.05 mol) was saponified by refluxing with 4 grams of NaOH in 50 ml of H_2O until a temperature of 100°C was reached. The material was diluted to 50 grams with H_2O to obtain 1 molal solution of

$$(Na)O_{1.5}Si(CH_2)_2SCH_2\underset{\displaystyle CH_2COONa}{\underset{|}{C}}HCOONa.$$

A titration indicated 0.02 equivalent of excess alkali.

Example 3: To a solution of 22 g (0.1 mol) of $(CH_3O)_3Si(CH_2)_3NH(CH_2)_2NH_2$ in 100 ml of water was added 28.5 g (0.3 mol) of $ClCH_2COOH$. To the stirred mixture was added 48 g (0.6 mol) of 50% aqueous NaOH with cooling to keep the temperature below 50°C. The course of the reaction was followed by titrating 5 ml of product (0.5 molal) against 0.5 molal $CaCl_2$ with ammonium oxalate indicator to a cloudy end point.

Time of Reaction	$CaCl_2$ (ml)	Equivalent Ca^{++} Product
After mixing	1	0.2
6 hours at 50°C	5	1.0

Chelation of 1 mol of Ca^{++} per mol of product indicated complete reaction to the compound

$$O_{1.5}Si(CH_2)_3N(CH_2COONa)CH_2CH_2N(CH_2COONa)_2$$

Example 4: To a solution of 144 g of acrylic acid in 364 g of water was added 200 g of technical grade $(CH_3O)_3Si(CH_2)_3NH(CH_2)_2NH_2$ and the mixture was refluxed for six hours. The course of the reaction was followed by thin layer chromatography on Adsorbasil-1 silica. A spot of 1% aqueous product was eluted with methanol and sprayed with bromcresol-purple indicator. The initial mix showed an immobile alkaline spot (purple in color) of unreacted amine and an eluted band of the acid (yellow color). After 6 hours reflux there was only an immobile acid spot indicating that the acid was not part of the silane. The product was a 1.37 molal solution of

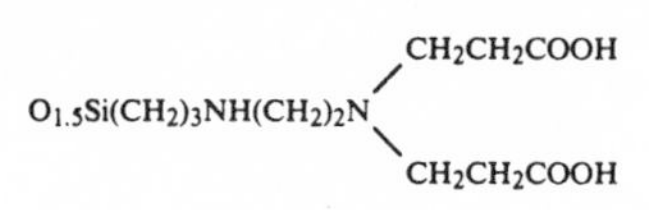

containing some

$$O_{1.5}Si(CH_2)_3N(CH_2CH_2COOH)(CH_2)_2NHCH_2CH_2COOH$$

having a density of 1.125 g/mol at 20°C, and a viscosity of 50 cs (0.5 Pa-s) at 25°C. A 73 g portion was diluted to 100 g with water to give a 1 molal solution of the acid adduct. 1 molal solutions of the acid were mixed with excess sodium silicate solutions in order to form the alkali metal salt in-situ. After aging one hour, the solutions were neutralized to pH 8.

Low-Foaming Compositions

It is an object of the process of *R. Widder, E. Getto and A. Hettche; U.S. Patent 4,344,862; August 17, 1982; assigned to BASF AG, Germany* to provide products which afford excellent corrosion protection, coupled with low foaming, in processes where iron, iron alloys, aluminum or aluminum alloys, or copper, zinc or their alloys, come into contact with water and especially with hard and saline water. This object is achieved with products which are obtained by reaction of an acid of the formula

$$R^1R^2{-}A{-}X{-}NR^3{-}C_6H_3(R^4)(COOH)$$

where R^1 and R^2 are hydrogen, fluorine, chlorine, bromine or alkyl or alkoxy of 1 to 4 carbon atoms, but the sum of the carbon atoms of R^1 and R^2 is not greater than 7 and preferably not greater than 3, A is a benzene, naphthalene or anthracene radical or a biphenyl structure, such as a diphenyl, diphenylmethane, diphenyl ether, diphenyl sulfide, diphenyl sulfoxide or diphenyl sulfone radical, R^3 is hydrogen, alkyl of not more than 4 carbon atoms, β-cyanoethyl or hydroxyalkyl of 2 to 4 carbon atoms, R^4 is hydrogen, fluorine, chlorine, bromine, alkyl or alkoxy of 1 to 4 carbon atoms, hydroxyl or carboxyl and X is $-SO_2-$ or $-CO-$, with an amine of the formula

$$N(R^5)(R^6)-(CH_2)_n-OH$$

where R^5 and R^6 are hydrogen, C_{1-4} alkyl, C_{2-4} hydroxyalkyl and/or

$$(CH_2)_n-N(R^8)-R^7,$$

R^7 and R^8 being hydrogen, C_{1-4} alkyl or C_{2-4} hydroxyalkyl and n being an integer from 2 to 4, in the ratio, expressed in terms of equivalents, of from 1:1 to 1:4.

This object is also achieved with mixtures of the above products with up to 85% by weight, based on the products, of one or more reaction products of orthoboric acid with one or more alkanolamines of 2 to 4 carbon atoms per alkanol group, the orthoboric acid and alkanolamine having been reacted in a molar ratio of from 1:1 to 1:4.

The compounds shown in the table were obtained in a conventional manner by reacting the particular acid chloride with the appropriate aminocarboxylic acid and then reacting the product with triethanolamine.

The corrosion test (cast iron/filter paper test) was carried out as follows: A Schwarzband circular filter paper is placed in a Petri dish of about 10 cm internal diameter, having a fitting cover dish; 5 to 10 g of coarse GG-20 cast iron filings are placed on the filter paper, by means of a suitable spoon, so as to create a uniform heap in the center, the heap being about 1.5 cm clear of the edge on all sides. The filings are about 5 to 8 mm long and must be produced from clean GG-20 cast iron without using any drilling oil or other cooling lubricant. All fines must be screened out.

5 ml of the solution or emulsion to be tested for corrosiveness are applied uniformly to the filings by means of a measuring pipette. The pH of the liquid being tested is recorded, since it is of substantial importance in assessing the results; it can, if desired, be adjusted to a particular standard value, for example, 8.5.

After the filings have been moistened, the cover is placed on the dish and the sample is left to stand for 2 hours under normal laboratory conditions at 23° to 25°C and about 70% relative atmospheric humidity. The cover is then removed and the filter paper is briefly inverted and floated on the surface of tapwater, which serves to remove the filings. Immediately thereafter, the filter paper, freed from the filings, is sprayed, and thus impregnated, with an indicator solution containing 1 g of potassium hexacyanoferrate(III) and 30 g of sodium chloride in 1 liter of water. The indicator is then allowed to act for 17 seconds, with the filter paper exposed to the air. Finally, the paper is thoroughly rinsed under running tapwater and is dried, exposed to air, in a moderately warm location. Depending on the corrosiveness of the medium, this procedure results in brownish yellow, yellow and/or bluish green stains, of varying intensity, on the filter paper.

The following is a suitable assessment scale:

Very poor (- -) – Intense, large, predominantly yellowish brown stains
Poor (-) – Intense, large stains with about equal proportions of yellowish brown and bluish green
Medium (+ -) – Pale medium-sized stains with about equal proportions of yellow and bluish green
Good (+) – Very pale, small (pinhead-sized) stains with a predominant proportion of bluish green
Very good (++) – No stains or at most very few, very small pale bluish green stains

A foam-whisking method, based on DIN 53,902, was used. A simple test procedure, in which the ram bearing the perforated plate is moved manually up and down 30 times in 30 seconds and is then carefully withdrawn (the IG whisking method) proved adequate. The foam volume is read off in ml, on the graduated foam-whisking cylinder, after 1, 5 and 10 minutes. The other important data are the temperature, concentration of corrosion inhibitor and water hardness. The results are shown in the following table in which the compounds tested in Examples 1 through 4 are:

Example 1 – Triethanolamine salt of N-phenylsulfonylanthranilic acid
Example 2 – Triethanolamine salt of N-(o-tolylsulfonyl)anthranilic acid
Example 3 – Triethanolamine salt of N-(xylylsulfonyl)anthranilic acid
Example 4 – Triethanolamine salt of N-benzoylanthranilic acid

Triethanolamine salt, in each case of 90% strength unless stated otherwise	Cast iron/filter paper test: water with an artificial 20° German hardness (DIN 51,360) 2%	Cast iron/filter paper test: tapwater, 23° German hardness 2%	Cast iron/filter paper test: tapwater, 23° German hardness 3%	Foaming (IG whisking method, 3% of inhibitor in distilled water, room temperature), in ml of foam after 10 min	pH 1% of inhibitor in water
Example 1	++	++	++	0	8.0
Example 2	++ to +	++	++	30	8.1
Example 3	++ to +	++	++	0	8.1
Example 4	+	+	++ to +	0	8.1
Comparative Example 1	+	+	++ to +	1,150	8.3

Agent Stable to Hard Water

In the field of metal processing and metal surface treatment, and in cooling cycles it is normal to use more or less strongly alkaline aqueous solutions containing corrosion-inhibiting additives for ferrous and nonferrous metals in order to prevent undesirable corrosion. This is valid for example for such widespread processes as cutting and noncutting metal shaping, cleaning of metal surfaces, or inner protection of streaming aqueous systems.

Inorganic salts such as sodium nitrite or chromates, for example, are well known and widespread corrosion-inhibiting additives; toxicological and ecological reasons, however, forbid their further use in these fields.

H. Fröhlich and R. Helwerth; U.S. Patent 4,348,302; September 7, 1982; assigned to Hoechst AG, Germany provide an anticorrosive agent stable to hard water. The agent comprises an alkali metal salt, alkaline earth metal salt, or amine salt of a compound of the formula shown below.

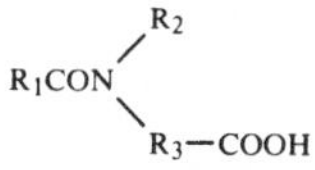

$$R_1CON\begin{matrix} \diagup R_2 \\ \diagdown R_3\text{—COOH} \end{matrix}$$

In the formula R_1 is branched C_{6-13} alkyl or C_{5-6} cycloalkyl, or polycycloalkyl having from 6 to 13 carbon atoms optionally substituted by 1 or 2 C_{1-4} alkyl groups, R_2 is hydrogen or C_{1-6} alkyl, and R_3 is C_{1-11} alkylene in linear or branched chain.

The salts of the above carboxylic acids have an excellent anticorrosive action with respect to iron, and they have an extremely low tendency to foaming, which is very important for practical application. They are furthermore substantially insensitive to the hardness-forming substances of water, and even under extreme electrolyte strain conditions, they leave deposits after drying which are of low viscosity and of oily consistency, so that they are not tacky and can be easily dissolved either with the service solution or with fresh water.

Example 1: *2-Ethylhexanoyl-ε-aminocaproic acid* — 113 g (1.0 mol) of ε-caprolactam are dissolved in 200 ml of water, and refluxed for 4 hours with 120 g (1.0 mol) of 33% sodium hydroxide solution. The batch is cooled to 20°C and 158.4 g (0.975 mol) of 2-ethylhexanoic acid chloride as well as simultaneously about 120 g of 33% sodium hydroxide solution (for maintaining a pH of 12) are added dropwise within 1 hour at 20° to 25°C. The solution is further stirred until no sodium hydroxide solution is consumed any longer, and subsequently acidified at 50°C with semiconcentrated hydrochloric acid in order to obtain a pH of 1. Separation is carried out in warm state, and the acid is washed with 350 ml of water. Subsequently, it is dehydrated in a rotation evaporator at 75°C/100 mm and separated as nearly colorless viscous oil, which solidifies to crystals after some time. Yield 233 g (93%). Acid number 225, water content 0.4%.

Example 2: *Mixture of 2-ethylhexanoyl-ε-aminocaproic acid and isononanyl-ε-aminocaproic acid* — 113 g (1 mol) of ε-caprolactam are hydrolyzed as described in Example 1 and subsequently reacted according to Example 1 with a mixture of 79.6 g (0.49 mol) of 2-ethylhexanoic acid chloride and 86.5 g (0.49 mol) of isononanoic acid chloride which can be prepared separately or from an equimolar mixture of 3-ethylhexanoic acid and isononanoic acid, in known manner. 238.6 g (90.4%) of a nearly colorless oil, acid number 216, are obtained.

Example 3: *Tricyclodecanoyl-ε-aminocaproic acid* — According to Example 1, 113 g (1 mol) of ε-caprolactam are hydrolyzed and reacted with 181.7 g (0.91 mol) of tricyclodecanoic acid chloride. The above acid is obtained as yellow, highly viscous oil with a yield of 236.1 g (88.5%), having an acid number of 194.

Example 4: *Isononanoyl-ε-aminocaproic acid* — According to Example 1, 113 g (1 mol) of ε-caprolactam are hydrolyzed and reacted with 172 g (0.975 mol) of isononanoic acid chloride. Workup yields 257.5 g (95%) of a nearly colorless viscous oil which solidifies to crystals after some time. Acid number 210.

For preparing an aqueous anticorrosive, 35 g each of the acids of Examples 1 through 4 were mixed with 50 g of triethanolamine and 15 g of water to give a clear, homogenous solution.

–3–

Aqueous/Nonaqueous System Applications

PHOSPHORUS ACID DERIVATIVES

Benzimidazolyl-2-Alkane-Phosphonic Acids

A. Botta, H.-J. Rother and G. Teichmann; U.S. Patent 4,278,791; July 14, 1981; assigned to Bayer AG, Germany provide new benzimidazolyl-2-alkane-phosphonic acids of the formula:

$$\text{(benzimidazole ring bearing } R^1, R^2 \text{ on the benzene ring and } R^3 \text{ on N)}-A-\overset{O}{\overset{\|}{P}}(OH)_2$$

where R^1 and R^2 are identical or different and denote hydrogen, lower alkyl, phenyl, halogen, trifluoromethyl, nitro or lower alkoxy, or together form a fused-on benzene ring, R^3 denotes hydrogen, lower alkyl or phenyl or benzyl which is optionally substituted by lower alkyl or halogen and A denotes a straight-chain or branched, saturated or unsaturated bivalent hydrocarbon radical with 1 to 15 carbon atoms, which can be substituted by phenyl which is optionally substituted by lower alkyl or halogen or by carboxyl or the phosphonic acid group, and their salts with inorganic or organic bases and acids.

The benzimidazolyl-2-alkane-phosphonic acids and their salts have a pronounced corrosion-inhibiting effect and can be used as corrosion inhibitors. For example, they can be employed as corrosion inhibitors in the heat transfer media of cooling cycles or heating cycles, cooling lubricants, motor oils or as pickling inhibitors. Corrosion of metals, especially of copper and alloys thereof, steel, cast iron, solder, aluminum and aluminum alloys, is prevented by adding the substances of the above formula according to the process and/or their salts to the media or circulating fluids mentioned.

Example 1: 534 g (2.22 mols) of triethyl 3-phosphonopropionate are allowed to run into a boiling solution of 259 g (2.4 mols) of o-phenylenediamine in 1,000 g of H_2O and 1,000 g of concentrated hydrochloric acid in the course of 30 min, while passing nitrogen over the mixture, and while stirring. The mixture is kept at the reflux temperature for a further 15 to 20 hr. After cooling the mixture, the solid is redissolved by carefully adding (cooling) 1,200 g of 45% strength sodium hydroxide solution and about 500 to 1,000 ml of H_2O. The brown solution is extracted by shaking several times with methylene chloride until the methylene chloride phase remains colorless.

The aqueous phase is then heated to the reflux temperature for 30 min using a large amount of active charcoal, and, after filtration, the filtrate is adjusted to pH 4.5 with hydrochloric acid. Light-colored to brownish crystals are obtained. The crystals are filtered off, suspended several times in warm water, filtered off and washed with water until free from chloride. On concentrating the mother liquors and wash waters in vacuo, further material is obtained, which is purified in the same manner. After drying the product in vacuo over P_2O_5, 405 g (80.4% of theory) of 2-[benzimidazol-2-yl]-ethanephosphonic acid are obtained as a light beige to colorless powder which does not melt up to a temperature of 300°C.

Examples 2 and 3: The following compounds are obtained according to the procedure of Example 1, using diethyl dimethylphosphonosuccinate:

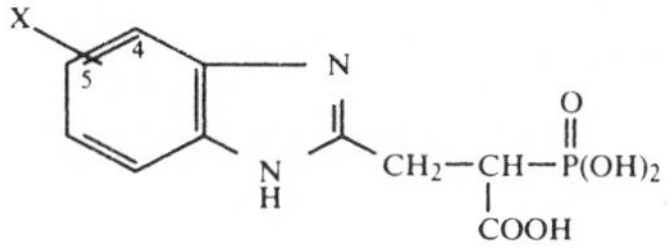

X, 4, 5, N, NH
$CH_2-CH(COOH)-P(O)(OH)_2$

Example	Starting Substances	X	Yield, %
2	o-Phenylenediamine	H	73.4
3	3,4-Diaminotoluene	5-CH_3	61.7

Example 4: *(Corrosion inhibitor test)* – 65 x 23 x 2 mm pickled and degreased copper sheets were used as test pieces; the test solution used was synthetic seawater according to ASTM D 665-IP 135, to which the particular corrosion inhibitor to be tested was added. Throughout the experimental period of 7.5 hr, the test pieces were immersed completely in the test solution, which was warmed to 55°C and into which about 100 ml of air were passed/min. After this test, the test pieces were cleaned for 15 sec in half-concentrated hydrochloric acid and washed with water and acetone. The dry test pieces were weighed before and after the experiment. The losses in weight thus obtained/m^2 of surface and the appearance of the test pieces and the test solution after the experiments are shown below.

Corrosion Inhibitor	Concentration Used, ppm	Loss in Weight Per Unit Area, g/m^2	Appearance . . . . After the Test	
			Test Pieces	Test Solution
None	–	3.25	highly corroded	severe precipitation of copper salts
1-Carboxy-2-[5-methylbenzimidazol-2-yl]-ethanephosphonic acid (Example 3)	50	0.66	slight dark tarnish color	clear

(continued)

Corrosion Inhibitor	Concentration Used, ppm	Loss in Weight Per Unit Area, g/m^2	Appearance .... After the Test .... Test Pieces	Test Solution
2-[Benzimidazol-2-yl]-1-carboxy-ethane-phosphonic acid (Example 2)	50	0.93	slight dark tarnish color	trace of precipitation of copper salts

Derivatives of Polyphosphoric Acid Partial Esters

T.P. Brady and H.G. Langer; U.S. Patent 4,301,025; November 17, 1981; assigned to The Dow Chemical Company provide derivatives of polyphosphoric acid partial esters of the formula:

$$M_z^{+n}H_y\left[\left(O-\overset{\overset{O}{\|}}{\underset{\underset{|}{O}}{P}}-O\right)\left(\overset{\overset{O}{\|}}{\underset{\underset{|}{O}}{P}}-O\right)_m\overset{\overset{O}{\|}}{\underset{\underset{|}{O}}{P}}-O\right)-R_q\right]_x$$

where R is each occurrence a remnant formed by removal of a hydroxyl from a monohydroxyl compound selected from:

(a) a (poly)alkylene glycol, monoether of the formula:

$$HO\left(CH_2-\underset{\underset{R_1}{|}}{CH}-O\right)_n R_2$$

where R_1 is hydrogen, methyl, or halomethyl, R_2 is C_{1-6} alkyl or haloalkyl, phenyl, halophenyl or methylphenyl; and n is an integer of from 1 to 4;

(b) a phenol or halophenol; or

(c) a C_{1-20} aliphatic or halogenated aliphatic compound;

provided that in at least one occurrence R is a remnant of (a); M is independently each occurrence an ammonium, substituted ammonium, or metal cation having valence n; m is an integer from zero to three; y is an integer equal to or greater than zero; and q, x and z are all integers greater than or equal to one selected such that $(z \cdot n) + y = x(m + 4 - q)$ and $q \leqslant m + 3$.

The compounds have been found to be useful corrosion inhibitors for use in functional fluids such as mechanical pressure transmission fluids, heat transfer fluids, metal cutting fluids and the like. More particularly, the compounds of the process have been found to effectively inhibit the corrosion of aluminum and iron and iron-containing metals such as steel or cast iron in the presence of functional fluids comprising water. It is also desirable to add the compounds to fluids which initially contain no water but are subject to possible water contamination during use.

Advantageously the compounds are combined with the remaining components of the functional fluid in minor proportions from about 0.1 to 10% by wt.

Example 1: To a reaction flask containing 500 ml CH_2Cl_2 under nitrogen atmosphere, phosphorus pentoxide (270 g, 1.9 mols) was added with stirring. Over approximately 2 hr, 2-n-butoxyethanol (425 g, 3.6 mols reagent grade) was added from a dropping funnel, causing a gentle reflux to occur. After complete addition only a small amount of unreacted P_2O_5 remained. The flask contained a clear yellow colored solution. Reaction for an additional 24 hr resulted in complete conversion of P_2O_5 and a darker colored solution. Analyses by ^{31}P nuclear magnetic resonance spectroscopy indicated the product comprised greater than 90% of the diphosphoric acid half ester with minor amounts of other partial esters of polyphosphoric acids plus monophosphates and full ester contaminants.

Example 2: A portion of the solution produced in Example 1 was added to a glass reaction vessel. Aluminum turnings were added a small amount at a time. Initial reaction was induced by adding a trace of water and heating to a temperature of about 148°C. After initiation of the reaction, more aluminum turnings were added and reacted until further hydrogen evolution ceased.

The viscous liquid reaction solution was decanted. Evaporation of the CH_2Cl_2 solvent left a white crystalline solid identified as the desired product by nuclear magnetic resonance spectroscopy.

Example 3: The reaction conditions of Example 1 were repeated except that the glycol ether utilized was 1-methoxy-2-propanol added to P_2O_5 in a molar ratio of about 1.9:1. The product recovered was primarily the diester of diphosphoric acid having the empirical formula $H_2P_2O_5[O{-}CH(CH_3)CH_2OCH_3]_2$.

Example 4: Potassium hydroxide (16.8 g, 0.3 mol) was dissolved in 20 ml water. To this solution was added 1-methoxy-2-propanol (200 g). The mixture was rapidly stirred while the diester of diphosphoric acid prepared in Example 3 (48.3 g, 0.15 mol), was slowly added. A clear homogeneous yellowish colored reaction mixture resulted containing 24.6% by wt of the alkali metal salt of the above diester of diphosphoric acid.

Example 5: The composition prepared in Example 4 was tested as an iron corrosion inhibitor. Testing was conducted according to the experimental procedure outlined in American Society of Testing Methods D-1384 glassware corrosion test. Accordingly, two solutions, one containing known corrosion-inhibiting compounds and the other the compound of Example 4, were prepared for comparison purposes. In both cases the base component consisted of a solution of 50% 1-methoxy-2-propanol and 50% deionized water. Formulation data and test results are shown below.

Fluid 1	Percent	Fluid 2	Percent
Base fluid inhibitors	99.4	Base fluid inhibitor	98.0
$NaNO_2$	0.5	Example 4	2.0
Tolyl triazole	0.1		

	Weight Loss Fluid 1, mg	Weight Loss Fluid 2, mg
Copper	1.2	4.3
Aluminum	5.4	3.9
Steel	5.2	1.4
Iron	642.3	5.2

Alpha-1,4-Thiazine Alkanephosphonic Acids and Their Quaternary Ammonium Salts

P.M. Quinlan; U.S. Patent 4,311,663; January 19, 1982; assigned to Petrolite Corporation describes a process of inhibiting corrosion of metals exposed to a corrosive liquid medium which comprises adding to the medium an effective amount of a compound of the formula:

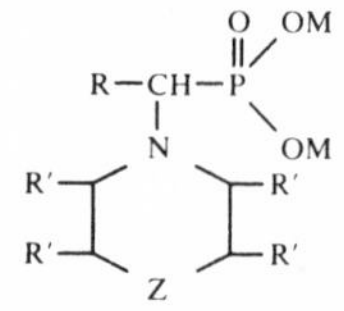

```
          O  OM
          ||/
   R—CH—P
      |    \
      N     OM
R'—⟨    ⟩—R'
R'—⟨    ⟩—R'
       Z
```

In the formula R is a member of the group selected from the group consisting of alkyl, alkenyl, cycloalkyl, aryl, alkaryl, aralkyl and hydroxyphenyl, and R' is hydrogen or alkyl, Z is S, SO or SO_2, and M is a hydrogen or a salt moiety.

The compounds of this process can be employed as corrosion inhibitors for acid systems, e.g., as illustrated by the pickling of ferrous metals, the treatment of calcareous earth formations and as a microbiocide in water flooding in the secondary recovery of oil and in hydrocarbon systems.

Loss of product, corrosion of the storage tank, clogging of filters and metering instruments, and fuel deterioration are among the harmful effects of bacteria growth in fuels. The activity of microorganism growth is often increased by the presence of rust. Not only do these microorganisms often encourage rust but rust encourages microorganism growth. Since microorganism growth appears to be considerably higher with kerosene than with gasoline, plugged filters experienced with jet fuels which contain large amounts of kerosene is a serious problem.

Example 1: To a solution of alpha-aminoethylphosphonic acid 12.6 g (0.1 mol) in 50 ml of a 50:50 (by volume) mixture of ethanol and water was slowly added, with stirring, divinyl sulfone 11.8 g (0.1 mol). The reaction mixture became warm and upon cooling deposited crystals. After cooling in an ice bath, the crystalline product was filtered and washed with cold ethanol. The product was recrystallized from aqueous ethanol. The product had the following structure which was characterized by NMR spectrum.

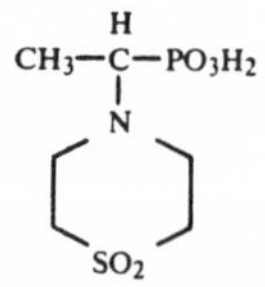

```
       H
CH3—C—PO3H2
       |
       N
     /   \
     \   /
      SO2
```

Analysis–Calculated for $C_6H_{14}O_5NPS$: P, 12.70%; N, 5.74%; S, 13.11%. Found: P, 12.59%; N, 5.68%; S, 13.21%.

Example 2: To a solution of alpha-aminopropylphosphonic acid 13.9 g (0.1 mol) in 50 ml of aqueous ethanol was slowly added, with stirring, divinyl sulfone 11.8 g (0.1 mol). The reaction mixture became warm and then deposited crystals upon cooling. The product was filtered and washed with cold ethanol. It was recrystallized from aqueous ethanol. The product had the following structure:

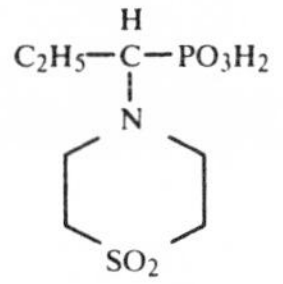

Analysis–Calculated for $C_7H_{16}O_5NPS$: P, 12.06%; N, 5.45%; S, 12.45%. Found: P, 12.00%; N, 5.38%; S, 12.33%.

Example 3: In a similar manner, alpha-aminobenzylphosphonic acid 18.7 g (0.1 mol) was reacted with divinyl sulfone 11.8 g (0.1 mol). The product was found to have the following structure:

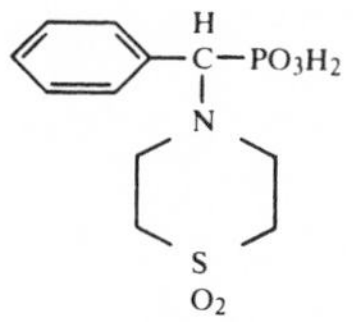

Analysis–Calculated for $C_{11}H_{16}O_5NPS$: P, 10.16%; N, 4.59%; S, 10.49%. Found: P, 10.24%; N, 4.61%; S, 10.65%.

In the corrosion test procedure described in U.S. Patent 4,248,796 above, this compound when used in the amount of 2,000 ppm in 15% HCl at 150°F showed a corrosion rate after 4 hr of 0.075 lb/ft^2/day. 1020 steel (AISI) coupons were used.

In related work *P.M. Quinlan; U.S. Patent 4,336,156; June 22, 1982; assigned to Petrolite Corporation* provides a process of inhibiting corrosion which comprises treating an acid system with a corrosion inhibiting amount of a compound of the formula:

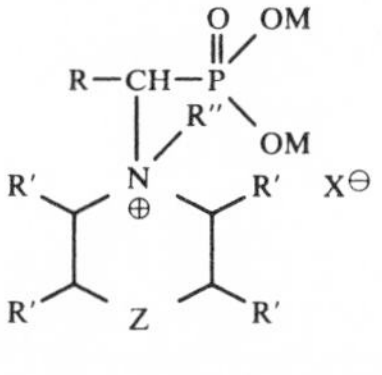

where R and R'' are members selected from the group consisting of alkyl, alkenyl, cycloalkyl, aryl, alkaryl, aralkyl, and hydroxyphenyl and from the group consisting of alkyl, aralkyl, cycloalkyl, alkenyl, and alkynyl, respectively, R' is hydrogen or alkyl, Z is S, SO or SO_2, M is hydrogen or a salt moiety and X is a halide, acetate, sulfonate or arylsulfonate.

Example 1: Into a flask equipped with a mechanical stirrer, heating jacket and reflux condenser are charged 14.2 g (0.1 mol) of methyl iodide, 50 ml of ethanol and 24.3 g (0.1 mol) of the compound shown on the following page.

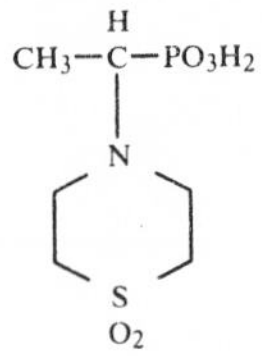

H
$CH_3-C-PO_3H_2$
N
S
O_2

The mixture was heated with stirring to reflux and held there for 8 hr. After the reflux period, the reaction mixture was cooled to room temperature and volatiles were stripped off on a rotary evaporator.

The resulting crystalline product was washed several times with acetone, filtered, and again washed several times with acetone. The product was recrystallized twice from aqueous ethanol.

The resulting product was identified by ^1H and ^{31}P spectrum as:

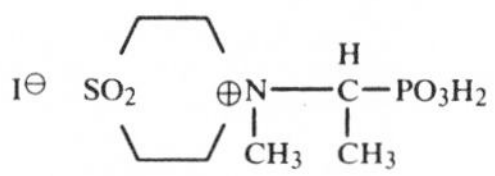

H
$I^{\ominus}$ SO_2 $\oplus N-C-PO_3H_2$
CH_3 CH_3

Analysis–Calculated for $C_7H_{17}O_5INPS$: N, 3.64%; P, 8.05%; I, 33.0%; S, 8.05%. Found: N, 3.55%; P, 7.89%; I, 32.7%; S, 7.98%.

Example 2: In a similar manner 24.3 g (0.1 mol) of the phosphonic acid used in Example 1 was reacted with 12.7 g (0.1 mol) of benzyl chloride in 50 ml of ethanol for 12 hr.

The product was identified as:

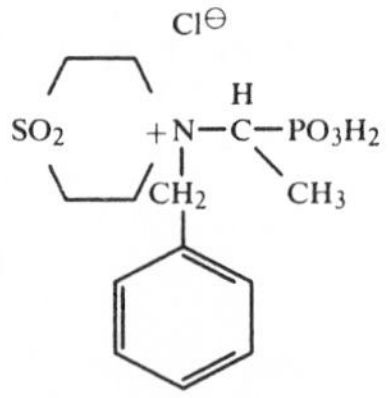

$Cl^{\ominus}$
H
SO_2 $+N-C-PO_3H_2$
CH_2 CH_3

Analysis–Calculated for $C_{13}H_{21}O_5ClNPS$: N, 3.78%; P, 8.38%; S, 8.42%; Cl, 9.59%. Found: N, 3.68%; P, 8.45%; S, 8.50; Cl, 10.0%.

Example 3: In a similar manner 25.7 g (0.1 mol) of:

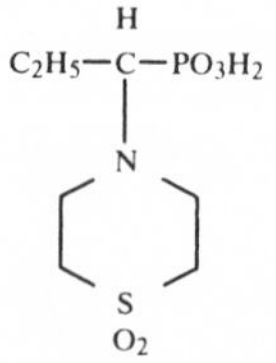

H
$C_2H_5-C-PO_3H_2$
N
S
O_2

and 11.9 g (0.1 mol) of propargyl bromide were reacted in 50 ml of refluxing ethanol for 8 hr. The product was isolated and purified in the usual manner. The structure of the product was:

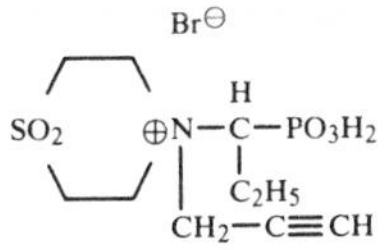

Analysis–Calculated for $C_{10}H_{19}O_5BrNPS$: N, 3.72%; Br, 21.27%. Found: N, 3.66%; Br, 20.98%.

The compounds of the process may be used as scale inhibitors, sequestering agents, as microbiocides in aqueous and hydrocarbon systems and as corrosion inhibitors for acid systems, e.g., in the pickling of ferrous metals, the treatment of calcarous earth formations, etc.

The corrosion inhibition in 15% HCl of 1020 steel (AISI) coupons were tested using the corrosion test procedure of U.S. Patent 4,248,796 with the following results.

Inhibitor	Parts per Million	Temperature (°F)	Time (hr)	Corrosion Rate ($lb/ft^2/day$)
Example 2	2,000	150°	4	0.054
Example 3	2,000	150°	4	0.040

In this process *P.M. Quinlan; U.S. Patent 4,309,383; January 5, 1982; assigned to Petrolite Corporation* uses, as a corrosion inhibitor, the diquaternary ammonium salts of alpha-1,4-thiazine alkanephosphonic acids which have the formula:

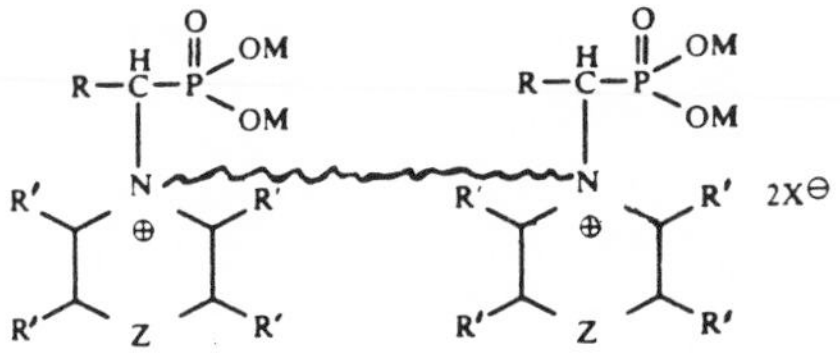

where R is a hydrocarbon or a substituted hydrocarbon group, e.g., alkyl, alkenyl, cycloalkyl, aryl, aralkyl, substituted aryl, etc., the R's are hydrogen or a substituted group such as a hydrocarbon group, i.e., alkyl, etc.; ~~ is a bridging group, e.g., a hydrocarbon group such as alkylene, alkenylene, alkaralkylene or hydrocarbon groups also containing other elements than carbon and hydrogen; Z is S, SO or SO_2; M is hydrogen, or a salt moiety such as alkali metal, alkaline earth metal, alkyl-ammonium, or ammonium; and X is an anion such as halide, acetate, sulfonate, aralkylsulfonate.

Example: Into a suitable reaction vessel were charged 24.4 g (0.1 mol) of 1,6-dibromohexane, 50 ml of water, 50 ml of ethanol and 51.4 g (0.2 mol) of the compound shown on the following page.

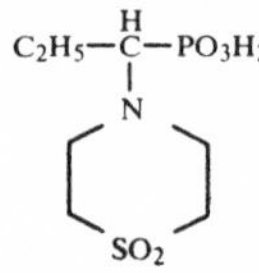

$$C_2H_5-\underset{\underset{\text{N (thiomorpholine-}SO_2\text{ ring)}}{|}}{\overset{H}{C}}-PO_3H_2,$$

The reaction mixture was heated to reflux and held there for 8 hr. The product was:

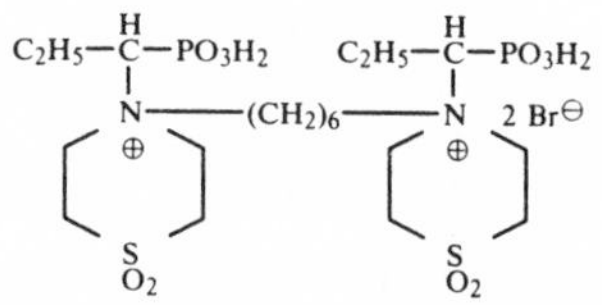

$$\left[C_2H_5-\overset{H}{C}(PO_3H_2)-\overset{\oplus}{N}(C_4H_8SO_2)-(CH_2)_6-\overset{\oplus}{N}(C_4H_8SO_2)-\overset{H}{C}(PO_3H_2)-C_2H_5\right]\ 2\,Br^{\ominus}$$

The diquaternary salts of this process have the same uses as the quaternary salts described above. In the same corrosion test procedure, the compound of the example under the same conditions showed a corrosion rate of 0.079 lb/ft^2/day compared to a rate for a blank of 0.220 lb/ft^2/day.

HETEROCYCLIC COMPOUNDS

Quaternized Derivatives of Polymerized Pyridines and Quinolines

The process of *P.M. Quinlan; U.S. Patents 4,297,484; October 27, 1981; and 4,341,657; July 27, 1982; both assigned to Petrolite Corporation* relates to alkylated derivatives of polymerized aromatic nitrogen heterocyclic compounds as illustrated by polymerized pyridine, quinoline and derivatives thereof.

These compositions are polymerized by treating the aromatic nitrogen heterocyclic compounds at elevated temperatures and pressures with catalytic amounts of alkyl halides and then alkylating the polymerized products formed.

Other examples of heterocyclic aromatic compounds that can be polymerized for utilization in this process include 2-methylquinoline, 4-methylquinoline, 2-methylpyridine, 4-methylpyridine, 4-phenylpyridine, 4-ethylpyridine, 2-hydroxypyridine, 2,4-lutidine and the like.

Alkylating agents include: methyl iodide, ethyl iodide, propyliodide, ethyl bromide, benzyl bromide, butyl bromide, dodecyl bromide, benzyl chloride, dodecyl benzyl chloride, ethyl bromoacetate, dimethyl sulfate and the like.

Example 1: Polyquinoline, 12.9 g (0.1 eq) with a molecular weight of 580 (Osmometer molecular weight in $CHCl_3$) was dissolved with heating and stirring in 40 ml of dimethylformamide. The solution was cooled and 14.2 g (0.1 eq) of methyl iodide was added. The resulting solution was heated at 100°C for 8 hr. The DMF was removed by heating under reduced pressure leaving a red-brown solid that was soluble in a 50:50, by volume, mixture of water and ethanol. Analysis–Calculated: I, 46.86%. Found: I, 47.21%.

The product consisted of the following structural species:

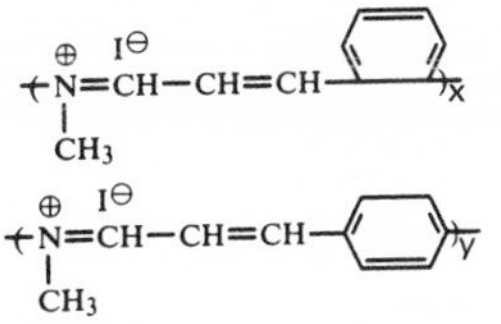

$$\left[\overset{\oplus}{N}(CH_3){=}CH{-}CH{=}CH{-}C_6H_4\right]_x \quad I^{\ominus}$$

$$\left[\overset{\oplus}{N}(CH_3){=}CH{-}CH{=}CH{-}C_6H_4\right]_y \quad I^{\ominus}$$

Example 2: In a similar manner, polyquinoline (molecular weight 580) 12.9 g (0.1 eq) was reacted with 17.0 g (0.1 eq) of propyliodide in 50 ml of dimethylformamide for 8 hr at 100°C. The product, soluble in a mixture of water and ethyl alcohol, had the following structural configuration:

$$\left[\overset{\oplus}{N}(C_3H_7){=}CH{-}CH{=}CH{-}C_6H_4\right]_x \quad I^{\ominus}$$

$$\left[\overset{\oplus}{N}(C_3H_7){=}CH{-}CH{=}CH{-}C_6H_4\right]_y \quad I^{\ominus}$$

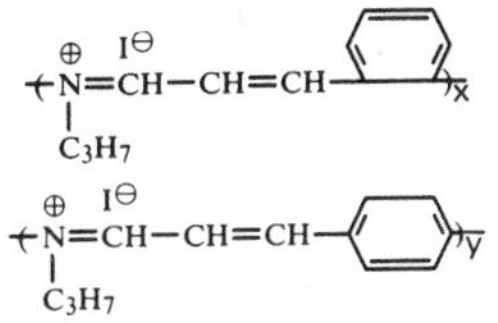

Example 3: In a similar manner, poly(2-methylquinoline) 14.3 g (0.1 eq) having a molecular weight of 820, was reacted with 27.7 g (0.1 eq) of tetradecyl bromide at 130°C in 50 ml of dimethylformamide for 24 hr. The product had the following structural configuration:

$$\left[\overset{\oplus}{N}(C_{14}H_{29}){=}C(CH_3){-}CH{=}CH{-}C_6H_4\right]_x \quad Br^{\ominus}$$

$$\left[\overset{\oplus}{N}(C_{14}H_{29}){=}C(CH_3){-}CH{=}CH{-}C_6H_4\right]_y \quad Br^{\ominus}$$

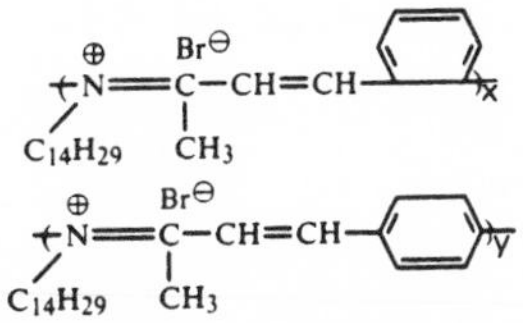

The compositions of this process can be used as corrosion inhibitors in acidizing media employed in the treatment of deep wells to reverse the production of petroleum or gas therefrom and more particularly to an improved method of acidizing a calcareous or magnesium oil-bearing formation.

The inhibitor compositions were employed to inhibit corrosion in 15% hydrochloric acid. In the corrosion test procedure described in U.S. Patent 4,248,796 above, using 1020 steel (AISI) coupons, the tests were run at 150°F for 4 hr. 0.2% by volume, inhibitor was employed. The results of the test are shown below.

Example No.	Corrosion Rate ($lb/ft^2/day$)
0	3.324
1	0.087
2	0.112
3	0.051

The compositions may also be used in metal cleaning and pickling baths which generally comprise aqueous solutions of inorganic acids such as sulfuric acid, hydrochloric acid, phosphoric acid and are useful in the cleaning and treatment of iron, zinc, ferrous alloys, and the like.

The compositions may also be used in inhibiting bacterial growth in the recovery of oil from oil-bearing strata by means of water flooding taking place in the presence of sulfate-reducing bacteria.

Microorganisms may bring about corrosion by acting on the metal structures of the wells involved, producing corrosive substances such as hydrogen sulfide, or producing conditions favorable to destructive corrosion such as decreasing the pH or producing oxygen.

In addition to being used as biocides in aqueous systems, the compounds of this process can also be employed as biocides in hydrocarbon systems, particularly when petroleum products are stored.

Poly(Oxyalkylated) Pyrazoles

H.W. Schiessl; U.S. Patent 4,306,986; December 22, 1981; assigned to Olin Corporation provides poly(oxyalkylated) pyrazoles of the formula:

$$\text{R}-\text{C}\langle\text{CH}=\text{C}-\text{R}'\rangle\text{=N}-\text{N}-(\text{CH}_2-\underset{\text{R}''}{\text{CHO}})_{\overline{n}}\text{H}$$

where R and R' are independently selected from lower alkyl groups having 1 to 4 carbon atoms; each R'' is individually selected from hydrogen and methyl; and n is from 2 to about 20. These compounds are effective corrosion inhibitors in corrosive liquids such as acids, antifreezes and hydraulic fluids.

Example 1: 3,5-dimethylpyrazole [120 g (1.25 mols)], toluene (300 ml), and powdered potassium hydroxide (6 g) were added to a stainless steel pressure vessel which was then flushed with N_2. Ethylene oxide [210 g (4.75 mols)] was then pressured with N_2 into the reaction vessel at a pressure up to 50 psig and a temperature of 130° to 140°C. This corresponds to a mol ratio of EO to pyrazole of 3.8. Addition time was 47 min with a post-addition time of 85 min; the temperature was kept in the desired range by periodic cooling by means of a water-cooled coil in the reactor. Toluene and unreacted ethylene oxide were then removed by stripping on a rotary evaporator using a vacuum pump and heating to 74°C. The residue weighed 307 g, a recovery of 91.3% based on the weight of ethylene oxide and 3,5-dimethylpyrazole added.

The elemental analysis of this product was: Found, % by wt, C, 57.07; H, 8.16, N, 10.91. Theory for DMP-3.8 EO, C, 57.45%; H, 8.43%, N, 10.64%.

Example 2: The apparatus and procedure of Example 1 was repeated, except using a molar ratio of propylene oxide to 3,5-dimethylpyrazole of 10:1. 403 g of final product were recovered, representing about 97% of the reactants fed.

Example 3: The compound prepared according to the method of Example 2 above was tested as corrosion inhibitor in glycol antifreezes according to test method ANSI/ASTM D 1384-70, "Corrosion Test For Engine Coolants in Glassware."

In carrying out this test method D-1384, several antifreeze solutions (each 750 ml) were formed which contained ethylene glycol (250 ml), corrosive water (500 ml), a potassium phosphate buffer [K_2HPO_4 (7.5 to 8 g)] and the compound of Example 2 (about 192 g with none for the blank test). The solutions were placed in 1,000 ml beakers.

A bundle of six different metal coupons (each coupon having already been weighed) was placed in the beaker and covered by the solution. The beakers were kept at 190°F for 336 hr, during which time the solutions were aerated. At the end of this time period, the bundles of coupons were removed from the beaker, disassembled, cleaned, reweighed, and the weight change determined. The weight change per square centimeter of each coupon was determined and is shown in the table below. As can be seen, the weight change with the inhibitor present was much less than the blank test, indicating the excellent protection provided by these compounds.

	. . . Weight Change, mg/cm^2. . . .	
Metal	**Blank**	**DMP-10 PO**
Copper	32.7	1.49
Solder	33.9	0.36
Brass	33.2	0.45
Steel	51.8	0.05
Cast Iron	42.5	0.03
Aluminum	3.2	0.01

The compound of Example 2 was further tested as corrosion inhibitor in aqueous acidic solutions according to the linear polarization method described in ASTM test method G5-72. The results of this test are given below.

By this method, the effectiveness of the compound as corrosion inhibitor was rated by, first, determining the linear polarization of a mild steel sample in an uninhibited 1.0 N H_2SO_4 solution, and second, in the same 1.0 N H_2SO_4 solution after the compound was added (the amount of compound added was equivalent to 0.25% by wt of the solution). From these linear polarization measurements, the percent protection afforded by inhibitor in this acid solution was determined by the following formula:

$$\text{Percent Protection} = \frac{LP_u - LP_i}{LP_u} \times 100$$

wherein LP_u is the linear polarization of the uninhibited sample and LP_i is the linear polarization of the sample placed in the acid solution containing the inhibitor compound.

The propylene oxide adduct of 3,5-dimethylpyrazole (DMP) was much better than 3,5-dimethylpyrazole by itself, giving 97.2% protection compared to 8.7% for DMP.

The compound (DMP-10 PO) prepared in Example 2 above, was tested as a corrosion inhibitor in a polyglycol-based hydraulic fluid according to the test method set forth in SAE-J1703f. This polyglycol-based fluid with the inhibitor included had the following formula:

	Percent by Weight
Triethylene glycol monomethyl ether (Poly-Solv)	75.8
Polypropylene glycol, molecular weight 1,000 (Poly-G 20-112)	20.00
Polyethylene glycol, molecular weight 300 (Poly-G 300)	3.0
Buffers and Antioxidants:	
Bisphenol A	0.2
Borax	0.2
Boric acid	0.2
Trimethylolpropane	0.2
DMP-10 PO	0.4

After this hydraulic fluid formation is made, a bundle of six different metal coupons (previously weighed) were placed in a test jar containing the fluid. All of the coupons were fully covered by the solution. After running the test at 100°C for 5 days, the coupons were removed, washed, dried and weighed. The weight change per square centimeter of each coupon was then determined. The results of this corrosion test in the hydraulic fluid containing the DMP-10 PO inhibitor versus the same uninhibited hydraulic fluid are given in the following table. As can be seen, the hydraulic fluid containing the inhibitor had a smaller weight change for some metals and, thus, offered protection against corrosion.

	. . . . .Weight Change of Coupons, mg/cm^2	
Metal Coupon	**Uninhibited Fluid**	**Fluid with DMP-10 PO**
Copper	0.67	0.40
Brass	0.69	0.45
Cast Iron	+0.11	+0.21
Aluminum	+0.01	0.005
Steel	+0.35	0.02
Tin	+0.01	0.03

Thiazine Quaternary Ammonium Salts of Epihalohydrin

The process of *P.M. Quinlan; U.S. Patents 4,316,007; Feb. 16, 1982; 4,331,554; May 25, 1982; and 4,371,497; February 1, 1983; all assigned to Petrolite Corporation* relates to thiazine quaternary ammonium salts of polyepihalohydrin, their preparation and uses. The compositions may be ideally expressed by the polymeric unit:

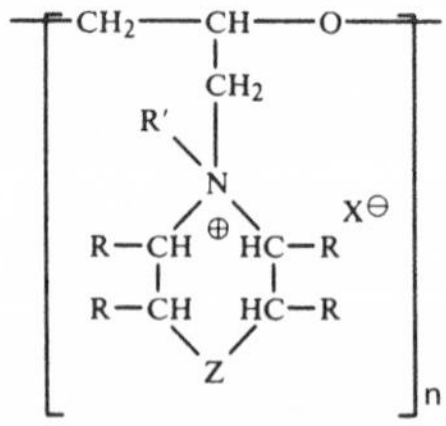

```
 ┌ CH2───CH───O ─┐
 │       │       │
 │       CH2     │
 │  R'   │       │
 │    \  │       │
 │       N       │
 │     / ⊕ \   X⊖│
 │ R─CH     HC─R │
 │   │       │   │
 │ R─CH     HC─R │
 │     \   /     │
 │       Z       │
 └               ┘n
```

where Z is S, SO, SO_2; the Rs are hydrogen or a substituted group such as a hydrocarbon group, i.e., alkyl, etc.; and R' is a hydrocarbon group such as alkyl, or a substituted alkyl, alkenyl, alkynyl, aryl, aralkyl, etc.; n may be from about 3 to 1,000, such as from 5 to 100 but preferably from 5 to 50. X is halogen.

The compounds are especially useful in the acidizing or treating of earth formations and wells traversed by a bore hole. They may also be used in metal cleaning and pickling baths which generally comprise aqueous solutions of inorganic acids and as biocides in water flooding in secondary recovery of oil and in hydrocarbon systems.

Example 1: Polyepichlorohydrin of molecular weight 1,300 (92 g; 1 eq) was heated at 150° to 160°C with

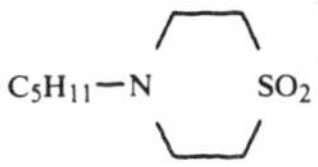

$C_5H_{11}-N$ SO_2

(205 g; 1 eq) and 297 ml of water in a closed reactor for 18 hr. At this time the ionic chloride content was found to be 5.8% (theoretical was 6.0%). The structure of the product was

CH_2-CH-O
CH_2
C_5H_{11} N ⊕ $Cl^{\ominus}$
S
O_2
$n \cong 14.1$

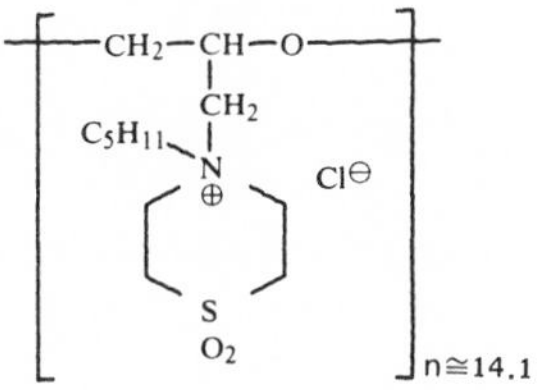

Example 2: Polyepichlorohydrin of molecular weight 1,300 (92 g; 1 eq) was heated in a closed pressure reactor with

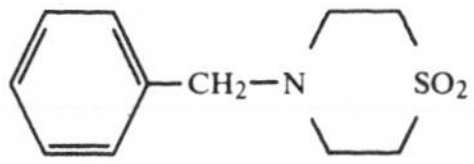

CH_2-N SO_2

(239 g; 1 eq) and a 1:1 water:isopropanol mixture (331 g) at 145° to 160°C for 20 hr. At this time the ionic chloride content was found to be 5.0% (theoretical was 5.4%). The structure of the product was

CH_2-CH-O
CH_2 $Cl^{\ominus}$
CH_2 N ⊕
S
O_2
$n \cong 14.1$

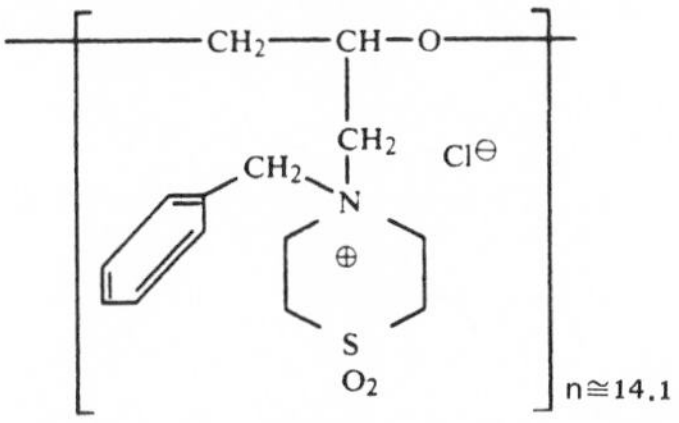

Example 3: Polyepichlorohydrin of molecular weight 1,300 (46 g; 0.5 eq) was heated in a closed reactor at 150° to 160°C with the compound of the structure shown on the following page (165 g; 0.5 eq) in a 1:1 mixture of water:isopropanol (211.0 g) for 24 hr.

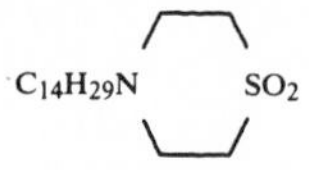

$$C_{14}H_{29}N\langle\!\!\!\;\;\rangle SO_2$$

The product had the structure

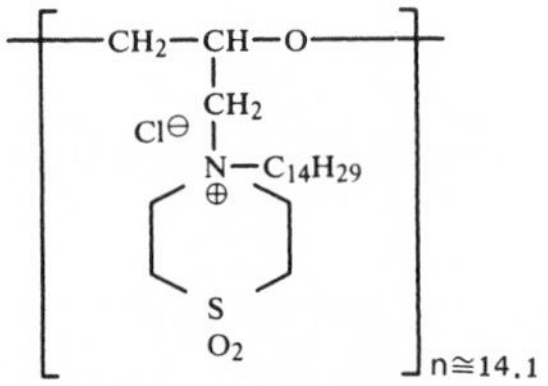

$$\left[\!-CH_2-\underset{\underset{\overset{\oplus}{N}(C_{14}H_{29})\text{-thiomorpholine-}S(O_2),\ Cl^{\ominus}}{|}}{\underset{CH_2}{\underset{|}{CH}}}-O-\right]_{n\cong 14.1}$$

Example 4: A mixture of polyepiiodohydrin with a molecular weight of 1,400 (92 g; 0.5 eq),

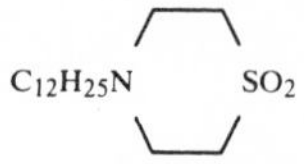

$$C_{12}H_{25}N\langle\!\!\!\;\;\rangle SO_2$$

(151 g; 0.5 eq) and a 1:1 mixture of water:isopropanol (243 g) was heated in a closed reactor at 140° to 150°C for 12 hr. The product had the following structure

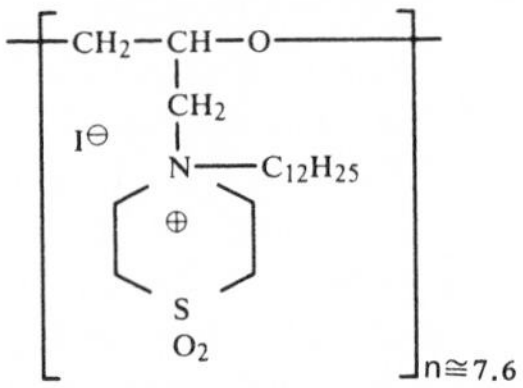

$$\left[\!-CH_2-\underset{\underset{\overset{\oplus}{N}(C_{12}H_{25})\text{-thiomorpholine-}S(O_2),\ I^{\ominus}}{|}}{\underset{CH_2}{\underset{|}{CH}}}-O-\right]_{n\cong 7.6}$$

In the corrosion test procedure described in U.S. Patent 4,248,796 1010 steel corrosion coupons were employed. The test conditions were as follows: concentration, 3,000 ppm; temperature, 150°F; time, 4 hr; HCl, 15%. The results of these tests are shown below.

Inhibitor	Corrosion Rate, $lb/ft^2/day$
Example 1	0.085
Example 2	0.054
Example 3	0.037
Example 4	0.030
Blank	2.02

Polyols of Tetrahydropyrimidines and Their Polymers

In U.S. Patent 4,145,545 there is described a method of preparing substituted 2,3,4,5-tetrahydropyrimidines (THP) of the formula:

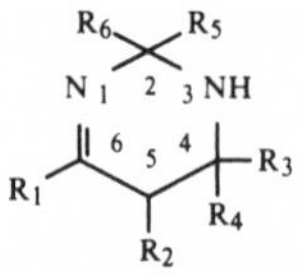

$$\text{Ring: } N_1{=}C_6(R_1)-C_5H(R_2)-C_4(R_3)(R_4)-N_3H-C_2(R_5)(R_6)-N_1$$

R_1, R_2, R_3, R_4, R_5 and R_6, which may be the same or different, are hydrogen or substituted group such as alkyl, aryl, cycloalkyl, alkaryl, aralkyl, heterocyclic, substituted derivatives thereof, etc. In addition, the R groups may be joined in a cyclic configuration which makes the THP structure a part of the substituted group.

B.A. Oude Alink and B.T. Outlaw; U.S. Patent 4,343,941; August 10, 1982; assigned to Petrolite Corporation have found that the THP's prepared in U.S. Patent 4,145,545 can be reacted with an aldehyde such as formaldehyde to yield methylol derivatives of THP.

The reaction is carried out by reacting the THP with formaldehyde under conditions which produce the desired product. In practice, aqueous formaldehyde (formalin) is employed. The reaction is carried out at about room temperature or higher, such as from about 25° to 140°C, e.g., from about 25° to 70°C, but preferably from about 25° to 40°C.

The stoichiometric ratio of THP to formaldehyde can vary widely from about 1 to 6, such as from about 1 to 5, but preferably from about 1 to 4.

The products are useful in the acidizing or treating of earth formations and wells traversed by a bore hole. They may also be used in metal cleaning and pickling baths and in oil field brines and oil- and water-based drilling fluids.

Example 1: *4,4,6-Trimethyl-2,3,4,5-tetrahydropyrimidine* – In a 2 ℓ three-necked round bottom flask equipped with a mechanical stirrer, a thermometer, and a reflux condenser was charged a mixture of 196.3 g of mesityl oxide and 400 ml of concentrated ammonium hydroxide. The mixture was stirred at room temperature for 18 hr. To this mixture was added 162.3 g of 37% formaldehyde (in H_2O) over 20 min. An ice-water bath was used to maintain the reaction temperature at 40°C. This mixture was stirred at room temperature for 18 hr. The excess ammonia was removed by distillation under reduced pressure. The remaining liquid was distilled to yield 176.7 g (68.5% of theory) of 4,4,6-trimethyl-2,3,4,5-tetrahydropyrimidine.

Example 2: Reaction of product from Example 1 with formaldehyde – In a 200 ml three-necked flask equipped with a mechanical stirrer, a thermometer, and a reflux condenser was charged 20 g of the product of Example 1 and 51.4 g of 37% formaldehyde solution. The mixture was stirred at room temperature for 5 hr. The solvent was removed by distillation under reduced pressure, to yield 33.2 g of a very viscous liquid (84.6% of theory). Analysis–Calculated for N: 11.38%. Found: 11.85%. Ratio of THP to formaldehyde is 1 to 4.

Example 3: In a similar manner to that in Example 2, the tetrahydropyrimidine of Example 1 (20 g) was reacted with 38.6 g of 37% formaldehyde (1:3 molar ratio). 36.2 g of viscous liquid was obtained (87.7% of theory). A mixture of products are possible, with ^{13}C NMR data favoring (A) as the major product.

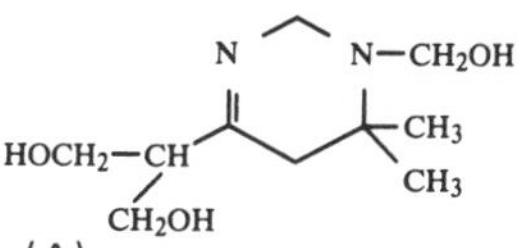

N, $N-CH_2OH$, $HOCH_2-CH$, CH_3, CH_3, CH_2OH
(A)

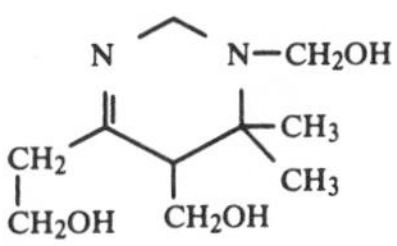

N, $N-CH_2OH$, CH_2, CH_3, CH_3, CH_2OH, CH_2OH
(B)

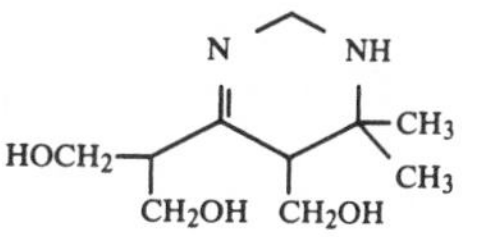

N, NH, $HOCH_2$, CH_3, CH_3, CH_2OH, CH_2OH
(C)

B.A. Oude Alink and B.T. Outlaw; U.S. Patent 4,343,930; August 10, 1982; assigned to Petrolite Corporation have found that the polyols of U.S. Patent 4,343,941 above can be polymerized with a deammoniation to yield polypyridines. The polypyridine units are believed to be joined by $-CH_2-X-CH_2-$ units where X is O or NH. The polymer may contain both O and NH groups.

Since the number of polyol units, i.e., $-CH_2OH$ groups on the monomer will vary, or the monomer may be a mixture of compounds containing varying amounts of $-CH_2OH$ groups, the structure of the final polymer will also vary. However, it is believed that during polymerization, ammonia is given off to yield a pyridine unit,

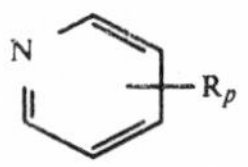

[pyridine ring with substituent R_p]

where the particular ring substitution (R) and the number of such substitutions (p) will vary with the particular starting materials, thus R represents substituted groups in p positions on the ring. The linking group in the polymer may be ideally represented as follows:

$$-CH_2OH \xrightarrow[\Delta]{\text{Lewis acid}} -CH_2-X-CH_2-$$

where X is oxygen or NH or both present in the same polymer. Thus, the polymer may be represented by the following:

$$R_p-\!\!(N)\!\!-(CH_2OH)_n \xrightarrow[\Delta]{\text{Lewis acid}} [R_p-\!\!(N)\!\!-(CH_2X)_n-]_m$$

where n = 2 or more and X is O or NH or both in the same polymer and (N) is a pyridine ring whose substitutions are represented by R_p which will vary with the starting materials.

The reaction is generally carried out in the presence of Lewis acid at a temperature sufficiently high to evolve ammonia and to yield the polymerized product. In general, the reaction is carried out at a temperature of at least about 170°C.

The products have the same uses as the products of the preceding patent.

Example: The product of Example 2 of U.S. Patent 4,343,941 was heated to 170° to 200°C with a trace of NH_4Cl. The product was a brittle solid that could be ground to a resinlike powder.

OTHER

Mixtures of Vicinal Aminoalkanols

The process of *W. Giede, K. Koch, G. Kolaczinski, W. Rupilius and W. Stein; U.S. Patent 4,302,354; November 24, 1981; assigned to Henkel Kommanditgesellschaft auf Aktien and Deutsche Gold- und Silber-Scheideanstalt, Germany* relates to mixtures of vicinal aminoalkanols. They are selected from the group consisting of (1) compounds having the formula:

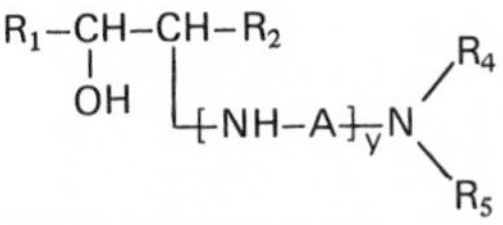

$$R_1-\underset{OH}{\underset{|}{CH}}-\underset{\lfloor}{CH}-R_2 \quad \{NH-A\}_y N \begin{matrix} R_4 \\ R_5 \end{matrix}$$

where R_1 and R_2 are alkyl having from 1 to 21 carbon atoms and the sum of the carbon atoms in $R_1 + R_2$ is from 6 to 22, A is a member selected from the group consisting of alkylene having from 2 to 6 carbon atoms and methylalkylene having from 3 to 7 carbon atoms, y is an integer from 1 to 3, and R_4 and R_5 are individually members selected from the group consisting of hydrogen, alkyl having from 1 to 4 carbon atoms and hydroxyalkyl having from 2 to 4 carbon atoms, and (2) their organic carboxylic acid salts. The mixtures must have at least two different and adjacent chain lengths of

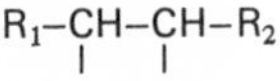

$$R_1-\underset{|}{CH}-\underset{|}{CH}-R_2$$

in the range of from 8 to 24 carbon atoms and the vicinal amino and hydroxyl substituents must be distributed statistically.

In the preparation of the vicinal aminoalkanol mixtures according to the process, mixtures of olefins with 8 to 24 carbon atoms containing statistically distributed nonterminal double bonds are employed as starting materials for the synthesis. These olefin mixtures are known per se and can be obtained, e.g., by catalytic dehydration or by chlorination/dehydrochlorination of paraffins having 8 to 24 carbon atoms and selective extraction of the nonterminal monoolefins obtained. However, it is also possible to use mixtures of such olefins with saturated hydrocarbons, as they are obtained in the preparation of these olefins. Preferred as mixtures of isomeric monoolefins are the fractions with a high content of linear C_{11-14} olefins or C_{15-18} olefins. The particularly preferred mixtures of nonterminal olefins have the following chain length distribution:

	Distribution, % by wt
Fraction C_{11-14} Olefins:	
C_{11}	~22
C_{12}	~30
C_{13}	~26
C_{14}	~22
Fraction C_{15-18} Olefins:	
C_{15}	~26
C_{16}	~35
C_{17}	~31
C_{18}	~6

The above preferred mixtures of C_{11} to C_{14} olefins and C_{15} to C_{18} olefins can also have deviations in the indicated chain length distributions. For the preparation of the products according to the process, the olefin mixtures are epoxidized by means of known methods, e.g., with peracetic acid. In the process, these nonterminal olefin epoxide mixtures are reacted with amines of the general formula:

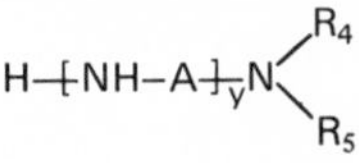

$$H\{NH-A\}_y N \begin{matrix} R_4 \\ R_5 \end{matrix}$$

where y, R_4 and R_5 are as described above. This reaction yields the vicinal aminoalkanol mixtures according to the process.

The amines are used in from 1.0 to 15 times the molar amount, based on the epoxide mixture and can serve at the same time as solvents, provided they are liquid at room temperature. If necessary, an additional catalytically-acting solvent can be used, preferably water.

The reaction between the olefin epoxide and the amine is carried out in a temperature range of 100° to 230°C, preferably at the reflux temperature of the amine or solvent used. If the boiling point of the diamine or triamine is below the required reaction temperature, or if a low-boiling catalytically-acting solvent is to be used, the reaction can also be carried out in the autoclave at temperatures of 150° to 230°C under pressure.

According to a preferred embodiment, the reaction with amines, wherein the above general formula, R_4 and R_5 are other than hydrogen, is also carried out under pressure if their boiling point is below the desired reaction temperature. The epoxide mixture is then charged with an amount of 0.1 to 0.5 mol of glycerin or ethylene glycol, based on the epoxide mixture, at a reaction temperature of 150° to 220°C, and the low boiling diamine is added slowly in an 1.0 to 1.5 molar amount, based on the epoxide mixture used, so that the temperature does not drop more than 10° to 20°C below the initially set reaction temperature. Subsequently the mixture can be stirred for 1 to 2 hr at reflux temperature. The catalytically-acting solvent is washed out with water or distilled off after the reaction.

The time required for the reaction can vary within a wide range, and usually takes about 1 to 50 hr, preferably 1 to 5 hr. The working up of the reaction mixture can be effected according to known methods, e.g., by distillation.

The amine salts of the mixtures of vicinal aminoalkanols with acids, particularly aliphatic carboxylic acids with 2 to 24 carbon atoms, such as alkanoic acids, alkenoic acids, hydroxyalkanoic acids, can be produced, if necessary, according to known methods, by neutralizing the aminoalkanol mixtures.

The aminoalkanol mixtures according to the process are suitable as corrosion inhibitor additives in fuels, oils and lubricants. They can also be used as solutions in organic solvents, and surprisingly also in water, particularly as 0.1 to 10% by wt solutions, as corrosion inhibitors.

Compared to the natural fatty amines otherwise used as corrosion inhibitors, the products, according to the process, show a much better inhibiting action and are also easier to dose, because of their low solidification temperatures.

Another advantage of the process is that the application of the vicinal aminoalkanol mixtures is much simpler, compared to the natural products, because they have only a weak amine odor and they are compatible with the skin. Therefore, they may be employed as corrosion inhibitors in cutting oils and as coating agents for metals handled by humans.

Compounds Containing Sulfur and Amino Groups

The process of *N.E.S. Thompson, D. Redmore, B.A. Oude Alink and B.T. Outlaw;*

U.S. Patent 4,332,967; June 1, 1982; assigned to Petrolite Corporation relates to the following: (1) an amino-containing composition characterized by polymercapto groups, (2) a polyamino-containing composition characterized by mercapto or polymercapto groups, and (3) a composition characterized by the presence of (A) mercapto or polymercapto groups, and (B) a nitrogen-containing group characterized by at least one of the following: an amido or a polyamido group, a cyclic amidine or a polycyclic amidine group, or an epihalohydrin-derived amino-containing group.

The presence of both sulfur-containing and amino-containing groups synergistically enhances the total effect as a corrosion inhibitor.

These compositions can be used in a wide variety of applications and systems where iron, steel and ferrous alloys are affected by corrosion. They may be employed for inhibiting corrosion in processes which require this protective or passivating coating as by dissolution in the medium which comes in contact with the metal and in pickling. They can be used in preventing atmospheric corrosion, underwater corrosion, corrosion in steam and hot water systems, corrosion in chemical industries, underground corrosion, etc.

The corrosion inhibitors find special utility in the prevention of corrosion of pipe or equipment which is in contact with a corrosive oil-containing medium, as, e.g., in oil wells producing corrosive oil or oil-brine mixtures, in refineries, and the like.

Example 1: In a flask, fitted stirrer, condenser, thermometer and addition funnel was placed octyl thiol (36.5 g; 0.25 mol) and Triton B (5 drops, catalyst) was added. To this thiol, methyl methacrylate (25 g; 0.25 mol) was added dropwise while stirring at a rate such that the temperature was maintained between 50° and 60°C. After stirring for 1 hr the addition reaction of the thiol to the methacrylate was complete. Infrared analysis showed the absence of thiol SH and unsaturated ester function and presence of saturated ester (1,740 cm^{-1}). To this thioether was added diethylenetriamine (26 g; 0.25 mol) and the mixture heated to 185° to 195°C. As the temperature reached 180°C methanol began to be liberated and distillation was continued for 1 to 1½ hr during which time 8 g of distillate was collected (0.25 mol). The product (80 g) was a yellow viscous liquid soluble in aqueous alcohol and water-dispersible.

Analysis gave: N, 13.20%; S, 10.15% (calculated: N, 13.25%; S, 10.09%). Infrared analysis showed C=O 1,650 cm^{-1} (amide). The product has the following structure:

$$CH_3(CH_2)_7SCH_2CH(CH_3)\overset{\overset{\large O}{\|}}{C}NH(CH_2)_2NH(CH_2)_2NH_2$$

Example 2: To dodecylthiol (40 g; 0.2 mol) containing 5 drops of Triton B was slowly added methyl methacrylate (20 g; 0.2 mol) so that the temperature was kept below 60°C. After heating for 1 hr at 50° to 60°C the reaction between thiol and methacrylate was complete. After cooling to room temperature diethylenetriamine (20 g; 0.2 mol) was added and the mixture heated to 180°C. Methanol formed in the condensation was allowed to distill from the reaction and was collected (6 g). The product was an amber viscous liquid dispersible in water. Infrared analysis showed C=O 1,650 cm^{-1} (amide). The product has the following structure:

$$CH_3(CH_2)_{11}SCH_2CH(CH_3)\overset{O}{\overset{\|}{C}}NH(CH_2)_2NH(CH_2)_2NH_2$$

Example 3: This example uses the same reactants as Example 1, but in different ratios.

To octyl thiol (14.6 g; 0.1 mol) containing 5 drops of Triton B as catalyst was added methyl methacrylate (40 g; 0.4 mol) dropwise during 45 min. After stirring for an additional 1 hr infrared analysis showed the absence of thiol SH and the presence of both saturated and unsaturated ester groups (1,740 cm^{-1} and 1,725 cm^{-1}). To this mixture diethylenetriamine (41.2 g; 0.4 mol) was added and the mixture heated with stirring. As the reaction temperature reached 160°C methanol was formed and removed by distillation. After heating at 180° to 185°C for 1 hr 13 g of methanol distillate had been collected leaving 75 g of a yellow viscous product. The product was readily water-soluble. Infrared analysis showed C=O 1,650 cm^{-1} (amide) and no residual ester groups. The product is an oligomeric amino-amide containing thioether groups.

Analysis–Calculated: N, 20.24%; S, 3.85%. Found: N, 20.96%; S, 3.97%.

Example 4: To octyl thiol (64 g; 0.44 mol) containing Triton B (5 drops) was slowly added acrylonitrile (30 g; 0.57 mol) during 45 min at less than 60°C. After stirring at ambient temperature for a further 1 hr aminoethylethanolamine (60 g; 0.57 mol) and thiourea (1 g) were added. Upon heating to 135°C ammonia evolution commenced and became rapid at 165° to 170°C. The ammonia evolution was essentially complete in 3 hr yielding a dark viscous water dispersible product.

The product is represented by the following formula:

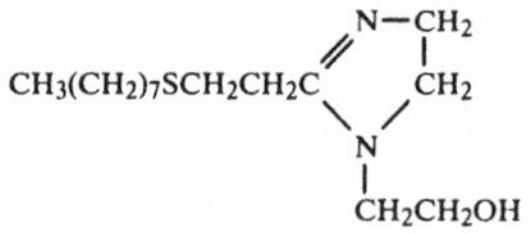

$$CH_3(CH_2)_7SCH_2CH_2C\begin{matrix} // N-CH_2 \\ \quad\;\; | \\ \quad\; CH_2 \\ \backslash N / \\ | \\ CH_2CH_2OH \end{matrix}$$

In order to compare the sulfur-amino compositions of this process with corresponding nonsulfur amino compositions, the following nonsulfur amino compositions were prepared and tested as corrosion inhibitors.

Example 5: To diethylenetriamine (52 g; 0.5 mol) cooled in an ice bath was added acrylonitrile (30 g; 0.56 mol) in 40 min at <60°C. Upon completion of this addition thiourea (0.8 g) was added as catalyst and the mixture heated to 150° to 165°C. Ammonia evolution proceeded rapidly and was complete after 4 hr. Upon cooling a viscous brown product was obtained readily soluble in water.

Example 6: Following the procedure of Example 5 aminoethylethanolamine (52 g; 0.5 mol) was condensed with acrylonitrile (30 g; 0.56 mol) in presence of thiourea as catalyst. The viscous product was water-soluble.

Corrosion Test Results: Corrosion tests were carried out at ambient temperature in 2% sodium chloride saturated with carbon dioxide. Corrosion rates were measured using PAIR meter of the type described in U.S. Patent 3,406,101. Inhibitors were injected after the electrodes had been allowed to corrode for 2 hr.

Protection is calculated in the usual manner from corrosion rate (R_1) of fluids without inhibitor and corrosion rate (R_2) in presence of particular inhibitor according to the formula:

$$[(R_1-R_2)/R_1] \times 100 = \text{Percent protection}$$

Blank corrosion rates under these conditions was 51 mpy. The corrosion test results are as follows:

	.... Percent Protection at Concentration		
Example	25 ppm	50 ppm	100 ppm
1	88	96	–
3	49	57	87
4	65	91	96
5	0	0	–
6	0	–	–

Acid Inhibitors Test in Hydrochloric Acid: 200 ml of 5% hydrochloric acid in a 300 ml beaker is heated to 165° to 170°F and the chemical to be tested is added at the appropriate concentration. Cleaned 1020 mild steel coupons (7/8 x 3¼ x 1/6 inches) are weighed and then placed in the acid for exactly 1 hr. The coupons are removed and washed with hot water, hot acetone, air dried and then reweighed.

Corrosion protection is calculated in the usual manner from the weight loss of the blank (W_1) and weight loss (W_2) in the presence of inhibitor according to the formula:

$$[(W_1-W_2)/W_1] \times 100 = \text{Percent protection}$$

The coupons used in corrosion experiments weighed 20.5 to 21 g and the typical weight loss without inhibitors was 1.3 g.

Compound	Concentration, ppm	Protection, %
Example 1	250	95
Example 2	250	95
Example 4	250	96
Example 6	250	77

5-Ketocarboxylic Acid Derivatives

The process of *D.R. Clark; U.S. Patent 4,366,076; December 28, 1982; assigned to Ciba-Geigy Corporation* relates to compositions comprising a functional fluid in contact with a ferrous metal and, as corrosion inhibitor, 5-ketocarboxylic acid derivatives. The corrosion inhibitor has the formula:

$$\begin{array}{l} R^1R^2R^3C \\ \qquad \searrow \\ \qquad\quad C{=}O \\ \qquad \nearrow \\ R^4R^5C\,CH_2CH_2\,COOX \end{array}$$

where X is H, alkali or alkaline earth metal, preferably Na, K, NH_4, the residue of a protonated amine, or the group –OX is the residue of an alkanol having from 1 to 20 carbon atoms; or of a di-, tri or tetraol having from 2 to 12 carbon atoms; R^1, R^2 and R^4 are the same or different and each is H or $-CH_2CH_2COOX$ where X has its previous meaning; R^3 and R^5 are the same or different and each is H, $-CH_2CH_2COOX$ where X has its previous meaning, or R^3 or R^5 is a straight or branched chain alkyl group having from 3 to 16, preferably 7 to 12 carbon atoms; or R^3 and R^5, together with the carbon atom to which they are attached, may form a cycloalkanone ring containing from 5 to 15 carbon atoms.

When X is the residue of a protonated amine it may be the residue of an alkylamine containing from 1 to 20 carbon atoms, or a heterocycle containing nitrogen, e.g., morpholine.

Preferably, the composition of the process contains from 0.001 to 5% by wt of the compound of the above formula, based on the total weight of the composition.

Examples of functional fluids which are useful in the compositions of the process include purely aqueous systems, e.g., cooling water systems; nonaqueous systems such as lubricants having a mineral oil or synthetic carboxylic acid ester base, greases, temporary protectants and hydraulic fluids based on mineral oils or phosphate esters; and mixed aqueous/nonaqueous systems, e.g., aqueous polyglycol/polyglycol ether mixtures, glycol systems, oil-in-water and water-in-oil emulsions, metal-working fluids having, as their base, mineral oils or aqueous systems, and aqueous-based paints.

In the following examples all parts and percentages are by weight unless otherwise specified.

Example 1: 3 parts potassium hydroxide dissolved in 7 parts methanol were added to a solution of 170 parts 2-undecanone in 200 parts tert-butanol. The mixture was cooled to 5°C whereupon a solution of 106 parts acrylonitrile in 120 parts tert-butanol was added over a period of 3 hr maintaining the temperatures of 5° to 10°C. When the addition was complete, the reaction was allowed to warm to room temperature and stirred for 4 hr. After addition of 5 parts concentrated hydrochloric acid (diluted with water to 15 parts) to neutralize the catalyst, the solvents and unreacted acrylonitrile were removed by distillation.

The residue was partitioned between ether and water, and the ethereal phase collected, dried over magnesium sulfate, the ether removed, and the residue distilled under vacuum.

After removal of 5 parts of unreacted ketone (boiling range 60° to 65°C/0.6 mm Hg) a fraction (A) distilling at 125° to 135°C/0.5 mm Hg was collected (38 parts; representing a yield of 17% based on ketone) and identified as 4-acetyldodecanitrile by NMR and elemental analysis. $C_{14}H_{25}NO$ requires: C, 75.28%; H, 11.28%; N, 6.27%. Found: C, 75.56%; H , 11.18%; N, 6.54%.

A second fraction (B), distilling at 209° to 212°C/0.3 mm Hg was collected (215 parts; representing a yield of 78% based on ketone) and identified as 4-n-octyl-4-acetylpimelonitrile by NMR and elemental analysis. $C_{17}H_{28}N_2O$ requires: C, 73.91%; H, 10.14%; N, 10.14%. Found: C, 73.41%; H, 10.41%; N, 10.52%.

138 parts of fraction B were added to a solution of 99 parts of potassium hy-

droxide pellets (assay 85%) in 600 parts water and the mixture heated to reflux with stirring. Afte refluxing for 8 hr the solution was cooled, neutralized with concentrated hydrochloric acid to pH 2, and the precipitated syrupy solid extracted with ether. After washing and drying the ether solution, the solvent was removed leaving a solid residue which was recrystallized from ethyl acetate/petroleum spirit giving 131 parts (90% yield) of 4-n-octyl-4-acetylpimelic acid (melting point 69° to 71°C) identified by NMR and elemental analysis. $C_{17}H_{30}O_5$ requires: C, 64.94%; H, 9.62%. Found: C, 65.15%; H, 9.71%.

Example 2: 22.3 parts of fraction A (described in Example 1) were added to a solution of 13.2 parts 85% potassium hydroxide in 100 parts water. The mixture was heated to reflux and maintained for 6 hr whereupon the solution was cooled and neutralized with concentrated hydrochloric acid to pH 2.

The precipitated oil was extracted with ether, the ether solution washed with water, dried over magnesium sulfate and the solvent removed. The residue solidified on standing and was recrystallized from 40° to 60° petroleum spirit giving 21.3 parts (88% yield) of 4-acetyldodecanoic acid (melting point 36° to 38°C) identified by NMR and elemental analysis. $C_{14}H_{26}O_3$ requires: C, 69.38%; H, 10.81%. Found: C, 69.02%; H, 10.58%.

When incorporated into an aqueous functional fluid in contact with a ferrous metal, 4-acetyldodecanoic acid exhibited excellent corrosion inhibitor activity.

Example 3: In a manner analogous to that described in Example 1, 99 parts of 2-tridecanone were treated with 53 parts of acrylonitrile in tert-butanol. After work-up, distillation afforded a fraction (A) distilling at 159° to 165°C/0.2 mm Hg (33 parts; representing a yield of 26% based on ketone) identified as 4-acetylmyristonitrile by NMR and elemental analysis. $C_{16}H_{29}NO$ requires: C, 76.49%; H, 11.55%; N, 5.58%. Found: C, 77.00%; H, 11.53%; N, 5.58%.

The residue was not further distilled but was identified by NMR, GLC and elemental analysis as being substantially (>95%) 4-n-decyl-4-acetylpimelonitrile (107 parts, 70% yield). $C_{19}H_{32}N_2O$ requires: C, 75.00%; H, 10.53%; N, 9.21%. Found: C, 74.52%; H, 10.19%; N, 9.41%.

The residue was hydrolyzed as described in Example 1 to yield 107 parts (75% yield) of 4-n-decyl-4-acetylpimelic acid, crystallized from ethyl acetate/petroleum spirit. $C_{19}H_{34}O_5$ requires: C, 66.63%; H, 10.01%; N, 23.36%. Found: C, 66.62%; H, 9.81%; N, 23.57%.

Example 4: 25 parts of fraction A described in Example 2 were hydrolyzed in a manner previously described to yield 23 parts (84% yield) of 4-acetylmyristic acid. $C_{16}H_{30}O_3$ requires: C, 71.11%; H, 11.11%. Found: C, 71.23%; H, 11.55%.

When incorporated into an aqueous functional fluid, 4-acetylmyristic acid exhibited excellent corrosion inhibitor activity.

–4–

Coatings and Films

FOR METAL SUBSTRATES

Thiofunctional Polysiloxanes

The compositions of *E.R. Martin; U.S. Patent 4,251,277; February 17, 1981; assigned to SWS Silicones Corporation* are effective as corrosion inhibitors and as release agents for metal substrates. They comprise (1) an organopolysiloxane fluid and (2) thiofunctional polysiloxanes which have at least one mercaptan group. Organopolysiloxanes employed in this composition may be represented by the general formula:

$$\left[-\overset{\displaystyle R}{\underset{\displaystyle R}{\overset{|}{\underset{|}{Si}}}}O- \right]_x$$

wherein R, which may be the same or different, represents monovalent hydrocarbon radicals or halogenated monovalent hydrocarbon radicals having from 1 to 18 carbon atoms and x is a number greater than 8.

Any linear, branched or cyclic organopolysiloxanes having an average of from 1.75 to 2.25 organic radicals per silicon atom may be employed. The organosiloxanes may be triorganosiloxy, alkoxy or hydroxy terminated; however, they should be free of aliphatic unsaturation. It is preferred that the polysiloxanes have a viscosity of between about 5 and 1,000,000 cs, and more preferably between about 50 and 300,000 cs at 25°C. Also, it is possible to combine high and low viscosity fluids to form a fluid having the desired viscosity. High molecular weight gums may also be employed, however, it is preferred that these gums be dissolved in an organic solvent before they are combined with the thiofunctional polysiloxanes.

The thiofunctional polysiloxanes employed in the composition may be prepared by reacting a disiloxane and/or a hydroxy or hydrocarbonoxy containing silane or siloxane with a cyclic trisiloxane in the presence of an acid catalyst in which at least one of the above organosilicon compounds contains a mercaptan group. Disiloxanes which may be used in this process may be represented by the formula:

$$\underset{(M)_aSi_2O}{\overset{(R)_{6-a}}{|}}$$

while the cyclic siloxanes are represented by the formula:

$$\underset{(M)_aSi_3O_3}{\overset{(R)_{6-a}}{|}}$$

wherein R, which may be the same or different, represents a monovalent hydrocarbon radical or a halogenated monovalent hydrocarbon radical having up to 18 carbon atoms, M is a mercaptan containing group represented by the formulas $R'(SR''')_y$ in which at least one R''' is hydrogen, and

R''
—SH
—R''

a is a number of from 0 to 5 and y is a number of from 1 to 3.

Catalysts which may be employed in affecting the reaction between a disiloxane and/or a hydroxy and/or hydrocarbonoxy containing silane or siloxane and a cyclic trisiloxane in which at least one of the reactants contains a mercaptan group are acid clays and organic and inorganic acids have a pK value less than 1.0 and more preferably below 0.7 in aqueous solutions. Suitable acid catalysts which may be employed are benzosulfonic acid, p-toluenesulfonic acid, sulfuric acid, sulfurous acid, nitric acid, perchloric acid, hydrochloric acid and acid clays such as Filtrol No. 13 and No. 24.

The compositions of this process can contain from 0.1 to 90% by wt of thiofunctional polysiloxanes and from 10 to 99.9% by wt of organosiloxanes.

Metals and alloys which may be treated with the composition of this process are those below and including magnesium in the electromotive series. The metals and alloys include aluminum, brass, bronze, copper, chromium, iron, magnesium, nickel, lead, silver, silverplate, sterling silver, tin, beryllium, bronze and zinc.

Example 1: A thiofunctional polysiloxane is prepared by adding 28.2 pbw of 3-mercaptopropyltrimethoxysilane, 0.5 pbw of water, and 25.5 pbw of Filtrol No. 13 acid clay to a reaction vessel containing 1,276 pbw of hexamethylcyclotrisiloxane heated to 70°C. The vessel is heated to 100°C and maintained at this temperature for 3 hr. The contents of the vessel are then cooled to 60°C and filtered. The volatiles are stripped off for 8 hr at 200°C at less than 1 torr. A clear transparent liquid is obtained having a viscosity of 100 cs at 25°C. Nuclear Magnetic Resonance (NMR) analysis shows that the product has a mol ratio of $CH_3O:HSC_3H_6:Si(CH_3)_2$ of 3:1:100. The SH content of the product is about 0.43%.

The resultant composition is applied as a thick film to metal panels and a layer of

Epoxical Urethane No. 1850 Foam "B" Pak (U.S. Gypsum) is then applied to the coated panels. The panels are then heated in a forced air oven at 100°C for 2 min, then removed from the oven and cooled. After repeated applications, a build up of residue is observed.

Example 2: About 100 parts of the mercaptofunctional fluid prepared in accordance with Example 1 are mixed with 900 parts of a trimethylsiloxy-endblocked dimethylpolysiloxane having a viscosity of 350 cs at 25°C. The resultant composition is applied as a film to metal panels and a layer of Epoxical Urethane No. 1850 Foam "B" Pack is then applied to the coated panels. The panels are then placed in a forced air oven at 100°C for 2 min, then removed from the oven and cooled. After repeated applications no residue build up is observed on the surface.

Example 3: Two copper panels are thoroughly cleaned with a commercial polishing compound. To one panel is applied the composition of Example 2 and the excess is removed. In this manner, the nontreated panel served as a control. Both panels are placed in a humid H_2S chamber which consists of an aqueous solution of sodium sulfide to which dilute formic acid is periodically added. Within 20 min, the untreated panel is badly discolored. The treated panel shows no evidence of discoloration.

Non-Newtonian Colloidal Disperse System and Phosphoric Acid Ester

L.S. Cech; U.S. Patent 4,264,363; April 28, 1981; assigned to The Lubrizol Corporation provides a coating composition for protecting metal from corrosion. The composition comprises the following:

(I) a mixture made by the process which comprises mixing at a temperature within the range of from about 25°C up to the decomposition temperature of (A) or (B):

- (A) from about 5 to 10 pbw of a non-Newtonian colloidal disperse system comprising
  - (1) solid, metal-containing colloidal particles selected from the class consisting of alkali and alkaline earth metal carbonate predispersed in
  - (2) a dispersing medium, and,
  - (3) as an essential third component, at least one organic compound which is soluble in the disperse medium, consisting of an alkaline earth metal salt of an acid selected from the class consisting of oil-soluble carboxylic and sulfonic acids with
- (B) from about 0.5 to 5 pbw of an acidic ester of a phosphoric acid wherein the alcoholic portions of the acidic ester are selected from the class consisting of hydrocarbyloxy and hydroxy-substituted hydrocarbyloxy compounds, and wherein the solid metal-containing colloidal particles and the third component, in combination constitute from about 10 to 70% by wt of the disperse system, and

(II) a hydrocarbon resin which is substantially insoluble in mixture (1) at temperatures below about 60°C.

In the following examples all parts and percentages are by weight unless explicitly stated to the contrary.

Example 1: An acidic acid ester of a phosphoric acid is prepared as follows. A polyisobutene-substituted phenol is prepared by mixing 940 parts of phenol and 2,200 parts of polyisobutene having a molecular weight of 350 at 50° to 55°C in the presence of 30 parts of boron trifluoride, and distilling off the unused phenol and other volatile substances by heating the alkylated phenol to 220°C/12 mm. The resulting alkylated phenol has a hydroxyl content of 3.7%.

A mixture of 1,089 parts of xylene and 524 parts of the above prepared polyisobutene-substituted phenol is heated to 50°C whereupon 523 parts of 1:1 (molar) copolymer of allyl alcohol and styrene having an average molecular weight of 1,100 is added over a period of 20 min at 50°C. Solution was complete after 1 hr at this temperature. Phosphoric anhydride (52 parts) is added over a period of 15 min at 50°C and the mixture is heated to the boiling point and to 145°C in 1.2 hr. The mixture is stirred and maintained at this temperature while removing a water-xylene azeotrope over a period of 6 hr. The residue is cooled to 40°C. The residue, a 50% solution in xylene, has a phosphorus content of 1.03% and a neutralization number (bromophenol blue) of 20 acid.

Example 2: Following the procedure of Example 1, to a mixture of 1,000 parts of aromatic hydrocarbon solvent and 595 parts of the polyisobutene-substituted phenol is added 357 parts of a styrene/allyl alcohol resin having a hydroxyl content of about 5.7 and an average molecular weight of 1,600 and phosphoric anhydride (47.6 parts) is added. The reaction mixture is heated at 145°C with nitrogen blowing. The product, a 50% solution in aromatic solvent, has a phosphorus content of 1.05% and a neutralization number of 16 acid.

Example 3A: A mixture of 520 parts of a mineral oil, 480 parts of a sodium petroleum sulfonate (molecular weight 480), and 84 parts of water is heated at 100°C for 4 hr. The mixture is then heated with 86 parts of a 76% aqueous solution of calcium chloride and 72 parts of lime (90% purity) at 100°C for 2 hr, dehydrated by heating to a water content of less than 0.5%, cooled to 50°C, mixed with 130 parts of methyl alcohol, and then blown with carbon dioxide at 50°C until substantially neutral. The mixture is then heated to 150°C to remove the methyl alcohol and water and the resulting oil solution of the basic calcium sulfonate filtered. The filtrate is found to have a calcium sulfate ash content of 16% and a metal ratio of 2.5.

Example 3B: Mineral oil (2,250 parts), 960 parts (5 mols) of heptylphenol, and 50 parts of water are introduced into a reaction vessel and stirred at 25°C. The mixture is heated to 40°C and 7 parts of calcium hydroxide and 231 parts (7 mols) of 91% assay paraformaldehyde is added over a period of 1 hr. The whole mixture is heated to 80°C and 200 additional parts of calcium hydroxide (making a total of 207 parts or 5 mols) is added over a period of 1 hr at 80° to 90°C. This mixture is heated to 150°C and maintained at that temperature for 12 hr while nitrogen is blown through the mixture to assist in the removal of water. If foaming is encountered, a few drops of polymerized dimethyl silicone foam inhibitor may be added to control the foaming. The reaction mass is then filtered for purposes of purification. The filtrate, a 33.6% oil solution of the desired calcium phenate of heptylphenol-formaldehyde condensation product, is found to contain 7.56% sulfate ash.

Example 3C: 1,000 parts of the carbonated calcium sulfonate complex of Example 3A, 75 parts of the calcium phenate of Example 3B and 325 parts of mineral oil are mixed together. Then, at 60°C are added, 17 parts methanol, 114 parts butanol and a solution of 1.6 parts $CaCl_2$ in 2 parts of water. To this mixture is added a total 667 parts lime in six increments at 50°C with carbonation after each increment until a final base number of 50 to 60 is obtained. The product is nitrogen blown at 150°C to remove water, 280 g additional oil is added, mixed, then filtered. The filtrate has a calcium sulfate ash content of 50% and a base number of 400.

Example 4: To a mixture of 5,000 parts of the dispersed system described in Example 3C and 1,665 parts of Rule 66 Stoddard solvent, at 50° to 55°C is added carbon dioxide at the rate of 2.0 scfh for 20 min. The carbonated mixture is then heated to 66°C and a mixture of 284 parts methanol and 216 parts of water is added over a 10 min period. The mixture is refluxed for 1.5 hr at which point it is very viscous. An additional 500 parts of Stoddard solvent is added and the mixture is stripped under nitrogen to remove the aqueous alcohol. The mixture is heated for 1.5 hr at 160°C under nitrogen. Then an additional 1,366 parts of Stoddard solvent is added and the mixture is heated to 110°C. The material described in Example 1 (655 parts) 50% active in a hydrocarbon solvent is slowly added to the stirred mixture and stirring is continued; the mixture is allowed to cool. The resulting product is the desired mixture of colloidal dispersion and acidic phosphoric acid ester.

Example 5: To 2,850 parts of the mixture described in Example 4 is added 150 parts of an ethylene derived polymer (Vybar 103). The polymer is cut in small pieces. This mixture is stirred for 2 hr while being heated to a temperature of between 115° and 135°C. The cooled mixture is then diluted with Stoddard solvent to form the desired coating composition.

Converting Pickle Oils to Nondripping Solid Coatings

F.P. Lochel, Jr.; U.S. Patent 4,264,653; April 28, 1981; assigned to Pennwalt Corporation has found a process that converts pickle oils, slushing oils and oily lubricants used for drawing and stamping operations into nondripping coatings on steel. This process provides for the inclusion of at least one long chain linear alcohol in the oils at elevated temperatures. The base oils for the pickle oil, slushing oil and oil-based lubricants are hydrocarbon oils. The long chain alcohols are selected from the group of C_{18} to C_{24} linear alcohols. The long chain linear alcohols constitute from about 3 to 30% by wt of the coating mixture of hydrocarbon oil and alcohol.

The process requires that the coating mixture be applied hot to the hot steel and thereafter the coated steel be cooled. A convenient way of cooling is to allow the coil to cool naturally by exposure to the surrounding air. At a temperature above ambient, the oily coating mixture becomes a nondripping mesosolid. This solid coating eliminates the fire and accided hazard previously encountered by oil drippings which accumulate on the floor of the steel mill. More important, the mesosolid coating significantly increases the corrosion protection of the steel.

Example 1: Steel strip continuously emerges from a hot rolling mill at a speed of about 500 fpm. The steel strip is then directed through a mechanical scale remover and while still hot is directed through a sulfuric acid pickling bath held at 160°

to 180°F. Immersion time in the pickle tank is about 2 to 3 sec. After leaving the pickle tank, the steel strip is rinsed twice with hot water and is then passed through a hot air dryer. In the hot air dryer, a blast of hot air literally blows the water off the surface of the steel strip. Any residual water is evaporated by the heat in the steel which will be at a temperature in the range of about 151° to 200°F.

The steel strip is then directed to the pickle oil station where the steel is contacted with a pickle oil coating mixture having a hydrocarbon oil base. The coating mixture will be held at a temperature in excess of about 150°F and is generally in the range of about 151° to 180°F. The pickle oil coating mixture has been previously prepared in a separate formulating tank and then it is pumped into the pickle oil coating storage tank.

The pickle oil coating mixture is continuously circulated by pump from the coating tank to a flooding station directly over the tank. At the flooding station, the hot strip is contacted with an excess of oily coating mixture which excess then drains back into the coating storage tank. Take-up rolls regulate the amount of coating mixture retained on the steel strip. Preferably, the pickle oil coating mixture is held at a temperature in the range of about 160° to 180°F. The pickling oil coating mixture is prepared according to the following formula:

	Percent by Weight
Mineral oil (naphthenic base) 300 SUS at 100°F	73.3
Behenyl alcohol (technical)	5.0
Emulsifier	1.43
Foam control agent	0.005
Pine oil (perfume)	0.005
Deodorant (perfume)	0.005
Germicide	0.005
Antioxidant	0.050
Phos-amine (EP additive)	4.7
Yellow grease	15.5

The pickle oil has a flash point of about 405°F, an acid number of 13.5 and a viscosity of 250 SUS at 100°F.

The coated steel strip is allowed to cool to ambient temperature while it is being recoiled. As soon as the temperature drops below 150°F, the coating mixture converts to a nondripping, corrosion resistant mesosolid. The recoiled strip is then transferred to the cold rolling mill or to a finishing operation.

Example 2: The steel strip as it leaves the cold rolling mill is coated with a slushing oil in the same manner as the pickle oil is applied in Example 1. The slushing oil has the following composition:

	Percent by Weight
Mineral oil (naphthenic base) 300 SUS at 100°F	78
Behenyl alcohol (technical)	15
Germicide (Dowicil)	0.005
Pine oil	0.05
Yellow grease	6.8
Surfactant	0.1

After the coating is applied to the steel strip, the strip is recoiled for transfer to manufacturing operations or to storage.

No-Rinse Chromate-Depositing Solution

No-rinse compositions and a process are provided by *W.D. Krippes; U.S. Patent 4,266,988; May 12, 1981; assigned to J.M. Eltzroth and Associates, Inc.* for inhibiting corrosion of ferrous or nonferrous metal surfaces and for producing a surface to which synthetic resin coating compositions will adhere so that the resultant coatings have satisfactory impact and bending resistance, together with resistance to creeping corrosion between the metal and the dried resin coating.

In accordance with the process a ferrous or nonferrous metal surfaced article is treated with an aqueous chromate-depositing solution containing hexavalent chromium but no trivalent chromium, together with fluoboric acid, hydrofluoric acid, sulfuric acid, with or without hydrofluosilicic acid. The solution also contains as an additive zinc oxide and/or magnesium oxide, and/or magnesium hydroxide, and/or aluminum sulfate, and/or aluminum hydroxide, the ratio of the additive to the total acids being such as to give a pH within the range of 1.5 to 3.6 at 22°C and a chromate concentration of 0.05 to 10.0 g/ℓ, as Cr.

In the following example, the proportions are in weight unless otherwise indicated.

Example: A concentrate was prepared by mixing together the following ingredients:

	Percent by Weight
CrO_3	12.0
ZnO (French processed)	0.8
H_2F_2 (48% concentration)	2.6
H_2SiF_6 (26% concentration)	0.4
HBF_4 (48% concentration)	0.8
H_2SO_4 (78 to 80% concentration)	0.8
H_2O	82.6

This concentrate has a specific gravity of approximately 1.11. Water is added to the above concentrate in sufficient amount to give a running bath having a concentration of 0.5 to 1% with a pH of approximately 1.8.

The metal to be processed or coated can be, e.g., cold rolled steel, aluminized and galvanized iron and steel, aluminum, aluminum-zinc alloys, magnesium or magnesium-aluminum alloys. Typical examples of cold rolled steel are SAE 1005 or 1010.

The concentration of 0.5 to 1.0% is given as Cr. The weight ratio of the amount of water added to the concentrate is approximately 15:1. This ratio may vary depending upon the desired concentration of the depositing solution but will usually be within the range of 3:1 to 50:1.

The metal to be coated is carefully cleaned with an alkaline cleaner at 160°F, hot water rinsed at 140° to 180°F and then coated in a coating bath containing a predetermined concentration of the above composition and having a predetermined pH which is adjusted by adding more or less of the zinc oxide or other additive previously described to the concentrates.

The coating weight on the metal will depend upon the particular metal and the pH of the coating bath. Thus, on aluminum, lowering the pH from 2.7 to 1.8 increases the coating weight with a given concentration of chromate as Cr, from 7 to 14.4 mg/ft², as Cr. Likewise on galvanized iron lowering the pH from 2.7 to 1.8 increases the coating weight from 2.8 to 12 mg/ft², as Cr. An optimum pH is 1.8 to 2.0.

On cold rolled steel lowering the pH reduces the coating weight. Thus, at a pH of 3 the coating weight is approximately 34.5 mg/ft², as Cr, and at a pH of 2 the coating weight is approximately 20.0 mg/ft², as Cr. As the Cr concentration is increased from 0.05 to 10.0 g/ℓ, with constant pH, the total coating weight may increase from 3 to 80 mg/ft².

The foregoing weights are based on an application of 3 sec contact time using a roll coater, dip, spray or other type of coating, followed by a squeegee to remove excess coating composition. Removal of excess composition is quite important. The application of the coating composition to the metal is preferably with a time period range of 1 to 10 sec. After coating, most of the excess is removed by passing the metal in strip form through a squeegee and it is desirable to dehydrate the coated metal as much as possible before painting. Paints are preferably baked on the metal at temperatures up to 550°F. Any kind of synthetic resin coating composition can be applied which dries to a water resistant film. Both corrosion resistance and adherence are enhanced.

Aluminum coated with a coating composition of the type described above is coated with a polyvinyl chloride primer and top coat baked on in the manner described above will withstand standard salt spray tests for at least 2,000 to 3,000 hr. Galvanized steel similarly coated will withstand standard salt spray tests at least 900 to 1,000 hr. Cold rolled steel similarly coated will withstand standard salt spray tests at least 600 hr.

Single-Package Zinc-Rich Coatings

T. Ginsberg and L.G. Kaufman; U.S. Patent 4,277,284; July 7, 1981; assigned to Union Carbide Corporation describe compositions consisting essentially of particulate zinc, an unhydrolyzed or a partially hydrolyzed organic silicate and a hardening amount of an hydrolyzable silicon compound. The hydrolyzable silicon compound is selected from the class consisting of (a) aminosilanes of the formula:

$$Z\!-\!\left(\underset{\displaystyle}{\overset{\displaystyle Q}{\overset{|}{N}}}\!-\!R^1\right)_t\!-\!\underset{\displaystyle M}{\underset{|}{N}}\!-\!Y$$

wherein t is an integer having values of 0 to 10; each of M, Y, Q and Z are R or

$$-R^1-\overset{\displaystyle R_b}{\overset{|}{Si}}-X_{3-b}$$

R is H, alkyl having 1 to 4 carbon atoms or hydroxyalkyl having 2 to 3 carbon atoms; R^1 is $-C_2H_4-$, $-C_3H_6-$ or $-R^2-O-R^2-$ and R^2 is an alkylene radical having about 1 to 8 carbon atoms; b is an integer having values of 0 to 2; with the proviso that at least one of M, Q, Y or Z is

$$-R^1-\overset{\displaystyle R_b}{\overset{|}{Si}}-X_{3-b}$$

X is an hydrolyzable organic group; (b) quaternary ammonium salts of the aminosilanes and (a); and (c) the hydrolyzates and the condensates of the aminosilanes in (a).

The compositions described above are stable for prolonged periods of time in a closed container. Thus separate packaging is not required. When applied on a ferrous substrate, the zinc-rich formulations dry rapidly with the result that a hard, continuous, smooth film is formed having excellent corrosion protecting properties. It is preferred to use about 15 to 45% by wt of hydrolyzable silicon compound.

In the following examples all parts and percentages are by weight unless otherwise specified.

Example 1: *Single-package zinc-rich coating with ethyl silicate 40 and γ-aminopropyltriethoxysilane* – A ferrous metal coating composition was prepared by mixing 45 g of partially hydrolyzed ethyl polysilicate containing 40% by wt of SiO_2, with 5 g of γ-aminopropyltriethoxysilane and 30 g of finely-divided zinc having a particulate size of about 2 to 15 μ (Asarco L-15). In addition, in order to maintain the mixture in an anhydrous state, 5 g of a water scavenging agent (Union Carbide Corp. molecular sieves 4A) were added and the composition was thinned with 50 g of a hydrocarbon solvent consisting of a mixture of 61% by vol of paraffinics and 39% by vol of naphthenics having a boiling range of about 158° to 196°C of Amsco Mineral Spirits 66-3 (American Mineral Spirits Co.). The resultant liquid protective coating or primer paint had a package stability of over 6 months.

When this paint was applied by spraying to sand blasted, cold-rolled steel panels measuring approximately 4 x 4 x ⅛", there was obtained a smooth film which dried in less than 10 min. The steel panel so coated was subjected for 1,000 hr to salt spray (ASTM Method B-117) and 1,000 hr in fresh water immersion (ASTM Method B-870). There was no evidence of corrosion or other signs of failure on the panel so coated.

Example 2: *Single-package zinc-rich coating with tetraethylorthosilicate and N-β-(aminoethyl)-γ-aminopropyltrimethoxysilane* – A ferrous metal protective composition was prepared by mixing 45 g of tetraethylorthosilicate with 5 g of N-β-(aminoethyl)-γ-aminopropyltrimethoxysilane and 300 g of Asarco zinc dust L-15, 5 g of molecular sieves 4A, and 50 g of Amsco Mineral Spirits 66-3. The resultant primer paint was stable in storage for over 6 months. When applied as a spray coating to a sand blasted steel panel, a dry film formed in less than 10 min. When panels were exposed as in Example 1 for 1,000 hr in the salt spray and water immersion test, there was no evidence of corrosion or other failure.

Example 3: *Single-package zinc-rich coating with Cellosolve silicate and γ-amino-*

propyltriethoxysilane – A ferrous metal protecting paint primer composition was prepared by mixing 45 g of partially hydrolyzed ethoxyethylpolysilicate containing 19% SiO_2, 5 g of γ-aminopropyltriethoxysilane, 300 g of Asarco L-15 zinc dust, 5 g of molecular sieves 4A and 50 g of Amsco Mineral Spirits 66-3. The resultant primer paint composition had a package stability of over 6 months. When applied as a spray over sand blasted panels, the composition dried to a hard film in less than 10 min. The panels, when subjected to the salt spray and water immersion test described in Example 1 for 1,000 hr, showed no evidence of corrosion or other failures.

Aqueous Alkaline Treating Solution for Zinc Surface

In accordance with a process of *E.R. Reinhold; U.S. Patent 4,278,477; July 14, 1981; assigned to Amchem Products, Inc.* there is provided an aqueous alkaline treating solution which has a pH of no greater than about 10.2 and which contains, as essential ingredients, one or more of the following metals in solution: cobalt, nickel, iron and tin; and an inorganic or organic complexing material which is effective in maintaining the metal in solution. In addition, the solution can include a reducing agent.

A preferred inorganic complexing material for use in the practice of this process is pyrophosphate and a preferred organic complexing material is nitrilotriacetic acid or a salt thereof.

A coating solution within the scope of the process can be used to treat a zinc surface in a manner such that there is formed on the surface a coating which is corrosion resistant and to which overlying coatings adhere excellently. In addition, the coating solution is effective in forming a coating which is readily visible by virtue of its being colored. This is important because it signals the user that the composition is indeed forming a coating on the surface.

The process provides several other important advantages. Excellent results can be achieved by the use of a composition which has a substantially lower pH than what the industry has been used to. The lower alkalinity mitigates handling problems and permits the use of conventional containers and other equipment. Also, a bath of the composition can be operated for prolonged periods of time without encountering sludge problems. And, in addition, a bath of the composition can be prepared utilizing a minimum of ingredients.

The coating solution of the process can be used to coat surfaces of pure zinc or of alloys in which zinc is present in a significant amount, including, e.g., zinc die castings, hot-dipped galvanized and electrogalvanized steel surfaces, a 50/50 Al/Zn alloy and galvanneal. It is believed that one of the widest uses of the coating solution will be in the coating of hot dipped and electrogalvanized steel coil.

Unless stated otherwise, each of the Zn surfaces treated with the compositions identified in Examples 1 through 6 below was a zinc panel of hot-dipped galvanized steel, 4 x 12" in size, which was subjected to the following sequence of steps:

(A) spray cleaned with an aqueous alkaline cleaning solution for 20 sec at 160°F (71°C);

(B) rinsed with a cold water spray for 2 to 3 sec at ambient temperature;

(C) treated with a composition of the examples at a temperature of 125°F (52°C) by immersing in a laboratory immersion cell for 15 sec;

(D) rinsed with a cold water spray for 2 to 3 sec at ambient temperature;

(E) treated with a 0.5 wt % Cr^{6+}/reduced Cr aqueous solution sold as Deoxylyte 41 by immersing for 5 sec, followed by squeegeeing through wringer rolls and air drying; and

(F) painted with a single coat of polyester paint (CWS 9039, Hanna Chemical Coatings Corp.) to a paint film thickness of about 0.8 to 1 mil, followed by baking for 75 sec in an oven having a temperature of 500°F (260°C) to a peak metal temperature of 420°F (216°C) and then quenching in cold water.

Corrosion resistant properties were evaluated by subjecting painted panels to salt spray conditions in accordance with ASTM B-117.

A test referred to herein as "T-Bend" was used to evaluate paint adhesion. In general, a rating of 1 or 2 is considered excellent and a rating of 4 or more is considered poor.

Examples 1 through 6: These examples show the use of a treating composition within the scope of this process and comprises an alkaline solution of 25 g/ℓ of $K_4P_2O_7$ and 2.5 g/ℓ of $Fe(NO_3)_3 \cdot 9H_2O$, and the use of modified forms of this composition. The modification encompasses and includes in the composition amounts of $Co(NO_3)_2 \cdot 6H_2O$ as indicated in the table below, which sets forth also the pH of the treating compositions and the results of paint adhesion tests. In these examples, the paint used was an acrylic paint (Durocron 630) and the thickness of the dry paint film was about 0.5 mil.

Example No.	$Co(NO_3)_2 \cdot 6H_2O$, (g/ℓ)	pH	Paint Adhesion, Number of T-Bends
1	–	10.2	2
2	0.1	10.0	2
3	0.2	10.0	2
4	0.5	9.9	2
5	1.0	9.8	3
6	2.0	9.8	2

It was observed that the salt spray corrosion resistance of the coated panels increased proportionately to the cobalt concentration up to 0.5 g/ℓ cobalt nitrate. Beyond that concentration, no further increase in corrosion resistance was evident. The color did, however, increase as the cobalt concentration increased to the limit tested.

Active Pigments Based on Calcium, Aluminum and Iron Oxides

Active anticorrosion pigments, which are of considerable practical importance, are red lead, zinc potassium chromate and zinc dust. By comparison with the substantially inactive iron oxide pigments used in large quantities for corrosion prevention, these active anticorrosion pigments either have a considerably higher specific gravity or are considerably more expensive.

An object of the process of *F. Hund and P. Kresse; U.S. Patent 4,285,726; Aug. 25, 1981; assigned to Bayer AG, Germany* is to develop active and anticorrosion pigments based on aluminum and iron oxide which are equivalent to known active anticorrosion pigments in regard to their anticorrosion effect but which do not have any of the disadvantages of known anticorrosion pigments.

The process provides anticorrosion pigments comprising about 30.0 to 2.0 mol % of CaO and about 70.0 to 98.0 mol % of Me_2O_3, where Me_2O_3 represents (1−x) Al_2O_3 and x Fe_2O_3 and x may assume values of from 0 to 1, preferably about 0.01 to 0.95, up to about 25 mol % of the Fe_2O_3 being replaceable by the corresponding quantity of Mn_2O_3, and having a specific surface area according to BET of about 0.1 to 200 and preferably about 1.0 to 150 m^2/g and a % wt loss/g of pigment, as determined by the Thompson corrosion test of less than about 0.05% and preferably of less than about 0.03%/g of pigment.

The pigments are produced by preparing intimate mixtures of suitable reactants calcining the mixtures thus prepared at temperatures in a certain range, intensively grinding the calcination product, optionally calcining it once more, followed by cooling and grinding.

The calcination temperature is of importance to the properties of the anticorrosion pigments formed. In the case of pigments having a low aluminum oxide content, e.g., below about 15 mol %, the calcination temperature is about 200° to 700°C. For pigments of relatively high aluminum oxide content, e.g., above about 15 mol %, the calcination temperatures selected are about 200° to 1100°C. The calcination time is generally about 0.1 to 20 hr and preferably about 0.5 to 10 hr.

The active anticorrosion pigments were tested in a long-oil alkyd resin based on tall oil fatty acid of low resinic acid content by comparison with standard commercial-grade zinc phosphate and zinc chromate. A pigment volume concentration (PVC) of 34% was selected for the test. The basic recipe in pbw consists of 167.00 long-oil alkyd resin, 60% in white spirit; 2.50 readily volatile oxime as antiskinning agent (Ascinin R55, a product of Bayer AG); 4.00 Co-, Pb-, Mn-octoate, 1:2 in xylene; 1.25 Ca-octoate, 4%; and 25.25 dilution white spirit/turpentine oil 8:2 quantity of pigment according to the PVC selected. Grinding is carried out for 5 hr in laboratory vibrating ball mills.

For testing the anticorrosion behavior of the pigments, the primers are sprayed onto bonderized steel plates in such a way that dry layer thicknesses of about 45 μ are obtained. After a minimum drying time of 7 days, the anticorrosion test is carried out by the salt spray test according to DIN 53167 or ASTM B 287-61 (permanent spraying with 5% sodium chloride solution at 35°±2°C).

The test plates are inspected after 3, 8, 11, 18, 24, 31, 38, 45 and 52 days. They are assessed by a marking system extending from 0 (= no damage) to 12 (= complete destruction of the paint film). This method of assessment is described by P. Kresse in the *XIII-FatipecKongressbuch* (Cannes 1976), pages 346-353. The sum of the individual marks on the abovementioned inspection days up to the 52nd day for the individual pigmenting systems is known as the "degree of corrosion." The greater this sum, i.e., the higher the degree of corrosion in a selected pigment-binder system, the weaker the corrosion-inhibiting effect of the anticorrosion pigment used.

By comparison with the commercial-grade zinc phosphate anticorrosion pigment with its relatively poor anticorrosion values, pigments number 2.2 and 2.3 containing 14.3% of CaO and 85.7 mol % of Al_2O_3 and calcined at 500° and 600°C are considerably better. Within the errors of the salt spray test, these nontoxic anticorrosion pigments according to the process are almost as good in their corrosion-inhibiting effect as the commercially available zinc chromate pigments containing chromium(VI)-ions. The following table shows the results of the anticorrosion test (salt spray test according to DIN 53167 or ASTM B 287-61).

Pigment	Degree of Corrosion*
Pigment No. 2.2 - 500°C 14.3 mol % CaO; 85.7 mol % Al_2O_3	16.0
Pigment No. 2.3 - 600°C 14.3 mol % CaO; 85.7 mol % Al_2O_3	21.5
Zinc phosphate	48.5
Zinc chromate	11.5

*After 1,248 hours under test.

Hydrolyzed Alkyl Silicate Binders for Zinc-Rich Coatings

K.H. Brown and K.M. Wolma; U.S. Patent 4,290,811; September 22, 1981; assigned to Rust-Oleum Corporation describe a method of preparing stable hydrolyzed alkyl silicate binders useful for producing protective coatings such as zinc-rich coatings. The method utilizes an ion exchange resin as a catalytic source of hydrogen ion instead of conventional acid catalysts. The ion exchange resin is removed after hydrolysis thereby greatly reducing the residual acidity. Lower residual acidity tends to reduce instability of the hydrolyzed alkyl silicate binders and increase the pot life of protective coatings made from the hydrolyzed alkyl silicate.

Generally the process involves dissolving an alkyl silicate in an organic solvent containing a strong acid form ion exchange resin and then slowly adding a quantity of water being in the range of 0.25 to 0.95 of an equivalent weight with respect to the quantity of alkyl silicate, or an amount sufficient to provide 0.125 to 0.475 mol of water for each alkoxy group carried by the alkyl silicate. When a partially hydrolyzed alkyl silicate is used in place of alkyl silicate the quantity of water must be adjusted so as not to exceed the equivalent weight of the unhydrolyzed alkyl silicate.

The alkyl silicates that may be employed in the process are alkoxy silicates such as tetraalkoxysilicate where the alkyl groups range from 1 to 4 carbon atoms. The most preferred silicate is tetraethylorthosilicate.

Partially hydrolyzed alkyl silicates may be used in place of the alkyl silicate when a higher degree of hydrolysis is desired. Commercially available partially hydrolyzed ethyl silicates such as Ethyl Silicate-40 (ES-40) are particularly preferred.

The following two examples illustrate the improvement over the prior art which is described in this process. Example 1 describes the method of manufacturing hydrolyzed ethyl silicate using a conventional acid catalyst. Example 2 describes the method of manufacturing hydrolyzed ethyl silicate using an anhydrous strong acid form ion exchange resin as the catalyst.

Example 1: A reaction vessel is flushed with dry inert gas. To the vessel are added 2,300 g of ES-40 and 600 g of anhydrous isopropanol. Agitation is then begun. A solution of 1.79 ml of concentrated hydrochloric acid in 261 ml of water is then added to the vessel over a two-hour period, with continuous agitation under the inert gas. The product is stirred for an additional hour and then filtered. The final product was 90% hydrolyzed ethyl silicate.

Example 2: A reaction vessel is flushed with dry inert gas. To the vessel are added 2,300 g of ES-40, 600 g of anhydrous isopropanol and 36 g of Amberlyst-15. Agitation is then begun. 261 ml of water is then added over a two-hour period, with continuous agitation under the inert gas. The product is stirred for an additional hour and then filtered to remove the Amberlyst-15. The final product was 90% hydrolyzed ethyl silicate.

Hydrolyzed ethyl silicate samples made by the methods of Examples 1 and 2 were tested for stability. Stability data are presented in the following table.

Acid Catalyst Type	Percent Hydrolysis	Residual Acidity (ppm as HCl)	Shelf Stability Gel Time (sec) 3 Day Reading	60 Day Reading
1. HCl	82.5	200	115-120	75-80
2. A-15	82.5	30-60	165	100
3. HCl	90.0	200	43	16
4. A-15	90.0	40-50	48	26

A-15 is Amberlyst-15. Samples 1 and 3 were prepared according to the prior art method and samples 2 and 4 were prepared according to the process. Percent hydrolysis of the ethyl silicate is determined by the amount of water used in the hydrolysis reaction. Residual acidity was measured by potentiometric titration with standard alcohol solution of base.

Gel time test consists of placing the sample of hydrolyzed ethyl silicate in a viscosity tube, adding a specific amount of morpholine, closing the tube and then turning the tube up, then down, so that the air bubble moves up the length of the tube. This is continued until the air bubble essentially stops moving. This time period is called the gel time and is usually expressed in seconds. Morpholine acts as a base for this test, and the further hydrolysis of the ethyl silicate is greatly accelerated in the presence of base.

The final zinc-rich product typically contains various pigments in addition to the zinc dust and the silicate binder. Many different pigments such as metal oxides are commercially available. Pigments are added to provide color and hiding and also some cost reduction.

In addition suspending agents such as waxes and clay may be added to provide proper pigment dispersion and rheology. The following table provides data on stability (measured by pot life) of pigmented zinc-rich coatings using hydrolyzed ethyl silicate binders.

Acid Catalyst Type	Percent Hydrolysis	Residual Acidity (ppm as HCl)	Pot Life After Activation of Fresh PHES with Zinc
1. HCl	82.5	200	9-12 Days
2. A-15	82.5	30-60	25-30 Days
3. HCl	90.0	200	6 Days
4. A-15	90.0	40-50	9-10 Days

PHES is pigmented hydrolyzed ethyl silicate. The pigmented hydrolyzed ethyl silicate is mixed with the proper amount of zinc dust and the time period until gellation is shown in the above table. The rows numbered 1, 2, 3 and 4 in the two tables above represent data from several experiments. Stability data may be averages of several experiments using the same system, or whenever possible, may be expressed as a range.

Production of Phosphate-Containing Pigments

The process of *A. Maurer, R. Adrian, K. Hestermann and G. Heymer; U.S. Patent 4,294,621; October 13, 1981; assigned to Hoechst AG, Germany* relates to the production of phosphorus-containing anticorrosive pigments with a particle size of at most 20 μ. The pigments are made by reacting one or more calcium or magnesium compounds with phosphoric acid or acid alkali metal or ammonium phosphates by intimately mixing an aqueous suspension or solution of the reactants inside a dispersing means rotating at a speed of 3,000 to 10,000 rpm. More particularly, fine pulverulent calcium and magnesium compounds are converted, with agitation, together with water and, if desired, a water-soluble alkanol to a homogeneous suspension or solution.

The suspension or solution is intimately mixed, inside the dispersing means at 0° to 40°C and in an approximately stoichiometric ratio, with the phosphoric acid or acid phosphate solution. Resulting and precipitated pigment is separated and dried under mild conditions while maintaining its content of water of crystallization.

The anticorrosive pigment is comprised of a homogeneous mixture and/or mixed crystals of which 10 to 95 mol % is calcium hydrogen phosphate dihydrate, the balance being magnesium hydrogen phosphate trihydrate. An at least 55 wt % proportion consists of particles with a size of up to 5 μ with a BET-surface of at least 1 m^2/g.

Example 1: 1,801.4 g of $CaCO_3$ and 80.6 g of MgO were introduced into a 30 ℓ vessel provided with a stirrer (500 rpm) and made with 20 ℓ of water into a uniformly concentrated suspension. By means of a pump, the suspension was metered into a funnel-shaped reaction zone which was provided with an axially disposed disperser. It was operated by the rotor/stator-principle. The dispersing means was a high efficiency disperser Ultraturrax. The disperser was operated at a speed of 10,000 rpm. Within 1 hr, the reaction zone was uniformly supplied with 2,305.2 g of 85 wt % phosphoric acid. The mixture was circulated by pumping for 30 min and then filtered off through a suction filter.

The calcium/magnesium/hydrogen-phosphate pigment so obtained was washed with water and acetone and dried for 3 hr at 60°C in a drying cabinet. The pigment had a particle size of at most 12 μ, and 60 wt % of the pigment had a size of less than 5 μ. The specific BET-surface area was 2.7 m^2/g. Analysis indicated that the pigment contained 21.0 wt % of Ca, 1.3 wt % of Mg and 18.0 wt % of P. 90.7 mol % was calcium hydrogen phosphate dihydrate and 9.3 mol % was magnesium hydrogen phosphate trihydrate.

Example 2: The procedure was the same as in Example 1, but phosphoric acid was metered into the reaction zone through a nozzle structure. A pigment with a particle size of at most 10 μ was obtained. 75 wt % of the pigment had a size of less than 5 μ. The BET-surface areas was 2.8 m^2/g. The pigment was analyzed and found to contain 21.0 wt % of Ca, 1.4 wt % of Mg and 17.9 wt % of P. 90.1 mol % was $CaHPO_4 \cdot 2H_2O$ and 9.9 mol % was $MgHPO_4 \cdot 3H_2O$.

Example 3: The procedure was the same as in Example 2, but 1,000.8 g of $CaCO_3$ and 403.2 g of MgO were used. A pigment with a particle size of at most 15 μ was obtained. 65 wt % of the pigment had a size of less than 5 μ. The BET-surface area was 2.5 m^2/g. The pigment was analyzed and found to contain 11.8 wt % of Ca, 7.0 wt % of Mg and 17.8 wt % of P. 50.6 mol % was $CaHPO_4 \cdot 2H_2O$ and 49.4 mol % was $MgHPO_4 \cdot 3H_2O$.

Example 4: The procedure was the same as in Example 2, but 400.3 g of $CaCO_3$ and 645.0 g of MgO were used. A pigment with a particle size of at most 15 μ was obtained. 60 wt % of the pigment had a size of less than 5 μ. The BET-surface area was 2.1 m^2/g. The pigment was analyzed and found to contain 5.0 wt % of Ca, 11.2 wt % of Mg and 17.9 wt % of P. 21.3 mol % was $CaHPO_4 \cdot 2H_2O$ and 78.7 mol % was $MgHPO_4 \cdot 3H_2O$.

Example 5: 100 kg of $CaCO_3$ and 10 kg of MgO were placed in a 1 m^3-vessel provided with a stirrer and made therein, with thorough agitation (100 rpm), and 500 ℓ of water into a homogeneous suspension. The suspension was allowed to flow freely into a screw pump which was used (a) for repumping the reaction mixture and (b) as the reaction zone. The screw pump was operated at a speed of 3,000 rpm. Immediately upstream of the screw pump, the suspension was admixed by means of a nozzle structure, within 3 hr, with 145 kg of 85 wt % phosphoric acid.

After the whole quantity of phosphoric acid had been added, the reaction mixture was circulated for 30 min and then filtered off with the use of a centrifuge. The calcium/magnesium/hydrogen-phosphate pigment so-obtained was water-washed inside the centrifuge and dried in a fluidized bed dryer, for 10 to 15 sec at 70°C. The pigment had a particle size of at most 8 μ and 80 wt % of the pigment had a size of less than 5 μ. The BET-surface area was 2.8 m^2/g. The pigment was analyzed and found to contain 18.8 wt % of Ca, 2.7 wt % of Mg and 18.0 wt % of P. 80.9 mol % was $CaHPO_4 \cdot 2H_2O$ and 19.1 mol % was $MgHPO_4 \cdot 3H_2O$.

Testing Pigments for Anticorrosive Efficiency: *Test formulation –*

	Parts by Weight
Alkyd resin (Alftalat AF 342)	38.0
Ethylene glycol	4.0
White spirit	4.0
n-Butanol	0.5
Antiskinning agent (Additol XL 297)	0.5
Dimethyl-dioctadecyl-ammonium montmorillonite (Bentone 34; 10% strength)	1.0

(continued)

	Parts by Weight
Talc	10.0
Barium sulfate	11.0
Titanium dioxide	13.0
Anticorrosive pigment	18.0

On the basis of the above formulation, an anticorrosive pigment with a pigment concentration by volume of 32 (hereinafter termed PCV) was made by varying the proportion of anticorrosive pigment. The term PCV as used herein denotes the ratio of pigment volume and filler volume to total volume of all nonvolatile lacquer or varnish ingredients.

Description of corrosion test – The pigments of the process and comparative pigments were applied to sheet metal specimens and the specimens were subjected to the following short time tests: Salt Spray Test (ASTM B 117-64; hereinafter referred to as SST); Condensed Moisture Test (DIN 50017; hereinafter referred to as CWT); and Kesternich Test (DIN 50018; hereinafter referred to as KT).

The specimens tested were inspected for: degree of corrosion (rust) (European scale for determining the degree of corrosion of anticorrosive paints); degree of blister formation (DIN 53 209); and corrosion of metal underlying pigment of cross-scratched specimen. The standardized magnitudes defining the degree of corrosion and formation of blisters, and the corrosion depth (in millimeters) of the underlying metal were assigned an evaluation score (ES) of 0 to 100. More specifically, the score (decreasing from 100 to 0) assigned to the individual specimens was lower the more serious the degree of corrosion or blister formation or corrosion of the underlying metal. Each of the above three tests provides for a maximum score of 300 to be assigned to a 100% corrosionproof specimen.

In order to identify the efficiency of the corrosion-inhibiting pigment by a numerical value, the three evaluation scores (ES) were converted to a single characteristic value (CV). The salt spray test is the best to reveal the protective efficiency under long term outdoor conditions. This is the reason why the factor 2 has been assigned to the salt spray test in calculating the characteristic value in accordance with the following formula:

$$CV = \frac{2 \times ES(SST) + ES(CWT) + ES(KT)}{1200} \times 100$$

As can be seen, a pigment affording a 100% corrosionproof effect can be assigned a maximum characteristic value of 100.

The results obtained in the dispersibility test and corrosion test are indicated in the following table.

Pigment	Evaluation Score Corrosion Test
$CaHPO_4 \cdot 2H_2O$	62
$MgHPO_4 \cdot 3H_2O$	57
Example 1	82
Example 2	83
Example 3	78
Example 4	75
Example 5	86

H.-D. Wasel-Nielen, R. Adrian, H. Panter, G. Heymer, A. Maurer and R. von Schenck; U.S. Patent 4,294,808; October 13, 1981; assigned to Hoechst AG, Germany provide a process which avoids the adverse effects that are associated with the grinding step and the use of an additional reactant in the production of phosphorus-containing anticorrosive pigments, especially zinc phosphate. More particularly, it is made possible by this process to produce, as early as during the precipitation step, e.g., zinc phosphate which presents the particle size necessary for use as an anticorrosive pigment and a regular constant particle size distribution, at least 90% of the particles having a size between the narrow limits of 0.1 to 8 μ. This is a product of improved dispersibility and improved anticorrosive efficiency of which the use entails a series of further beneficial effects.

It was found that zinc phosphate which combines in itself dispersibility with a particularly narrow particle size distribution is unexpectedly obtained by converting zinc oxide, with agitation and with the use of water or mother liquor, to a concentrated suspension, placing the suspension in a vessel with stirrer and maintaining its concentration constant by thorough agitation, and reacting the suspension with concentrated phosphoric acid in a rapidly rotating dispersing means (3,000 to 10,000 rpm), e.g., a screw pump. By the structure selected for the dispersing means and by this process it is ensured that the reactants are intimately intermixed and undergo reaction within a minimum period of time, e.g., while passing through the screw pump.

Zinc oxide and phosphoric acid unexpectedly undergo complete reaction substantially without any undesirable crystal growth later during the process, irrespective of the extremely short sojourn time of the reactants in the reaction zone. To ensure this, care should be taken to avoid the presence of any excess of phosphoric acid, which may adversely affect the quality of the resulting product, in the reaction zone. To this end, it is good practice to use zinc oxide and phosphoric acid from the onset in a molar ratio greater than 1.5 for preparation of the zinc oxide suspension. Failing this, care should be taken that the reaction zone, i.e., the dispersing means, is fed with exactly metered quantities of phosphoric acid, which may be supplied, e.g., with the use of a nozzle. Only in this manner is it possible to ensure complete and more especially rapid reaction in the dispersing means and the formation of zinc phosphate of very narrow particle size distribution and dispersibility.

A further technically beneficial effect of this process resides in the fact that use is made of a concentrated suspension and that very limited quantities of mother liquor are produced. This has favorable effects especially on the period necessary for filtration of the zinc phosphate and on the dimensions to be selected for the reactors.

The method of this process and its technically beneficial effects are of assistance not only in the production of $Zn_3(PO_4)_2 \cdot XH_2O$, but also in the manufacture of a wide variety of other phosphorus-containing anticorrosive pigments. In this latter case, it is possible for the metallic reaction component to be used in the form of the respective oxide or hydroxide or salt, e.g., chloride or acetate, the phosphorus-containing reaction components being selected from orthophosphoric acid, pyrophosphoric acid, higher condensed phosphoric acids, phosphorous acid or hypophosphorous acid or an alkali metal or ammonium salt of these oxygen acids of phosphorus. It is also possible to produce these alkali metal or ammonium salts in situ by means of an alkali liquor or ammonia and an oxygen acid of phosphorus.

Example 1: *Preparation of $Zn_3(PO_4)_2 \cdot 4H_2O$* – 100 kg of ZnO and 300 ℓ of water were thoroughly stirred into a suspension which was circulated for 1 hr by means of a screw pump rotating at a speed of 6,000 rpm so as to intensify the suspending effect. Next, the zinc oxide suspension was admixed over 2 hr inside the screw pump with 94.4 kg of orthophosphoric acid of 85 wt % strength which was admitted in metered proportions and underwent spontaneous reaction therein.

The ZnO/H_3PO_4-molar ratio was stoichiometric, equal to 1.5. The temperature rose from 20° to 80°C. After the phosphoric acid had been added, the zinc phosphate suspension was circulated by pumping for about 30 min more and then stored in an intermediate container. Next, zinc phosphate was separated from its mother liquor which was recycled to a container receiving the ZnO-suspension. The solid matter so-obtained was dried as usual at 85°C in a fluid bed dryer. A zinc phosphate which had the following particle size distribution (determined by means of a whizzer air separator) was obtained: 0.05 to 8 μ = 94.9%; 8 to 12 μ = 4%; and 12 to 15 μ = 1%.

Example 2: *Preparation of $Zn_3(PO_4)_2 \cdot 4H_2O$* – 183 kg of ZnO was suspended in 600 ℓ of water with thorough agitation and circulation by pumping with the aid of a screw pump rotating at a speed of 6,000 rpm. A uniformly concentrated suspension was obtained which was admixed within 2 hr and by means of the screw pump with 150 kg of phosphoric acid of 85 wt % strength. The reaction occurred spontaneously.

The ZnO/H_3PO_4-molar ratio was overstoichiometric, equal to 1.73. The temperature rose to 70°C. After reaction the zinc phosphate suspension was worked up and mother liquor was recycled to the container receiving ZnO-suspension, in the manner described in Example 1. A zinc phosphate pigment which had the following particle size distribution was obtained: 0.05 to 8 μ = 94.9%; 8 to 12 μ = 4%; and 12 to 15 μ = 1%.

Example 3: *Preparation of $MgHPO_4 \cdot 3H_2O$* – 50 kg of phosphoric acid of 85 wt % strength was placed in a stirring vessel, mixed therein with 60 kg of sodium hydroxide solution of 50 wt % strength, and the whole was diluted with 200 ℓ of water. The solution so-obtained was cooled down to about 20°C and circulated by means of a screw pump rotating at a speed of 6,000 rpm. Next, 121.5 kg of a 34 wt % solution of $MgCl_2$ was added in metered proportions within 2 hr by means of the screw pump.

The $MgCl_2/H_3PO_4$ molar ratio was stoichiometric, equal to 1:1. The whole was also admixed with 4.7 kg of a 10 wt % solution of NaOH, which was necessary to ensure stoichiometric conversion. The reaction temperature varied between 20° and 50°C. After all had been added, the suspension was circulated for 1 hr. Next, dimagnesium phosphate (magnesium hydrogen phosphate) was filtered off and dried at 50°C. A dimagnesium phosphate pigment which had the following particularly narrow particle size distribution was obtained: 0.05 to 8 μ = 90.8%; 8 to 12 μ = 6.5%; and 12 to 15 μ = 2.5%.

This process of *A. Maurer, R. Adrian and G. Heymer; U.S. Patent 4,346,065; August 24, 1982; assigned to Hoechst AG, Germany* relates to the manufacture of a finely dispersed, sparingly soluble salt of an oxyacid of phosphorus, having a maximum particle size of 20 μm.

The salt is prepared by reacting a compound of a divalent metal with the oxyacid of phosphorus or an alkali metal or ammonium salt thereof, which comprises reacting an at least 5% by wt aqueous solution or suspension of a compound of a divalent metal selected from the group Mg, Ca,Sr, Ba, Mn, Zn, Cu, Pb, Sn, Co and Ni, while maintaining a pH range of between 3 and 9 and while stirring, with an at least 2 mol % solution of at least one trivalent metal selected from the group Al, Fe and Cr in the oxyacid of phosphorus, and optionally, with an aqueous solution of an alkali metal or ammonium salt of the oxyacid of phosphorus, with the resultant formation of finely dispersed salt consisting to an extent of at least 90% of particles with a size between 0.05 and 7 μm, and separating off and drying the precipitated salt in known manner.

The finely dispersed, sparingly soluble salts so produced are used preferably as anticorrosion pigments.

Example 1: A solution of 27.0 g (1 mol) of Al in 2,305 g (20 mols) of H_3PO_4 (85% by wt strength) was metered into a well stirred suspension of 1,902 g (19 mols) of $CaCO_3$ in 20 ℓ of water in the course of 2 hr. The solution had been prepared by dissolving the corresponding amount of $Al(OH)_3$ in boiling H_3PO_4. The temperature during the reaction of the calcium carbonate with the solution of Al in H_3PO_4 was 25°C and the pH was between 9 and 4.5. A precipitate was obtained which was filtered off, washed with water and acetone and dried in a drying cupboard for 1 hr at 70°C. A white pigment was obtained. X-ray analysis indicated that it was calcium hydrogen phosphate dihydrate (JCPDS card index No. 9-77 and 11-293). Its average particle size was 4 μm and the maximum particle size 10 μm. The BET surface area was 2.7 m^2/g.

Example 2: The procedure was the same as in Example 1 except that a solution of 10.8 g (0.4 mol) of Al in 2,305 g (20 mols) of H_3PO_4 (85% by wt) was metered into a suspension of 1,962 g (19.6 mols) of $CaCO_3$ in 20 ℓ of water. A white pigment was again obtained. It had an average particle size of 4 μm and a maximum particle size of 10 μm. The BET surface area was 2.5 m^2/g.

Example 3: The procedure was the same as in Example 1 except that a solution of 55.85 g (1 mol) of Fe in 2,305 g (20 mols) of H_3PO_4 (85% by wt) was metered in. A yellowish pigment was obtained. X-ray analysis identified it as calcium hydrogen phosphate dihydrate. It had an average particle size of 4 μm and a maximum particle size of 10 μm. The BET surface areas was 2.4 m^2/g.

Example 4: The procedure was the same as in Example 1 except that a solution of 52.0 g (1 mol) of Cr in 2,305 g (20 mols) of H_3PO_4 (85% by wt) was metered in. The reaction product was a greenish pigment. By x-ray analysis, it was identified as calcium hydrogen phosphate dihydrate. It had an average particle size of 5 μm and a maximum particle size of 12 μm. The BET surface area was 2.4 m^2/g.

The process of *K. Hestermann, A. Maurer, J. Kandler, G. Mietens and H. Beumling; U.S. Patent 4,337,092; June 29, 1982; assigned to Hoechst AG, Germany* relates to a corrosion-inhibiting active pigment for use in the surface protection of iron and iron alloys. The pigment consists of a mixture of calcium hydrogen phosphate dihydrate, magnesium hydrogen phosphate trihydrate and zinc oxide. More specifically, the mixture contains 2 to 98 wt % of ZnO and 98 to 2 wt % of phosphate components. These in turn contain 3 to 97 wt % of $CaHPO_4 \cdot 2H_2O$ and 97 to 3 wt % of $MgHPO_4 \cdot 3H_2O$. The active pigment is suitable for use in painting compositions or lacquers.

A pigment mixture of $MgHPO_4 \cdot 3H_2O$, $CaHPO_4 \cdot 2H_2O$ and ZnO had corrosion-inhibiting properties which compare favorably with those of products consisting of $MgHPO_4 \cdot 3H_2O$ and $CaHPO_4 \cdot 2H_2O$, or ZnO alone.

It has also been found that these corrosion-inhibiting pigments have no inherent coloration which would adversely affect the coloration of the compositions made therefrom.

The pigments were tested as described in U.S. Patent 4,294,621 above using the same test formulation except that the active pigment was 18.0 parts by weight of $CaHPO_4 \cdot 2H_2O + MgHPO_4 \cdot 3H_2O + ZnO$.

Examples 1 through 17: The materials were made into a homogenized primer with the use of an agitator-provided mixer. On the basis of the test formulation, various primers with a pigment volume concentration of 36 were made by varying the quantitative composition of the active pigment. The corrosion test results are indicated in the following table.

Example 18: *(Comparative Example)* – The formulation was the same as that in U.S. Patent 4,294,621 except that zinc phosphate was used as the active pigment. A pigment volume concentration of 36 was established. The result obtained is indicated in the following table.

Example 19: *(Comparative Example)* – The formulation was the same as that in U.S. Patent 4,294,621 except that zinc oxide was used as the active pigment. A pigment volume concentration of 36 was established. The result obtained is indicated in the following table.

Example 20: *(Comparative Example)* – The formulation was the same as that in U.S. Patent 4,294,621 except that a mixture of $CaHPO_4 \cdot 2H_2O$ and $MgHPO_4 \cdot 3H_2O$ in a ratio by weight of 60:40, which was free from zinc oxide was used as the active pigment. A pigment volume concentration of 36 was established. The result obtained is indicated in the following table.

Example	.... Weight Percent Proportion in Active Pigment				Characteristic Value
	$CaHPO_4 \cdot 2H_2O$	$MgHPO_4 \cdot 3H_2O$	ZnO	$Zn_3(PO_4)_2 \cdot 2H_2O$	
1	87.3	2.7	10.0	–	82
2	81.0	9.0	10.0	–	90
3	72.0	18.0	10.0	–	88
4	58.5	31.5	10.0	–	85
5	2.7	87.3	10.0	–	81
6	77.6	2.4	20.0	–	82
7	72.0	8.0	20.0	–	91
8	64.0	16.0	20.0	–	92
9	52.0	28.0	20.0	–	89
10	2.4	77.6	20.0	–	81

(continued)

Example	.... Weight Percent Proportion in Active Pigment				Characteristic Value
	$CaHPO_4{\cdot}2H_2O$	$MgHPO_4{\cdot}3H_2O$	ZnO	$Zn_3(PO_4)_2{\cdot}2H_2O$	
11	48.5	1.5	50.0	–	81
12	45.0	5.0	50.0	–	89
13	40.0	10.0	50.0	–	87
14	32.5	17.5	50.0	–	90
15	1.5	48.5	50.0	–	79
16	88.2	9.8	2.0	–	81
17	1.6	0.4	98.0	–	79
18	–	–	–	100	65
19	–	–	100.0	–	72
20	60.0	40.0	–	–	71

Hydrolyzed Ethyl Silicate Binders for Zinc-Rich Primers

Typical ethyl silicate based inorganic zinc-rich primers consist of two separate components which are mixed together just prior to use. Upon mixing the product will be useful for a limited period of time. Normally one of the components consists solely of zinc dust and the other component contains a mixture of pigments in a solution of hydrolyzed ethyl silicate. These pigments are present primarily to lower the cost of this otherwise relatively expensive product.

It has been found that pigmentation need not be 100% zinc dust to give equivalent corrosion protection. In fact, choice of the proper type and amount of extender pigments can actually improve performance. Other reasons for using these additional pigments are product appearance and hardness. For economic reasons primarily, these pigments are almost invariably included in the binder component rather than dry blended with the zinc dust.

The binder in this product is partially hydrolyzed ethyl silicate. It is produced by partially hydrolyzing tetraethyl orthosilicate. This hydrolysis can be stopped at any desired level of hydrolysis by adjustment of the amount of water used in the reaction. Typically, a range of 80 to 95% hydrolysis is desired for use in zinc-rich primers. The final product, however, has inherent stability limitations. Within a certain period of time a reaction process will continue making the product unusable as a coating. Certain conditions are known to accelerate this inherent instability. Among the most critical are presence of excess moisture and heat.

The extender pigments such as wet ground mica which are used in zinc-rich primers are not soluble in the solvent used in the system. (See *Pigment and Resin Technology,* pp. 40-43, January, 1972, and *Modern Paint and Coatings,* pp. 67-69, November, 1975.) Typical solvents are Cellosolve (ethylene glycol monoethyl ether) and alkanols such as ethanol and isopropanol. For optimum effectiveness, the pigments must be dispersed to their minimum particle size. This dispersion process must be carried out in typical manufacturing equipment.

Dispersion carried out directly in the partially hydrolyzed ethyl silicate binder results in the likelihood of accelerated instability of the binder due to possible moisture contamination and the heat generated during pigment dispersion due to friction. Since ethyl silicate reacts with the metal lining of typical paint manufacturing equipment, special manufacturing equipment is also necessary when dispersion is carried out in the ethyl silicate binder. Therefore, it is desirable to disperse the pigments in a solvent compatible with the ethyl silicate and not directly in the ethyl silicate. The dispersed pigment/solvent slurry can then be added to the partially hydrolyzed ethyl silicate solution just prior to packaging.

Although it is desirable to disperse the pigments in a compatible solvent rather than in the ethyl silicate, itself, it is not possible to do so effectively unless the viscosity of the solvent is increased. Dispersion in solvent alone does not allow pigment particle agglomerates to be reduced to their minimum size. Solvent in itself is simply too thin to be an effective dispersion medium. Merely adding pigments to thicken the solvent creates a thixotropic or "puffy" rheology which is equally ineffective for true dispersion. A means was needed to thicken the solvent without creating puffiness.

D.C. Dulaney and J.A. Bowman; U.S. Patent 4,294,619; October 13, 1981; assigned to Rust-Oleum Corporation describe a method of preparing pigmented partially hydrolyzed ethyl silicate binders used in protective coatings such as zinc-rich primers. The method utilizes ethyl hydroxyethyl cellulose to aid in the dispersion and suspension of pigments in the binder solution. The addition of ethyl hydroxyethyl cellulose to the solvent enables the pigments to be dispersed in the solvent rather than in the ethyl silicate solution, thereby preserving product stability and avoiding additional manufacturing costs incurred from dedicating equipment for manufacturing the coatings. The separate pigment/solvent blend can then be added to the partially hydrolyzed ethyl silicate solution just prior to packaging.

Example: 1 pbw ethyl hydroxyethyl cellulose is added to a solution of 31.7 pbw Cellosolve and 26.7 pbw anhydrous isopropanol. To the solution are added 17.3 pbw calcined clay, 10.8 pbw wet ground mica and 10.8 pbw chromium oxide. Approximately 1.8 pbw castor wax type thixotrope may be added to aid suspension. The slurry is then ground in a typical paint manufacturing grinder such as a high speed grinder. When the proper pigment size and dispersion is achieved the slurry may be added to the hydrolyzed ethyl silicate solution.

Chromium-Free Passivating Film

In the process of *J.L. Greene; U.S. Patent 4,298,404; November 3, 1981; assigned to Richardson Chemical Company* substrates, especially those having plated metal surfaces, are subjected to passivation treatments in baths that incorporate one or more film-forming agents at least one of which does not require chromium and includes anions or cations of elements other than chromium. Typically, the anions or cations are introduced as bath-soluble salts which react with the plated surface of the substrate to form an adherent, coherent passivation surface film.

Also present within these baths are a source of hydrogen ions and a bath-soluble carboxylic acid or derivative activator for enhancing the rate of the passivation reaction. Articles passivated in baths incorporating these film-forming agents have a hydrophobic surface that exhibits corrosion resistance and that typically has a bright finish.

Example 1: A powdered film-forming composition was prepared by blending together 3.7 g of potassium titanium fluoride, 0.8 g of boric acid, 1.2 g of sodium sulfate, 1.0 g of sodium nitrate, and 1.0 g of sodium heptagluconate. This composition was dissolved in 1 ℓ of water to which had been added 0.25 vol % of nitric acid, the resulting solution having a pH of 1.85. A freshly zinc-plated steel panel was rinsed in water to remove adhering zinc plating solution and was then immersed in the passivation bath at room temperature for 25 sec, after which it was removed from the solution, rinsed in cold water, and blown dry in a stream of warm air. The surface of the panel was covered by a uniform adherent blue bright film that showed definite hydrophobic characteristics and that was adequate to protect the underlying zinc surface for 24 hr in a 5% neutral salt spray in accordance with ASTM test method B-117.

Example 2: A powdered blended mixture of 1.5 g of sodium aluminum fluoride, 0.8 g of boric acid, 1.2 g of sodium nitrate, 1.0 g of sodium sulfate, and 1.5 g of sodium heptagluconate was dissolved in 1 ℓ of water containing 0.25 vol % of nitric acid in order to form a bath having a pH of 1.95. Upon immersion for 25 sec, a panel of zinc-plated steel developed a pale reddish-green film that protected the underlying zinc during a 16 hr ASTM B-117 salt spray exposure.

Example 3: The zinc surface of a panel formed a good, uniform blue bright film and was protected against an ASTM B-117 salt spray test for 16 hr upon being immersed for 25 sec within 1 ℓ of water to which had been added 0.40 vol % of nitric acid, together with a powdered blended mixture of 3.5 g of sodium orthovanadate, 3.7 g of sodium silicofluoride, 1.0 g of sodium heptagluconate, 1.2 g of sodium nitrate, and 0.8 g of boric acid, the solution having a pH of 1.7.

Example 4: A bath was prepared by adding to water a composition consisting of 52 wt % of potassium titanium fluoride, 11 wt % of boric acid, 13 wt % of sodium sulfate, 14 wt % of sodium nitrate, and 10 wt % of sodium oxalate. About 0.25 wt % of a powdered dye was also added, and the pH was adjusted to about 2.1 with nitric acid, approximately 5.5 ml of nitric acid having been added for each gal of bath. Electroplated zinc substrates were dipped in the bath for up to about 50 sec to provide a blue bright passivation coating on the zinc.

Example 5: Dissolved into 1 ℓ of 0.15 vol % nitric acid in water was a blended powder mixture of 3.1 g of potassium titanium fluoride, 0.8 g of boric acid, 1.0 g of sodium sulfate, 1.0 g of sodium nitrate, and 0.7 g of sodium oxalate, the resulting solution having a pH of 1.96. Freshly zinc-plated steel panels that had been rinsed to remove adhering plating solution were immersed into the passivation bath, and within 15 to 20 sec, a blue bright passivation film was formed.

Example 6: The procedure of Example 5 was substantially repeated, except the sodium oxalate was replaced in the formulation with 0.7 g of sodium malonate, and substantially the same result was achieved.

Example 7: About 0.7 g of sodium succinate was used in place of the sodium oxalate of Example 5, and a substantially identical blue bright film was formed.

Example 8: Powdered formulations generally in accordance with Example 1 were dissolved in various baths employing several concentrations of acids other than nitric acid, between 0.005 M for immersion time periods up to 5 min and up to 0.20 M for immersion time periods less than 1 min. The acids used were sulfuric acid, sulfamic acid, and phosphoric acid, and each of these baths produced acceptable passivation films.

Barium Salts of Carbonyl-Group-Containing Compound

It is an object of the process of *E. Kuehn; U.S. Patent 4,304,707; December 8, 1981; assigned to ICI Americas Inc.* to provide a pigment which inhibits corrosion when incorporated in a film-forming resin binder paint formulation. Pigment blends coprecipitated on inert substrates are also included.

This is realized by applying to metals paint systems containing an effective amount, about 5 to 95% by wt of a barium salt of a carbonyl-group-containing organic compound having at least one acidic hydroxy hydrogen. Representative of such compounds are substantially water-insoluble hydrous and anhydrous barium salts of organic acids, such as citric acid, tartaric acid, salicylic acid, alizarine, quinizarine, chloranilic acid, alizarincarboxylic acid, and blends thereof. Such barium salts when incorporated in films which contact metal surfaces, particularly iron and steel surfaces, offer a method for inhibiting corrosion equivalent to or better than most toxic lead and chromium-containing pigments.

In addition to barium salts of the compounds mentioned above, barium salts of other materials, such as glutaric acid, glycolic acid, glyceraldehyde, glyceric acid, malic acid, gluconic acid, and β-hydroxypropanoic acid, to name a few, are additional examples of materials which would be effective either as anhydrous or hydrated salts of barium as sole ingredients, blends or deposits on inert or active substrates.

Preparation 1–Barium Citrate: Into a glass beaker is intermixed citric acid and barium hydroxide 8 hydrate [$Ba(OH)_2 \cdot 8H_2O$] in a mol ratio of 1/1.5 by intermixing 300 g of distilled water, 96 g citric acid and 236.6 g of barium hydroxide 8 hydrate heated to a temperature of 58°C. Upon precipitation of the barium citrate, 100 g of distilled water is added whereupon the white pasty pigment slurry is then held for an additional hour at 70° to 90°C. A total of 294.6 g of a white fluffy powder is obtained which corresponds to a yield of 92% of hydrated barium citrate. Analysis: 47.45% $Ba_3C_{12}O_{14}H_{10}$.

Preparation 2–Barium Tartrate: Hydrated barium tartrate is prepared by intermixing 315.5 g barium hydroxide 8 hydrate, 150 g of tartaric acid and 400 g of water with stirring at 50°C. Upon thickening, another 100 g of distilled water is added and further heated to about 80° to 95°C for 1 hr. The slurry is cooled to 1°C and filtered. Filter cake resulting from vacuum drying overnight at 90° to 95°C was 288 g of fine powder $Ba_2(C_4H_4O_6)_2 \cdot 2H_2O$.

Preparation 3–Barium Salicylate: Hydrated barium salicylate is prepared by the intermixing of 157.75 g of barium hydroxide 8 hydrate, 69.06 g of salicylic acid and 350 g of distilled water. Barium salicylate precipitates as a white slurry upon heating and stirring for 1 hr at 75° to 95°C. Upon cooling, filtering and drying in a vacuum oven at 90° to 95°C overnight 140.5 g of hydrated barium salt is obtained, $Ba(C_6H_4OHCO_2)_2 \cdot H_2O$.

Preparation 4–Barium Alizarate: The barium salt of alizarine (1,2-dihydroxy-9,10-anthraquinone) is prepared by intermixing a 2/1 mol ratio of alizarine and barium hydroxide octahydrate (240.22 g of alizarine and 175.75 g barium hydroxide 8 hydrate) in 600 ml of water in a 2 ℓ glass beaker. The mixture turns purple upon the formation of a pasty mixture. The slurry is stirred for an additional 2 hr at 90°C and filtered at room temperature, after which the filter cake is dried at 90° to 95°C overnight in a vacuum oven to yield 237.2 g of fine free-flowing powder $Ba[C_6H_4(CO)_2C_6H_2O_2]_2$.

Preparation 5–Barium Quinizarate: The barium salt of quinizarine (1,4-dihydroxy-9,10-anthraquinone) is prepared by dissolving in 350 g of water, 84.5 g of barium hydroxide 8 hydrate and 115.5 g of quinizarine at a temperature of 40°C. A dark purple precipitate forms immediately. The slurry is then stirred at 95°C for a period of 1 hr after which it is filtered, washed, and dried. 151.2 g of a dark purple free-flowing pigment is obtained. Analysis: 23.9% Ba.

Preparation 6–Barium Salt Coated Substrate: It may be desirable to precipitate active corrosion resistant barium compounds on to inert carriers. A pigment consisting of 60% by wt alumina trihydrate (Alcoa C-331), 30% by wt barium citrate and 10% by wt barium alizarate is made as follows.

72.4 g of barium hydroxide 8 hydrate is dissolved in 300 ml of distilled water in a 1 ℓ beaker at 50°C. Into this is stirred 103 g of alumina trihydrate. This slurry is heated to 70°C and thereafter 50 g of an aqueous solution containing 4.32 g of citric acid and 15 g of alizarine are added simultaneously with vigorous stirring. The color of the slurry turns purple and an increase in viscosity is noted. Another 200 ml of water is added and heated to a temperature of 90° to 95°C and held for 1 hr. The pigment slurry is then cooled and filtered, washed repeatedly with water, and dried overnight at 90°C. After grinding the pigment in a mortar and pestle, 134 g of a uniformly purple colored free-flowing pigment is obtained. The pigment can be employed in corrosion resistant coatings.

Salt spray test results indicated that commercially available pigments are relatively poor as compared to the pigments of the process. The barium salts are nearly successful in preventing rust entirely.

Alkyl Esters of Amino Acid as Vapor Phase Inhibitors

H. Sato and K. Osada; U.S. Patent 4,308,168; December 29, 1981; assigned to Idemitsu Kosan Co., Ltd., Japan have found that alkyl esters of amino acids are effective vapor phase corrosion inhibitors. The method of inhibiting corrosion of a metal comprises contacting the metal with an amount of a volatile alkyl ester of an amino acid sufficient to inhibit corrosion, the alkyl group of the acid containing not more than 7 carbon atoms. The amino acid is at least one selected from the group consisting of glycine, α-alanine, valine, leucine, isoleucine, N-methylglycine, cysteine, cystine, methionine, lysine, arginine, glutamic acid, β-alanine, γ-aminobutyric acid, phenylalanine, tyrosine, histidine, tryptophan, proline and hydroxyproline.

The vapor phase inhibitors of this process can be applied in different ways depending on its formulation. For example, when the vapor phase inhibitor is prepared by adding the alkyl ester of the amino acid to oil to form an oil-base composition, the composition can be applied to the surface of various products (usually metal) such as the inner wall of metal pipes by coating.

The metal surface coated with the above vapor phase inhibitor is maintained rust-free, even if the coating is incompletely applied or the oil film is broken, because the vaporized alkyl ester of the amino acid protects the metal surface. When a small engine is run-in at the factory and then, after removing the oil, transported to delivery, the empty engine surfaces are liable to rust. However, when the vapor phase corrosion inhibitor of the process is incorporated in the run-in oil, the engine is protected from rust.

The vapor phase inhibitor of this process is distinguished by its high anticorrosive ability and long diffusion distance of the gaseous (vaporized) alkyl ester of the amino acid. Furthermore, the vapor phase inhibitor of this process corrodes neither iron nor nonferrous metals and, therefore, can be applied to any system in which different metals are alloyed or are in contact, e.g., plated, or clad or coated.

The anticorrosion inhibitors and compositions of this process prevent or at least minimize corrosion attack by the effect of the contact of the alkyl ester of the amino acid with the metal being protected. This contact may solely be contact of the vapor (gaseous) alkyl ester of the amino acid with the metal which may be provided by placing a volatilizable alkyl ester of the amino acid in the vicinity of the metal to be protected, e.g., placing it in a closed container with the metal, e.g., admixed with a porous carrier such as a silica gel or zeolite. It may also be impregnated into a porous packaging material such as kraft paper, etc.

The alkyl ester of the amino acids will also be incorporated into oil base compositions (using the term oil to include greases) which are applied to the metal which is being protected. Such oils include lubricating oils, light oils, base oils for brake oil, base oils for antifreeze or synthetic lubricating oils such as synthetic hydrocarbon oils, hindered polyol esters and the like, ethylene glycol, various glycol ethers and organic silicone oils, etc.

The amount of the alkyl ester of the amino acid incorporated in the oil may vary within wide limits depending upon conditions such as the type of oil, the object of the application and the like, but in general, the amount should be at least 0.1 g, preferably 0.3 to 3.0 g/ℓ of oil. The alkyl ester of the amino acid is added in an amount of about 0.5 to 1.5 g to 1 ℓ of hydraulic oil.

Example 1: 5 ml of concentrated sulfuric acid was added to a suspension of 5 g of phenylalanine in 50 ml of dioxane. The mixture was added to about an equivalent amount of liquefied isobutylene and then allowed to stand at room temperature in an autoclave for a whole day and night. The resulting solution was poured into an excess of a cold solution of 2 N sodium hydroxide to form the tert-butyl ester of phenylalanine which is thereafter extracted with ether. After distilling out the ether, the distillation of the residue under reduced pressure produced 4.7 g of phenylalanine-tert-butyl ester (yield 70%) having a boiling point of 96°C/1 mm Hg.

Examples 2 and 3: Following the procedure described in Example 1 but substituting N-methylglycine and proline, respectively, for phenylalanine, N-methylglycine-tert-butyl ester (Example 2) and proline-tert-butyl ester (Example 3) were obtained.

Example 4: To 960 ml of tert-butyl acetate, 5.7 g (64 mmol) of α-alanine and then 11.76 g (70.4 mmol) of 60% perchloric acid were added and thereafter stirred at room temperature for 4 days. The resulting solution was cooled to 0°C and extracted four times with 160 ml of 0.5 N aqueous solution of hydrochloric acid. The extract was neutralized with 6 N sodium hydroxide solution and was extracted twice with 400 ml of ethyl ether. The ethereal layer was dried over anhydrous sodium sulfate and ether was distilled out. By distilling under reduced pressure, 5.22 g of α-alanine-tert-butyl ester (yield 56%) was obtained. The boiling point of the obtained substance was 61°C/21 mm Hg.

Example 5: Following the procedure described in Example 5 but substituting glycine for alanine, the tert-butyl ester of glycine was obtained.

Example 6: A laboratory dish containing 20 mg of each sample obtained in the preceding examples and several other compounds of the process was added to a 5 ℓ glass vessel containing 50 ml of 30% aqueous solution of glycerin.

A piece of polished carbon steel (JIS G-4051) was arranged at the upper part of the vessel, at a distance of 25 cm from the dish. After keeping the vessel at 20°±2°C during 1 to 5 days, water droplets were condensed on a surface of steel from moisture by pouring cold water into an aluminum pipe, and after 5 hr the presence of the rust on the steel piece was observed. The results are shown in the following table.

Substance	After Leaving for 1 Day*	After Leaving for 5 Days*
Phenylalanine-tert-butyl ester	o	o
N-methylglycine-tert-butyl ester	o	o
Proline-tert-butyl ester	o	o
α-Alanine ethyl ester	o	o
α-Alanine-tert-butyl ester	o	o
Glycine-tert-butyl ester	o	o
β-Alanine-tert-butyl ester	o	o
Dichan (prior art)	x	x

*o = The amount of the rust is less than that in the blank test.
x = The amount of the rust is the same as that in the blank test.

Low Temperature Curing

The process of *G.A. Collins, Jr. and J.M. Klotz; U.S. Patent 4,319,924; March 16, 1982; assigned to Coatings for Industry, Inc.* relates to improved coatings (e.g., corrosion resistant coatings) which can be formed by curing an aqueous coating composition at a relatively low temperature.

There is provided an aqueous coating composition comprising in coating forming proportions: dissolved phosphate; dissolved dichromate; dissolved aluminum; and dispersed solid particulate material, preferably an aluminum-containing material; this composition being capable of being heat-cured at elevated temperatures into a water-insoluble material; and further comprising diethanolamine in an amount at least sufficient to reduce the temperature at which the composition can be cured into the water-insoluble material.

Amounts of the above ingredients to be used are as follows: at least about 1 mol/ℓ of dissolved phosphate; at least about 0.1 mol/ℓ of dissolved dichromate; at least about 0.5 mol/ℓ of dissolved aluminum; at least about 20 g/ℓ of inorganic solid particulate material having a particle size of at least about 1 μ and capable of being bonded to a metallic surface by phosphate bonding; and at least about 0.02 mol/ℓ of diethanolamine.

It is believed that the coating composition of the process will have its widest applicability of use in coating metallic substrates such as iron and ferrous alloys and other metals which are subject to corrosion or which require a specialty coating due to requirements of use.

There is described below an aqueous binding solution (Solution A) to which can be added solid particulate materials to form coating compositions within the scope of the process. Solution A contains 25 ml H_2O, 100 ml aluminum phosphate and CrO_3 solution, and 3 g $(HOCH_2CH_2)_2NH$. The aluminum phosphate/CrO_3 solution was prepared by combining 300 g of hydrated alumina ($Al_2O_3 \cdot H_2O$) with 558 ml of 75% phosphoric acid and thereafter high speed mixing. After standing overnight the solution was decanted from the insoluble $Al(OH)_3$ which had settled to the bottom. The resulting solution was diluted with water to yield a 60% by wt aluminum phosphate solution. To 100 ml of this solution was added 12 g of CrO_3.

Solution A is an example of an aqueous binding solution that can be described as being versatile in that it can be used very effectively in binding a wide variety of particulate materials, both refractory and nonrefractory, to metallic or other substrates. Such coating compositions prepared from Solution A can be cured effectively at relatively low temperatures (e.g., at about 180°F for 4 to 6 hr or at 215°F for 1 hr or at 300°F for ½ hr) into coatings which maintain their corrosion resistant properties after being exposed to corrosive environments. Solution A with pigment had no shelf life problem after 6 months.

Example 1: This example is illustrative of a coating composition containing a metal oxide. Such a composition can be used in forming coatings, which have properties that make them particularly effective as electrical insulation coatings. Example 1 contains 100 ml of Solution A and 120 g of aluminum oxide powder (-400 mesh).

Example 2: This example is illustrative of a coating composition containing a titanate. Such a composition can be used in forming coatings which have properties that make them particularly effective as thermal insulative coatings with high K factors. Example 2 contains 100 ml of Solution A and 100 g of barium titanate (-400 mesh).

Example 3: This example is an illustration of a coating composition containing a metal alloy and contains 100 ml of Solution A and 90 g of aluminum-manganese alloyed metal powder, 50% Al-50% Mn (-325 mesh).

The use of the aluminum-manganese alloy as the particulate material in the coatcomposition of Example 3 affords various advantages over the use of pure aluminum powder as particulate material. Coatings including the alloy have a higher melting point than coatings including aluminum metal only. The presence of the aluminum in the alloy provides excellent corrosion protection for metal substrates coated with the composition of Example 3 which substrates are exposed alternately to high and low temperatures (e.g., -100°F to 1800°F), and to salt spray environments. (An example of such an application is a coating for hot section parts of gas turbine engines such as turbine blades.)

In an application where the coated substrate is exposed to sulfur, such as in applications where the substrate is exposed to sulfur-bearing gases, the manganese in the alloy functions to tie up the sulfur as MnS, thus reducing the susceptibility of the metal substrate to corrosive attack by the sulfur.

Example 4: This example is illustrative of a coating composition containing metal powder as the particulate material, and contains 100 ml of Solution A and 90 g of aluminum metal powder (-400 mesh).

Each of the coating compositions of Examples 1 through 4 above was applied to a steel panel by conventional paint spraying equipment. The coating compositions were then cured by placing the coated panels in an oven at 300°F for 30 min. The coatings had thicknesses of about 1 to 2 mils. Thereafter the corrosion resistant properties of coated substrates were evaluated by subjecting the substrates to a 5% salt spray test (ASTM-117B) for 168 hr. It was found that there was no loss of adhesion and the coatings could not be removed readily.

Beta-Diketone-Epoxy Resin Reaction Products

N.T. Castellucci and J.F. Bosso; U.S. Patent 4,321,305; March 23, 1982; assigned to PPG Industries, Inc. provide compositions of matter comprising a water-dispersible reaction product of beta-diketones and epoxy material. The beta-diketones useful herein are those which enolize in amounts that render them sufficiently acidic and reactable with epoxy groups of the resin. The reaction products can be dispersed in water and they can be formulated optionally with curing agents. Ferrous metals treated with the compositions of this process exhibit excellent corrosion-resistant properties.

Preferably, the water-solubilizing group is a precursor of an onium salt. Accordingly, the water-dispersible compositions of the process comprise an onium salt. Such compositions contain the reaction product of: (A) an epoxy material; (B) a beta-diketone of the formula:

$$R_1-\overset{\overset{\large O}{\|}}{C}-\overset{\overset{\large R_3}{|}}{CH}-\overset{\overset{\large O}{\|}}{C}-R_2$$

wherein R_1, R_2 and R_3, each independently, is a hydrogen or a hydrocarbyl group preferably containing from about 1 to 18 carbon atoms, selected from the group consisting of alkyl or aryl; and (C) an onium salt precursor which is an acid salt of a tertiary amine, a tertiary phosphine-acid mixture, or a sulfide-acid mixture.

Aqueous compositions of the reaction product and a curing agent can be applied to the surface of the substrate in any convenient manner such as by immersion, spraying, or wiping the surface either at room temperature or at elevated temperature. When desirable, the aqueous compositions of the combination can be electrodeposited on a variety of substrates. After the application, the substrate can be baked at temperatures such as 90° to 210°C for about 1 to 30 min.

All percentage compositions in the following examples are parts by weight unless otherwise stated.

Example 1: This example illustrates the preparation of the composition of the process, and the quaternary ammonium salt derivative thereof. The following charge was used in the preparation:

Ingredients	Parts by Weight, g
Epon 1001*	500.0
1,3-Diphenyl-1,3-propanedione	70.8
2-Ethylhexanol	70.0
Ethyl triphenyl phosphonium acetate	1.3
Dimethylethanolamine lactate	69.2
Deionized water	23.4
Deionized water	156.7

*Epon 1001 is a polyglycidyl ether of bisphenol A having an epoxy equivalent of about 500 and a molecular weight of about 1000 (Shell Chemical Company).

The Epon 1001, the diphenyl propanedione and the 2-ethylhexanol were charged to a properly equipped reaction vessel and heated under a nitrogen sparge for about 1 hr and 25 min, to a temperature of 92°C. The reaction mixture was allowed to cool, and at 32°C the ethyl triphenyl phosphonium acetate (catalyst) was added.

Thereafter, the reaction mixture was heated to 135°C; there was an exotherm and a resulting temperature rise to 155°C. After a slight temperature drop to 151°C, the reaction mixture was heated to about 175°C. The temperature was maintained over the range of 170° to 175°C for 1 hr to a Gardner-Holdt viscosity of E, measured as a 44% resin solids solution in 2-butoxyethanol at 25°C.

At 90°C, a solution of the dimethylethanolamine lactate in the first portion of water was introduced into the reaction vessel and the temperature was maintained over the range of 80° to 90°C for 2 hr. A clear brown solution was obtained which was thinned with the second portion of water. The resulting reaction product had a solids content of 72.5%.

The following illustrates the preparation of an electrodeposition bath using the quaternary ammonium salt group-containing polymer prepared as described above, use of the bath for electrodeposition of ferrous metal substrates, and the evaluation of the corrosion resistance of the coated substrates.

The electrodeposition bath was prepared by mixing at room temperature the following mixture of ingredients (pbw, g): The quaternary ammonium salt group-containing polymer, 310.3; and deionized water, 1,190.0.

The electrodeposition bath (15% resin solids) was then used to electrocoat iron phosphate pretreated steel substrates, zinc phosphate pretreated steel substrates, and untreated steel substrates.

The substrates were electrocoated at standard electrocoating conditions. The coatings were baked and scribed with an "X" and then placed in a salt spray chamber at 38°C (100°F) at 100% relative humidity atmosphere of 5% by wt aqueous sodium chloride, for a period of 14 days.

The coating and baking schedules, and the evaluation of the substrates in terms of the measurement of scribe creepage due to corrosion are reported in the examples of the following table.

Substrate	Voltage at Which Electrocoated for 90 seconds at 80° F. (27° C.)	Appearance of Film After Baking at 350° F. (177° C.) for 30 minutes	Film Thickness in mils	Scribe Creepage (in mm.)
zinc phosphate pretreated steel	200	clear and glossy	0.06–0.08	0.0
zinc phosphate pretreated steel	250	"	0.10–0.15	—
iron phosphate pretreated steel	250	glossy	0.3–0.5	—
iron phosphate pretreated steel	200	"	0.35	0.0
untreated steel	200	glossy	1.2	0.0–0.6
untreated steel	150	"	0.6	—
zinc phosphate pretreated steel	250	"	0.55	—
untreated steel	150	"	0.6	—

Example 2: This example illustrates the use of the quaternary ammonium salt group-containing polymers of the process as pretreatment agents for steel substrates. The quaternary ammonium salt group-containing polymer was prepared in essentially the same manner as described in Example 1 and dispersed in deionized water to form a 10% resin solids dispersion.

Ferrous metal articles were dipped in the dispersion at room temperature for 2 min, blown dry with air, baked at 300° and 400°F (149° and 204°C) for 5 min and then coated with a thermosetting acrylic coating composition (Duracron 200, PPG Industries, Inc.). Coating was accomplished by drawing down to about 1 mil thickness with a draw bar. The coated sample was then baked for 10 min at 400°F (204°C), scribed with an "X" and placed in a salt spray chamber for salt spray corrosion testing as described in Example 1.

For the purposes of comparison, there were evaluated control panels in the form of substrates which were dipped in an art-known pretreatment agent which is an onium salt comprising the reaction product of Epon 1001 and an acid salt of a tertiary amine (no beta-diketone moiety), and then coated with Duracron 200 as described above.

Composition	Substrate	Scribe Creepage, mm
Test Panels:		
Quaternary ammonium salt of composition (10% resin solids)	untreated steel	2.0
Quaternary ammonium salt of composition (20% resin solids)	untreated steel	2.5
Control Panels:		
Quaternary ammonium salt of Epon 1001 (10% resin solids)	untreated steel	3.0
Quaternary ammonium salt of Epon 1001 (20% resin solids)	untreated steel	3.5

In related work by *N.T. Castellucci; U.S. Patent 4,321,304; March 23, 1982; assigned to PPG Industries, Inc.* the water dispersible composition is described as follows. It is a blend of: (A) the reaction product of (1) an epoxy material, (2) a beta-diketone as described in the preceding patent, (3) an onium salt precursor also as described above; and (B) at least 2% by wt, based on the total weight of

(1) and (2) of a monomeric or polymeric phosphonium salt which is different from (A).

Example: This example illustrates the improvement in the corrosion resistance properties obtained when the blends of the process comprising the phosphonium salts and the epoxy material-beta-diketone products are used. Onium salt (15% by wt) prepared essentially in the manner of Example 1 of the preceding patent was blended with 2% by wt of ethyl triphenyl phosphonium acetate. The blend was dispersed in water to form a 17% resin solids dispersion.

In the manner of Example 2 of the above patent, the dispersion was applied to the surface of ferrous metal substrates which were then evaluated for their corrosion resistance properties. The method of evaluation was as described above and the results are reported in the table below.

For the purpose of comparison, there were evaluated control panels in the form of ferrous metal substrates which were similarly treated with 15% dispersions comprising onium salts prepared essentially in the manner of Example 1 above.

			Salt Spray Results: Scribe-Creepage After			
Pretreatment Agent	Substrate	Pretreatment Conditions	96 hrs.	240 hrs.	336 hrs.	408 hrs.
Quaternary ammonium salt group-containing polymer prepared in the manner of Example 1 (control)	Bare cold rolled steel (degreased)	Room temperature dipped for 2 minutes, drip-dried and baked at 400° F. (204° C.) for 5 minutes	1.2 mm	1.5 mm	4.5 mm	7.0 mm
The blend prepared in the manner of Example	Bare cold rolled steel (degreased)	Room temperature dipped for 2 minutes, drip-dried and baked at 400° F. (204° C.) for 5 minutes	0.5 mm	1.0 mm	2.5 mm	4.5mm

Zinc or Lead Salts of Heterocyclics

H. Eschwey, J. Galinke, R. Galinke, H. Linden, B. Wegemund and N. Wiemers; U.S. Patent 4,329,381; May 11, 1982; assigned to Henkel KGaA, Germany describe a method for providing corrosion protective coatings for metal surfaces. The coatings contain conventional coating components to which are added at least one zinc or lead salt of five- or six-membered heterocyclic compounds substituted by at least one hydroxyl or mercapto group and the ring includes at least one nitrogen atom and at least two conjugated double bonds, of which at least one had the formation:

(1) $\diagdown N{=}C\diagup$ with $-OH$ on C   or   (2) $\diagdown N{=}C\diagup$ with $-SH$ on C

as corrosion inhibitors.

Further, coating compositions including such salts have excellent stability with respect to their sedimentation behavior.

In addition, such corrosion inhibitor compounds have further advantages, compared to the electrochemically-active inhibitors known to the state of the art. The extremely low water-solubility of the salts according to the process [the solubility of the zinc salts in water (20°C) is 0.1% or lower, the solubility of the lead salts is 0.01% or lower], provides an improved stability and extended protection of the corrosion protective coatings or lacquer films prepared with them, since these salts cannot easily be washed out of the lacquer film when formed. This minimal solubility of the inhibitors, according to the process, is a positive advantage with respect to their environmental behavior under leaching conditions. The drained waters have much lower concentrations of toxic materials.

The coatings, according to the process, are usually at least 5 μm in thickness but, preferred, are coatings in the range 10 to 100 μ and containing the zinc, lead or zinc-lead salts as defined.

Preparation of the Zinc and/or Lead Salts: The preparation of the zinc and/or lead salts of s-triazine derivatives is easily accomplished by different methods. In the following, these methods are illustrated based on cyanuric acid.

Preparative method 1 – Lead and/or zinc carbonate is reacted with cyanuric acid in boiling water. The corresponding metal cyanurates are formed with evolution of carbon dioxide.

Preparative method 2 – Lead and/or zinc nitrate, chloride, or acetate in aqueous solution is reacted at room temperature with an aqueous solution of trisodium or tripotassium cyanurate. The appropriate zinc or lead cyanurate precipitates from the aqueous solution.

Preparative method 3 – Zinc and/or lead oxide is reacted with cyanuric acid in boiling water, preferably in the presence of a small amount of acetic acid as catalyst. The corresponding metal cyanurates are formed. The preparation of other zinc or lead salts with other moieties is exemplified by their preparation utilizing mercaptobenzothiazole as a representative reactant.

Preparative method 4 – Lead or zinc carbonate is reacted with mercaptobenzothiazole in boiling water. The corresponding zinc or lead salts of mercaptobenzothiazole are formed with evolution of carbon dioxide.

Preparative method 5 – Lead or zinc nitrate, chloride, or acetate is treated, in aqueous solution at room temperature, with an aqueous solution of the sodium or potassium salt of mercaptobenzothiazole. The corresponding zinc or lead salts of mercaptobenzothiazole precipitates from the aqueous solution.

Preparative method 6 – Zinc or lead oxide is reacted with mercaptobenzothiazole in boiling water, preferably in the presence of a small amount of acetic acid as catalyst. The corresponding zinc or lead salts of mercaptobenzothiazole are formed.

The salts according to the process are easily incorporated by mixing with conventional binder formulations to prepare the corrosion protective coating compositions of this process. Generally, small amounts of these salts suffice to obtain the described anticorrosive effects; i.e., the salts are added to the binders in amounts of from 0.5 to 10% by wt, preferably 1 to 3% by wt, based on the total formulation. Suitable binders for such primer lacquer compositions are the resins conventionally used for such purposes, dissolved or emulsified and dispersed in suitable solvent systems and treated with the usual pigments and fillers used in such compositions. The following example demonstrates the effectiveness of the corrosion protective compositions and the resultant coatings according to the process, with regard to the corrosion protective values (CV) obtained with them, as well as with regard to the improved dispersion of these agents in such lacquers. The following testing methods were used to obtain the values reproduced in the example.

(A) Corrosion Resistance Testing: The corrosion protective coatings prepared according to the example were applied on rustfree, degreased steel sheets (150 x 70 x 1 mm) by a lacquer-film centrifugal device (type 334/II, Erichsen, Germany). The resulting film thickness was standardized to measure between 30 and 33 μm. After drying the lacquer films so formed for 2 hr at 60°C, then 8 days at room temperature, two of the steel sheets, each coated with the corrosion protective coating, were subjected to the salt spray test according to ASTM B 117-64 (continuous spray with a 5% sodium chloride solution at 35°C) for 200 hr for each corrosion inhibitor composition tested. One of the test sheets was unmarked, the other was provided, prior to the salt spray test, with a so-called Andreas cross (Andreas cross = crosscut with a blade of 0.1 mm) through the lacquer film to the substrate.

Two additional coated steel sheets, per each corrosion inhibitor composition under test, were subjected to the Kesternich test according to DIN 50,018 without Andreas cross for the duration of 12 cycles (1 cycle = 8 hr at 40°C in a humid atmosphere containing 0.2 ℓ sulfur dioxide; followed by an additional 16 hr without load). The stressed test sheet evaluation was based on a rust degree scale according to DIN 53,210. The CV was then calculated according to the method described by Ruf [*Farbe and Lack* 75 (1969) 943-949] from the results obtained.

(B) Sedimentation Behavior: To test the sedimentation behavior of the pigments, i.e., the corrosion inhibitors in the compositions of this process, 250 ml wide-neck flasks were filled with samples of the corrosion protective coating compositions described in the example. By careful manual probing with a 3 mm wide steel spatula, the structure of the sediment in question, was determined at regular intervals. The rating of the structure of the sediment was based on an evaluation scale of 0 to 4. This evaluation scale was based on the following key:

Number	Sediment	Is the pigment stirrable?	Is the lacquer fit for practical application?
0	none	—	yes
1	slight	easily	yes
2	moderate	moderately	conditionally
3	strong	difficult	barely
4	very strong	no, cemented	no

Example: Composition of the binder formulation:

	Parts by Weight
Short oil, resin-modified alkyd resin with 44% linseed oil/wood oil, 60% solution in xylene	34.0
Xylene	9.0
Benzine, BP 145° to 200°C	5.6
Ethylene glycol ethyl ether	2.3
Decalin	2.3
Calcium naphthenate, 4% Ca	0.2

(continued)

	Parts by Weight
Cobalt naphthenate, 6% Co	0.4
Lead naphthenate, 24% Pb	0.1
Methyl ethyl ketoxime	0.3
Titanium dioxide, rutile	6.8
Microtalc	4.5
Barium sulfate	23.7
Zinc phosphate	8.5
Lacquer base	97.7

To each 97.7 parts of this lacquer base were added 2.3 pbw each of the corrosion inhibitors tested. The Examples 1.1 through 1.12 are the salts according to the process. Examples 1.13 through 1.16 contain as corrosion inhibitors, compounds according to the state of the art. In Example 1.16, zinc phosphate was present from the basic composition without any additional inhibitor.

Example	Corrosion Inhibitor	CV in %	Sediment 3 days	after-21 days
1.1	lead salt of mercapto-pyridine (44.0% Pb)	95	0	0
1.2	zinc salt of mercapto-pyridine (22.6% Zn)	89	0	0
1.3	lead salt of 3-cyano-6-hydroxy-4-methylpyridone-2 (56.4% Pb)	88	0	0
1.4	zinc salt of 3-cyano-6-hydroxy-4-methylpyridone-2 (30.2% Zn)	80	0	0
1.5	lead salt of barbituric acid (64.5% Pb)	88	0	0
1.6	zinc salt of barbituric acid (35.2% Zn)	78	0	0
1.7	lead salt of 5-nitroorotic acid (57.9% Pb)	90	0	0
1.8	zinc salt of 5-nitroorotic acid (28.8% Zn)	81	0	0
1.9	lead salt of hydantoin (56.9% Pb)	83	0	0
1.10	zinc salt of hydantoin (40.0% Zn)	75	0	0
1.11	lead salt of mercapto-benzothiazole (37.3% Pb)	96	0	0
1.12	zinc salt of mercapto-benzothiazole (16.4% Zn)	88	0	0
1.13	zinc salt of 5-nitro-isophthalic acid (44.1% Zn)	73	2	4
1.14	zinc/lead mixed salt of 5-nitro-isophthalic acid (31.0% Zn/; 19.1% Pb)	83	1	2
1.15	zinc potassium chromate, lead containing acc'g to DIN 55902	70	0	0
1.16	barium sulfate (i.e., zinc phosphate alone)	55	0	0

(CV = corrosion protection value)

The PVC values of the corrosion protective coatings of Examples 1 through 16 are in the range of from 39.4 to 39.8%. These examples of the formulation show the excellent corrosion-inhibiting effect of the salts of mercaptopyridine as well as those of mercaptobenzothiazole. Salts of mercaptobenzothiazole particularly show excellent activity at remarkably low contents of lead or zinc. Thus, the lacquer composition with 2.3% by wt mercaptobenzothiazole salt contains, in the case of the lead salt (37.3% Pb), only 0.86% lead and in the case of the zinc salt (16.4% Zn), only 0.38% zinc, when the small lead and/or zinc content of the other components of the binder formula are disregarded.

Micro-Throwing Alloy Undercoating

J. Hyner, S. Gradowski and T.F. Maestrone; U.S. Patent 4,329,402; May 11, 1982; assigned to Whyco Chromium Co., Inc. provide micro-throwing alloy undercoatings and a method for improving the corrosion resistance of a ferrous metal substrate by utilization of a micro-throwing alloy as the initial layer, or undercoating, applied directly thereover.

It has been discovered that the micro-throwing alloy undercoatings provide a reliable, uniform coating of corrosion resistant metal plating, most notably over ferrous metal articles having surface defects, pits, cracks, laps or the like. It is believed that this substantial improvement stems from the micro-throwing power of these alloys and their demonstrated ability to coat, or even fill in, the surface defect areas, thus providing a uniformly receptive surface for subsequently applied conventional platings and/or coatings. A coating for such a substrate comprises:

(a) a layer of nickel-cadmium alloy comprising 95 to 99.9% by weight nickel and 0.1 to 5.0% by weight cadmium, the alloy being characterized by the ability to be electrodeposited upon the substrate to form a layer which is thicker inside of surface defects thereon than on the plane surface in which the defect is formed, the layer of nickel-cadmium alloy being applied directly over the substrate and having a thickness which ranges from about 0.005 to 0.00005 inches, and

(b) at least one coating which contributes to further improving the corrosion resistance of the ferrous metal substrate, the coating being applied over the layer of nickel-cadmium alloy and being selected from (1) a galvanically protective metal or alloy selected from the group consisting of cadmium and tin, zinc or zinc alloy, (2) alloys of (1), (3) paints, (4) metal dyes, or (5) a chromate film.

The following bath formulations are among those which can be used, as required, to plate the desired metal or alloy layer.

Nickel-Cadmium Alloy Bath	
$NiSO_4 \cdot 7H_2O$	350 g/ℓ
$NiCl_2 \cdot 6H_2O$	45 g/ℓ
Boric acid	40 g/ℓ
Gelatin	5 g/ℓ
Cadmium sulfate	1.08-3.6 g/ℓ
Operating Conditions	
Temperature	57°C
Current density	16 amp/dm^2
pH	about 6.0
Cadmium Bath	
Cadmium oxide	31.5 g/ℓ
Sodium cyanide	142.3 g/ℓ
Plating Conditions	
Temperature	23.9°-32.2°C
Current density	0.5-16.2 amp/dm^2
Zinc Bath (Kenlevel II, 3M Co.)	
Concentrated zinc chloride	101.86 g/ℓ
Potassium chloride	224.7 g/ℓ

(continued)

Zinc Bath (Kenlevel II, 3M Co.)	
Boric acid	33.7 g/ℓ
Kenlevel II TB	29.96 g/ℓ
Kenlevel II TM	0.26 ml/ℓ
Plating Conditions	
Temperature	26.7°C
pH	5.0
Current density	3.2 amp/dm^2
Tin Bath	
Potassium stannate	104.86 g/ℓ
Potassium hydroxide (free)	39.7 g/ℓ
Sodium hydroxide (free)	14.98 g/ℓ
Plating Conditions	
Temperature	71°C
Current density	3.2 amp/dm^2

In accordance with the method, several steel fasteners were electroplated with an initial layer, or undercoating, of nickel-cadmium micro-throwing alloy. The steel fasteners were made cathodic and electroplated using the aforementioned nickel-cadmium plating bath. The resulting layer of nickel-cadmium alloy comprised between about 2.5% by weight of cadmium and was electroplated to a thickness of about 0.00025 inch. A series of these undercoated fasteners were then subsequently plated with the respective layer or layers of galvanically protective metals and/or organic coatings and subjected to a 5% Neutral Salt Spray resistance test (ASTM B117). These results were compared with similarly coated fasteners which lacked the initial undercoating layer of micro-throwing alloy.

Example 1: Several steel fasteners having micro-throwing alloy undercoating were electroplated with 0.00030 inch of zinc, using the aforementioned conventional zinc plating bath and operating conditions (i.e. Kenlevel II). Several "control" fasteners, (i.e. without an undercoating layer of micro-throwing alloy), were likewise plated with 0.00030 inch of zinc using the same bath and plating conditions.

Both sets of fasteners were subjected to 5% Neutral Salt Spray testing, with the results set forth in the table below. A substantially superior degree of corrosion resistance was clearly demonstrated by the fasteners which were undercoated with the micro-throwing alloy undercoatings of the process.

Example 2: Example 1 was repeated, except that instead of a layer of zinc, a 0.00030 inch layer of cadmium was applied, using the aforementioned cadmium bath. The comparative performance of these fasteners in 5% Neutral Salt Spray testing, is also set forth in the table below. Again, the fasteners having micro-throwing alloy undercoatings exhibited a superior level of resistance to corrosion.

Example 3: Example 1 was again repeated, except that a uniform layer of a thermosetting, phenolic paint, (Polyseal, R.O. Hull Co.) was applied over the plated zinc layers on both the fasteners plated with the micro-throwing alloy and zinc and the control fasteners plated with zinc alone. The superior performance of the fasteners undercoated with a layer of micro-throwing alloy in accordance with the process is likewise set forth in the table below.

Example 4: Example 3 was repeated, except that a chromate film was applied over both sets of fasteners. The chromate film was applied from a commercially available bath (Kenvert No. 5). The superior performance of the fasteners undercoated with a layer of micro-throwing alloy is likewise documented in the table.

Example 5: Example 2 was repeated, except that a 0.00005 inch layer of tin was electroplated over the cadmium platings on both sets of fasteners. The aforementioned conventional cadmium bath was used. The superior performance of the fasteners undercoated in accordance with the process is also set forth in the table below.

Example No.	Sequence of Coatings Applied	Time to Red Rust (Hrs.)
1	Nickel-Cadmium micro-throwing alloy/Zinc	340
	Zinc alone	160
2	Nickel-Cadmium micro-throwing alloy/Cadmium	265
	Cadmium alone	80
3	Nickel-Cadmium micro-throwing alloy/Zinc/paint	550
	Zinc/paint	240
4	Nickel-Cadmium micro-throwing alloy/Zinc/Chromate	418
	Zinc/Chromate	172
5	Nickel-Cadmium micro-throwing alloy/Cadmium/Tin	650
	Cadmium/Tin	194

Paint-Receptive Coating Not Requiring Rinsing

A composition and process useful for imparting corrosion resistance and paint receptivity to metal surfaces are provided by *K. Yashiro and Y. Moriya; U.S. Patent 4,341,558; July 27, 1982; assigned to Hooker Chemicals & Plastics Corp.* The process involves application of an aqueous wet film which is dried in place and thus avoids effluent problems from conventional processes involving rinsing. The composition contains a water-soluble titanium or zirconium compound, an inositol 2 to 6 phosphate ester and silica.

The amount of the water-soluble titanium or zirconium compound that is added to the metal surface coating agent of this process is determined by the wet film thickness applied to the metal surface and on the basis of Ti or Zr they are added to the coating agent so as to give surface deposits of 1 to 200 mg/m^2 and desirable 5 to 50 mg/m^2.

The inositol 2 to 6 phosphate ester that are used are inositol diphosphate ester, inositol triphosphate ester, inositol tetra-phosphate ester, inositol pentaphosphate ester and inositol hexaphosphate ester or salts in which the hydrogen atoms of the inositol 2 to 6 phosphate esters have been replaced with alkali metals or with alkaline earth metals. Inositol hexaphosphate ester is generally called phytic acid. Furthermore, the inositol di-penta-phosphate esters are obtained mainly by the hydrolysis of phytic acid and so industrially phytic acid is the most useful. The amount used is determined by the amount that is to be deposited on the metal surface, as in the case of Ti and Zr and is added to give a surface concentration of 1 to 500 mg/m^2 and desirably 5 to 200 mg/m^2.

The amount of SiO_2 used is determined by the amount that is to be deposited on the metal surface so as to give a surface concentration of 20 to 2000 mg/m^2 and desirably 50 to 1000 mg/m^2.

In order to adjust the pH of the metal surface coating agent, any conventional alkaline bases such as ammonia, ethylamine or caustic soda and the oxides, hydroxides or carbonates of metals such as Al, Zn, Ni can be used and the pH is adjusted to 1 to 7.

Example: A metal surface coating agent was prepared in the way outlined below.

Composition of the Metal Surface Coating Agent

	Parts by Weight
Ammonium fluotitanate	10
Phytic acid (50% aqueous solution)	16
Silica (Aerosil #200)	30
PVA (Degree of polymerization 1400)	50
Deionized water	894

The pH was adjusted to 5.3 using concentrated aqueous ammonia.

After cleaning the surface of a commercially available electrozinc plated steel sheet by wiping with acetone, it was coated with the abovementioned metal surface coating agent. The coating was carried out with a roll coater and the system used a mesh roll for the metering roll for reverse coating and the amount of material coated in the wet coat was set at 10 g/m^2 by adjusting the space between the rolls. Immediately after coating the sample was dried for 35 seconds at 120°C in a hot air circulating oven. At this time, the temperature of the sheet was 70°C.

As a reference example, identical sheet was cleaned in same manner, rinsed with water, and treated with a chromate treatment liquid with the composition shown below:

Composition of the Chromate Treatment Liquid

	Parts by Weight
Anhydrous chromic acid	10
Phosphoric acid	1
Silicon hydrofluoride	2
Chromium carbonate	1
Sulfuric acid	0.5
Water	985.5

The pH was adjusted to pH 2.0 with zinc carbonate.

The conditions of the chromate treatment involved treating for 10 seconds with a spray method at a temperature of 40°C and then rinsing immediately with water and drying. The amount of Cr deposited was measured by X-ray fluoresence and was 28 mg/m^2.

The electrozinc plated steel sheets that had been treated as above were then coated with an alkyd-melamine based paint by the bar coating method and after setting for 20 minutes they were baked for 25 minutes in a hot air circulating oven at an air temperature of 140°C to yield painted sheets with a paint film thickness of 30±2 microns. These painted sheets were then subjected to paint film adhesion tests and salt spray tests and the results of these tests are shown in the following table.

Electrozinc plated steel sheets that had been treated as above but not painted were cleaned by spraying for 3 minutes with a commercially available alkaline cleaning solution at 20 g/ℓ at 60°C. The samples were then rinsed with water for 15 seconds and dried in a drier. These specimens were then painted in exactly the same way as noted above and the results of paint film adhesion tests carried out on these painted samples are also shown in the following table.

	Paint Film Adhesion Test		Salt Spray Test	
Alkaline Cleaner	No	Yes	No	Yes
Example 1	5	4	4	3
Reference example 1	3	1	3	1

Paint Film Adhesion Test: The paint film was subjected to convex impact of 7 mm with an Erikesen tester, tape pulled and an assessment made on the basis of the 5 point scale given below:

5 points	No peeling of the paint film
4 points	Peeling of the paint film less than 5%
3 points	Peeling of the paint film 6 to 25%
2 points	Peeling of the paint film 26 to 50%
1 point	Peeling of the paint film more than 50%

Salt Spray Test: A cross cut was made on the painted sheet down to the underlying metal using an NT cutter and after carrying out a salt spray test in accordance with JIS-Z-3271 for 120 hr the samples were washed with water and dried and the paint surface was tape pulled and the paint peeling from the cross cut part was assessed as follows:

5 points	No peeling
4 points	Width of peeling on both sides of the cross cut. Less than 1 mm
3 points	Width of peeling on both sides of the cross cut. Less than 2 mm
2 points	Width of peeling on both sides of the cross cut. Less than 5 mm
1 point	Width of peeling on both sides of the cross cut 6 mm or more

Reducing Hexavalent Chromium

L. Schiffman; U. S. Patent 4,341,564; July 27, 1982 describe corrosion inhibiting pigments which are obtained by partial reduction of a hexavalent chromium compound in aqueous solution to yield a reaction product containing 5 to 95% of the chromium in lower valence state. In the aqueous reaction mixture there is incorporated a water-soluble or water-dispersible polymeric material capable of reacting with chromate to form an insoluble product therewith and the obtained mixture evaporated to dryness under controlled conditions, followed by subdividing the obtained dry mass to suitable size.

Example 1: In the following table are listed some preparations for chromium chromate pigments of this process. An equivalent of 25 parts CrO_3 to 100 parts H_2O (w/w) is used as the hexavalent chromium oxidant. The reductant to oxidant (calculated as CrO_3) mol ratio is given in column 3. The % hexavalent chromium in the reacted solution is given in Column 4. The reported analysis is based on reacted solution—approximately 8 hr after "apparent" end of reaction—that is, when reactants have been completely added and gassing has ceased.

Oxidant (Hexavalent Cr Compound)	Reductant	Ratio Oxidant: Reductant	% Cr^{+6} Content
CrO_3	H_2O_2 (30%)	3:1	88
CrO_3	HCHO (37%)	3:1	55.4
CrO_3	CH_3OH	3:1	65.6
CrO_3	furfural	3:1	42.4
CrO_3	citric acid	3:1	25
$(NH_4)_2Cr_2O_7$	HCHO (37%)	3:1	80
CrO_3	H_2O_2 (30%)	1:2	70

The reductant (in aqueous solution or dispersion) is carefully added to the oxidant CrO_3 solution, with the rate of addition controlled so that excessive rate of reaction is avoided. After the reaction has largely subsided, the reaction mix is transferred to evaporating trays for drying. The dried product is then ground to pigment particle size.

Since the extent of reduction of the hexavalent chrome depends not only upon the concentration of the reactants, but also on the specific reducing agent employed, or upon combination of reducing agents, pigments having a specific hexavalent to reduced chrome ratio can be obtained, or tailor made. Although corrosion inhibition properties exist even with as much as 95% chromium reduced, most effective performance occurs when the hexavalent chromium content is in the 25 to 75% range or inversely the reduced chromium content is in the 75 to 25% range.

Example 2: *Typical Pigment Preparation* – 25 kg of chromium trioxide are dissolved in 100 kg of water. To this solution is added slowly and with vigorous stirring, a solution prepared by diluting 6.8 kg of 37% formalin with 3 kg water. The addition rate is controlled so that the reaction temperature does not exceed 95°C. When the reaction subsides, the reaction solution is permitted to cool down to room temperature. To this solution is then added slowly and with stirring 1 kg of Acrysol A-1 (25% by weight polyacrylic acid solids). The obtained liquid is transferred to evaporating trays for drying, which is carried out at temperatures of about 110°C. After the product is visibly dry it is heated for another 2 hr at 150°C.

The dried product is then ground to pigment particle size, with maximum 1% coarse particles retained on #325 standard sieve.

Non-Petroleum-Based Composition

According to the process of *A.J. Conner, Sr.; U.S. Patent 4,342,596; August 3, 1982* there is provided a metal corrosion inhibiting composition which is a water-based solution of:

(1) a $C_{8\text{-}20}$ aliphatic monobasic acid;
(2) a lubricant;
(3) an aminoalkylalkanolamine;
(4) an aromatic mono- or polycarboxylic acid; and
(5) an amine which forms a water-soluble salt with the acids.

The composition can be applied to the metals by spraying or rolling.

A coating of the solution inhibits oxidation of metal surfaces, provides lubricity and need not be removed from a metal surface prior to painting.

The composition contains a minor amount of a lubricant which may be either a petroleum or a non-petroleum product. Any of the petroleum oils presently employed in petroleum based corrosion inhibiting compositions for steel are believed to be useful in the present composition. Good results have been obtained using a 100 SSU viscosity petroleum oil. In lieu of a petroleum oil, esters such as butyl stearate, dioctyl sebacate, butyl benzoate, or any of the light alkyl esters with boiling ranges above 350°F can be used as the lubricant. In a particularly preferred embodiment a petroleum oil is used as the lubricant.

The preparation of typical 55 gallon batches of a concentrated solution of the non-petroleum based corrosion inhibitor is described below (approximate weights are in parenthesis):

(1) Pump 30 gal of water (250 lb) at 120°F into tank and agitate. Add 10 gallons of a tall oil fatty acid/rosin mixture (80 lb) (Unitol-DT-40, Union Camp) and 1 or 2 gallons of 100 SSU viscosity petroleum oil (7-14 lb). The oil will dissolve in the tall oil-rosin mixture, but neither the petroleum oil nor the tall oil fatty acid/rosin mixture will dissolve in water. While agitating add 1 gallon of aminoethylethanolamine (8 lb). An oil in water emulsion will form. This emulsion is milky and completely opaque. Add 8 gallons of monoethanolamine (64 lb) and the mixture will become clear and stable. Add 100 lb of benzoic acid and the mixture will become hazy because of the portion of the benzoic acid which has not been neutralized to a soluble salt. To complete neutralization of the benzoic acid, add more monoethanolamine (or morpholine, cyclohexylamine, etc.) until the solution is completely clear and has a pH of 8.0 to 9.5. Continue mixing for 30 minutes and recheck pH. If pH drops below 8.0, add more monoethanolamine to bring pH to 9.0.

(2) Dump 30 gallons of water (250 lb) at 110° to 120°F into a tank, add 10 gallons of tall oil fatty acids containing 8 to 12% rosin acids. While agitating, add 1 quart of aminoethylethanolamine. The tall oil/rosin mixture will emulsify (solution will be milky). Then add 2½ gallons of diethanolamine and the solution will clear and thicken. While agitating slowly add 45 lb of terephthalic acid. The solution will remain clear and the viscosity will drop. Dilute up to 55-58 gallons with water and continue agitating until all the terephthalic acid has dissolved. The viscosity of the finished solution at 100°F will be about the same as a 30 wt commercial grade lubricating oil.

For use at the mills or manufacturing plants, one part of a composition prepared as described above is diluted with up to 5 parts of water and applied as either a rust preventative and/or lubricant. The recommended dilution ratio is 1 part concentrate to about 4 parts water.

Improved Automotive Powder Primer

The process of *R.-T. Khanna; U.S. Patent 4,351,914; September 28, 1982; assigned to E.I. Du Pont de Nemours and Company* is related to an improved powder coating composition and, in particular, to an automotive powder primer surfacer exhibiting substantially improved corrosion resistance under a powder topcoat.

The improved powder coating composition is of the kind consisting essentially of finely divided powder particles of a blend of epoxy functional film-forming resin, an approximately stoichiometric amount of a curing agent that is polyhydroxyl functional, polycarboxyl functional, or polyamino functional, and optionally, a flow control agent, pigment, and filler particles. The improvement comprises:

(A) 12-45% by weight, based on the weight of the film-forming resin, of a compound selected from the group consisting of zinc oxide, a zinc oxide-containing complex, and mixtures of these; and

(B) a dicarboxylic acid in an amount by weight equal to 0.5-10% of the total weight of (A), this amount being independent of the amount of any dicarboxylic acid which may be present as the curing agent.

An example of a zinc oxide-containing complex suitable for use in the process is a zinc phosphooxide complex marketed as Nalzin SC-1.

Although any dicarboxylic acid can be used in the improvement, a preferred acid is dodecanedioic acid. For purposes of the process, the acid is micronized, i.e., the particle size is 5 microns or less.

Although it is not known what the exact mechanism is by which the zinc oxide (or modified zinc oxide) and acid combine to inhibit corrosion, it is hypothesized that providing zinc ions in "slowly" leachable form from a water insoluble organic polymer acid ionomer creates a corrosion-inhibiting environment around steel.

The following example illustrates the process. In the example, the components will be referred to according to the following numbering system:

(1) Epoxy resin having the formula

$$\overset{O}{\widehat{CH_2-CH}}-CH_2-\left[-O-C_6H_4-\underset{CH_3}{\overset{CH_3}{\underset{|}{\overset{|}{C}}}}-C_6H_4-O-CH_2-\overset{OH}{\overset{|}{CH}}-CH_2-\right]_n-O-$$

$$-C_6H_4-\underset{CH_3}{\overset{CH_3}{\underset{|}{\overset{|}{C}}}}-C_6H_4-O-CH_2-\overset{O}{\widehat{CH-CH_2}}$$

where n is sufficiently large to provide a resin having a Gardner-Holdt viscosity of H-S and an epoxide equivalent weight of 400-850.

(2) Phenol-modified epoxy resin which is an epoxy resin (epoxide equivalent weight 186-192) of the formula of (1) above reacted with bisphenol-A and phenol in an epoxy-resin/bisphenol-A/phenol equivalent weight ratio of 1.82/1.0/0.5 to provide a phenol-modified epoxy resin having an epoxide equivalent weight of 590-630.

(3) Curing agent which is a combination of:

(a) a mixture of compounds of the general formula

$$HO{-}Ar{-}[{-}O{-}CH_2{-}\underset{\displaystyle OH}{\underset{|}{CH}}{-}CH_2{-}O{-}Ar{-}]_x{-}OH$$

where Ar =

$$-C_6H_4-\underset{\displaystyle CH_3}{\overset{\displaystyle CH_3}{\overset{|}{\underset{|}{C}}}}-C_6H_4-$$

and x is 0 or a positive number. providing a mixture of compounds having an equivalent weight of 230-1000; and

(b) 0.67% by weight, based on the weight of (a), of 2-methylimidazole.

Example: The following components are blended as described:

Ingredient	% by Weight (Based on Total Weight of all Ingredients)
Component 1	22.48
Component 2	15.23
Zinc oxide	8.94
Titanium dioxide pigment	2.37
Carbon black	0.11
Micronized Nalzin SC-1 (National Lead) (max. diameter of 5 microns)	8.91
Micronized dodecanedioic acid (max. diameter of 5 microns)	1.18
Fluorad (flow control agent)	0.37
Triethanolamine	0.59
Component 3	10.16
Barium sulfate	29.67

Charge the entire mixture into a Marion mixer and mix for 1 hr. This blend is then extruded twice in a Cokneader at the highest speed and the lowest possible temperature settings. This is followed by pin milling twice. Afterward, the blend is classified at 600 rpm, to remove coarse particles. Finally, the powder is passed through a 325-mesh kason sieve.

The powder is sprayed as a primer onto 20-gauge panels using Ransburg electrostatic powder guns, and the coated panels are then heated in a gas-fired oven for approximately 45 minutes at 160° to 170°C.

Hydroxybenzotriazole plus Aliphatic Dicarboxylic Acid

In the process of *K. Tanikawa, T. Obi, S. Otsuka, I. Manabe, A. Inubushi and C. Maeda; U.S. Patent 4,354,881; October 19, 1982; assigned to Nippon Steel Corporation, Japan* a steel stock such as hot-rolled pickled steel plate, cold-rolled steel plate, cast iron or the like is subjected to antirust treatment with an aqueous solution or an emulsion. The composition consists mainly of 0.01 to 10% by weight 1-hydroxybenzotriazole compound represented by the general formula:

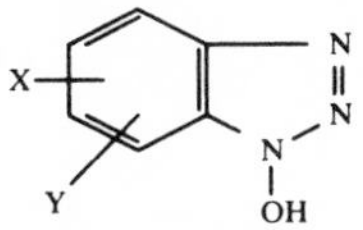

where X and Y represent hydrogen atoms and hydroxy, alkyl, carboxyl, nitro and sulfonic groups. The composition further contains 0.01 to 5% by weight one or more of aliphatic C_{8-13} dicarboxylic acids added thereto, and has a pH within the range of 7-10.

The effects of the process are not limited to antirust ones in a particular environment. The antirust effects are excellent not only in water but also in the air, and in an acidic atmosphere or in an atmosphere of high temperature and high humidity where water droplets are present between the contacted treated steel plates or in other various environments where a temper rolling solution is scattered or its vapor attaches as water droplets on steel plates rolled in a high speed temper rolling step. Also no work in removing the anticorrosive agent before the next step is necessary, and painting or other required treatments may be conducted directly, so the workability can be improved. Further, a problem of toxicity resulting from the treatment has recently arisen in the case of the conventional treatment involving the use of sodium nitrite, whereas in case of this process, such toxicity is extremely low, and in the welding or like operations, there is no environmental pollution caused by fuming, nor any lowering of workability caused by staining of the electrodes.

In the following examples which illustrate the process, all percentages are by weight.

Example 1: An aqueous solution containing 0.5% of 1-hydroxybenzotriazole and 0.3% of azelaic acid was adjusted to pH 8 by the addition of monoethanol amine as the neutralizing agent to give a treating solution. A steel material was subjected to the conventional temper rolling, and water was purposely applied to a part of the steel material at the exit, which was coiled as it was.

Example 2: An aqueous solution prepared by adding 0.2% azelaic acid and further 0.03% of a nonionic surfactant (a polyoxyethylene alkyl ether) to 0.4% of 1-hydroxybenzotriazole is adjusted to pH 8 by the addition of monoethanol amine as the neutralizing agent to give a treating solution. This treating solution is sprayed continuously over a surface-cleaned cold-rolled steel plate. Immediately thereafter, the plate is treated with rubber rolls and dried with a drier.

Example 3: An aqueous solution prepared by adding 0.3% of sebacic acid and 0.1% of an acrylic resin, a water-soluble high molecular material, to 0.4% of 1-hydroxybenzotriazole is adjusted to pH 8 by the addition of triethanolamine as the neutralizing agent to prepare a treating solution. This treating solution is sprayed continuously over a surface-cleaned cold-rolled steel plate, and immediately the plate is treated by rubber rolls and dried with a drier.

Example 4: An aqueous solution, prepared by adding 0.2% of sebacic acid and 0.03% of a nonionic surfactant (a polyoxyethylene alkyl ether) to 0.5% of 1-hydroxybenzotriazole, is adjusted to pH 9.0 by the addition of monoethanolamine as the neutralizing agent to give a treating solution. It is sprayed continuously over a surface-cleaned cold-rolled steel plate. Immediately thereafter, the plate is treated with rubber rolls and dried with a drier.

The results of antirust tests using steel plates treated according to the above method of the process and controls are shown in the table.

Results of Antirust Tests

	Housing test in wet box*				
	Stacking condition	Stacked after dropping of treating solution (1)	Stacked after dropping of tap water (2)	Indoor exposure for 10 days	In 0.6 N HCl atmosphere for 24 hours (3)
Present invention					
Example 1	Coiled and allowed to stand in plant for 1 month	—	—	◎	◎
Example 2	◎	◎	◎	◎	◎
Example 3	◎	◎	◎	◎	◎
Example 4	◎	◎	◎	◎	◎
Controls**					
Ammonium sebacate	○	○	Δ	X	X
Triethanol amine azelate	○	○	Δ	X	X
1-hydroxy-benzotriazole ammonium	◎	◎	Δ ~	○	○
Commercially available product A (4)	○	○	Δ	X	X
No treatment	XX	—	XXX	XX	XXX

In the table,

*Ten test pieces of 10 x 10 cm subjected to a 7-day housing test in a wet box (50°C, 98% RH) were bound tightly by the use of a miniature vise and then subjected to the tests.

**The concentration of all the treating solutions used as the controls was 1%, and the treatment using them was made by the same procedures as in the case according to this process.

(1) The aqueous treating solution was dropped on the treated steel plates, and the treated steel plates were stacked for the tests. Discoloration and rust condition of the portions on which the solution was dropped were evaluated.

(2) Tap water was dropped on the treated steel plates, and the treated steel plates were stacked and tested. Rust condition of the portions on which the water was dropped was evaluated.

(3) In the bottom of a desiccator, there was placed a 0.6 N aqueous HCl solution, and the treated steel plates were placed on a perforated plate in the desiccator by the capping. Rust conditions after 24 hr was evaluated.

(4) The commercially available product A was a sodium nitrite system. Evaluations: ◎. . . No change; ○. . .Slight discoloration; Δ . .Rust is noticeable; X. . .About 10% rust; XX. . .About 30% rust; XXX. . .More than 60% rust.

Hydroxybenzyl Amines

D. Frank and L.D. Metcalfe; U.S. Patent 4,357,181; November 2, 1982; assigned to Akzona Incorporated provide a method for inhibiting corrosion of a metal surface. The method comprises contacting the metal surface with a corrosion inhibitor of the formula

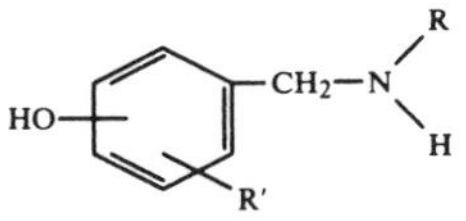

$$\text{HO}-C_6H_3(R')-CH_2-N(R)(H)$$

wherein R is selected from the group consisting of aliphatic radicals containing from about 6 to about 22 carbon atoms, $R_1 \leftarrow C_m H_{2m} \rightarrow$ wherein R_1 is alkoxy containing from 6 to 22 carbon atoms and m is an integer of from 2 to 6, and

$$R_2 \leftarrow OCH_2\overset{\displaystyle R_3}{\overset{|}{C}}H \rightarrow_x$$

wherein R_2 is alkyl containing from 1 to 20 carbon atoms, x is an integer of from 1 to 10 and each R_3 is independently hydrogen or methyl; and R' is selected from the group consisting of hydrogen, C_{1-12} alkyl, and C_{1-12} alkoxy, at a temperature and for a period of time sufficient to inhibit corrosion of the metal surface.

Examples 1 through 5: Seven steel coupons (3" x 6") were dipped into chloroform solutions containing 1% of a compound in accordance with this process, except for Examples 3 and 4, which were for comparative purposes and contained no additive. Examples 1 and 2 contained 1% of N-(2-hydroxybenzyl)octadecylamine and Example 5 contained 1% of N-(2-hydroxylbenzyl)cocoamine.

The coupons were dipped into the solution for 2 minutes, removed, and the solvent allowed to evaporate. The panels were then slightly wiped with a paper towel. A black acrylate was then sprayed twice onto the coupons and was cured for 72 hr.

The coupons were supported at an angle on plastic racks in a salt spray apparatus and sodium chloride 99.97% pure dissolved in distilled water at a concentration of 5% was used at a temperature of 93°F and a pH of 6.8-7.0. The test procedure was in accordance with ASTM B-117-73. The results of the testing are summarized in the following table.

Example	Hours Tested	Results
1	24	Scattered blisters
	48	Slight increase in blistering
	72	Rust in blisters
	96	No further change
	120	No further change
	150	Blistering over 15%
2	24	A few scattered blisters
	48	Blistering slightly increasing
	72	Rust running from top
	96	No further change
	120	No further change
	150	Blistering over 25%

(continued)

Example	Hours Tested	Results
3	24	A few scattered blisters
	48	No further change
	72	Rust in blisters
	96	No further change
	120	No further change
	150	Blistering over 50%
4	24	A few scattered blisters
	48	Blistering increasing
	72	Rust in blisters
	96	No further change
	120	No further change
	150	Blistering over 50%
5	24	A few scattered blisters
	48	No further change
	72	Rust in blisters
	96	A few more blisters
	120	No further change
	150	Blistering over 10%

Sprayed Glass Flake Composition

A process by *M. Hoshino, T. Shinohara, H. Tanabe, T. Taki and S. Nakayama; U.S. Patent 4,363,889; December 14, 1982; assigned to Dai Nippon Toryo Co., Ltd., Japan* provides for the formulation of anticorrosive coatings. The process comprises spray-coating in a coating thickness of 200 to 1500 microns a composition comprising (a) 100 parts by weight of an unsaturated polyester resin, (b) 10 to 100 parts by weight of a glass flake having an average thickness of 0.5 to 5 microns and an average particle size of 100 to 400 microns, or 10 to 70 parts by weight of the glass flake and 10 to 150 parts by weight of scaly metal pigment, (c) 0.1 to 1.0 part by weight of a ketone peroxide and (d) 0.5 to 2.0 parts by weight of a hydroperoxide and/or a peroxy ester by using a spray gun having a spray tip diameter of 0.5 to 3 mm.

In this process, the glass flake is laminated in a plurality of layers in the coating in parallel to the substrate, and serves to increase the strength of the resin and, simultaneously, to prevent permeation and transmission of ambient materials such as vapor and water. This preventive effect is ordinarily enhanced when the thickness of the glass flake is small and the diameter is large. This tendency is conspicuous in a highly corrosive atmosphere or environment.

If the thickness of the glass flake is smaller than 0.5 micron, the strength of the glass flake is low and the glass flake cannot be used in practice. On the other hand, if the thickness of the glass flake is larger than 5 microns, the glass flake is seldom arranged in parallel with the substrate.

Furthermore, if the size of the glass flake is smaller than 100 microns, the glass flake is seldom arranged in parallel with the substrate, and the strength and corrosion resistance of the coating cannot easily be improved. If the size of the glass flake exceeds 400 microns, the adaptability of the coating composition to the spray-coating operation is reduced and use of a glass flake having such a large size is not preferred.

Example 1:

Formula 1–Main Ingredients

	Parts by Weight
Unsaturated polyester resin (Rigolac 150 HR, Isophthalic acid type)	37
Organic bentonite (Bentone #34)	2.0
Styrene	35.5
Cobalt naphthenate (containing 6% of metallic cobalt)	0.5
Glass flake (average particle size = 200 microns, average thickness = 2-3 microns)	15
Talc	5
Titanium oxide	5
Total	100
Curing agents:	
Methyl ethyl ketone peroxide	0.1
Cumene hydroperoxide	0.3

The unsaturated polyester resin was first mixed with the organic bentonite and the mixture was kneaded by a roller. The remaining main ingredients were added to the mixture, and the mixture was stirred with a disperser to form a main ingredient composition. A soft steel plate (JIS G-3141) having a size of 150 x 50 x 1.6 mm was subjected to shot blasting to remove scale, rust and oil completely. The main ingredient composition of Formula 1 in which the curing agents had been incorporated was coated on this steel plate to a dry coating thickness of 500±50 microns by compressed air spraying (tip diameter = 1 mm). The coating was dried at 20°C for 7 days, and the coated steel plate was subjected to comparison tests.

Example 2:

Formula 2–Main Ingredients

	Parts by Weight
Unsaturated polyester resin (Rigolac LP-1, Bisphenol type)	40
Organic bentonite (same as used in Example 1)	3.0
Paraffin wax	0.3
Styrene	11.2
Cobalt naphthenate (same as used in Example 1)	0.5
Glass flake (same as used in Example 1)	25
Talc	15
Titanium oxide	5
Total	100
Curing agents:	
Methylethyl ketone peroxide	0.15
tert-Butyl peroxyphthalate	0.3

A coating composition of Formula 2, formed by kneading in the same manner as described in Example 1, was coated on the same type of steel plate sample used in Example 1, and the coated steel plate was subjected to comparison tests.

Example 3:

Formula 3–Main Ingredients

	Parts by Weight
Unsaturated polyester resin (Repoxy R-800, Epoxy Acrylate type)	50
Organic bentonite (same as in Example 1	1.0
Paraffin wax	0.3
Styrene	23.2
Cobalt naphthenate (same as in Example 1)	0.5
Glass flake (same as in Example 1)	20
Titanium oxide	5
Total	100
Curing agents:	
Methyl ethyl ketone peroxide	0.3
tert-Butyl peroxyphthalate	0.7

A composition of recipe Formula 3, which was formed by kneading in the same manner as described in Example 1 was airless-spray coated on the same type of steel plate sample as used in Example 1, so that the dry coating thickness was 500±50 microns (the tip diameter was 1 mm). The coated steel plate was dried at 5°C and subjected to comparison tests.

Comparative Example 1:

Formula 4–Main Ingredients

	Parts by Weight
Unsaturated polyester resin (same as used in Example 1)	50
Organic bentonite (same as used in Example 1)	2.5
Styrene	26.7
Cobalt naphthenate (same as used in Example 1)	0.8
Glass flake (average particle size = 40 microns, average thickness = 2-3 microns)	20
Total	100
Curing agent:	
Methyl ethyl ketone peroxide	1.5

A coating composition of Formula 4 was air-spray-coated on the same type of steel plate sample as used in Example 1 so that the dry coating thickness was 500±50 microns. The coated sample was dried at room temperature for 7 days and subjected to comparison tests.

The results of the comparison tests made on coatings formed in the foregoing Examples and Comparative Example are shown in the following table.

	Example 1	Example 2	Example 3	Comparative Example 1
Salt spray resistance test[1]	no change	no change	no change	rusting of coating from cut portion
Adhesion test (kg/cm^2)[2]	65	60	63	40
Weather resistance test[3]	no change	no change	no change	partial peeling at periphery of cut portion
Service water dip resistance test[4]	29 days	30 days	35 days	18 days
Alkali resistance test[5]	no change	no change	no change	partial blistering and rusting on coating
Acid resistance test[6]	no change	no change	no change	partial blistering and rusting on coating
Humidity resistance test[7]	no change	no change	no change	partial blistering and rusting on coating
Impact resistance test[8]	no change	no change	no change	cracking

[1] Square cross cuts were formed in the coating and after 1000 hr of salt spray, the coating was peeled and a check was made to determine whether or not rusting had advanced below the coating from the cut portion.
[2] The coating was tested by a Tensilon tester and the strength per cm^2 necessary for breakage of the coating was measured.
[3] Cuts were formed in the coating, the coating was tested for 500 hr by a weatherometer, the coating was folded at an angle of 20° and the state of the coating was examined.
[4] The sample was immersed in service water maintained at 60°C and the number of days for the appearance of any change in the coating were recorded.
[5] The sample was immersed in a 5% aqueous solution of sodium hydroxide for 5 months.
[6] The sample was immersed in a 20% aqueous solution of hydrochloric acid for 5 months.
[7] The sample was allowed to stand at a relatively humidity of 100% for 1000 hr.
[8] The sample was tested by a Du Pont impact tester under conditions of 1 kg x ¼" R x 50 cm.

SPECIAL APPLICATIONS

Rust-Preventive Compatible with Phosphate Esters

Phosphate esters are finding increasing use as fire resistant lubricating and hydraulic fluids. These ester lubricants have the desirable properties of low flammability, high lubricity and long service life.

New machinery or machinery in storage or transport which is designed for use with phosphate esters is frequently rust-proofed with conventional petroleum based compositions or other formulations which are not compatible with phosphate esters. This situation often necessitates an extensive cleaning of rust preventive treated machinery before operation with phosphate ester based fluids may be begun.

J.F. Anzenberger; U.S. Patent 4,263,062; April 21, 1981; assigned to Stauffer Chemical Company provides improved corrosion inhibiting compositions which are compatible with phosphate ester lubricants and are effective in controlling rust.

The compositions consist of:

(A) from 65 to 90 weight percent of an aryl phosphate ingredient

(B) from about 0.1 to about 5.0 weight percent of an oil soluble calcium petroleum sulfonate ingredient; and

(C) from about 20 to about 30 weight percent of a liquid polyolester ingredient.

Metal surfaces (especially, ferrous metal surfaces) are protected from corrosion by applying a coating of the rust-preventive composition of this process. The composition may be applied by any conventional means such as spraying, dipping, brushing, flushing, etc. Since the rust-preventive composition has the ability to adhere to metal surfaces, it is only necessary to contact the metal with the composition to deposit a rust-preventive effective coating.

Example: *Test procedure* – A steel paint panel (approx. 7.62 cm x 12.7 cm) was completely immersed in a rust-preventive formulation. The panel was allowed to drip dry for 30 minutes. The panel was then laid flat in a container slightly larger than the panel and 80 ml of distilled water was poured into the container to completely cover the panel. The container was allowed to stand at room temperature until all of the water evaporated. The test panel was then examined and evaluated for rust formation.

Rust-preventive compositions (all percentages by weight) –

Sample A: Trixylyl phosphate (Fyrquel 220) 79 weight percent; oil-soluble calcium sulfonate detergent additive (Tergol 8BH) 1% liquid polyolester, a reaction product of pentaerythritol with C_7 and C_4 alkanoic acids (Base Stock 874) 20%;

Sample B: Trixylyl phosphate (ingredient used in Sample A) 78%; oil-soluble calcium sulfonate detergent additive (ingredient used in Sample A) 2%; liquid polyolester (ingredient used in Sample A) 20%;

Sample C: Trixylyl phosphate (ingredient used in Sample A) 100%;

Sample D: Trixylyl phosphate (ingredient used in Sample A) 99%; calcium sulfonate detergent additive (Tergol 180H) 1%;

Sample E: Trixylyl phosphate (ingredient used in Sample A) 99%; calcium sulfonate detergent additive (ingredient used in Sample A) 1%;

Sample F: Trixylyl phosphate (ingredient used in Sample A) 99%; dilauryl acid phosphate rust inhibitor (Ortholeum 162) 1%;

Sample G: Trixylyl phosphate (ingredient used in Sample A) 99%; fatty imidazoline tertiary amine inhibitor (Unamine C) 1%.

Each of the above sample compositions was employed in the described test procedure. Experimental results are displayed in the table below.

Sample	Test Results
A*	No rust
B*	No rust
C	Heavy rust
D	Medium rust
E	Light rust
F	Medium rust
G	Heavy rust

*These samples conform to the rust-preventive compositions of this process.

Tolyltriazole to Protect Mirror Metal Layer

Glass mirrors are generally made by applying a layer of silver or copper or both to one side of a sheet of glass so as to provide a reflective surface. Generally this coating of metal is coated with an outer protective layer in the form of an organic paint system which covers the metal and the side of the glass as well. However, even with such a protective layer, the metal coatings may be subject to deterioration from oxidation or corrosion by the environment or by residual chemicals which remain on the mirror after completion of the "mirroring" step. This deterioration is very difficult if not impossible to stop over a long period of time by any of the practices known to date. In addition, many mirrors are cut from a larger mirrored sheet which cut portions expose the edges of the metal mirroring material which can then be attacked by atmospheric corrosion or oxidation or by cleaning chemicals, such as ammonia.

F.M. Workens; U.S. Patent 4,255,214; March 10, 1981; assigned to Falconer Plate Glass Corporation has found a method of protecting mirrors against all of these problems by passivating the metal coating prior to applying an organic protective coating such as paint. The passivating can be accomplished by applying a solution of tolyltriazole to the metal surface, either during or after application of the metal.

Preferably a 0.25% or more solution of tolyltriazole is applied to the metal surfaces of the mirror immediately after their application on the mirroring line, followed by rinsing and drying. However, the solution of tolyltriazole may be applied after the mirror is completed either by spraying on the solution or by wiping the solution on by hand or machine.

In the case of small mirrors cut from a large mirror sheet, each cut edge must be wiped with the solution as well as the metal surface itself. The maximum upper limit of tolyltriazole which may be used is limited only by economics and by foaming problems which may arise when the concentration is too high in certain types of equipment. The presence of tolyltriazole in the coolant solutions used in glass grinding and drilling solutions will not only protect the mirror being ground or drilled but will also inhibit corrosion of the grinding machinery itself. Tolyltriazole may be incorporated in the organic protective paint used to cover the metal with similar reduction in corrosion of the mirror.

Heat Reflective Enamel for Fuel Tanks

M.A. Shtern, N.I. Levit, A.V. Bondarenko and A.K. Kaschentseva; U.S. Patent 4,272,291; June 9, 1981 provides a heat-reflective enamel comprising a weather resistant vehicle, a pigment, an extender, and a solvent. The enamel contains additionally the product of thermal interaction between titanium dioxide or metatitanic acid, a nickel compound thermally decomposed to yield nickel oxide, antimony trioxide, and ammonium dichromate or an alkali metal dichromate, taken in the following amounts, in parts by weight:

titanium dioxide or metatitanic acid based on titanium dioxide: 100

nickel compound thermally decomposed to yield nickel oxide, based on nickel oxide: 3 to 32

antimony trioxide: 15 to 35

ammonium dichromate or an alkali metal dichromate, based on chromium oxide: 2 to 5

In one enamel the constituents have the following proportions expressed in parts by weight:

vehicle: 100

pigment: 30 to 80

extender: 1 to 60

solvent: 90 to 840

product of thermal interaction between titanium dioxide or metatitanic acid, nickel compound thermally decomposed to yield nickel oxide, antimony trioxide, and ammonium dichromate or an alkali metal dichromate: 15 to 60.

The heat-reflective enamel formulated as above has higher heat reflectance characteristics compared to a prior-art enamel, as attested by the reverse-side temperature of the coated specimen, equal to 66°C and being thus 6° lower than that of the prior-art heat reflectance enamel.

The coating life of the heat reflectance enamel formulated as proposed is three months longer over that of the prior-art dark colored enamel, the protective physical properties being retained over the service period.

The thermal-interaction product is obtained as follows. The starting mixture comprising:

100 kg of titanium dioxide or metatitanic acid based on an equivalent amount of titanium dioxide, 3 to 32 kg of nickel sulfate or some other nickel compounds which can be thermally decomposed to give nickel oxide, taken in equivalent amounts based on nickel oxide, 15 to 35 kg of antimony trioxide, 2 to 5 kg of ammonium dichromate or an alkali metal dichromate based on equivalent amounts of chromium oxide, and 25 ℓ of water, is ground by a conventional technique, dried, calcined within a temperature range of 1150° to 1200°C, and wet milled in a ball mill, a vibrating mill, or any other milling equipment.

This product features high lightfastness, chemical resistance (insolubility in acids or alkalis), and water resistance, combined with high pigment properties.

The pigment properties of the product are assessed in terms of the following characteristics: color, hiding power, oil absorption, and water-soluble salts content. The product is yellow in color, has a hiding power of 65 g/m^2, an oil absorption of about 17 to 20 g of oil per 100 g of the pigment product, and a water-soluble salts content of 0.2%.

The heat reflectance properties of the coatings are determined under laboratory conditions using the following procedure.

A specimen metal plate is coated on one side with 4 layers of the enamel to be tested, to a total thickness of 100 to 120 microns. Each enamel layer is dried for 1 hr at 120°±5°C.

The specimen with the dry coating is placed under an incandescent lamp rated at 500° W, mounted horizontally in a plastic foam cell sized 25 x 140 x 260 mm and fitted in the bottom with a hole of 20 mm diameter to enable measurements of the reverse side temperature of the specimen to be taken. The distance between cell and lamp is to be such that the specimen temperature on the reverse side coated with a black enamel be equal to 80°±2°C.

Temperature measurements are taken 4 minutes after the initiation of heating, using a miniature-sized electrical thermometer.

Example 1: A green colored heat-reflective enamel formulation, in parts by weight:

Copolymer of butyl acrylate, styrene, methacrylate and methacrylic acid: 100

Chromium oxide: 38

Zinc oxide: 12

Dolomite: 30

Product of thermal interaction between titanium dioxide, nickel nitrate, antimony trioxide and potassium dichromate: 26

Xylene: 65

Butyl acetate: 65

The heat reflectance and weather resistance characteristics of the enamel are presented hereinunder in the table.

Example 2: A green colored heat-reflective enamel formulation, in parts by weight:

Polyvinyl acetate: 100

Zinc oxide: 24

Chromium oxide: 44

Talc: 20

Product of thermal interaction between metatitanic acid, nickel carbonate, antimony trioxide and ammonium dichromate: 15

Acetone: 300

The heat reflectance and weather resistance characteristics of the enamel are presented hereinunder in the table.

Example 3: A brown colored heat-reflective enamel formulation, in parts by weight:

Polyvinyl butyral: 100

Iron oxide red: 60

Zinc oxide: 20

Micronized baryte: 20

Product of thermal interaction between titanium dioxide, nickel sulfate, antimony trioxide and ammonium dichromate: 38

Butanol: 250

Ethanol: 340

Acetone: 250

The heat reflectance and weather resistance characteristics of the enamel are presented hereinunder in the table.

Characteristics	Unit of Measure	1	2	3
Reverse side temperature of coated specimen	°C	65	65	67
Exposure to atmosphere	month	18	22	12

The process can most advantageously be used for protection against atmospheric effects and to reduce heat build-up in metal surfaces, specifically, for coating the exterior surfaces of fuel tanks in aircraft and other applications, those of stationary and transportation tanks for volatile liquid products, and the decks of tankers used to carry liquid fuel.

Polyepoxide-Cashew Oil Reaction Product for Can Coating

B. Passalenti, G. Giuliani and O. Fiorani; U.S. Patent 4,272,416; June 9, 1981; assigned to Societa Italiana Resine S.I.R. SpA, Italy describe a resinous substance for use as film-forming component in anticorrosive and can-coating paint and varnish compositions. The substance consists of the product obtained by reacting a normally solid polyepoxide with cashew nut oil in such amounts as to ensure equivalence between the phenolic hydroxyl groups of the cashew nut oil and the epoxy groups of the polyepoxide, and modifying the resulting reaction product with fatty acids having at least 8 carbon atoms per molecule, using a molar ratio of from 0.1:1 to 0.8:1 between the fatty acids and the cashew nut oil used in the reaction with the polyepoxide.

This resinous substance can be used as film-forming component in compositions which: dry in air and display, even on rusted supports, a resistance to corrosion at least equal to that of the traditional compositions based on epoxy and alkylphenolic resins; when applied to tinned plate, have the peculiar property of requiring hardening temperatures about 50° to 60°C lower than those required for the traditional compositions, while having the same mechanical characteristics and resistance to various types of food substances.

Example 1: A resin is prepared from cashew nut oil, a polyepoxide and fatty acids from dehydrated castor oil. More particularly, cashew nut oil is loaded into a flask provided with an agitator, a system for passing in inert gas and means for controlling the temperature, and the mass is heated to 140°C. An atmosphere of nitrogen is maintained in the flask, while the polyepoxide, consisting of the product of the reaction of bisphenol-A with epichlorhydrin and having a molecular weight of 950, is added in an amount of 0.5 mol for every mol of cashew nut oil. The mass is then heated to 260°C and is maintained at this temperature until the reaction mixture has a Gardner viscosity of U, measured at 25°C in a 55% by weight solution in a solvent mixture formed from white spirit and solvent naphtha in a weight ratio of 80:20.

The mass is then cooled to 180°C and fatty acids of dehydrated castor oil are added in an amount of 0.5 mol for every mol of cashew nut oil loaded in initially. The mass is heated to 240°C and maintained at this temperature until the acid value of the resulting product is equal to or less than 3. The resin thus obtained is dissolved in a solvent mixture of white spirit and solvent naphtha (weight ratio of 80:20) until a concentration of 50% by weight is achieved (Resin A).

Example 2: *(Comparative)* – A resin according to the known art is prepared from cashew nut oil, formaldehyde, epoxy resin and linseed oil.

More particularly, a condensation product of cashew nut oil and formaldehyde in a molar ratio of 1:0.6 is prepared in a flask provided with an agitator, a system for passing in an inert gas and means for controlling the temperature.

The formaldehyde is fed in as 96% by weight paraformaldehyde and the condensation is carried out at 90°C and at a pH of 5 obtained by the addition of sulfuric acid, until the reaction product has a Gardner viscosity of E-F, measured at 25°C in a 66% by weight solution in xylene. The reaction water is removed by distillation at atmospheric pressure, after the pH of the mass has been brought to a value of 7 by means of the addition of aqueous sodium hydroxide. The polyepoxide, consisting of a reaction product of bisphenol-A with epichlorhydrin, having a molecular weight of 1500, and the linseed oil are then added.

Both the epoxy resin and the linseed oil are used in an amount of 0.1 mol for every mol of cashew nut oil fed in initially. The mass is heated to 220°C and maintained at this temperature until the Gardner viscosity is K, measured at 25°C in a 50% by weight solution in a solvent mixture of white spirit/solvent naphtha in a 80:20 weight ratio. At the end of the reaction, the mass is cooled and the resin obtained is dissolved in the solvent mixture mentioned above until a concentration of 50% by weight is obtained (Resin B).

The anticorrosive primer formulation (a) is prepared from the Resin A of Example 1. Similarly an anticorrosive primer formulation (b) is prepared from the Resin B of Example 2.

In the following table the parts and percentages are given by weight. In the formulations the following commercial products are used: finely ground talc MT 120 (Talco Grafite Co.), red iron oxide M81 (Montedison Co.), asbestine 1634 (Massimiliano Massa Co). Bentone 38 and Eskin 2 (both Urai Co.) and soya lecithin (Comiel Co.).

Moreover, by drier is meant a toluene solution of metal salts containing cobalt naphthenate (6%), manganese naphthenate (8%) and lead naphthenate (30%).

Finally in the table by solvent is meant the white spirit/solvent naphtha mixture in a 50/50 weight ratio.

	Formulation (a)	Formulation (b)
Resin A	44.0	–
Resin B	–	44.0
Talc MT 120	18.8	18.8
Iron oxide M 81	15.0	15.0
Asbestine 1634	10.0	10.0
Bentone 38	0.4	0.4
Eskin 2	0.2	0.2
Soya lecithin	0.3	0.3
Drier	1.1	1.1
Solvent up to a viscosity of 100" in Ford Cup No. 4 (25°C)	10.0	10.0

The formulation (a) and the formulation (b) are dried in the form of films on sheet metal supports of the Unichim 5867 type which have been completely and perfectly cleaned and on similar supports which have rusted. The anticorrosive compositions of the process have been shown to adhere to a metallic surface even if it is rusted and exposed to saline fog for 600 hours.

One-Step Cleaning and Priming Composition

The process of *H.R. Choung; U.S. Patent 4,281,037; July 28, 1981; assigned to DAP, Inc.* provides a composition for cleaning and priming a surface in one step, thereby saving considerable labor over the standard two-step procedure. Furthermore, compositions of the process cure more quickly, allowing rapid painting of the primed surface without the long wait required by prior priming compositions. Additionally, the composition are surprisingly more stable than those of the prior art.

The cleaning and priming composition comprises a solution of (a) 0.1 to 10 weight percent titanium acetylacetonate, (b) 20 to 50 weight percent alkanol having 1 to 4 carbon atoms, (c) 20 to 50 weight percent selected alkanone, and (d) 5 to 40 weight percent water. All percentages used herein are weight percents based on total weight unless otherwise stated.

Titanium acetylacetonate is sold in solution with about 25% isopropanol (Tyzor AA).

Preferred alkanols having 1 to 4 carbon atoms are methanol, ethanol, propanol and isopropanol. Isopropanol is most preferred.

The alkanones are selected from the group consisting of methyl ethyl ketone (MEK), methyl isobutyl ketone (MIBK), acetone, and mixtures thereof. MEK is most preferred.

The most preferred composition contains about 1% titanium acetylacetonate, about 39% isopropanol, about 30% methyl ethyl ketone, and about 30% water.

Although not necessary for operability, 0.1 to 0.5% of a perfume may be added to the composition to mask the odor of the alkanone and/or alcohol. Other acceptable additives would be dye to improve visability during application and corrosion inhibitor, such as magnesium chromate.

The cleaning and priming composition can be applied to wood, metal, plastic, insulating board and wallboard surfaces which may be finished or unfinished. The method of using the composition will be dependent on the surface to be worked. Contemplated modes of application would include applying the solution with a cloth, brush, preimpregnated pad, pump spray, aerosol spray, or using a dipping tank. After application and the composition, the surface should be rubbed with an absorbant material such as a sponge, newspaper, cloth, etc. to remove dirt, grease, oil, and the like.

As with prior art cleaners, more than one application may be necessary for particularly dirty surfaces. But unlike the prior art, the present composition require no second step. A thin layer of the composition is simply allowed to cure on the surface. After a brief cure, which takes only a few minutes, the surface is ready for the application of paint, sealer, adhesive, or the like. The prior art would require a second step involving drying the cleaner, applying a primer, and curing, which could take several hours.

Keeping in mind the ease of application and short cure time, use of the composition is envisioned in the aircraft and automotive industries where the cleaning and priming of metal surfaces is a primary factor before the application of finishing coats

Composition for Use with Thermal Insulation

M.T. Orillion; U.S. Patents 4,298,657; Nov. 3, 1981; and 4,349,457; Sept. 14, 1982; both assigned to The Dow Chemical Co. provides a corrosion protection cover for metal surfaces. It comprises: (a) thermal insulation, to reduce the flow of heat between the metal surface and its surroundings; and (b) a composition useful for the corrosion-protection of metal surfaces, disposed in or on the metal-contacting surface of the thermal insulation, which includes a borate salt of an alkali metal, a nitrite salt of an alkali metal, and a molybdate salt of an alkali metal. The process likewise provides compositions useful for corrosion protection, whether disposed on the metal-contacting surface of or within the thermal insulation or applied to the metal surface by one or more of several means well-known to the art.

In a preferred method of practicing this process the corrosion inhibitor compositions just described are impregnated in or disposed on the metal-contacting surface of the thermal insulation by well-known means such as soaking, brushing or spraying. The thermal insulation is most preferably fiberglass, foamglass, or calcium silicate. The thermal insulation, including the inhibitor compositions, is then applied to the metal surface to be protected by using well-known techniques. In utilizing this embodiment of the process, the metal surfaces may thereby be protected against corrosion at relatively high temperatures.

The process is further illustrated by the following examples, in which three sets of carbon-steel pipe were subjected to a corrosion environment with and without the protection of a corrosion-inhibitor composition prepared according to this process. One set of pipes was tested without any type of preconditioning, a second set was preconditioned by sandblasting, and a third set was zinc-primed.

Thereafter one-foot lengths of each specimen were wrapped in fire-clad paper wherein holes and open seams were preset to permit penetration of moisture. The specimens were then exposed to outdoor weather conditions.

Example 1: A corrosion-inhibitor composition was prepared from 40 g of borax, 200 g of sodium nitrite, 200 g of sodium molybdate, 1200 ml of an epoxy resin containing about 50% zinc chromate by weight, 600 ml of an epoxy hardener, and 1,000 ml of a thinning agent which comprised a mixture of diacetone alcohol, methyl isobutyl ketone, and toluene. This composition was applied by brush to several of the steel pipe specimens; and also applied by brush to several sets of fiberglass, foamglass, and calcium silicate. These coated insulation materials were then used to insulate other specimens of steel pipe.

Example 2: A corrosion inhibitor composition was prepared from 20 g of borax, 50 g of sodium nitrite, 50 g of sodium molybdate, 300 ml of an epoxy resin, 150 ml of an epoxy hardener, and 150 ml of a thinning agent which comprised a mixture of diacetone alcohol, methyl isobutyl ketone and toluene. This composition was applied to six sections of sandblasted carbon-steel pipe and to six sections of fiberglass insulation, which were thereafter used to insulate six other sections of pipe. The composition was applied by brush to the exterior of the pipe and to the inside surface of the insulation.

Example 3: A corrosion inhibitor composition was prepared from 30 g of borax, 100 g of sodium nitrite, 100 g of sodium molybdate, and 7,000 ml of water. Several sections of fiberglass insulation were immersed in the composition and then dried in an oven. Thereafter the fiberglass insulation was used to insulate several sections of steel pipe.

Example 4: A corrosion inhibitor composition was prepared from 10 g of borax, 20 g of sodium nitrite, 20 g of sodium molybdate, 550 ml of an epoxy resin, 275 ml of an epoxy hardener, and 200 ml of a thinning agent which comprised a mixture of diacetone alcohol, methyl isobutyl ketone, and toluene. In applying this composition to carbon steel pipe, three sections of pipe were first sandblasted prior to applying the composition. Thereafter, two sections of pipe were coated with the composition by brush and one section was left untreated for testing.

The specimens in Examples 1 through 4 were examined for evidence of corrosion after one month, after six months, and after one year. Those specimens not protected by application of the corrosion inhibitor compositions specified in Examples 1 through 4 were in all cases found to display evidence of corrosion attack. The degree of severity was observed to increase as the time of exposure increased from one month to six months to one year.

In significant contrast, the specimens that had been protected by application of the corrosion-inhibitor compositions specified in Examples 1 through 4 showed little or no evidence of corrosive attack. There was, in fact, no significant evidence of any surface corrosion after one month for the protected specimens. After six months and even after a year these specimens showed very little or no surface corrosion. Where slight indications of surface corrosion were found for the protected specimens, such corrosion was very slight and in all cases substantially less than for any of the untreated specimens.

Treating Interior Enamel of Steel Food Containers

In the food industry, the interior of the enamel coated steel containers is attacked by the food product resulting in the oxidation of the steel container and/or the formation of salt deposits which discolors the steel, and are clearly visible to the consumer. There is a need for further protecting the steel after conventional protective coatings have been applied to improve the corrosion resistance characteristics of the steel.

A. Khayat and K.P. Redenz; U.S. Patent 4,310,575; January 12, 1982; assigned to Ralston Purina Company provide a method of treating the interior enamel of steel containers suitable for packing food.

In accordance with the process, a silyation agent is applied on the enamel coated steel to improve the corrosion resistance thereof. The silyation agent is then bonded, usually by heating, to render the enamel nonpolar and resistant to later corrosion of the steel.

The silyation agent can be any one that is commercially available. It should also be the type which can be easily attached to the epoxy phenolic resins found in enamels used for coating the interior of steel food containers. The term "silyation" is usually used to abbreviate "trimethylsilyation." It is also used to designate the attachment of similar groups such as dimethylsilyl [$-SiH(CH_3)_2$] or chloromethyldimethylsilyl [$-SiCH_2Cl(CH_3)_2$].

Example: Number 307 x 112 cans were formed from rolled flat steel previously coated with an epoxy phenolic resin base enamel supplied by Mobil Oil Co. The cans were cleaned by washing with heptane in order to free them from any lubricant or grease. The cans were then washed with a soap solution followed by distilled water and then they were dried. A few containers were then set aside to serve as a control. Into the remaining containers, the silyation agent, N,O-bis-(trimethylsilyl)acetamide, was poured. The silyation agent was applied on the enamel surface of each container in an amount of about 8 micrograms per square inch. The containers were then drained. Next, the containers were allowed to set at a temperature of 50°C for 60 minutes. A specially prepared test solution had been formulated beforehand. This test solution was formulated to show in a relatively short period, about 3 days, if there were any lack of resistance to corrosion of the interior coating. The test solution had the following composition:

Sodium chloride--2.5%
Acetic acid–1.2%
Glucose–1.0%
Methionine--0.05%
Sodium sulfide–0.05%
Water–balance

Both the cans with the surfaces treated with silyation agents and the cans left untreated as controls were rinsed with the abovedescribed test solution. Then, the cans were filled with test solution and retorted at 232°F for 132 min. After retorting the containers were after 3 days inspected by photomicrographs showing the gas phase portion of a container. By gas phase, it is meant that space normally not occupied by food content. There were no gas pockets, rust spots, or black deposits for these cans which were treated with silyation agents. In the gas phase portion of a container which was not treated with silyation agents, the large rust spots and gas pockets were clearly visible.

Protecting Superalloys from High Temperature Corrosion

The process of *L.E. Dardi and S. Shankar; U.S. Patent 4,313,760; February 2, 1982; assigned to Howmet Turbine Components Corporation* is concerned with coatings adapted to significantly improve the elevated temperature corrosion resistance of articles composed of iron-, cobalt- or nickel-based superalloys whereby more satisfactory performance and longer life for such articles can be obtained.

Elevated temperature exposure of metal articles is experienced in many situations. Metal components are subjected to such conditions, for example, in various aerospace applications and in land and marine operations such as in the case of components utilized in gas turbine engines.

In such applications, it is important to provide some means for preventing undue oxidation/sulfidation of the components involved since such corrosion can materially shorten the useful life of the components. Deterioration of components can also create significant performance and safety problems.

Various alloys, including most superalloys, are characterized by a degree of corrosion resistance, however, such resistance is significantly decreased when unprotected superalloy components are exposed at the operating temperatures involved in certain systems. For that reason, such components have been provided with coatings, such as aluminide coatings, which increase the corrosion resistance at elevated operating temperatures.

Aluminide coatings are applied by methods such as the pack cementation process. In this process, the substrate chemistry and the processing temperature exert a major influence on coating chemistry, thickness and properties. Specifically, the coatings comprise a hard, brittle outer layer and a hard, brittle multiphase sublayer that can crack when subjected to mechanically or thermally induced strain. This leads to poor fatigue properties, and the cracks can also materially reduce the corrosion resistance of the coated components.

Another class of coatings is the MCrAlY overlay coatings where M stands for a transition metal element such as iron, cobalt or nickel. MCrAlY coatings have been shown to have an advantage over aluminide coatings in providing extended life to turbine components. Specifically, MCrAlY coatings generally demonstrate greater corrosion resistance than aluminide coatings and also greatly superior ductility.

The coating compositions of this process are particularly resistant to oxidation-sulfidation at elevated temperatures, are otherwise highly efficient in their performance at these temperatures, and are well suited for application to substrates by plasma spraying. In the broadest sense, the coatings consist essentially of from 10 to 50% chromium, 3 to 15% aluminum, a metal mixture from the group consisting of tantalum, tungsten and manganese and combinations thereof, and the balance selected from the group consisting of nickel, cobalt and iron and combinations thereof. The mixture of tantalum, tungsten and manganese is present in amounts from 1 to 15%, and tantalum comprises at least one-fifth by weight of the mixture, or at least 0.5% of the total coating weight, whichever is greater. The balance of the mixture consists of at least 0.5% manganese or tungsten, or combinations thereof.

Optionally, the coating may have up to 5% reactive metal from the group consisting of lanthanum, yttrium and the other rare earths. Also the addition of rare earth and/or refractory metal oxide particles to the aforementioned coating composition is contemplated; these ingredients preferably being individually utilized in amounts from about 0.05 up to about 2.0% by weight. Such additions can be beneficial to the overall protective response of the coating because the metal oxide particles assist in pinning protective oxide scales. This pinning phenomenon results in superior adherence (less spalling) of the protective scale, thus increasing the overall coating life. Additions of titanium up to about 5% and of noble metals up to about 15% are also contemplated.

Example: A typical nickel-base superalloy of the type used in gas turbine engines, known as IN738, having a nominal composition of 0.09% C, 16.0% Cr, 8.5% Co, 1.7% Mo, 2.5% W, 1.7% Ta, 3.5% Ti, 3.5% Al, 0.01% B, 0.03% Zr and the balance Ni, was provided as one substrate. A typical cobalt-base superalloy of the type used in gas turbine engines, known as Mar-M509 and having a nominal composition of 0.60% C, 23.4% Cr, 10.0% Ni, 7% W, 3.5% Ta, 0.23% Ti, 0.01% B, 0.45% Zr, 1.5% Fe, 0.10% Mn, 0.40% Si and the balance Co, provided a second substrate for testing.

A first series of coatings was produced by plasma spraying prealloyed powders. These powders were sprayed in a plasma arc (>Mach 3 velocity) using argon and helium as primary and secondary gases, respectively. Spraying was performed in a chamber maintained at a dynamic pressure of 55 torr. The process parameters were:

Gun to workpiece distance	16 in.
Primary gas (argon)	370 CFH at 225 psi
Secondary gas (helium)	150 CFH at 250 psi
Voltage	50-52 volts
Current	1400-1440 amps
Powder flow	0.07 lb/min
Carrier gas (argon)	25 CFH at 100 psi
Time for deposition	45 sec.

The articles were then subjected to heat treatment in a vacuum for 4 hours at 1975°F.

The following table illustrates the compositions, tested and the test results.

Properties of MCrAlY Coatings

COATING	COMPOSITION (WI %)									AVERAGE LIFE[1] (HOURS)	
SYSTEM	Ni	Co	Cr	Al	Ta	Mn	W	La	Y	IN738 SUBSTRATE	MAR-M509 SUBSTRATE
UTC[3]											
NiCoCrAlY	Bal	23	18	13					0.3	100	190
MDC-35A	Bal	15	20	12	2.5			0.5		107	
MDC-34H	Bal	10	20	17					0.6	186	
MDC-1A		Simple Aluminide								23[2]	
LDC-2E		Platinum Aluminide								135	
MDC-35B	Bal	15	20	12	2.5		1.5	0.5		124	237
MDC-35C	Bal	15	20	12		2.5		0.5		110	
MDC-35D	Bal	15	20	12	2.5	2.5		0.5		175	238
MDC-35E	Bal	15	20	12	2.5		1.5		0.5	125	
MDC-35M	Bal	21	16	12	2.5	1.7		1.0		230	

[1] Rig Cycle: 2100°F/2 min + 1750°F/4 min + 2100°F/2 min + Cool/2 min (5 ppm salt).
[2] Result from one test.
[3] Composition conforming to United Technologies Pat. No. 3,928,026.

Electron microprobe analysis showed that the coating chemistry was not very much different from that of the chemistry of the powder.

The performance of the articles coated pursuant to this example was evaluated using a 0.7 Mach burner rig. The testing cycle was 2100°F/2 min; 1750°F/4 min; 2100°F/2 min; air cool/2 min. Five (5) ppm salt solution was injected into the combustion products of JP5 fuel containing 0.2% sulfur. This cycle closely simulates the gas turbine engine environment for turbine vanes and blades, it highlights the oxidation phenomenon, and it imposes significant thermal stresses on the protection system.

Easily Removable Temporary Coating

Temporary corrosion inhibitors are employed in the automotive industry and machine construction for the protection of lacquered and bright metal surfaces. In contra-distinction to antirust lacquers and chemical passivation processes the protective film is meant to protect only temporarily and thereafter to be removed easily and completely. Generally there are used solvent-containing wax dispersions and the wax coatings have to be removed with the aid of solvents or surfactant-containing water-solvent mixtures prior to putting the protected objects into service. When being applied and removed, the evaporating solvents are not only annoying for the workers, but generally represent an environmental problem which is to be avoided.

A. Hereth, K. Rieger and J. Wildgruber; U.S. Patent 4,315,957; February 16, 1982; assigned to Hoechst AG, Germany provide a process for temporarily protecting a metal or lacquered surface with a wax coating which is easily removable by treatment with a water-steam mixture. The process comprises applying a liquid, aqueous wax-containing preservative onto the surface. The preservative consists of

(a) from 4 to 20% by weight of an acid wax based on montan wax with a drop point of 80° to 90°C and an acid number of from 100 to 150, or an ester wax based on montan wax with a drop point of 78° to 88°C, an acid number of from 20 to 50 and a saponification number of from 100 to 150 or of a mixture of the waxes,

(b) from 0 to 12% by weight of a natural wax selected from the group consisting of carnauba wax, candelilla wax, ouricury wax and Japan wax,

(c) from 10 to 20% by weight of a hydrocarbon wax selected from the group consisting of cake paraffin with a softening point of from 50° to 56°C, synthetic paraffin with a softening point of from 102° to 104°C, high-molecular 1-olefin with a softening point of from 73° to 75°C, microcrystalline wax with a softening point of from 60° to 70°C, polyolefin wax with a softening point of from 100° to 130°C and polyolefin wax oxidate with a drop point of from 100° to 115°C and an acid number of from 15 to 30, or a mixture of the waxes,

(d) from 0.02 to 0.5% by weight of lithium or potassium hydroxide,

(e) from 0.5 to 8% by weight of an alkane sulfonate as emulsifier, and

(f) water in an amount so that the total of (a) through (f) is 100% by weight.

Finally, corrosion inhibitors, for example alkanolamine salts of nitrogen-containing condensation products, or substances serving as processing auxiliaries, for example from 0.001 to 0.05% by weight (calculated on the basic emulsion) of a wetting or levelling agent, for example a fluorine surfactant, and/or from 0.5 to 3% by weight of an oxyethylated fatty alcohol and/or an alkylphenol, may also be added to the preservative emulsion.

For preparing the preservative of the process, the waxes are melted at a temperature of from about 100° to 130°C, then the saponifying agent is added, and the mixture is maintained at this temperature for about 5 to 15 minutes. Thereafter the emulsifier is added, and the batch is emulsified by mixing it with water, while stirring. The water is optionally added to the wax melt, or the latter is introduced by stirring into the water being present.

In this manner, emulsions of up to 50% strength may be obtained as required. With emulsions of 25 to 30% strength, which are preferred, films having a thickness of the dry film of from about 10 to 15 microns may be obtained. The emulsions should be applied with spray guns operating according to the combination spraying process (material pressure=airless+additional air coating at the spray nozzle). If the spraying is effected according to the airless method or with compressed air only, irregular films might be obtained.

The preservative of the process has the following advantages:

(1) No autoclave is required for the emulsification which may be carried out in the open vessel.

(2) The emulsions are very finely divided and stable to storage for several weeks.

(3) They form on freshly lacquered surfaces as well as on bright metal parts uniformly closed films which show a thickness of the dry layer of 15 microns when 30% emulsions are applied.

(4) After a drying period of 30 to 60 minutes at room temperature or 10 minutes in a current of warm air of about 40°C, the films are generally waterproof.

(5) The wax films may be removed with a water-steam mixture of about 90° to 95°C without additives. The same is true also for films having been exposed to a weather test—about 6 weeks in the open air—and for artificially aged films which have been washed for 2 hr (water pipe) and tempered for 2 hr at 75°C.

(6) Upon complete removal of the wax films, the lacquers do not show any damage, such as a swelling, dulling, blooming, etc.

Chromium-Free Coating for Tin Cans

N. Oda and H. Terada; U.S. Patent 4,339,310; July 13, 1982; assigned to Hooker Chemicals & Plastics Corp. are concerned with a method for the surface treatment of tin plated steel sheet (referred to below as tin plate) and drawn and ironed (DI) cans of tin plated steel sheet (referred to below as tin cans). It is an object to form a coating on the tin surface which is corrosion resistant and oxidation resistant and exhibits good paintability either chemically or electrolytically.

In the past aqueous solutions of phosphoric acid and chromic acid or chromic acid salts in aqueous solution have been used as surface treatment baths for tin surfaces. Surface treatment methods based on chromic acid salts of this sort are excellent surface treatment methods for tin but there is a disadvantage in that effluent treatment is needed to prevent pollution and there are further disadvantages in connection with environmental health and with operability, etc. Furthermore, tin plate cans are often employed as food or beverage containers where the presence of chromium is not desirable.

It has been found that it is possible to provide a film that is better than the surface treatment films of the chromic acid salt system and to form a coating which has improved corrosion resistance, oxidation resistance and paintability for DI tin cans and tin plate. The tin surface is given either a chemical or an electrolytic treatment in an aqueous solution at pH 1.0-10 which contains as its main components (1) at least one soluble compound of a metal selected from the group consisting of titanium and zirconium; (2) at least one pyrazole comound of the formula:

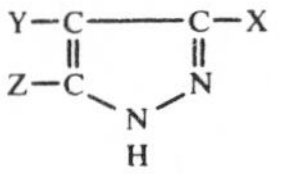

```
Y—C———C—X
   ‖      ‖
Z—C      N
    \   /
      N
      H
```

(Where X, Y and Z are independently selected from the group consisting of hydrogen atom, hydroxyl group, alkyl group of up to 5 carbon atoms, amino group, and nitro group.)

(3) at least one myoinositol phosphate ester having 2-6 phosphate groups per molecule or a salt thereof and; (4) at least one silicon compound.

The total concentration of the titanium and zirconium compounds is 0.001 to 10 g/ℓ and desirably 0.01 g/ℓ. The total concentration of the pyrazole derivatives is 0.01 to 20 g/ℓ and desirably 0.1 to 5 g/ℓ.

The myoinositol hexaphosphate ester is commonly named phytic acid. Furthermore, since the myoinositol di-penta phosphate esters are obtained mainly by the hydrolysis of phytic acid, phytic acid is the most useful industrially.

The overall concentration of the myoinositol 2-6 phosphate esters calculated as phosphoric acid is 0.005 to 50 g/ℓ and desirably 0.01 to 10 g/ℓ. The total concentration of the silicon compound(s) calculated as silicon is 0.001 to 10 g/ℓ and preferably 0.005 to 1 g/ℓ.

Example 1: 3-Methyl-6-hydroxypyrazole (10 g) was dissolved in tap water (8 ℓ) and phytic acid (5 g) and 20 wt% in water of zirconium hydrofluoride (25 g) were then dissolved successively in this solution. Then, 5 g of γ-aminopropyltriethoxysilane and 2 g of 55% hydrofluoric acid were added and dissolved and then, after the addition of 2 g of 30% aqueous hydrogen peroxide the total volume was made up to 10 ℓ by adding tap water. The pH of this aqueous solution was 2.7.

After cleaning a tin plate can by degreasing with a solution of 10 g/ℓ of a conventional alkaline degreasing agent and rinsing with water, a coating was formed by spraying the surface for 30 seconds at 1.0 kg/cm^2 (gauge pressure) with the abovementioned aqueous solution heated to 45°C after which the residual aqueous solution was removed by first rinsing with tap water for 20 seconds and then by spraying with deionized water with a specific resistance in excess of 500,000 ohm-cm the can was dried for 3 minutes in a hot air convection oven at 200°C. The results of testing the treated can painting are shown in the table.

Example 2: 3-Methyl-5-hydroxypyrazole (12 g) was dissolved in tap water (5 ℓ) and then phytic acid (10 g) was added to make up solution A. Potassium fluotitanite (10 g) and 40% silicon tetrafluoride (20 g) were then dissolved in 4 ℓ of tap water to make up solution B. Then after mixing solutions A and B and adding and dissolving sodium nitrate (10 g) the total volume was made up to 10 ℓ with tap water. The pH was adjusted to 9.0 using aqueous ammonia to yield the aqueous solution.

A coating was then formed on a tin can that had been cleaned using the same method as in Example 1 by spraying for 10 seconds at 0.8 kg/cm^2 (gauge pressure) with the abovementioned aqueous solution that had been heated to 40°C and the residual aqueous solution was then removed by first rinsing for 10 seconds in tap water and then 10 seconds in deionized water (resistance in excess of 300,000 ohm-cm) followed by drying in a hot air convection oven at 150°C. The results of testing are shown in the table.

Example 3: After cleaning a tin can in the same way as in Example 1, the can was immersed for 10 seconds in the same aqueous solution as used in Example 1 and dried for 3 minutes after removal from the aqueous solution in a hot air convection oven at 120°C without first rinsing with water. The results of testing are shown in the table.

Comparative Example: For purposes of comparing Examples 1 to 3 with chromic acid based treatments, after cleaning a tin can in the same way as in Example 1, the treatment was as follows:

Composition of the Aqueous Solution:	
Anhydrous chromic acid	40 g) made up to 10 ℓ
Phosphoric acid (75%)	20 g) with water
Aqueous solution temperature	30°C
Treatment Conditions:	
Spray treatment (spray pressure:	0.5 kg/cm^2 gauge pressure)
Spray time	30 sec

Results of testing are shown in the table.

Example	Corrosion Resistance	Moisture Resistance
1	5 points	5 points
2	5	5
3	5	5
Comparative	4.5	5

Test Methods: *Corrosion resistance test* – The tin can specimen was placed base up and introduced into a salt water spray tester and after testing for 30 minutes as prescribed in JIS-Z-2371 the tarnished condition of the surface of the tin can was assessed. The scale of assessment was as follows: No tarnishing 5 points, tarnishing over the whole surface 1 point with intermediate scores for intermediate states of tarnishing.

Moisture resistance test – The tin can specimen was placed base up and introduced into a moisture tester in accordance with JIS-Z-0228, the state of tarnishing of the surface of the tin can was assessed after a 3 hr test. The assessment was made on the same basis as the corrosion resistance test.

Multilayered Coatings for Chimney Liners

The process of *W.R. Slama, R.B. Washburn and D.J. Semanisin; U.S. Patent 4,347,277; August 31, 1982; assigned to General Signal Corp.* provides coated articles such as chimney liners, flue ducts, and scrubbers, which possess improved corrosion resistance when placed in contact with corrosive substances such as strong acids, particularly sulfuric acid. The multilayered, coated articles comprise a substrate, a primer coating layer, one or more inner coating layers, an intermediate coating layer, and an outer coating layer.

The coated articles are prepared as follows. A primer coating layer of a thermosetting polymer is placed over the surface to be protected. Then, one or more inner coating layers of a thermosetting polymer and a siliceous filler are overlaid upon the primer coating layer. The chemical properties of the thermosetting polymers employed in the primer coating layer and in the inner coating layer or layers are such that bonding occurs between the primer coating layer and the surface being coated and between the primer coating layer and the adjacent inner coating layer as well as between contiguous inner coating layers if more than one such layer is present. Over the outermost, inner coating layer there is placed a nonsiliceously filled, intermediate layer of a thermosetting polymer capable of bonding to the inner coating layer.

Finally, an outer coating layer of a fluoroelastomer and a nonsiliceous filler is applied over the intermediate coating layer. The chemical properties of the fluoroelastomer employed are such that, upon curing, a chemical bond forms between the intermediate coating layer and the fluoroelastomeric outer coating layer.

In the preferred embodiment, a coated article is prepared which possesses improved corrosion resistance and comprises a substrate, a primer coating layer of polyester 1-5 mils in thickness, an inner coating layer of glass flake-filled vinyl ester 20-80 mils in thickness, an intermediate coating layer of carbon-filled vinyl ester 10-15 mils in thickness, and an outer coating layer of a carbon-filled fluoroelastomer 10-50 mils in thickness.

The underlying substrates which are protected may be metals, especially steel, concrete, or brick.

Coating for Fabricated Steel Parts

In the manufacture of automobiles and the like, there is a tendency for fabricated metal parts, such as steel parts, to corrode even after painting especially in the inside areas where portions of the metal are folded into close proximity to other portions and not painted.

Accordingly, an objective of the process of *N. Kaliardos; U.S. Patent 4,356,036; October 26, 1982* is to provide anticorrosion compositions for metal which will protect the metal surfaces against corrosion even when portions of the part are folded on one another.

The anticorrosion coating composition comprises a mixture of aluminum paste or zinc paste containing aluminum or zinc flakes and a solvent, microcrystalline wax, aromatic hydrocarbon resin having a softening point, R and B, of 70° to 140°C, chlorinated rubber having a combined chlorine content of at least 66%, neutral barium sulfonate, chlorinated solvent, aromatic hydrocarbon solvent and mineral spirits.

Various compositions of the process have been tested and have been found to provide good corrosion resistance on bare steel panels in a salt cabinet for over 360 hr, humidity resistance for over 500 hr, no blistering, partial immersion salt solution resistance for at least 260 hr, partial immersion detergent solution resistance for at least 260 hr, and good adhesion.

The compositions utilizing aluminum may have the following ranges of composition by weight:

	Percent
Aluminum paste	21-28
Microcrystalline wax	1.3-3.5
Aromatic hydrocarbon resin	7-23
Neutral barium sulfonate	0.5-1
Chlorinated solvent	7-23
Aromatic hydrocarbon solvent	23-42
Mineral spirits	1.3-3.5

In the above aluminum compositions, the chlorinated solvent is preferably perchloroethylene and the aromatic hydrocarbon solvents are preferably toluene ranging from 8 to 14% by weight and xylene ranging from 15 to 28% by weight.

The compositions utilizing zinc may have the following ranges of compositions by weight:

	Percent
Zinc paste	15-43
Microcrystalline wax	0.5-4
Aromatic hydrocarbon resin	10-18
Neutral barium sulfonate	0.5-1
Chlorinated solvent	7-23
Aromatic hydrocarbon solvent	23-42
Mineral spirits	1.9-4

In the above zinc compositions, the chlorinated solvent is preferably perchloroethylene and the aromatic hydrocarbon solvents are preferably toluene ranging from 11 to 16% by weight and xylene ranging from 11 to 16% by weight.

Hot Corrosion Resistance at Moderate Temperatures for Superalloys

A coated superalloy article is described by *J.A. Goebel, R.H. Barkalow and N.E. Ulion; U.S. Patent 4,371,570; February 1, 1983; assigned to United Technologies Corporation* as is the method of producing the article. The coated article has a resistance to hot corrosion in the relatively low temperature range of 1200° to 1700°F and finds particular application in gas turbine engines.

The coating comprises an overlay coating selected from the group consisting of MCr, MCrAl, MCrAlY, MCrAlHf alloys (where "M" is Ni, Fe, Co and mixtures thereof). The outer surface portion of this coating is substantially enriched in silicon and contains from about 10 to about 50% silicon.

The silicon rich portion of the coating may be produced by chemical means such as pack cementation process or by vapor deposition means such as electron beam vapor deposition. The silicon rich layer has a thickness of from 10 to 40% of the total coating thickness.

Example 1: A nickel base superalloy was coated with a NiCoCrAlY 4 mil coating material containing 15% chromium, 25% cobalt, 11.5% aluminum, 0.7% yttrium, balance nickel. The electron beam vapor deposition technique was used to apply silicon to the surface of this NiCoCrAlY coating. Conditions were such that a 1 mil thick surface layer containing about 22% silicon was produced.

A test specimen having this coating system was tested at 1300° to 1350°F in a hot corrosion test using a burner rig with additions of sea salt and SO_2 to accelerate corrosion. Also tested were an unsiliconized NiCoCrAlY (18% Cr, 24% Co, 14% Al, 0.2% Y, balance nickel) coated superalloy sample and an unsiliconized superalloy sample coated with a silicon containing NiCoCrAlY composition (18% Cr, 24% Co, 10.3% Al, 0.1% Y, 2.6% Si, balance Ni) containing 2.6% silicon homogeneously dispersed throughout as described in U.S. Patent 4,034,142. The results were as follows: after 470 hr both of the unsiliconized NiCoCrAlY specimens had completely failed. Corrosion attack penetrated the entire 4 mil thickness of the coatings and caused substrate damage. After 1800 hr, the siliconized NiCoCrAlY displayed only about 0.5 mil of attack.

Considering both the amount of corrosion attack and the time involved, siliconizing reduced the degree of corrosion by about 340%. This clearly demonstrates the surprising and unexpected advantage of this process at resisting hot corrosion at intermediate temperatures.

Example 2: Cast specimens of Ni 30Cr 8Al alloy were obtained. The cast Ni 30Cr 8 Al alloy sample represented the overlay coating and there was no superalloy substrate. One sample was siliconized using a physical vapor deposition process to produce a silicon rich surface layer having a thickness of 0.6 to 0.9 mil and a silicon concentration of 10 to 16%. This siliconized sample was tested in a burner rig at 1650°F along with an unsiliconized sample. Additions of sea salt and SO_2 were made to increase the severity of the test. The results were as follows: after 353 hr of testing, the unsiliconized sample had corrosion attack to a depth of 80 mils. The siliconized sample showed only 0.7 mil of attack after 1521 hr of testing. Taking the testing time and degree of attack into account, the siliconized surface layer is seen to provide a reduction of 400 to 500X in corrosion attack under these test conditions.

Example 3: A superalloy test specimen having a nominal composition of 8%, Cr, 10% Co, 1.0% Ti, 6% Al, 1% Mo, 4.3% Ta, 0.1% C, balance Ni, was provided with an overlay coating having a composition of 15.5% Cr, 10.7% Al, 0.4% Y, balance Co. Another test specimen of the same alloy was provided with a similar overlay coating having a composition of 18% Cr, 12% Al, 0.5% Y, balance Co. Both overlay coatings were applied by an electron beam physical vapor deposition process.

A second coated sample was siliconized using a pack cementation process. A pack material consisting of 19% Si powder, 1% NH_4Cl, balance Al_2O_3 was employed. The overlay coated sample was embedded in this pack material and was heated at 1600°F for 6 hr. The sample was then removed from the pack and heated to 1800°F for 4 hr in an argon atmosphere. The resultant silicon rich surface layer contained about 25% Si and was about 1 mil thick.

These coated samples were evaluated in a burner rig operated at 1300° to 1350°F with additions of sea salt and SO_2 to accentuate corrosion. The following results were observed: after 36 hr the unsiliconized sample showed 2.2 mils of corrosion attack while after 599 hr the siliconized sample displayed only 1.75 mils of attack. In this example, the siliconized surface provided a 20X reduction in the corrosion attack.

Alkaline Resistant Primers for Automotive Steel

The process of *R.A. Dickie, J.W. Holubka and S.B.A. Qaderi; U.S. Patent 4,374,213; Feb. 15, 1983 and R.A. Dickie and J.W. Holubka; U.S. Patent 4,374,965; Feb. 22, 1983; both assigned to Ford Motor Co.* relates to compositions that retard corrosion of susceptible substrates in the absence of conventional amounts of inhibiting pigments as chromates and includes, especially, primer compositions adapted for use on automotive steels and other such substrates. This process particularly includes crosslinking compositions that cure into coatings on bare and treated steels and resist spread of corrosion, particularly from coating surface defects.

Briefly, the crosslinking compositions include (1) the reaction product of (a) epoxy reactant and (b) amine reactant, (2) amino resin crosslinking agent, and (3) optionally, a di- or polyhydroxy compound, wherein the amine reactant comprises secondary or primary and secondary amine and, also, is mostly, on a molar basis, hydroxy amine. In an alternative embodiment, (a) and (b) need not be fully reacted prior to application onto the substrate. Surprisingly such compositions may be cured within commercially acceptable curing schedules without volatilization of the amine to provide coatings that are exceptionally resistant to degradation/displacement under corrosive conditions and may advantageously be cured at even lower temperatures in the presence of catalyst.

In one embodiment, the crosslinking compositions may be formulated into solvent-based coatings as spray primers that exhibit corrosion resistance considerably in excess of certain conventional primers. Moreover, the crosslinking compositions may be formulated into water-based primers that exhibit desirable properties.

In another embodiment, certain reactive catalysts have also been found that reduce cure temperature required for primers containing the crosslinking compositions while maintaining and improving film characteristics compared to usual catalysts.

The discovery that corrosion prevention may be accomplished in the absence of inhibiting pigments provides an alternative means of preventing corrosion that is believed to utilize a mechanism different than when inhibiting pigments as chromates are included. More particularly, available evidence indicates that cured coatings, rather than slowing the overall rate of corrosion by retarding anodic dissolution of iron as a means to retard corrosion, provide an adherent crosslinked network which is exceptionally resistant to degradation/displacement by cathodically produced hydroxide and thus limit the total area over which the corrosion reactions can occur.

The following examples are intended as illustrating some of the more preferred aspects. All degrees are in degrees Celsius and parts in parts by weight unless specified otherwise.

Example 1: A heat curable coating composition suitable for automotive application is prepared from a bis-phenol A-epichlorohydrin epoxy resin, alkanolamine, and an amino resin crosslinking agent in the manner hereinafter set forth:

Part A – Preparation of epoxy-alkanolamine resin:

Materials	Parts by Weight
Epon 1004*	625
Diethanolamine	70
Methyl amyl ketone	625
Butanol	200
Reactor Charge	
Diethanolamine	70
Methyl amyl ketone	625

*A product of Shell Chemical Co., which is a reaction product of epichlorohydrin and bis phenol-A and described such as being a solid with melting point 95° to 105°C; Gardner-Holdt viscosity at 25°C of Q-U (40% w soln in butyl Dioxitol); epoxide equivalent 875-1,000 (grams resin per gram-equivalent of epoxide); equivalent weight (grams resin to esterify 1 mol acid) 175. (Data Sheet SC: 69-58).

The reactor charge is heated to 70° to 80°C in a reaction vessel equipped with a stirrer, reflux condenser, and thermometer. The Epon 1004 is added over a 4 hr period. The temperature is maintained at 70° to 80°C throughout the addition of the epoxy resin and for 4 hr thereafter. The epoxy-alkanolamine resin so formed is then cooled to 50°C and diluted with butanol. The resin is then cooled to room temperature, filtered and formulated into a primer as hereinafter described.

Part B – Formulation of primer:

Materials	Parts by Weight
Butylated melamine resin (RN 602)*	46
Urea resin (RN 512)**	6.7
Titanium dioxide pigment	15
Carbon black pigment	15
Silica pigment	11
Barytes pigment	102
Xylene	50
Epoxy-alkanolamine resin from Part A	399

*Product of Mobil Chemical Co. that is described as being butylated melamine formaldehyde resin with viscosity of V-X; 58% non-volatiles; Acid number 2 maximum; solvent n-butanol.

**Product of Mobil Chemical Co. that is described as being butylated urea formaldehyde resin; viscosity W-Y; non-volatiles 50%±2%; Acid number 3.5-4.5; Color Gardner/max; Solvent 80% n-butanol and 20% ethyl-benzene.

An unpigmented primer is prepared by dissolving the epoxyalkanolamine resin, the butylated melamine and the urea resin in xylene and butanol. A mill base is then prepared using one-third of the unpigmented resin and the pigments shown above. The remaining unpigmented primer is then added and thoroughly dispersed with the mill base. The resulting fully formulated primer is filtered, applied to cold rolled, unphosphated steel panels by spraying and cured at 180°C for 20 minutes. The cured coating displays excellent corrosion resistance in salt spray, good solvent resistance to xylene and good flexibility and hardness.

Example 2: (a) Heat curable coating compositions suitable for automotive application are prepared in the manner discussed in Example 1 with the exception that from 2.5% by weight of phosphoric acid is added to the primer. The inclusion of small amounts of acid catalyst lowers the cure temperature.

(b) A heat curable coating composition suitable for automotive application is prepared from a bis-phenol-A-epichlorohydrin epoxy resin, alkanolamine, and an amino resin crosslinking agent in the manner hereinafter set forth:

Part A – Preparation of epoxy-alkanolamine resin:

Materials	Parts by Weight
Epon 1004*	625
Diethanolamine	70
Methyl amyl ketone	625
Butanol	200
Reactor Charge:	
Diethanolamine	70
Methyl amyl ketone	62

*Product of Shell Chemical Co., see Example 1.

The reaction charge is heated to 70° to 80°C in a reaction vessel equipped with a stirrer, reflux condenser, and thermometer. The Epon 1004 is added over a 4 hr period. The temperature is maintained at 70° to 80°C throughout the addition of the epoxy resin and for 4 hr thereafter. The epoxy-alkanolamine resin so formed is then cooled to 50°C and diluted with butanol. The resin is then cooled to room temperature, filtered and formulated into a primer as hereinafter described.

Part B – Formulation of primer:

Materials	Parts by Weight
Epoxy-alkanolamine resin from Part A	39.9
Butylated melamine resin (RN602)*	4.6
Urea resin (RN512)*	0.67
Titanium dioxide pigment	15
C-Black pigment	15
Silica pigment	11
Barytes pigment	102
Xylene	50
Butanol	75
Reaction product of 2-ethyl-1,3-hexanediol with P_2O_5 (1% based on resin solids)**	0.23

*Products of Mobil Chemical Co., see Example 1.

**Prepared as follows: In a three-necked round bottom flask equipped with a stirrer dropping funnel and a thermometer are placed 2500 g of dry (dried over molecular sieves)-2-ethyl-1,3-hexandiol. Phosphorus pentoxide (450 g) is added portionwise with continuous stirring and an exothermic reaction occurs. The addition of phosphorus pentoxide is regulated to maintain the temperature at 50°C. After the addition is complete, the reaction mixture is stirred at 50°C for one more hour and then filtered. The acid equivalent weight by tritation with medium hydroxide solution is found to be 357.

An unpigmented primer is prepared by dissolving the epoxy-alkanolamine resin, the butylated melamine, the urea resin, and the reaction product of 2-ethyl-1,3-hexanediol with P_2O_5, in the butanol and xylene. A mill base is then prepared using one-third of the unpigmented primer and the pigments shown above. The remaining clear primer is then added and thoroughly dispersed with the mill base. The resulting fully formulated primer is filtered, applied to cold rolled, unphosphated steel panels by spraying and cured at 140°C for 20 min. The cured coating displays excellent corrosion resistance in salt spray, good xylene solvent resistance, flexibility and hardness.

Example 3: A heat curable water-based coating composition suitable for automotive application is prepared from epoxy-amine reaction product and amino resin crosslinking agent in the manner as hereinafter set forth:

Part A – Preparation of epoxy alkanol amine reaction product:

Materials	Parts by Weight
Epon 1004*	405.27
Diethanol amine	45.41
Propasol P**	299.31

*See Example 1, a product of Shell Chemical Co.
**A product of Union Carbide Corp.–propylene glycol monopropylether.

The reactor charge is heated to 80° to 85°C in a reaction vessel equipped with a stirrer, reflux condenser and thermometer. The Epon 1004 is added over a 2 hr period. The temperature is maintained at 80° to 85°C throughout the addition and for 16 hr thereafter. The epoxy alkanol amine reaction product so formed is then cooled to room temperature and formulated into a water-based primer as hereinafter described.

Part B – Preparation of millbase:

Materials	Parts by Weight
Epoxy alkanol amine reaction product from Part A	160.00
Propasol P	41.04
Barytes pigment	115.64
Titanium dioxide pigment	15.84
Iron oxide	15.84
Silica	11.08

To 160 parts of the epoxy alkanol amine reaction product, 41.04 parts of Propasol P is added and then shaken for 30 minutes. A homogeneous solution is obtained. To this then the pigments as shown above are added and then shaken for 6 hr. A Hegmann gauge reading of 6 to 7 is obtained. This is then used for preparation of fully formulated pigmented primer.

Fully formulated water-based primer:

Materials	Parts by Weight
Millbase from Part B	79.6
Butylated melamine resin (RN602)*	2.4
Acetic acid	1.04
KR-55**	0.40
Water (distilled)	69.30

*See Example 1, a product of Mobil Chemical Co.
**Ken-React TTMDTP-55 titanate coupling agent.

The millbase is weighed into a stainless steel container. To this millbase resin is added and stirred well using a disperator. Then the KR-55 coupling agent is added and mixed well. The acetic acid is weighed into 10 parts of water in a glass container and then added to the millbase, melamine resin and KR-55 mixture under stirring. The balance of water is then added in 10 to 15 minutes after vigorous stirring. The whole mixing is accomplished in 15 to 20 minutes. The resulting fully formulated primer is filtered, applied to cold rolled, unphosphated steel panels by spraying and cured at 180°C for 20 minutes. The cured coating displays excellent corrosion resistance in salt spray, good solvent resistance to xylene and good flexibility and hardness.

Composition for Industrial Chimney Interiors

The coating composition of the process of *J. Blitstein and D. Kathrein; U.S. Patent 4,374,874; February 22, 1983; assigned to T.C. Manufacturing Co., Inc.* includes a temperature resistant fluoroelastomer containing 3 to 50% by volume hollow glass microballoons. The coating composition is particularly desirable for coating the interior surfaces of industrial chimneys, particularly those chimneys where the interior surfaces are exposed to hot boiler flue gases containing sulfur, for protecting metal, interior surfaces against chemically induced corrosion.

The hollow glass microspheres or microballoons substantially increases the corrosion resistance of the fluoroelastomeric liquid composition, yielding thermal and chemical resistance, particularly when the coating composition is applied in two or more layers.

Example: One hundred grams of a copolymer of hexafluoropropylene and vinylidene fluoride is dissolved in 300 g of acetone to form the fluoroelastomer solution. Then 10 g of hollow sodium borosilicate glass microspheres having the following particle size distribution and density are hand stirred into the fluoroelastomer solution:

Particle Size Range, Microns (% by weight)	>175	(5)	100–125	(12)
	149–175	(10)	62–100	(44)
	125–149	(12)	44–62	(10)
			<44	(7)
Average Particle Diameter, Microns (weight basis)	80			
Density	.311 grams/cc.			

The resulting composition contains about 24.4% by weight fluoroelastomer and about 2.4% by weight hollow microspheres representing approximately 30% by volume of the composition after solvent removal.

A steel substrate is first grit blasted and then degreased. A first coating of the fluoroelastomer composition is brush-applied and allowed to dry for 45 minutes at ambient temperature. A second coating is then brush-applied over the dried first coating and the second coating is allowed to dry for 12 hr. The coated substrate is then baked at 400°F for 8 hr to affect a complete cure of the fluoroelastomer.

Comparative testing was performed to determine the effect of incorporation of the hollow glass microspheres by comparing the corrosion protecting capabilities of the fluoroelastomer composition with hollow glass microspheres to the fluoroelastomer composition without the hollow glass microspheres. Two identical fluoroelastomer solutions were prepared by dissolving equal amounts of a copolymer of hexafluoropropylene and vinylidene fluoride into acetone. Hollow glass microspheres were then hand stirred into one of the fluoroelastomer solutions in an amount of 30% by volume of fluoroelastomer plus microspheres. The two fluoroelastomer compositions then were coated on a steel plate using approximately equal amounts of fluoroelastomer in each coating, and the coatings were dried and cured. Acid tests of the two coatings show that the hollow glass microspheres used in accordance with the process approximately double the acid resistance of the steel substrate coated, as shown in the following table.

	Copolymer of Hexafluoro-propylene and Vinylidene Fluoride (Without Microspheres)	Copolymer of Hexafluoro-propylene And Vinylidene Fluoride And 30% By Volume Microspheres
Coats	2	2
Total Thickness	2 Mil	4 Mil
Bake	1 Hr. at 400° F.	1 Hr. at 400° F.
Acid Test Temp.	500° F.	500° F.
Acid Drops ($90\%\ H_2SO_4$)	10	14
Time To Failure	4 Hours	6½ Hrs.
Acid Failure Diameter	17.5 mm	12.0 mm
High Temp.	Over 600° F.	Over 600° F.
High Temp. Failure Area	100.0 mm^2	53.0 mm^2

–5–

Lubricant, Fuel and Hydraulic Fluid Additives

FOR OILS AND GREASES

Rust-Inhibiting Glycerol Ethers

An object of the process of *S. Kimura and N. Ishida; U.S. Patent 4,247,414; January 27, 1981; assigned to Nippon Oil Company, Japan* is to provide a rust inhibitor not having the drawbacks usually encountered in conventional rust inhibitors. The rust inhibitor comprises glycerol-1-alkylphenoxy-3-ethylene glycol ether of the formula

$$R\text{-}C_6H_4\text{-}O-CH_2-\underset{OH}{\underset{|}{CH}}-CH_2-O-CH_2-\underset{OH}{\underset{|}{CH_2}}$$

wherein R is an alkyl group of C_{6-18}. The alkyl group is preferably in the sequence of nonyl, dodecyl, pentadecyl, hexyl and octadecyl, and the most preferable alkyl group is nonyl group of C_9.

A further object is to provide a composition, obtained by adding the aforesaid rust inhibitor to a mineral oil. The mineral oil is the one generally used as a base oil of a lubricant or a rust preventive oil, for example, lubricant fraction of naphthenic or paraffin oils having 20 to 150 cs of viscosity at 37.8°C, refined products thereof, petroleum solvents, petrolatum and the like.

The addition amount of the rust inhibitor for preparing a lubricant is in the range of 0.01 to 1.0 part by weight based on 100 parts by weight of a mineral oil used, preferably 0.03 to 0.5 part by weight; the one for preparing a rust preventive is in the range of 0.1 to 20 parts by weight per 100 parts by weight of the mineral oil, preferably 0.5 to 5 parts by weight.

Preparation Example: Into a flask, were put 27.6 g (1 mol) of p-nonylphenylglycidyl ether and 62 g (1 mol) of ethyleneglycol and the mixture was gradually heated in the presence of 1.15 g (0.05 mol) of metallic sodium to react them for 6 hours at 180°C under nitrogen atmosphere. Then, the reactants were cooled.

Following the extraction of the resulting product with 1 liter of ethyl ether, the extract was washed with water and distilled at 180°C/1 mm Hg to remove unreacted substances and water, giving 28.1 g of transparent and liquid glycerol-1-p-nonylphenoxy-3-ethyleneglycol ether.

In the same manner as described above, glycerol-1-alkylphenoxy-3-ethyleneglycol ethers of C_{12}, C_{15}, C_{16} and C_{18} were obtained.

The resulting ethers were evaluated by the following performance tests on the basis of the Japanese Industrial Standard as shown below.

JIS K 2510 (Testing method for rust-preventing characteristics of turbine oils) – Into a mixture of a sample and water, a test piece of iron is immersed and the mixture is stirred at 60°C for 24 hours. Thereafter, the presence or absence of rust appearing on the surface of the test piece is investigated.

JIS K 2517 (Testing method for steam emulsion number of lubricating oils) – Steam is blown into a sample until the total volume of the sample and condensed water becomes 52-55 ml and then the separated state of the sample and condensed water is observed. The demulsibility is determined as number of seconds required by the time where the separation amount of the oil phase becomes 20 ml. It is regarded as "above 1200 seconds" if the amount of the separated oil does not reach 20 ml after 20 minutes.

JIS Z 0236 (Testing method for rust preventing oils) – A test piece of steel, coated with a sample, is hung in a humidity cabinet and rotated for 200 hours at 49°±1°C in above 95% relative humidity. Thereafter, the degree of rusting on the test piece is measured to evaluate same with "A (superior) through E (inferior)".

JIS K 2520 (Testing method for emulsion characteristics of lubricating oils) – A sample is mixed with water and kept at 54°±1°C with stirring to observe the state, separated into water and oil phases, of the resulting emulsion. The evaluation of the state is shown as follows: oil phase (ml); water phase (ml); emulsion phase (ml); [time passed (min)].

The process will be further illustrated in more detail by way of the following examples and controls.

Examples 1 through 5 and Control 1: The lubricants used in Examples 1 through 5 were prepared by respectively adding glycerol-1-p-nonylphenoxy-3-ethyleneglycol ether, glycerol-1-p-dodecylphenoxy-3-ethyleneglycol ether, glycerol-1-p-pentadecylphenoxy-3-ethyleneglycol ether, glycerol-1-p-hexylphenoxy-3-ethyleneglycol ether and glycerol-1-p-octadecylphenoxy-3-ethyleneglycol ether made according to the manner as described in the foregoing preparation. Example of a purified lubricant fraction obtained from Minas crude oil, having a viscosity of 56 cs at 37.8°C and the lubricant used in Control 1 was prepared by adding commercially available 2-hydroxypropyl-2'-alkenylmonosuccinate of the formula

$$\begin{array}{l} C_{12}H_{23}-\underset{\displaystyle CH_2COOH}{\underset{|}{CH}}-COOCH_2CH(OH)CH_3 \end{array}$$

as a rust inhibitor to the above lubricant fraction.

The results of the test for steam emulsion number of lubricating oils and the test for rust preventing characteristics of turbine oils with the above lubricants are shown in the table below.

	Lubricant composition (parts by wt.)		Test result	
	Mineral Oil	Rust inhibitor	Test for rust preventing characteristics of turbine oils	Test for steam emulsion number of lubricating oil
Example 1	lubricant fraction obtained from Minas crude oil (100)	glycerol-1-p-nonylphenoxy-3-ethyleneglycol ether (0.05)	pass	49
2	same as above (100)	glycerol-1-p-dodecylphenoxy-3-ethyleneglycol ether (0.05)	"	70
3	same as above (100)	glycerol-1-p-pentadecylphenoxy-3-ethyleneglycol ether (0.05)	"	98
4	same as above (100)	glycerol-1-p-hexylphenoxy-3-ethyleneglycol ether (0.05)	"	129
5	same as above (100)	glycerol-1-p-octadecylphenoxy-3-ethyleneglycol ether (0.05)	"	210
Control 1	same as above (100)	alkenyl-monosuccinate (0.05)	"	>1200

The compositions used in Examples 1 through 5 had excellent performance in each test. On the contrary, the composition used in Control 1 stood the turbine oil anticorrosive test but showed an unfavorable result in the test for steam emulsion number of lubricating oils. It seems that the unfavorable result is due to the hydrolysis of the ester bonds contained in the alkenyl monosuccinate.

Examples 6 through 10 and Control 2: In Examples 6 through 10 and Control 2, rust-preventive oil made with glycerol-1-p-nonylphenoxy-3-ethyleneglycol ether, glycerol-1-p-dodecylphenoxy-3-ethyleneglycol ether, glycerol-1-p-octadecylphenoxy-3-ethyleneglycol ether, glycerol-1-p-pentadecylphenoxy-3-ethyleneglycol ether and glycerol-1-p-hexylphenoxy-3-ethyleneglycol ether and with commercially available sorbitan monooleate were used to carry out the test for rust-preventive oils and the test for emulsion characteristics of lubricating oils. The results obtained are shown in the table below.

	Rust preventing oil composition (parts by wt.)		Test for rust preventive oils	Test for emulsion characteristics of lubricating oils
	Mineral oil	Rust inhibitor		
Example 6	Purified spindle oil obtained from Arabian crude oil (100)	glycerol-1-p-nonylphenoxy-3-ethyleneglycol ether (0.5)	A	40-40-0 (5)
7	same as above	glycerol-1-p-dodecylphenoxy-3-ethyleneglycol ether (0.5)	A	40-40-0 (6)
8	same as above	glycerol-1-p-octadecylphenoxy-3-ethyleneglycol ether (0.5)	A	40-40-0 (9)
9	same as above	glycerol-1-p-pentadecylphenoxy-3-ethyleneglycol ether (0.5)	A	40-40-0 (7)
10	same as above	glycerol-1-p-hexylphenoxy-3-ethyleneglycol ether (0.5)	A	40-40-0 (9)
Control 2	same as above	sorbitan monooleate (0.50)	A	41-38-1 (10)

As will be seen from the above table, with the compositions according to the process, there can scarcely be found an emulsified layer and they are excellent in demulsibility as compared with that of a conventional rust-preventive oil and shows excellent performance as a rust-preventive oil.

S. Kimura and N. Ishida; U.S. Patent 4,247,415; January 27, 1981; assigned to Nippon Oil Company, Ltd., Japan provide a rust inhibitor consisting of glycerol alkyl phenylether of the formula

$$R-C_6H_4-O-CH_2-CH(OH)-CH_2(OH)$$

wherein R is an alkyl group of C_{6-18}.

A composition having rust preventing result may be obtained by adding the rust inhibitor to a mineral oil. The amount of the rust inhibitor to be added to a mineral oil is in the range of 0.01 to 20 parts by weight based on 100 parts by weight of the mineral oil.

Preparation Example: Into a 500 ml flask were put 110 g (0.5 mol) of p-nonylphenol and 110.5 g (1 mol) of glycerol-α-monochlorohydrin and the mixture was gradually heated in the presence of 24 g (0.6 mol) of sodium hydroxide in nitrogen atmosphere to react them for 5 hours at 100°C. Then the reactants were cooled.

Following the extraction of the resulting product with 1 liter of ethyl ether, the extract was washed with water and distilled for 3 hours at 170°C/1 mm Hg to remove unreacted substances and water, giving 120.4 g of transparent glycerol-p-nonylphenyl ether.

In the same manner as described above, glycerol alkylphenyl ethers of C_{10-18} could be obtained. The resulting ethers were evaluated by the performance tests on the basis of the Japanese Industrial Standard described in U.S. Patent 4,247,414 above.

Examples 1 through 5 and Control 1: The lubricants used in Examples 1 through 5 were prepared by respectively adding glycerol-p-hexylphenyl ether, glycerol-p-nonylphenyl ether, glycerol-p-pentadecylphenyl ether, glycerol-p-dodecylphenyl ether and glycerol-p-octadecylphenyl ether made according to the manner described in the foregoing Preparation Example to a purified lubricant fraction, obtained from Minas crude oil, having a viscosity of 56 cs, at 37.8°C and the lubricant used in Control 1 was prepared by adding the commercially available 2-hydroxy-2'-alkenyl-monosuccinate described in U.S. Patent 4,247,414 above.

The results of the test for steam emulsion number of lubricating oils and the test for rust-preventing characteristics of the turbine oils carried out with the above lubricants are shown in the table below.

The compositions used in Examples 1 through 5 had excellent performance in each test. On the contrary, the composition used in Control 1 stood the test for rust preventing characteristics of turbine oils but showed unfavorable results in the test for steam emulsion number of lubricating oils.

	Lubricant composition (parts by weight)		Test Result	
	Mineral oil	Rust inhibitor	Test for rust preventing characteristics of turbine oils	Test for steam emulsion number of lubricating oil
Example 1	lubricant fraction obtained from Minas crude oil (100)	glycerol-p-hexylphenyl ether (0.10)	pass	110
2	same as above	glycerol-p-nonylphenyl ether (0.10)	pass	93
3	same as above	glycerol-p-pentadecylphenyl ether (0.10)	pass	105
4	same as above	glycerol-p-dodecylphenyl ether (0.10)	pass	100
5	same as above	glycerol-p-octadecylphenyl ether (0.10)	pass	120
Control 1	same as above	alkenyl monosuccinate (0.10)	pass	above 1200

Examples 6 through 10 and Control 2: In Examples 6 through 10 and Control 2, rust-preventive oils made with glycerol alkylphenyl ethers prepared according to the foregoing Preparation Example and made with commercially available sorbitan monooleate as a control were used to carry out the test for rust-preventive oils and the test for emulsion characteristics of lubricating oils. The results obtained are shown in the table below.

	Rust preventing oil composition (parts by weight)			
	Mineral oil	Rust inhibitor	Test for rust preventive oils	Test for emulsion characteristics of lubricating oils
Example 6	Purified spindle oil obtained from Arabian crude oil (100)	glycerol-p-nonylphenyl ether (0.50)	A	40-40-0 (5)
7	same as above (100)	glycerol-p-dodecylphenyl ether (0.50)	A	40-40-0 (7)
8	same as above (100)	glycerol-p-pentadecylphenyl ether (0.50)	A	40-40-0 (7)
9	same as above (100)	glycerol-p-hexylphenyl ether (0.50)	A	40-40-0 (9)
10	same as above (100)	glycerol-p-octadecylphenyl ether (0.50)	A	40-40-0 (9)
Control 2	same as above (100)	sorbitan monooleate (0.50)	A	41-38-1 (10)

A rust inhibitor comprising glycerol-1-alkylphenoxy-3-glycerol ether of the formula

$$R-C_6H_4-O-CH_2-\underset{\displaystyle OH}{\underset{|}{CH}}-CH_2-O-CH_2-\underset{\displaystyle OH}{\underset{|}{CH}}-\underset{\displaystyle OH}{\underset{|}{CH_2}}$$

wherein R is an alkyl group of C_{6-18} is described by *S. Kimura and N. Ishida; U.S. Patent 4,273,673; June 16, 1981; assigned to Nippon Oil Company, Ltd., Japan.*

A composition having rust preventing effect may be obtained by adding the rust inhibitor to a mineral oil. The amount of the rust inhibitor to be added to a mineral oil is in the range of 0.01 to 20 parts by weight based on 100 parts by weight of the mineral oil.

Preparation Example: Into a flask, were put 27.6 g (0.1 mol) of p-nonylphenylglycidyl ether and 46.1 g (0.5 mol) of glycerin and the mixture was gradually heated in the presence of 1.15 g (0.05 mol) of metallic sodium to react them for 5 hours at 180°C under nitrogen atmosphere. Then, the reactants were cooled.

Following the extraction of the resulting product with 500 ml of ethyl ether, the extract was washed with water and then, ethyl ether was removed, giving 24.1 g of transparent, colorless and viscous glycerol-1-p-nonylphenoxy-3-glycerol ether in liquid form.

In the same manner as described above, glycerol-1-alkylphenoxy-3-glycerol ether of C_{12}, C_{15}, C_{16} and C_{18} were obtained. The resulting ethers were evaluated by the tests on the basis of the Japanese Industrial Standard described in U.S. Patent 4,247,414.

Examples 1 through 5 and Control 1: The lubricants used in Examples 1 through 5 were prepared by respectively adding glycerol-1-p-nonylphenoxy-3-glycerol ether, glycerol-1-p-dodecylphenoxy-3-glycerol ether, glycerol-1-p-pentadecylphenoxy-3-glycerol ether, glycerol-1-p-hexylphenoxy-3-glycerol ether and glycerol-1-p-octadecylphenoxy-3-glycerol ether made according to the manner as described in the foregoing Preparation Example to a purified lubricant fraction, obtained from Minas crude oil, having a viscosity of 56 cs at 37.8°C and the lubricant used in Control 1 was prepared by adding commercially available 2-hydroxypropyl-2'-alkenylmonosuccinate described in U.S. Patent 4,247,414.

The results of the test for steam emulsion number of lubricating oils and the test for rust preventing characteristics of turbine oils carried out with the above lubricants are shown in the table below.

	Lubricant composition (parts by wt.)		Test result	
	Mineral oil	Rust inhibitor	Test for rust preventing characteristics of turbine oils	Test for steam emulsion number of lubricating oils
Example 1	lubricant fraction obtained from Minas crude oil (100)	glycerol-1-p-nonylphenoxy-3-glycerol ether (0.10)	pass	53
2	same as above	glycerol-1-p-dodecylphenoxy-3-glycerol ether (0.10)	"	74
3	same as above (100)	glycerol-1-p-pentadecylphenoxy-3-glycerol ether (0.10)	"	103
4	same as above (100)	glycerol-1-p-hexylphenoxy-3-glycerol ether (0.10)	"	140
5	same as above (100)	glycerol-1-p-octadecylphenoxy-3-glycerol ether (0.10)	"	220
Control 1	same as above (100)	alkenyl monosuccinate (0.10)	"	>1200

Examples 6 through 10 and Control 2: In Examples 6 through 10 and Control 2, rust-preventive oil made with glycerol-1-p-nonylphenoxy-3-glycerol ether, glycerol-1-p-dodecylphenoxy-3-glycerol ether, glycerol-1-p-octadecylphenoxy-3-glycerol ether, glycerol-1-p-pentadecylphenoxy-3-glycerol ether and glycerol-1-p-hexylphenoxy-3-glycerol ether and with commercially available sorbitan monooleate were used to carry out the test for rust-preventive oils and the test for emulsion characteristics of lubricating oils. The results obtained are shown in the table below.

	Rust preventing oil composition (parts by wt.)		Test for rust	Test for emulsion characteristics of
	Mineral oil	Rust inhibitor	preventive oils	lubricating oils
Example 6	Purified spindle oil obtained from Arabian crude oil (100)	Glycerol-1-p-nonylphenoxy-3-glycerol ether (0.50)	A	40-40-0 (5)
7	same as above	glycerol-1-p-dodecylphenoxy-3-glycerol ether (0.50)	A	40-40-0 (6)
8	same as above	glycerol-1-p-pentadecylphenoxy-3-glycerol ether (0.50)	A	40-40-0 (7)
9	same as above	glycerol-1-p-hexylphenoxy-3-glycerol ether (0.50)	A	40-40-0 (9)
10	same as above	glycerol-1-p-octadecylphenoxy-3-glycerol ether (0.50)	A	40-40-0 (9)
Control 2	same as above	sorbitan monoleate (0.50)	A	41-38-1 (10)

Oil-Free Basic Calcium Alkyl Salicylate for High Performance Grease

According to *W.D. Carswell, A. Youd and A.C.B. MacPhail; U.S. Patent 4,251,431; February 17, 1981; assigned to Shell Oil Company* an improved high performance grease especially suitable for lubrication of high speed bearings comprises a base fluid comprising a major amount of a linear perfluoroalkyl polyether fluid together with a minor amount of a methyl polysiloxane fluid and a thickener. The grease of this process can be further improved in its performance by the addition of a highly basic mineral oil-free dispersant/antirust additive.

Example 1: A grease was prepared using a mixture of linear perfluoroalkyl polyethers (Krytox 143) together with methylphenylsilicone fluid, and thickened with polytetrafluoroethylene.

Example 2: A grease was prepared similar to Example 1 but having added a highly basic mineral oil-free antirust additive.

Example 3: *Antirust additive* — Although the grease formulation according to the process and exemplified by Example 1 meets the requirements for a high performance grease, evaluation of its antirust properties in the ASTM D 1743 Rust Test showed them to be poor. This means that although the grease is suitable for use in stainless steel bearings, it cannot be recommended for use in a wider range of applications with complete confidence. In order to improve the antirust properties of the grease, several additives were evaluated in the same antirust test in the grease. The results are shown in the table below. Where considered appropriate, the starting and running torques in the IP 186/64 Low Temperature Torque Test were also determined and these too are included in the table.

From the table below it can be seen that two of the additives, basic calcium alkyl salicylate and barium dinonyl naphthalene sulfonate, proved successful in improving the antirust ability of the basic grease formulation sufficiently for it to meet the requirements of the ASTM D 1743 Test. However, when grease containing these two additives were evaluated for Low Temperature Torque by the method of IP 186 at -54°C, unacceptably high values were obtained. It was considered that these be caused by the presence of mineral oil diluent in the additive packages as obtained and further tests were made using additives freed of this diluent. The oil-free version of basic calcium alkyl salicylate was found to be particularly effective and the grease containing it passed the IP 186/64 test.

Anti-rust additive added	Performance in ASTM D 1743 Rust Test	IP 186/64	
		Starting g.cm	Running g.cm
None	Fail	1530	80
5%m basic Ca Sulphonate	Fail	—	—
5% m basic Ca alkyl salicylate	Pass	5360	380
5%m Ba dinonyl naphthalene sulphonate	Pass	8670	260
2.5%m oil-free basic Ca alkyl-salicylate	Pass	3600	130
5%m oil-free basic Ca alkyl-salicylate	Pass	4200	450
5% oil-free Ba dinonyl naphthalene sulfonate	Fail	2900	150

It was envisaged that the basic oil-free additive can be incorporated as the usual additive concentration stage of between 0.1% m and 10% m. In particular, 5% m of the additive was used in the example tested.

Base fluids suitable for use in the improved grease formulations of the process can be prepared from a range of combinations of perfluoroalkyl polyether fluids and methyl silicone fluids containing between 55% m and 95% m of the perfluoroalkyl polyether fluid together with between 45% m and 5% m of the methyl silicone fluid. Preferred combinations are between 75% m and 90% m of the perfluoroalkyl polyester fluid and between 25% m and 10% m of the methyl silicone fluid. The exact proportions used depends upon the actual components selected and their respective viscosities and average molecular weight.

The base fluids can be thickened to any required consistency using a suitable thickener. A commercially available thickener is Fluon L 170. In the above examples between 20% m and 30% m polytetrafluoroethylene were used.

5-Aminotriazole-Succinic Anhydride Reaction Product for Diesel Lubricant

The process of *R.L. Sung and B.H. Zoleski; U.S. Patent 4,256,595; March 17, 1981; assigned to Texaco Inc.* relates to an antioxidant and corrosion inhibitor for diesel crankcase lubricant composition. More especially, this process relates to a diesel crankcase lubricant composition which satisfies the onerous criteria of the Union Pacific OxidationTest (UPOT) for oxidative stability and corrosion resistance.

The diesel lubricating oil composition comprises a diesel lubricating oil and the reaction product of a hydrocarbyl succinic anhydride in which the hydrocarbyl radical has from about 12 to about 30 carbon atoms and 5-aminotriazole.

The diesel lubricating oil composition contains the reaction product in an amount of between 0.25 and 2.0 weight percent, preferably between 0.5 and 1.5 weight percent, especially 0.75 to 1.5 weight percent and more especially at least 1% by weight.

It has been surprisingly found that the reaction product of a hydrocarbyl succinic anhydride in which the hydrocarbyl radical has from about 12 to 30 carbon atoms and, specifically, 5-aminotriazole functions as an exceptionally effective antioxidant and corrosion inhibitor for diesel lubricating oil compositions, especially when the product is present in the composition in an amount of at least 1.0 weight percent. This is especially surprising considering that a position isomer thereof, namely, the reaction product of a hydrocarbyl succinic anhydride in which the hydrocarbyl radical has 12 to 30 carbon atoms and 4-aminotriazole does not function as a corrosion inhibitor for a diesel lubricant. For instance, whereas the hydrocarbyl succinic anhydride-5-aminotriazole composition at the 1.0 weight percent level inhibits corrosion to the extent that the weight loss of the strips is only 35.1 mg, the hydrocarbyl succinic anhydride-4-aminotriazole reaction product permits the removal of material from the test strip in an amount of 225 mg or approximately 175 mg beyond that which can be tolerated in accordance with the standards of UPOT. The 3-aminotriazole compound is also not effective as a corrosion inhibitor.

In the preparation of the antioxidant/corrosion inhibitor, the 5-aminotriazole is reacted with the hydrocarbyl substituted succinic acid broadly at a temperature ranging from room temperature to about 150°C until a substantial completion of the reaction. The reaction is conducted in the absence of any catalyst, but generally in the presence of a solvent or diluent to facilitate the reaction. The solvent or diluent is one compatible with the components of the diesel crankcase lubricating oil. Of the 5-aminotriazoles used, 5-aminotriazole (1,3,4) is preferred.

Example 1: 0.2 mol of polyisobutenylsuccinic anhydride of molecular weight of about 335 was allowed to react with 0.25 mol of 5-aminotriazole in 30 ml benzene. The reaction mixture was refluxed for 36 hours, filtered and stripped.

In a similar manner, 0.2 mol of tetrapropenylsuccinic acid anhydride of molecular weight 168 is allowed to react with 0.26 mol 5-aminotriazole in 300 benzene. Similarly, the reaction mixture is refluxed for 36 hours, filtered and stripped.

Benzotriazole as Silver Corrosion Inhibitor for Diesel Lubricant

The process of *R.L. Sung and B.H. Zoleski; U.S. Patent 4,278,553; July 14, 1981; assigned to Texaco Inc.* relates to a diesel lubricant composition with improved ability to protect the silver plated areas of a railway diesel engine such as the piston pin insert bearing-carrier combination. The diesel lubricating oil composition comprises a diesel lubricating oil containing a silver corrosion inhibiting amount of a benzotriazole compound of the formula:

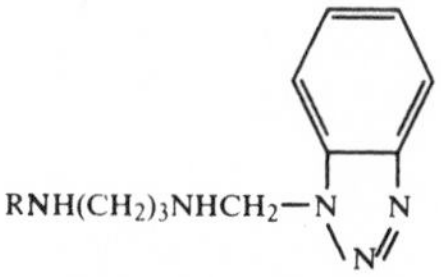

$RNH(CH_2)_3NHCH_2$—N N
N

wherein R is a straight chain alkyl having 12 to 18 carbon atoms and preferably is oleoyl or tallow.

The composition contains the above additive in an amount from about 0.5 to 2.0 weight percent and especially at least 1% by weight.

In the preparation of the additive, an N-alkyl-1,3-propanediamine wherein the alkyl group has from 12 to 18 carbon atoms, is condensed with formaldehyde and a benzotriazole.

Suitable N-alkyl-1,3-propanediamines are those available commercially under the name of Duomeen, wherein the alkyl group is straight chain with an average of 12 to 18 carbon atoms and is attached to the nitrogen through the second carbon in the chain. Examples of suitable N-alkyl-1,3-propanediamine include N-oleoyl-1,3-propanediamine (Duomeen O), N-stearyl-1,3-propanediamine and N-tallow-1,3-propanediamine (Duomeen T).

Example: *Preparation of (1-Duomeen T-methyl) benzotriazole* – To a mixture of 0.2 mol benzotriazole, 0.2 mol of Duomeen T, 60 parts methanol and 60 parts xylene, at reflux temperature, 0.2 mol of formaldehyde was added dropwise. After addition was completed, the mixture was refluxed for 1¾ hours more, let cool and 100 parts xylene was added. The mixture was filtered and stripped under vacuum. The reaction product was analyzed by IR, NMR and elemental analysis.

Samples of a base oil and of the same oil containing different amounts of the additive prepared above were tested in the Texaco Modified Silver Disc Friction Test. This procedure is a laboratory test for determining the antiwear properties of a lubricant oil. The test machine comprises a system wherein a one-half inch diameter 52100 steel ball is placed in assembly with three one-half inch silver discs of like size and of a quality identical to that employed in the plating of the silver pin insert bearing of railway diesel engines manufactured by the Electromotive Division of General Motors, Inc.

These discs are disposed in contact with one another in one plane in a fixed triangular position in a reservoir containing the oil sample to be tested for its silver antiwear properties. The steel ball is positioned above and in contact with the three silver discs. In carrying out these tests, the ball is rotated while it is pressed against the three discs at the pressure specified and by means of a suitable weight applied to a lever arm. The test results are determined by visual reference, using a low power microscope, to the scars on the discs, the scar texture, whether scored or smooth, for example, and coloration, in a rating system using a standard for comparison and a classification of "poor," "fair", "good" and "excellent".

The rotation of the steel ball on the silver discs proceeds for a period of 30 minutes at 600 rpm under a 60 kg static load. Each oil is tested at 300°, 400°, 450° and 500°F.

Under these test conditions (see table below), the fully formulated diesel engine oil "A" which contained no silver corrosion inhibitor was characterized as "poor". Oil "B" which contained 0.05 weight percent of the additive of the process was rated "borderline". Oils "C" and "D" containing 1.00 and 2.00 weight percents of the additive, respectively, were characterized as "excellent".

Run Identification Composition, Wt. %	A	B	C	D
Naphthenic Mineral oil of 300 visc.	3.00	—	—	—
Paraffin Mineral oil	60.00	—	—	—
Naphthenic Mineral Oil of 75/80 viscosity	29.70	—	—	—
Mono (3-hydroxyethyl) alkene thiophosphonate	3.20	—	—	—
Sulfurized calcium alkylphenolate (188TBN) detergent	4.10	—	—	—
Silicone Foam inhibitor ppm	150	—	—	—
Ex. 1 Product	—	0.05	1.00	2.00
Oil A	—	99.95	99.00	98.00
Silver Disk Fricton Test Rating	Poor	Borderline	Excellent	Excellent

Oxyalkylated Trimethylol Alkane-Oxyethylated Alkylphenol Combination

It has been found by *E.F. Zaweski and C.S. Harstick; U.S. Patent 4,278,555; July 14, 1981; assigned to Ethyl Corporation* that an improved crankcase lubricating oil can be formulated by including a small amount of an oxyalkylated trimethylol alkane in combination with an oxyethylated alkylphenol. Tests have been carried out which demonstrate that the combination provides rust proteccion beyond that provided by either individual component.

A preferred embodiment of the process is a lubricating oil composition comprising a major amount of lubricating oil and a minor corrosion inhibiting amount of the combination of (a) an oxyalkylated trimethylol alkane having a MW of about 1,000 to 8,000 in which an initial hydrophobic oxyalkylene block is bonded to trimethylol alkane and a terminal hydrophilic oxyalkylene block is bonded to the other end of the hydrophobic block, the hydrophobic block consisting essentially of propyleneoxy units and optionally containing random ethyleneoxy units in an amount such that the average oxygen-to-carbon atom ratio in the hydrophobic block does not exceed 0.4, the hydrophilic block consisting essentially of ethyleneoxy units and optionally propyleneoxy units such that the oxygen-to-carbon atom ratio in the hydrophilic block is in excess of 0.4, the hydrophilic block forming 5 to 90 wt % of the oxyalkylated trimethylol alkane, (b) an ethoxylated C_{4-12} alkylphenol containing an average of about 2 to 10 ethyleneoxy units and (c) an overbased metal detergent selected from the group consisting of alkaline earth metal alkylbenzene sulfonates, petroleum sulfonates, phenates, sulfurized phenates and salicylates.

Representative trimethylol alkanes include trimethylol ethane, propane or butane. In a most preferred embodiment the trimethylol alkane is trimethylol propane.

The oxyalkylated trimethylol propanes of this process are known compounds. They are described in U.S. Patent 3,101,374. These additives can be obtained from BASF Wyandotte Corporation under the name "Pluradot". They are available in various molecular weights. Pluradot HA-510 has an average MW of 4,600. Pluradot HA-520 has an average MW of 5,000 and Pluradot HA-530 has an average MW of about 5,300. All of these are very effective in the present combination. The most preferred alkylphenol is nonylphenol.

The most preferred coadditive is an ethoxylated nonylphenol containing an average of about 4 oxyethylene groups. Such additives are commercially available. One such additive is known as "Sterox ND" (Monsanto Co.).

The amount of each additive used need only be an amount such that the combination provide adequate corrosion and rust protection in an engine. A useful range is about 0.005 to 0.3 weight pecent of the oxyalkylated trimethylol alkane and 0.01 to 0.5 weight percent of the oxyethylated alkylphenol.

Example 1: In a blending vessel place 1,000 gallons 150 SUS solvent refined mineral oil. To this add 100 gallons 1-decene oligomer containing mainly trimers and tetramers. Add sufficient zinc isobutylamyldithiophosphate to provide 0.07 weight percent zinc. Add overbased (TBN 300) calcium alkylbenzene sulfonate in an amount to provide 0.15 weight percent calcium. Add 30 gallons of a poly-laurylmethacrylate VI improver. Add sufficient polyisobutyl succinimide of tetraethylene pentamine to provide 3 weight percent active dispersant. Add 0.03 weight percent Pluradot HA0510 and 0.3 weight percent Sterox ND. Blend the oil until homogenous and package for distribution.

In many cases the additive combination of this process is first packaged in an additive concentrate formulated for addition to lubricating oil. These concentrates contain conventional additives such as those listed above in addition to the oxyalkylated trimethylol alkane and ethoxylated alkylphenol. The various additives are present in a proper ratio such that when a quantity of the concentrate is added to lubricating oil the various additives are all present in the proper concentration. The additive concentrate also contains mineral oil in order to maintain it in liquid form. The following example illustrates formulation of an additive concentrate formulated for addition to lubricating oil to provide an effective crankcase lubricant.

Example 2: In a blending vessel place 1,199.5 pounds of 100 SUS mineral oil, 784 pounds of polyisobutyl succinimide of tetraethylene pentamine, 2,352 pounds of ethylene/propylene copolymer VI improver, 254.8 pounds of zinc di-(2-ethylhexyl)dithiophosphate, 245 pounds of Pluradot HA-520 and 58.8 pounds of Sterox ND. Blend until homogenous and then package. The addition of 2,450 pounds of the above concentrate to 1,000 gallons of 150 SUS mineral oil will provide an effective crankcase lubricant.

Tests were carried out which demonstrate the corrosion protection provided by the present additive combination. These tests have been found to correlate with the Multi Cylinder Engine Sequence IId tests. In the test an oil blend is prepared containing a commercial succinimide ashless dispersant, a zinc dialkyldithiophosphate, an ethylene/propylene copolymer VI improver and 0.14% calcium as a 300 TBN overbased alkylbenzene sulfonate.

30 ml of 0.01 N hydrochloric acid is placed in a cell and heated to 50°C. Then 20 ml of the test oil blend is added and the mixture stirred for 10 minutes while measuring pH. The gradual increase in pH is measured over a 10-minute period and the results stated in terms of pH change per minute. It has been found that an increase of at least 0.4 pH units per minute indicates that the oil will pass the standard IId engine test. The following results were obtained in a series of these tests.

Additive	Conc (wt %)	pH shift (units/min)
1. Sterox ND	0.3	0.48
2. Pluradot HA-510	0.3	0.50
3. Sterox ND	0.15	0.59
Pluradot HA-510	0.15	

As the above results show, the combination produces synergistic results in that their combined effect is greater than either component could contribute individually at the same concentration.

Diphosphatetraazacyclooctatetraenes for Perfluorinated Fluids

Present interest in the utilization of perfluorinated fluids for high temperature lubricant applications has provided an impetus in a research effort directed toward the discovery of antioxidant and anticorrosive additives. Because of their thermal stability, the perfluorinated fluids have a great potential for use in lubrication and hydraulic applications. However, there is a serious drawback in their use resulting from the fact that certain metals present in aircraft components tend to corrode and degrade the fluids being used as lubricants. This occurs at temperatures above 550°F in an oxidative environment.

It would be highly desirable, therefore, to provide an additive for the perfluorinated fluids that would overcome the degradation and corrosion problems associated with their use. As a consequence, a considerable research effort has evolved in an attempt to provide such additives and an efficient and convenient method for their synthesis. As a result of this effort, it has been found by *K.J.L. Paciorek, R.H. Kratzer, T.I. Ito and J.H. Nakahara; U.S. Patent 4,281,185; July 28, 1981; assigned to the U.S. Secretary of the Air Force* that symmetrical diphosphatetraazacyclooctatetraenes can be obtained by effecting a reaction between a perfluoroalkylether amidine and a trihalophosphorane. The corrosion problems encountered when using perfluorinated fluid lubricants are obviated by using the additives of this process.

Example: Under nitrogen bypass, 2.89 g (8.81 mmol) of a solution of the amidine, $C_3F_7OCF(CF_3)C(=NH)NH_2$ and 3.9 ml (27.96 mmol) triethylamine in 20 ml Freon-113 was added over a period of 1.7 hours to 5.48 g (18.8 mmol) of a solution of diphenyltrichlorophosphorane in 50 ml benzene at 50°C. The mixture was then stirred at 50°C for 87 hours. After removal of solvents under reduced pressure, the residue was treated with Freon-113 (5 x 15 ml) and filtered through a 1.5 x 5 cm column of Woelm neutral alumina. The product (2.17 g, 47% yield) was distilled in vacuo giving the desired product 1,5-bis(diphenylphospha)-3,7-bis-[$C_3F_7OCF(CF_3)$]-2,4,6,8-tetraazacyclooctatetraene (1.46 g, 31.5%); BP 138° to 139°C/0.001 mm Hg; MP 77.5° to 79°C.

Analysis – Calculated for $C_{36}H_{20}F_{22}N_4O_2P_2$: C, 42.37%; H, 1.98%; F, 40.96%; N, 5.49%; P, 6.07%; O, 3.14%; MW 1,020.49. Found: C, 42.56%; H, 2.03%; F, 41.14%; N, 5.53%; P, 5.42%; MW 1,130.

The prepared 1,5-bis(diphenylphospha)-3,7-bis[$C_3F_7OCF(CF_3)$]-2,4,6,8-tetraazacyclooctatetraene was found to effectively inhibit oxidation of perfluoroalkylether fluids, e.g., fluids of the type disclosed in U.S. Patent 3,393,151 and to prevent corrosion of various metals by these fluids. For example, a 1% by weight solution of this diphosphatetraazacyclooctatetraene decreased oxygen consumption to zero and volatile products formation by a factor of ~17 during a 24 hour exposure to oxygen at 600°F as compared to an identical treatment of the fluid in the absence of the additive. In addition, an M-50 coupon test surface in the presence of the additive appeared unchanged, whereas in the absence of any additive, under otherwise identical conditions, the surface becomes covered with deeply colored irregular deposits.

It has also been found by *K.J.L. Paciorek, R.H. Kratzer, T.I. Ito and J.H. Nakahara; U.S. Patent 4,297,510; October 27, 1981; assigned to the U.S. Secretary of the Air Force* that unsymmetrical diphosphatetraazacyclooctatetraenes can be obtained by effecting a reaction between a perfluoroalkylether imidoylamidine and an imido-tetraaryl-diphosphinic acid trichloride. The resulting 8-membered ring compounds are excellent additives for use in perfluorinated lubricants and overcome the corrosion problem associated with the use of these lubricants in high temperature, aerospace applications.

Example 1: Under nitrogen bypass, to a solution of 1.79 g (3.65 mmol) of imido-tetraphenyl-diphosphinic acid trichloride in 22 ml acetonitrile at 50°C was added a solution of 2.0 g (3.13 mmol) of an imidoylamidine, of the particular formula, $C_3F_7OCF(CF_3)C(=NH)N=C(NH_2)CF(CF_3)OC_3F_7$, and 1.38 ml (9.89 mmol) triethylamine in Freon-113 (7 ml). The mixture was then stirred at 50°C for 111 hours. Following solvent removal, the Freon-113 soluble 1,3-bis(diphenylphospha)-5,7-bis[$CF(CF_3)OC_3F_7$]-2,4,6,8-tetraazacyclooctatetraene (3.19 g, 80% yield) was crystallized from Freon-113/acetone/pentane; MP 126° to 126.5°C. Analysis – Calculated for $C_{36}H_{20}F_{22}N_4O_2P_2$: C, 42.37%; H, 1.98%; F, 40.96%; N, 5.49%; P, 6.07%; O, 3.14%; MW, 1,020.49. Found: C, 42.64%; H, 2.09%; F, 41.33%; N, 5.45%; P, 6.07%; MW, 1,050.

Example 2: Under nitrogen bypass, to a solution of the imido-tetraphenyl diphosphinic acid trichloride (1.17 g, 2.38 mmol) in acetonitrile (10 ml) at 50°C was added dropwise over 1.75 hr a solution of the imidoylamidine, of the particular formula, $C_3F_7OCF(CF_3)CF_2OCF(CF_3)C(=NH)N=C(NH_2)CF(CF_3)OCF_2CF(CF_3)OC_3F_7$, (2.05 g, 2.11 mmol) and triethylamine (0.98 ml, 7.03 mmol) in Freon-113 (18 ml). The resulting mixture was heated at ~50°C for 142 hours. After solvent removal the residue was titrated with Freon-113 (4 x 10 ml) and then filtered through a 1.5 x 5 cm column of Woelm neutral alumina. 1,3-bis(diphenylphospha)-5,7-bis-[$CF(CF_3)OCF_2CF(CF_3)OC_3F_7$]-2,4,6,8-tetraazacyclooctatetraene thus obtained (2.80 g, 98% yield) was distilled in vacuo, BP 146° to 148°C/0.001 mm Hg.

Analysis – Calculated for $C_{42}H_{20}F_{34}N_4O_4P_2$: C, 37.30%; H, 1.49%; F, 47.76%; N, 4.14%; P, 4.58%; O, 4.73%; MW, 1,352.55. Found: C, 38.29%; H, 1.55%; F, 48.82%; N, 4.23%; P, 3.94%; MW, 1,400.

The above cyclooctatetraene exhibited good thermal oxidative stability as shown by 1% and 4% volatilization after 24 hour heat treatment in air at 235° and 316°C, respectively.

The 1,3-bis(diphenylphospha)-5,7-bis[$CF(CF_3)OCF_2CF(CF_3)OC_3F_7$]-2,4,6,8-tetraazacyclooctatetraene was found to effectively inhibit oxidation of perfluoroalkylether fluids of the type disclosed in U.S. Patent 3,393,151 and to prevent corrosion of various metals by these fluids. For example, a 1% by weight solution of this cyclooctatetraene in such a fluid decreased oxygen consumption to zero and volatile products formation by a factor of 330 during a 24 hour exposure to oxygen at 600°F as compared to an identical treatment of the fluid in the absence of the additive. In addition, an M-50 coupon surface in the presence of the additive appeared unchanged, whereas in the absence of any additive, under otherwise identical conditions, the surface becomes covered with deeply colored irregular deposits.

Sulfurized Molybdenum-Containing Compositions

In the process of *L. deVries and J.M. King; U.S. Patent 4,283,295; August 11, 1981; assigned to Chevron Research Company* antioxidant additives for lubricating oil are prepared by combining a polar promoter, ammonium tetrathiomolybdate, and a basic nitrogen compound complex to form a sulfur- and molybdenum-containing composition.

The polar promoter used is one which facilitates the interaction between the ammonium tetrathiomolybdate and the basic nitrogen compound. A wide variety of such promoters may be used. Typical promoters are 1,3-propanediol, 1,4-butanediol, diethylene glycol, butyl Cellosolve, propylene glycol, 1,4-butylene glycol, methyl Carbitol, ethanolamine, dimethylformamide, N-methyl acetamide, dimethyl acetamide, methanol, ethylene glycol, dimethylsulfoxide, hexamethyl phosphoramide, and tetrahydrofuran and water. Preferred are water and ethylene glycol. Particularly preferred is water.

While ordinarily the polar promoter is separately added to the reaction mixture, it may also be present, particularly in the case of water, as a component of non-anhydrous starting materials or as waters of hydration of the molybdenum compound. Water may also be added as ammonium hydroxide.

The basic nitrogen compound must have a basic nitrogen content as measured by ASTM D-664 or D-2896. Typical of such compositions are succinimides, carboxylic acid amides, hydrocarbyl monoamines, hydrocarbon polyamines. Mannich bases, phosphonamides, thiophosphonamides, phosphoramides dispersant viscosity index improvers, and mixtures thereof. Any of the nitrogen-containing compositions may be after-treated with, e.g., boron using procedures well known in the art so long as the compositions continue to contain basic nitrogen. These after-treatments are particularly applicable to succinimides and Mannich base compositions.

The lubricating oil compositions containing the additives of this process can be prepared by admixing, by conventional techniques, the appropriate amount of molybdenum-containing composition with a lubricating oil. The selection of the particular base oil depends on the contemplated application of the lubricant and the presence of other additives. Generally, the amount of the molybdenum containing additive will vary from 0.05 to 15% by weight and preferably from 0.2 to 10% by weight.

The lubricating oil which may be used in this process includes a wide variety of hydrocarbon oils, such as naphthenic bases, paraffin bases and mixed base oils as well as synthetic oils such as esters and the like. The lubricating oils may be used individually or in combination and generally have a viscosity which ranges from 50 to 5,000 SUS and usually from 100 to 15,000 SUS at 38°C.

In many instances it may be advantageous to form concentrates of the molybdenum containing additive within a carrier liquid. These concentrates provide a convenient method of handling and transporting the additives before their subsequent dilution and use. The concentration of the molybdenum-containing additive within the concentrate may vary from 0.25 to 90% by weight although it is preferred to maintain a concentration between 1 and 50% by weight. Another embodiment of this process is a lubricating oil concentrate composition comprising an oil of lubricating viscosity and from 15 to 90% by weight of the molybdenum-containing additive.

The final application of the lubricating oil compositions of this process may be in marine cylinder lubricants as in crosshead diesel engines, crankcase lubricants as in automobiles and railroads, lubricants for heavy machinery such as steel mills and the like, or as greases for bearings and the like. Whether the lubricant is fluid or a solid will ordinarily depend on whether a thickening agent is present. Typical thickening agents include polyurea acetates, lithium stearate and the like.

Example 1: To a 500 ml flask was added 145 g of a solution of 45% concentrate in oil of the succinimide prepared from polyisobutenyl succinic anhydride and tetraethylene and having a number average molecular weight for the polyisobutenyl group of about 980 and 75 ml of hydrocarbon thinner. The mixture was heated to 75°C and then 0.1 mol (26 g) of $(NH_4)_2MoS_4$ in a pulverized state was added to the reaction mixture along with 35 ml of water. A nitrogen atmosphere was maintained in the reaction mixture which was heated at 65°C for 45 minutes.

The temperature was then increased to 95°C and water was removed. A heavy evolution of hydrogen sulfide and ammonia was observed. The temperature was increased to reflux at 155°C and maintained for one hour. The mixture was then filtered through diatomaceous earth and the filtrate stripped to 160°C at 20 mm Hg to yield a product containing 4.78% molybdenum, 4.00% sulfur and 1.79% oxygen.

Example 2: In the Oxidator B test the stability of the oil is measured by the time required for the consumption of 1 ℓ of oxygen by 100 g of the test oil at 340°F. In the actual test, 25 g of oil is used and the results are corrected to 100-g samples. The catalyst which is used at a rate of 1.38 cc per 100 cc oil contains a mixture of soluble salts providing 95 ppm copper, 80 ppm iron, 4.8 ppm manganese, 1,100 ppm lead, and 49 ppm tin. The results of this test are reported as hours to consumption of 1 ℓ of oxygen.

The copper strip test is a measure of corrosivity toward nonferrous metals and is described as ASTM Test Method D-130. Antiwear properties are measured by the 4-ball wear tests. The 4-ball wear test is described in ASTM D-2266.

The formulation tested was neutral oil containing 3.5% of a 50% solution of succinimide in oil, 22 mmol/kg of the product of the process, 20 mmol/kg sulfurized calcium phenate, 30 mmol/kg overbased magnesium sulfonate, 5.5% viscosity index improver and, if necessary, additional succinimide to bring the total nitrogen content of the finished oil to 2.14%. Results are as follows: Oxidator B, 8.6 hours; ASTM D-2266, 0.36 mm; D-130, 2A.

In the process of *L. deVries and J.M. King; U.S. Patent 4,285,822; August 25, 1981; assigned to Chevron Research Company* antioxidant additives for lubricating oil are prepared by (1) combining a polar solvent, an acidic molybdenum compound and an oil-soluble basic nitrogen compound to form molybdenum-containing complex and (2) contacting the complex with carbon disulfide to form a sulfur- and molybdenum-containing composition.

The molybdenum compounds used to prepare the additives for compositions of this process are acidic molybdenum compounds. By acidic is meant that the molybdenum compounds will react with a basic nitrogen compound as measured by ASTM test D-664 or D-2898 titration procedure. Typically these molybdenum compounds are hexavalent and are represented by the following compositions: molybdic acid, ammonium molybdate, sodium molybdate, potassium molybdate and other alkaline metal molybdates and other molybdenum salts such as hydrogen salts, e.g., hydrogen sodium molybdate, $MoOCl_4$, MoO_2Br_2, $Mo_2O_3Cl_6$, molybdenum trioxide or similar acidic molybdenum compounds. Preferred acidic molybdenum compounds are molybdic acid, ammonium molybdate, and alkali metal molybdates. Particularly preferred are molybdic acid and ammonium molybdate.

The polar promoters and basic nitrogen compounds used in this process are those described in the preceding patent. The lubricant compositions containing the additive are also prepared as described in the preceding patent using the same quantities of materials and having the same applications.

Example 1: To a 500 ml flask was added 290 g (0.1 mol active) of a solution of 45% concentrate in oil of the succinimide prepared from polyisobutenyl succinic anhydride and tetraethylene pentaamine and having a number average molecular weight for the polyisobutenyl group of about 980. This mixture was heated to 140°C and to it was added dropwise a solution containing 28.8 g (0.2 mol) of molybdenum trioxide dissolved in approximately 100 ml of concentrated ammonium hydroxide. The addition took place over a period of 2 hours and was accompanied by heavy foaming. The reaction mixture was then heated to 170°C to remove the water, and a small amount of xylene was added to remove the remaining amount of water from the solution. The reaction was filtered through diatomaceous earth and approximately 8.34 g of molybdenum trioxide was removed on the filter pad.

The product was then dissolved in 300 ml of xylene and heated to 70°C. Slowly, 60 ml carbon disulfide was added, the heat was increased to 105°C (reflux) and held for four hours. Hydrogen sulfide gas evolved. Heating was continued at 115°C for 2 hours until no more hydrogen sulfide gas evolved. The reaction mixture was filtered through diatomaceous earth to yield a product containing 1.36% sulfur, 4.61% molybdenum, 2.88% oxygen and 1.82% nitrogen.

Example 2: To a 1 ℓ flask containing 290 g of the succinimide described in Example 1 and heated to 140°C was added dropwise under nitrogen 28.8 g (0.2 mol) of molybdenum trioxide dissolved in 100 ml of concentrated ammonium hydroxide. The foaming of the product was very heavy and it took 2 hours to add about one-third of the molybdenum trioxide solution. Five drops of foam inhibitor was added and the remainder of the molybdenum solution was added over a period of one hour. To this mixture was added, 400 ml toluene and then the solvent was stripped at 120° to 125°C. To this mixture was added 500 ml hexanes and the solution was filtered through diatomaceous earth. The hexanes were removed, 200 ml toluene was added and then at 70°C, 60 g of carbon disulfide was added.

The reaction mixture was heated to 105°C and maintained at this temperature for 5 hours. Heating was continued for 2 hours at 120°C and carbon disulfide was removed with distillation. This mixture was treated with hydrogen sulfide at room temperature for 3 hours using a hydrogen sulfide sparge to give a light positive pressure. Toluene was removed at 140°C to yield a composition containing 4.51% molybdenum, 1.75% oxygen, 1.73% nitrogen and 3.75% sulfur.

Example 3: Lubricating oil compositions containing the additives prepared according to this process have been tested in a variety of tests. Reported below are results from certain of these tests which are described below. The Oxidator B test is described in the preceding patent.

The anticorrosion properties of compositions can be tested by their performance in the CRC L-38 bearing corrosion test. In this test, separate strips of copper and lead are immersed in the test lubricant and the lubricant is heated for 20 hours at a temperature of 295°F. The copper strip is weighed and then washed with potassium cyanide solution to remove copper compound deposits. It is then reweighed. The weight losses of the two strips are reported as a measure of the degree of corrosion caused by the oil.

The copper strip test is a measure of corrosivity toward nonferrous metals and is described as ASTM Test Method D-130. Antiwear properties are measured by the 4-ball weld tests. The 4-ball wear test is described in ASTM D-2266 and the 4-ball weld test is ASTM D-2783.

The coefficient of friction of lubricating oils containing additives of this process was tested in the Kinetic Oiliness Testing Machine. The procedure used in this test is described by G.L. Neely, *Proceeding of Midyear Meeting, American Petroleum Institute* (1932), pp. 60-74 and in *ASLE Transactions,* vol 8, pp. 1-11 (1965) and *ASLE Transactions,* vol. 7, pp. 24-31 (1964). The coefficient of friction was measured under boundary conditions at 150° and 204°C using a 1 kg load and a molybdenum-filled ring on a cast-iron disk. The data for some of the tests run on compositions of this process is reported in the table below. The particular formulations tested are as described in Example 2 of the preceding patent.

Product of Example*	Oxidation B (hours)	ASTM D-2266 (mm)	ASTM D-2783 (kg)	....L-38.... Cu (mg)	Pb (mg)	D-130	Coefficient of Friction 150°C	204°C
1	13.4	0.41	170	51.3	4.7	C	0.089	0.036
2	9.75	0.34	187	18.3	3.3	C	0.052	0.026

*Neutral oil formulation containing 3.5% of a 50% concentrate of succinimide, 20 mmol/kg sulfurized calcium phenate, 30 mmol/kg overbased magnesium sulfonate, 5.5% viscosity index improver, and 22 mmol/kg product. (If necessary, additional succinimide was added to bring the total nitrogen content of finished oil to 2.14%.)

N-Alkylaminomethyl-5-Amino-1H-Tetrazole as Silver Corrosion Inhibitor

In accordance with the process of *R.L. Sung and B.H. Zoleski; U.S. Patent 4,285,823; August 25, 1981; assigned to Texaco Inc.* there is provided a diesel lubricating oil composition comprising a diesel lubricating oil containing a silver corrosion inhibiting amount of at least one N-alkylaminomethyl-5-amino-1H-tetrazole.

The composition of this process contains the above additive in an amount from about 0.5 to 2.0 weight percent, preferably between 0.5 and 1 weight percent and especially at least 1% by weight.

The preparation of the reaction product used in a diesel lubricating composition according to the process is relatively uncomplicated and can be economically conducted. An N-alkyl-1,3-propanediamine wherein the alkyl group has from 12 to 18 carbon atoms, formaldehyde and 5-amino tetrazole are reacted in substantially equimolar amounts to form the desired products.

Suitable N-alkyl-1,3 propanediamines known as Duomeens (Armak Co.) are available commercially, wherein the alkyl group is straight chain with an average 12 to 18 carbon atoms and is attached to the nitrogen through the second carbon in the chain. Examples of suitable N-alkyl-1,3-propanediamines include N-oleyl-1,3-propanediamine (Duomeen O), N-stearoyl-1,3-propanediamine and N-tallow-1,3-propanediamine (Duomeen T).

Example 1: *Preparation of (1-Duomeen T methyl)-5-amino-1H-tetrazole* — To a mixture of 0.2 mol of 5-amino-1H-tetrazole, 0.2 mol of Duomeen T, 60 parts methanol and 60 parts of xylene, at reflux temperature, 0.2 mol of formaldehyde was added dropwise. After addition was completed, the mixture was refluxed for 1¾ hr more, allowed to cool and 100 parts of xylene was added. The mixture was filtered and stripped under vacuum. The reaction product was identified and analyzed by IR, NMR and elemental analysis.

Example 2: *Preparation of (1-Duomeen O methyl)-5-amino-1H-tetrazole* —

0.2 mol of formaldehyde was added dropwise to a boiling mixture of 0.2 mol Duomeen O, 0.2 mol of 5-amino-1H-tetrazole, 60 parts of methanol and 60 parts of xylene. The mixture was refluxed for 1¾ hr after addition of formaldehyde was completed, let cool and 100 parts of xylene was added. The mixture was filtered and stripped under vacuum. The residue was analyzed by elemental analysis, IR and NMR.

Samples of a base oil and of the same oil containing different amounts of the additive proposed in Example 1 were tested in the Texaco Modified Silver Disc Friction Test described in U.S. Patent 4,278,553 above. The results are shown in the table below.

	Blends			
	A	B	C	D
Components (wt %)				
Naphthenic mineral oil of 300 viscosity at 100°F	3.00	–	–	–
Paraffin mineral oil	60.00	–	–	–
Naphthenic mineral oil of 75 to 80 viscosity at 210°F	29.70	–	–	–
Phosphosulfurized polyisobutylene dispersant	4.10	–	–	–
Calcium sulfurized alkyl phenolate detergent	3.20	–	–	–
Silicone foam inhibitor	150	–	–	–
Oil A	–	99.5	99.0	98.0
Product of Example 1	–	0.5	1.0	2.0
Silver Disc Function Test Rating	Poor	Excellent	Excellent	Excellent

From the data above, it is seen that the present additives are effective corrosion inhibitors for diesel lubricating oil compositions.

More specifically, in the table above, blend A is a fully formulated diesel engine oil except that it does not contain a silver corrosion inhibitor. Under the test conditions, additions thereto of 0.5 to 2.0% of the additive of the process to form blends B, C, and D, cause the rating of the oil to pass from poor to excellent.

Oil-Soluble Metal (Lower) Dialkyl Dithiophosphate Succinimide Complex

Metal dihydrocarbyl dithiophosphates are useful for a variety of purposes known in the art. The zinc dialkyl dithiophosphates in particular are employed as oxidation and corrosion inhibitors in lubricating oil compositions. There is a problem, however, with metal (lower) dialkyl C_{2-3} dithiophosphates, in that they are essentially insoluble in lubricating oil compositions.

It has been found by *E.S. Yamaguchi; U.S. Patent 4,306,984; December 22, 1981; assigned to Chevron Research Company* that an oil-insoluble metal C_{2-3} dialkyl dithiophosphate may be made oil-soluble by forming a complex between the dithiophosphate and an alkenyl or alkyl mono- or bis-succinimide.

Example 1: *(A) Diisopropyl dithiophosphoric acid* – To a two liter three-necked flask, equipped with a stirrer, nitrogen inlet, dropping funnel and condenser was charged under nitrogen 288.6 g (1.3 mols) P_2S_5 and 600 ml of toluene. To this slurry was added 312 g (5.2 mols) of isopropyl alcohol over a period of about 17 minutes. After stirring for about 30 minutes, the reaction mixture was heated to reflux and maintained at reflux for about 2.5 hours. The clear yellow solution containing the reaction product was decanted off (1,050 g) leaving a small amount of black solids; acid No. 241.6; 238.4 mg KOH/g equivalent weight 234.

(B) Zinc diisopropyl dithiophosphate – To a two liter three-necked flask equipped with a stirrer, nitrogen inlet, and a Dean-Stark trap/condenser was added under nitrogen, 526.5 g (2.25 equivalents) of diisopropyl dithiophosphoric acid and about 600 ml toluene. To this solution was added 146.5 g zinc oxide (60% equivalent excess) at which point the temperature rose to about 74°C. The reaction mixture was heated to reflux and maintained at reflux for about 4 hours. A total of about 20 ml of water was collected, after which the reaction mixture was further diluted with 200 ml toluene and filtered hot, two times through Celite. The clear filtrate was stripped in a Rotary-Evaporator under full pump vacuum and a water-bath temperature up to 72°C. The product obtained weighed 592.3 g and was a soft, white, crystalline solid; zinc, 14.00%, phosphorous 13.83%.

In a similar manner, following the procedures described above, there was obtained zinc di-n-propyldithiophosphate and zinc diethyl dithiophosphate.

Example 2: Various oil blends were prepared as indicated in the table below using Mid Continental Paraffinic base oil (CC100N) and containing zinc diisopropyl dithiophosphate with and without the solubilizing polyisobutenyl succinimide dispersant component (prepared by reacting polyisobutenyl succinic anhydride wherein the number average molecular weight of the polyisobutenyl was about 950 and triethylenetetramine in a mol ratio of amine to anhydride of 0.87).

Component	Amount (mmol/kg)	Observation
Zinc diisopropyl dithiophosphate	3	Solids present in oil
Zinc diisopropyl dithiophosphate	9	Solids present in hazy oil
Zinc diisopropyl dithiophosphate plus polyisobutenyl succinimide of triethylenetetramine (3.5%)*	12	Bright and clear oil solution
Zinc diisopropyl dithiophosphate plus polyisobutenyl succinimide of triethylenetetramine (3.5%)*	18	Bright and clear oil solution

*The zinc diisopropyl dithiophosphate and the succinimide were first dissolved in chloroform, the chloroform evaporated off and the complex blended into the oil.

Example 3: Formulated oils containing the additives shown in the table below were prepared and tested in a Sequence V-D Test method Phase 9-L (according to candidate test for ASTM). This procedure utilizes a Ford 2.3 liter four cylinder engine. The test method simulates a type of severe field test service characterized by a combination of low speed, low temperature "stop and go" city driving and moderate turnpike operation. The effectiveness of the additives in the oil is measured in terms of the protection provided against sludge and varnish deposits and valve train wear.

Formulation 1 was prepared by mixing the components together at 135°C until homogeneous. The complex thus formed was added to the oil. Formulations 2 and 3 were prepared by adding each of the components directly to the oil.

The comparisons were made in a formulated base oil Cit-Con 100N/Cit-Con 200N at 55%/45% containing 30 mmols/kg of a magnesium sulfonate, 20 mmols/kg of a calcium phenate and 8.5% of a polymethacrylate V.I. improver.

		Cam Lobe Wear × 10^{-3}		Varnish	Sludge
Entry	Formulation	SF Spec. Max. (2.5)	SF Spec. Avg. (1.0)	SF Spec. Avg. (6.6)	SF Spec. Avg. (9.4)
1	8.1 mmoles/kg zinc diisopropyl dithiophosphate + 3.5% succinimide of Example 2	*2.0	*0.7	7.5	9.7
2	8.1 mmoles/kg zinc di(2-ethylhexyl) dithiophosphate + 3.5% succinimide of Example 2	10.8	5.8	8.8	9.3
3	8.1 mmoles/kg zinc di(isobutyl/mixed primary hexyl) dithiophosphate + 3.5% succinimide of Example 2	7.8	4.0	7.0	8.0

*Average of two runs.

As indicated from the results shown in the table above, the combination of zinc diisopropyl dithiophosphate and succinimide dispersant gave superior wear performance relative to the zinc dithiophosphates derived from the primary alcohols (Entries 2 and 3).

Reducing Corrosivity of Phenol Sulfides

According to *M. Braid; U.S. Patent 4,309,293; January 5, 1982; assigned to Mobil Oil Corporation* sulfurized phenols, e.g., phenol sulfides, disulfides or polysulfides, oligomers thereof or mixtures of same when treated with alkyl vinyl ethers provide excellent metal anticorrosivity characteristics without significant reduction of antioxidant, antiwear or other desired properties when incorporated into organic media such as lubricants.

The additives of this process or mixtures thereof may be used in mineral oils, synthetic oils or mixtures of mineral and synthetic oils of lubricating viscosity. Amounts from about 0.1 to about 5 weight percent of the total composition are highly effective for the intended purpose.

Example 1: A typical solvent refined mineral oil base stock having a viscosity of 200 SUS at 100°F.

Example 2: *Reaction of 2-methyl-4-tert-butylphenol with sulfur monochloride* – To a solution of 2-methyl-4-tert-butylphenol (82.2 g) in petroleum ether (200 ml) cooled to 8°C there was added over about five hours sulfur monochloride (33.7 g). The temperature was maintained at 6° to 8°C. After an additional 20 hours the reaction mixture cooled in an ice bath was treated with dilute ammonium hydroxide and extracted with benzene. Removal of solvent from the washed and dried benzene extract left the sulfurized phenol, 93 g of reddish oil. Elemental analysis gave C, 67.66%; H, 7.92%; S, 15.0%; and Cl, 0.1%.

Example 3: *Reaction of sulfurized 2-methyl-4-tert-butylphenol with butyl vinyl ether* – To a solution of sulfurized 2-methyl-4-tert-butylphenol (17.5 g) prepared as described in Example 2 in benzene (200 ml), containing one drop of glacial acetic acid as catalyst and heated at reflux temperature, there was added a solution of butyl vinyl ether (40 g) in petroleum ether (50 ml). The addition required 0.75 hr; the temperature was maintained at 76°C during addition and for an additional 2 hr reaction period. The reaction mixture was washed with water, neutralized with sodium bicarbonate solution, washed again with water and dried. Solvents and unreacted butyl vinyl ether were stripped off in a rotary film evaporator at reduced pressure leaving the reaction product as a moderately viscous amber oil.

Example 4: *Sulfurized 4-tert-octylphenol* — Sulfur monochloride (50.6 g) was added over five hours to a stirred solution of 4-tert-octylphenol (154.7 g) in n-octane (150 ml) while the temperature was maintained at 125° to 127°C. After an additional 0.25 hr of heating the reaction mixture was allowed to cool to room temperature. The reaction mixture was then poured while stirring into a solution of ammonium hydroxide (150 ml). The resulting mixture was extracted with benzene. The extracts were washed with water and dried. Solvent was removed in a rotary film evaporator under reduced pressure leaving the sulfurized 4-tert-octylphenol as dark viscous oil containing 14.5% of sulfur.

Example 5: *Reaction of sulfurized 4-tert-octylphenol with ethyl vinyl ether* — To a solution of sulfurized 4-tert-octylphenol (44.3 g) prepared as in Example 4 in xylene (200 ml) heated to 100°C there was added during 3 hours while stirring, ethyl vinyl ether (50 g). The rate of addition was controlled so as to maintain the temperature for one hour more, and then solvent and unreacted vinyl ether were removed by rotary distillation at reduced pressure. The treated sulfurized phenol was obtained as a hazy oil residue which contained substantial hydroxyl absorption in the infrared spectrum. It was filtered to remove a minor amount (0.3 g) of solids melting above 300°C.

Certain of the examples were then subjected to the Copper Strip Test after being incorporated into the above-referred to base oil (Example 1). The test data clearly demonstrates the excellent anticopper corrosion characteristics of the additives disclosed herein.

The test employed for this purpose was a standard ASTM Test D-130 which, in general, comprises immersion of a polished copper strip in the material to be tested for a period of three hours at a temperature of 250°F. At the end of this period the copper strip is removed, washed, and rated for degree of corrosion by comparison with the ASTM standard strips.

After alkyl vinyl treatment, all of the examples tested showed the excellent copper corrosivity rating of 1 A, the lower the rating the better the anticorrosion properties.

Benzimidazoles for Fluorinated Polysiloxane and Polyfluoroalkyl Ether Base Fluids

It has been found by *J.B. Christian and C. Tamborski; U.S. Patent 4,324,671; April 13, 1982; assigned to the U.S. Secretary of the Air Force* that the addition of a small amount of certain benzimidazoles to a fluorinated polysiloxane base fluid and a thickener therefor provides a grease having unexpectedly outstanding properties. Thus, the resulting grease composition inhibits rust formation when utilized as a lubricant for ferrous metals under mild temperature and high humidity conditions. Furthermore, the grease inhibits corrosion when used as a lubricant for ferrous metals under high temperature conditions.

More specifically, the grease composition consists essentially of (1) about 60 to 65 weight percent of base fluid, (2) about 33.5 to 39.5 weight percent thickener, and (3) about 0.5 to 1.5 weight percent benzimidazole, based upon a total of 100 weight percent.

As a thickener, it is generally preferred to use a fluorinated ethylene-propylene copolymer or polytetrafluoroethylene.

The benzimidazole antirust and anticorrosion additives used in the grease compositions have the following structural formula:

$$\text{C}_6\text{H}_4\begin{matrix}\text{N}\\ \text{N(H)}\end{matrix}\text{C—R}$$

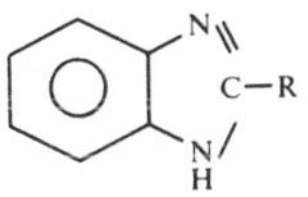

wherein R is H, hydrocarbon alkyl, hydrocarbon aryl, perfluoroalkyl or perfluoroalkyl or perfluoroalkylene ether.

Example 1: A series of runs was conducted in which grease compositions of this process were formulated and tested. As a base fluid there was used a fluorinated polysiloxane having the following formula:

$$\text{(A)} \quad CF_3CH_2CH_2Si(CH_3)_2O\left[\begin{matrix}Si(CH_3)O\\ |\\ CH_2CH_2CF_3\end{matrix}\right]_n Si(CH_3)_2CH_2CH_2CF_3$$

where n is an integer having a value such that the fluid has a viscosity of 75 cs at 100°F. The base fluid was known as FS-1265 (Dow Corning Corp.). The thickener used was a fluorinated copolymer of ethylene and propylene having a molecular weight of 150,000.

The benzimidazole additives used in the formulations had the following structural formula shown above in which R was one of the following: H, C_6H_{13}, C_6H_5, $CF(CF_3)OCF_2CF(CF_3)OC_3F_7$, and $CF(CF_3)[OCF_2CF(CF_3)]_4OC_3F_7$.

In preparing each of the greases, the components were mixed and stirred until a uniform mixture was obtained. The amounts of base fluid used ranged from 60 to 65 weight percent while the amounts of thickener used ranged from 34 to 39 weight percent. Each grease composition contained 1.0 weight percent of the abovedefined benzimidazole additives. Each mixture was further blended to a grease consistency by passing it two times through a 30 roll mill with the rollers set at an opening of 0.002" at about 77°F.

The various grease compositions were tested according to several test procedures. The penetration test was conducted in accordance with Federal Test Method Standard 791a, Method 313.2. The rust preventive properties test was carried out in accordance with Method 4012 of the same standard. The high temperature corrosion was determined in accordance with the method set forth in Technical Documentary Report AFML-TR-69-290. The results of the test are set forth in the table below.

Example 2: A control run was conducted in whch greases were prepared, utilizing the base fluid and thickener of Example 1. The greases consisted of 65 weight percent base fluid and 35 weight percent thickener and did not contain any of the benzimidazole additives. The greases were formulated and tested according to the procedures described in Example 1. The results of the test are included in the table below.

	Greases of Example 1	Grease Based Formula A Fluid, No Additive*
Penetration range of greases, dmm	264-290	302-309
Rust preventive properties	Pass**	Marginal***
High temperature corrosion, 450°F		
72 hours	Pass	Fail†
52-100 steel	Pass	Fail
440C steel	Pass	Fail
M-10 steel	Pass	Fail
M-59 steel	Pass	Fail

*Control run.
**Pass = no rusting or corrosion, a maximum of three spots allowed.
***Marginal = The maximum allowable number of rust spots were present at the end of the test.
†Fail = more than three rust or corroded spots or pitting and etching.

The following is an example of a specific grease formulation utilizing a benzimidazole additive of this process: 65% fluid given in Example 1; 34% thickener given in Example 1; and 1% benzimidazole wherein R is H.

In this process *J.B. Christian and C. Tamborski; U.S. Patent 4,324,673; April 13, 1982; assigned to the U.S. Secretary of the Air Force* provide an antirust, anticorrosion grease composition comprising a major amount of a polyfluoroalkylether base fluid, a minor amount of a fluorocarbon polymer thickening agent, and a rust and corrosion inhibiting amount of a benzimidazole.

More specifically, the grease composition consists essentially of (1) about 65 to 72 weight percent of base fluid, (2) about 26.5 to 34.5 weight percent thickener, and (3) about 0.5 to 1.5 weight percent benzimidazole, based upon a total of 100 wt %. In general, any suitable polyfluoroalkylether can be used as a base fluid in formulating a grease of this process. As a thickener, it is usually preferred to utilize a fluorinated ethylene-propylene copolymer or polytetrafluoroethylene. The benzimidazole antirust and anticorrosion additives used in the grease compositions are identical to those described in the preceding patent.

Example 1: A series of runs was carried out in which grease compositions of this process were formulated and tested. As a base fluid there was used a polyfluoroalkylether having the following formula:

$$C_3F_7O(\underset{\displaystyle CF_3}{\underset{|}{C}}FCF_2O)_nC_2F_5$$

where n is an integer having a value such that the fluid has a kinematic viscosity of 270 cs at 100°F. The base fluid was Krytox 143AC fluid, a product of E.I. du Pont de Nemours & Co. The thickener used was a fluorinated copolymer of ethylene and propylene having a molecular weight of about 150,000. The benzimidazole additives used in the formulations had the following structural formula:

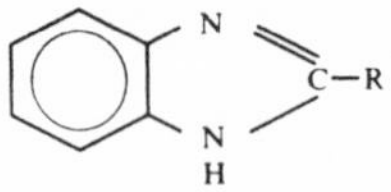

[Structure: benzimidazole ring, N= and N–H bonded to C–R]

in which R was one of the following particular formulations: H, C_6H_{13}, C_6H_5, $CF(CF_3)OCF_2{-}CF(CF_3)OC_3F_7$, and $CF(CF_3)[OCF_2CF(CH_3)]_4OC_3F_7$.

In preparing the greases, the components were mixed and stirred until a uniform mixture was obtained. The amounts of base fluid used range from 65 to 72 weight percent while the amounts of thickener ranged from 27 to 34 weight percent. Each grease composition contained 1.0 weight percent of one of the above-described benzimidazole additives. Each mixture was further blended to a grease consistency by passing it two times through a 3-roll mill with the rollers set at an opening of 0.002" at about 77°F.

The several grease compositions were tested according to several standard test procedures as described in Example 1 of U.S. Patent 4,324,671, above. The results of the tests are set forth in the table below.

Example 2: A series of runs was conducted in which greases were prepared, utilizing, as described in Example 1, the same thickener and benzimidazole additives and amounts thereof as well as the same amounts of a polyfluoroalkylether base fluid. However, the polyfluoroalkylether had the following structural formula:

$$XO(C_3F_6O)_p(CF_2O)_q(C_2F_4O)_sY$$

where X and Y are CF_3, C_2F_5, or C_3F_7 and p, q and s are integers such that the fluid has a kinematic viscosity of about 90 cs at 100°F. The base fluid used was Fomblin Y fluid, a product of Montedison, SpA, Italy.

The greases were formulated and tested according to the procedures described in Example 1. The results of the tests are shown below in the table.

Example 3: Control runs were carried out in which greases were prepared, utilizing the base fluids and thickeners of Examples 1 and 2. The greases consisted of 70 weight percent base fluid and 30 weight percent thickener and did not contain any benzimidazole additives. The greases were formulated and tested according to the procedures described in Example 1. The results of the tests are included below in the table.

Greases	Range of Penetration Values (dmm)	Rust Preventive Properties	High Temperature Corrosion, 450°F, 72 hours 52-100 Steel	440C Steel	M-10 Steel	M-50 Steel
Example 1	264-277	Pass	Pass	Pass	Pass	Pass
Example 2	283-300	Pass	Pass	Pass	Pass	Pass
Control Runs						
Based on 270 cs fluid	298-300	Fail	Fail	Fail	Fail	Fail
Based on 90 cs fluid	310-310	Fail	Fail	Fail	Fail	Fail

Transition-Metal-Containing Calcium Phenol Sulfides

According to *M. Braid; U.S. Patent 4,330,421; May 18, 1982; assigned to Mobil Oil Corporation* lubricant compositions containing minor effective amounts of calcium phenol sulfides, containing controlled amounts of a transition metal wherein the transition metal is derived by a direct exchange reaction for calcium provide stabilizer/detergent properties thereto and protect the compositions thereof against oxidative deterioration.

The exchange reaction is effected by reaction of the neutral or overbased calcium phenate with a solution of the sulfate salt of the transition metal in an appropriate solvent system. The calcium phenate may be reacted as a solution in diluent oil or it may be free of oil and reacted in solvents such as benzene, hexane, petroleum ether, isopropyl alcohol as mixtures of these and similar solvents. The transition metal sulfates as hydrates may be reacted in aqueous solution or in alcohol solvents such as methanol, ethanol, isopropyl alcohol and the like. When the reaction system is heterogeneous for example as in the reaction of an oil diluted calcium phenol sulfide with an aqueous solution of nickel sulfate an interphase reaction promoter such as ethanol or isopropyl alcohol in minor amounts can be used to expedite the reaction.

The calcium atoms in the calcium phenol sulfide are not equivalent in ease or rate of direct exchange and this is the basis for controlling the selectivity and extent of exchange. Reactant ratios, choice of solvent systems, length of reaction period, and temperature can be used severally and jointly in effecting the exchange to achieve the desired balance of stabilizer/detergent properties. While it is not the sole controlling factor in determining the properties of the additives in this process the ratios of calcium to transition metal in terms of weight percent is useful in judging the stabilizer effectiveness and the preferred ranges vary from about 0.2 to about 80 and the most preferred range is from about 1.5 to about 10. The resultant calcium/transition metal compounds contain at least about 0.01 to 10 weight percent of transition metal and preferably from about 0.025 to 2.5 weight percent based on the weight of the total composition.

The additive compounds are highly useful in mineral oils, mineral oil fractions, synthetic oils and in mixtures of mineral and synthetic oils in providing increaased resistance to degradation and deterioration resulting from oxidation. The synthetic oils or fluids include synthetic hydrocarbon oils derived from long chain alkanes or olefin polymers, ester oils obtained from polyhydric alcohols and monocarboxylic acids or monohydric alcohols and polycarboxylic acids.

The additive compounds in accordance with this process may be effectively used in concentrations of from about 0.01 to about 5 weight percent based on the total weight of the composition. Preferred are concentrations of from about 0.25 to about 5 weight percent. Other additives may be effectively used in the compositions for their intended purposes.

Example 1: This is an oil diluted commercial overbased calcium alkylated phenol sulfide typically having a calcium content of 5.25% and a total base number of about 140 to 150. This commercial calcium phenate may be manufactured as disclosed in the patent literature by reaction of an alkylated phenol in diluent oil solution with a basic calcium reactant in the presence of a catalyst or promoter and thereafter carbonated to the final degree of overbasing. The alkylated phenols usually contain from 4 to 32 carbon atoms total or 4 to 16 carbon atoms in any one side chain.

Example 2: *Separation of diluent oil from overbased calcium phenol sulfide* – To an aliquot of commercial oil diluted overbased (total base number, 140) calcium phenol sulfide described in Example 1 (29.4 g) stirred magnetically there was added rapidly at room temperature, 2-propanol (400 ml) which caused the precipitation of solids. After several hours of stirring the precipitated solids were collected and air dried to give 7.2 g which gave the following elemental analysis: C, 55.10%; H, 7.53%; S, 6.70%; Ca, 10.3%. Additional treatment with 2-propanol (300 ml) afforded a second fraction of 6.9 g of solids with the following elemental analysis: C, 55.29%; H, 7.69%; S, 6.50%; Ca, 9.9%.

Removal of the 2-propanol from the liquid residue left mainly the diluent oil which the infrared spectrum showed to contain phenol and/or phenol sulfide which had not been converted to calcium derivatives in the original processing.

Example 3: *Nickel exchanged overbased calcium phenol sulfide; effect of prolonged reaction time at higher temperature in the presence of nickel sulfate* – To a clear solution of the commercial overbased (total base number, 140) calcium thiobis-(alkylphenate) of Example 1, an approximately 57% concentrate in diluent oil (183.9 g) in xylene (500 ml) there was added a solution of nickel sulfate hexahydrate (52.5 g) in water while stirring at room temperature. The resulting reaction mixture was heterogeneous but free of solids.

Denatured ethanol (50 ml) was added and after a few minutes the temperature was raised to 35°C for 0.5 hr and then to 80° to 90°C as precipitating solids caused substantial thickening. Additional xylene (200 ml) was added as water was removed by azeotropic distillation during a total 5 hr reaction period as the temperature rose to 140°C. The resulting mixture was filtered hot and the filtrate was stripped of solvent xylene under reduced pressure. The oil concentrate of the nickel-exchanged product was obtained as a greenish dark brown moderately viscous liquid with the following elemental analysis: S, 3.39%; Ca, 0.51%; Ni, 3.72%.

Example 4: *Nickel exchanged overbased calcium phenol sulfide; 2-propanol/methanol (all alcohol) solvent system* – To a mixture of the overbased calcium phenol sulfide of Example 1 (150 g) and isopropyl alcohol (400 ml) there was added while stirring over a period of one hour a solution of nickel sulfate hexahydrate (53 g) in methanol (150 ml). During the addition the temperature was raised from 45° to 70°C. An additional quantity of isopropyl alcohol (400 ml) was added to disperse balled solids which agglomerated during the addition and the mixture was refluxed for six hours. From this reaction there was obtained the oil diluted nickel exchange product as a greenish brown viscous oil with the following elemental analysis: S, 2.20%; Ni, 1.36%; Ca, 0.57%.

The compounds thusly prepared were thereafter tested in the following manner. The catalytic oxidation test was used to evaluate the oxidative stabilization properties of the calcium-transition metal exchanged phenol sulfides. In this test the transition metal interchanged products were tested in a catalytic oxidation test in which the stabilizer agent in solution in a neutral solvent refined base oil having a viscosity at 100°F of 130 SUS is subjected to heating at 325°F in the presence of lead, iron, copper, and aluminum metal specimens for 40 hr while air is passed through at a rate of 5 liters per hour. The extent of oxidative deterioration is assessed by the increase in acidity, change in the neutralization numbers (ΔNN) as measured by ASTM D-974 and by the observed increase in viscosity (percentage change, %ΔKV). The results in the table clearly show the effectiveness of these additives in stabilizing oils of lubricating viscosity against oxidation, and differentiate strikingly the degree of effectiveness of these additives compared with the reactant calcium phenates.

Test Oil	Conc. of Additive (wt %)	ΔNN	ΔKV, %
Base oil, no additive	–	17	334
Base oil plus additive of Example 1	1	12.4	224
Base oil plus additive of Example 2*	1	5.8	53
Base oil plus additive of Example 3	1	4	13
Base oil plus additive of Example 4	1	7.5	39

*100% active, no diluent oil.

Phenylated Benzylamines and Phenylated Tetrahydro Naphthylamines

B.K. Bandish, F.C. Loveless and W. Nudenberg; U.S. Patent 4,335,006; June 15, 1982; assigned to Uniroyal, Inc. provides a method for stabilizing lubricating oils utilizing phenylated benzylamines or phenylated tetrahydro naphthylamines, which may be substituted or unsubstituted, either alone or in conjunction with a metal deactivator and a metal compound. This stabilizer system provides a surprisingly high degree of resistance to oxidative breakdown of lubricating oils as well as resulting in dramatic reductions in sludge formation.

The method for stabilizing a lubricating oil comprises utilizing as the stabilizer amino compounds having the general formula:

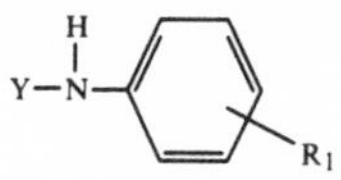

wherein Y is: or

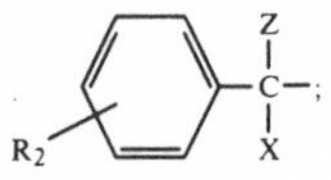

(1) (2) (3)

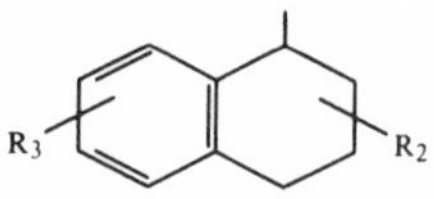

When Y is the moiety of formula (2), Z and X are each independently selected from the group consisting of H or C_{1-3} alkyl; R_2 selected from the group consisting of hydrogen, C_{1-12} alkyl, C_{1-12} alkoxy; C_{2-18} carbalkoxy, halogen or nitro and R_1 is hydrogen, C_{1-12} alkyl, C_{1-12} alkoxy, C_{2-18} carbalkoxy, halogen, amino or nitro. When Y is the moiety of formula (3), R_1 and R_3 are each independently selected from the group of moieties set forth as R_1 above and R_2 is as previously defined.

The preferred antioxidants of this process are N-(alpha-methyl-p-octylbenzyl)-aniline and N-(alpha-methyl benzyl)-p-nonylaniline.

Example: This example shows the unexpected ability of the antioxidant of this process to protect polyester based lubricating oils against oxidative degradation. The oil used was a commercially available polyolester fluid, Hercolube A (Hercules Inc.) and believed to be one prepared from pentaerythritol and a mixture of monocarboxylic acids, e.g., valeric acid and pelargonic acid.

Experiments were carried out in order to evaluate the effectiveness of the antioxidant. The oil sample used in runs B and C was prepared by adding N-(alpha-methyl-benzyl)aniline as stabilizer in the amount set forth in the table to 100 g of the polyester based oil. Sample D was similarly prepared containing phenyl alpha-naphthylamine, a commercially available stabilizer. The amounts used in each case are set forth in the table below.

Each of the samples was tested according to the following test procedures: A 100 ml sample having the compositions set forth in the table is poured into a pyrex glass test cell and aged by inserting one end of a glass air delivery tube into the test cell while the remaining 25 ml portions of each original oil sample is set aside and analyzed for neutralization number and kinematic viscosity in centistokes at 100°F. Around this glass air delivery tube immersed in the oil was placed from zero to four metal washers (Mg, Cu, Ag and Fe) as identified in the table.

When more than one washer was used, they were separated from each other by glass spacers. These remained in the oil during the aging process and served to indicate the extent of corrosion of the oil oxidative decomposition products on the metal. The test cell was then fitted with a reflux condenser. The assembly was placed in a constant temperature aluminum block. An air hose was then attached to the other end of the air delivery tube and the air flow was adjusted so that five liters of air per hour was bubbled through the oil. This aging test was carried out for 48 hours at 125°F. After aging, the oil was filtered hot and the amount of sludge developed was collected and was determined and recorded in milligrams per 100 ml of the oil. The filtered oil was then analyzed to determine changes in neutralization number and kinematic viscosity at 100°F.

The neutralization number was determined by the color-indicator titration method according to ASTM procedure D974-55T.

The kinematic viscosity was determined according to ASTM procedure D445-53T. The metal washers, which were weighed initially, were then carefully washed and weighed again to determine the weight change in grams.

The data in the table below dramatically show that when an amine such as N-(alpha-methylbenzyl)aniline is added to a polyolester based lubricating oil the aged properties of oil samples B and C are excellent as noted by very little change in the viscosity or neutralization number, very low sludge and essentially no weight change in the metals. The amine antioxidant of this process clearly provides better all around protection than a commercially available stabilizer (sample D).

It should be noted that sample C, wherein the test was conducted in the presence of copper, was slightly more deteriorated than sample B, where no copper was present.

							WEIGHT CHANGE (in grams) OF WASHERS			
RUN#	OIL*	GRAMS OF STABILIZER	% V_{100}**	N.N.***	SLUDGE (in mg.)	OIL*	Mg	Fe	Cu	Ag
A	100	—	94.9†	26.2	Undetermined	100	−0.1366	−0.0052	−0.0091	−0.0001
B	100	2.0	13.68	2.7	5.1	100	−0.0001	+0.0003	††	−0.0017
C	100	2.0	18.81	2.64	19.7	100	+0.0001	−0.0004	−0.0011	−0.0001
D	100	2.0	31.5	2.2	135.7	100	−0.0130	0.0000	−0.0024	−0.0001

*Hercolube A, commercially available.
**Percent change in viscosity at 100°F.
***Neutralization number of aged oil.
†Aged oil was so viscous that it could not be completely filtered to determine the amount of sludge formed. Neutralization number and viscosity of the aged oil are actually the properties of a small sample that could be filtered.
††No metal included.

Phenol-Containing Dithiophosphates

S. Rosenberger; U.S. Patent 4,349,445; September 14, 1982; assigned to Ciba-Geigy Corporation found a class of phenol-containing dithiophosphates in which the dithiophosphate part is linked in a quite specific manner to the phenolic molecule. These compounds impart to lubricants, in addition to having therein a good antioxidation and anticorrosion action, excellent extreme pressure and antiwear properties. Furthermore, the compounds are characterized by negligible formation of sediment and by no formation of ash.

Example 1:

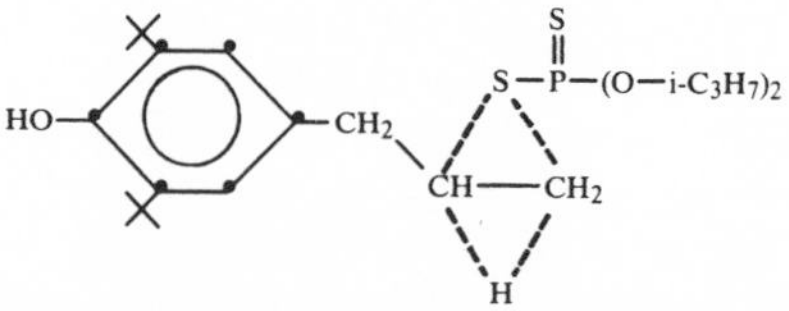

214 g of dithiophosphoric acid-O,O-diisopropyl ester are added under nitrogen to 246 g of 3-(3,5-di-tert-butyl-4-hydroxyphenyl)-prop-1-ene, and with the addition of 1 g of azodiisobutyric acid nitrile in several portions the mixture is heated, with stirring, for 8 to 10 hours at 90° to 110°C. At the end of the reaction period, there is practically no further olefin detectable (thin-layer chromatography).

The substance is freed in vacuo (~15 mm Hg/80°C) from readily volatile impurities. The isomeric mixture thus obtained forms a slightly brownish viscous oil, and can be used directly for application according to this process as a lubricant additive.

Example 2:

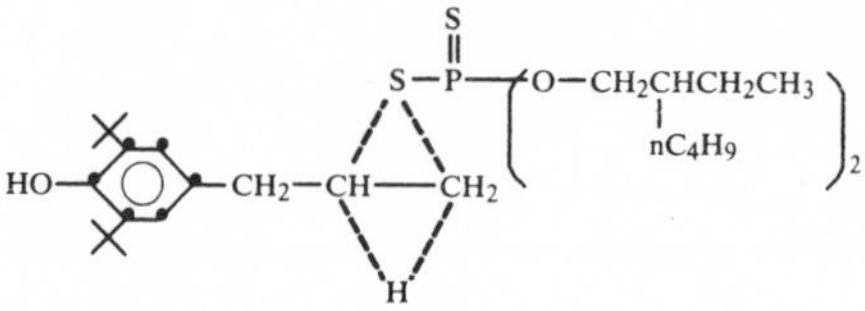

By replacing the diisopropyl ester of Example 1 by the corresponding amount of dithiophosphoric acid-di-2-ethyl-n-hexyl ester, with otherwise the same procedure, there is obtained the above isomeric mixture in the form of a viscous, almost colorless oil.

Example 3:

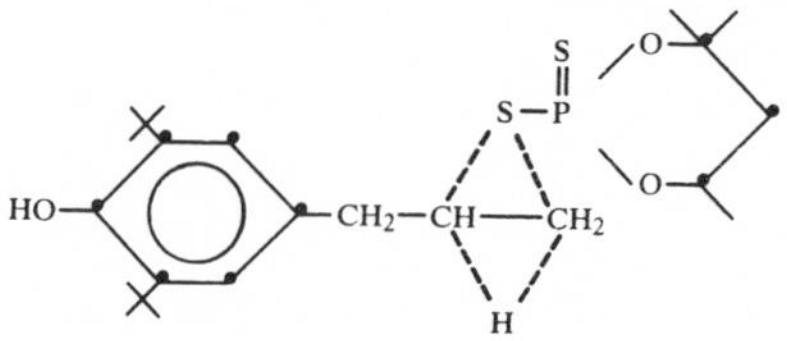

With use of the corresponding cyclic O,O-diester of dithiophosphoric acid with 1,1,3-trimethyl-1,3-dihydroxypropane, there is obtained, analogously to Example 1, the above-given isomeric mixture in the form of a viscous, almost colorless oil.

FOR FUELS

Alkenyl Succinic Acid and Trialkanolamine Triester

The process of *R.L. Godar, C.C. Hendricks and K.R. Roux; U.S. Patent 4,253,876; March 3, 1981; assigned to Petrolite Corporation* relates to a composition comprising: (1) an alkenyl succinic acid or anhydride (ASAA), and (2) the reaction product of ASAA and a trialkanol amine such as triethanolamine (TEA) where ASAA (3 mols) is reacted with TEA (1 mol) to yield the triester; and to the use thereof in corrosion inhibition.

The weight ratio of (1) ASAA to (2) triester can vary widely depending on the particular reactants, the particular systems in which it is employed, etc.

In general, the weight ratio of (1) to (2) is about 60 to 40, such as from about 40 to 60, for example from about 80 to 20, but preferably from about 20 to 80, with an optimum of about 50 to 50.

The compositions of this process are particularly useful as rust or corrosion inhibitors such as in refined petroleum products such as in gasoline, jet fuels, turbine oils, fuel oils, etc. They may be employed in any amount capable of inhibiting rust or corrosion preferably 25 to 50 ppm.

The methods of preparing the alkenyl succinic acid anhydrides are well known to those familiar with the art. The most feasible method is by the reaction of an olefin with maleic acid anhydride. Since relatively pure olefins are difficult to obtain, and when thus obtainable, are often too expensive for commercial use, alkenyl succinic acid anhydrides are usually prepared as mixtures by reacting mixtures of olefins with maleic acid anhydride. Such mixtures, as well as relating pure anhydrides, are utilizable herein. Corresponding alkyl succinic anhydrides can also be employed, i.e., where the alkenyl group is saturated, the preparation of alkyl succinic acids and anhydrides thereof is well known to the art.

A typical procedure for the preparation of the triester is the following. The desired substituted succinic anhydride and trialkanol amine charges are weighed into a suitable jacketed acid-resisting vessel equipped with agitation temperature recording means and a reflux condenser. The charge is heated to about 95° to 100°C while agitating and held thereat and the progress of the reaction observed from time to time by withdrawing a sample and determining the neutralization number thereof. Heating is continued until the desired neutralization number is obtained or until it remains substantially constant.

Oxazolonium Hydroxides

The process of *R.L. Sung and P. Dorn; U.S. Patent 4,257,780; March 24, 1981; assigned to Texaco Inc.* provides oxazolonium hydroxides having utility as fuel additives which are represented by the following structure:

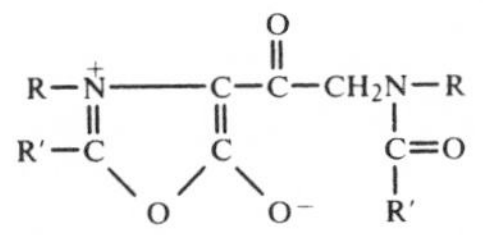

```
                       O
       +               ‖
  R—N———————C—C—CH2N—R
       ‖        ‖          |
  R'—C       C          C=O
        \    / \          |
         O     O⁻        R'
```

wherein R stands for a hydrocarbyl radical having from 1 to 5 carbon atoms and R' is a hydrocarbyl radical having from 10 to 20 carbon atoms. Preferred compounds are those where the substituent R group has from 1 to 3 carbon atoms and R' has from 12 to 13 carbon atoms in the chain.

Both the R and R' radicals can be straight chain or branched and may be substituted with one or more typical noninterfering substituents such as halogen, cyano, trifluoromethyl, nitro or alkoxy.

The process also provides a motor fuel composition comprising a mixture of hydrocarbons in the gasoline boiling range containing a minor detergent and corrosion inhibiting amount of at least one of the above compounds. Preferably, the amount ranges from 20 to 200 parts per thousand barrels of fuel to provide both detergency and corrosion inhibiting effects. As little as 2.5 parts per 1,000 barrels will provide corrosion inhibition.

Example 1: *Preparation of anhydro-2-lauroyl-3-methyl-4-(N,N′-methyl-lauroyl-glycyl)-5-hydroxy-1,3-oxazolonium hydroxide* – 45 parts of N,N'-methyl lauroyl glycine were dissolved in 125 parts nitromethane. To the mixture, 35 parts of dicyclohexyl carbodiimide in 125 parts nitromethane was added. The mixture was heated at 43°C for 2 hours, then filtered. The filtrate was stripped to yield the desired product. The structure was confirmed by IR and NMR.

Example 2: The procedure of Example 1 was repeated using N,N'-methyl cocoyl glycine yielding anhydro-2-cocoyl-3-methyl-4(N,N'-methyl-cocoyl-glycyl)-5-hydroxy-1,3-oxazolonium hydroxides. Other compounds where R and R' are as above stated are prepared in similar fashion.

Any gasoline suitable for a spark-ignited, internal combustion engine can be used in the practice of this process. In general, the base fuel will consist of a mixture of hydrocarbons in the gasoline boiling range, i.e., boiling from about 75° to 450°F. The hydrocarbon components can consist of paraffinic, naphthenic, aromatic and olefinic hydrocarbons. This gasoline can be obtained naturally or it can be produced by thermal or catalytic cracking and/or reforming of petroleum hydrocarbons. The base fuel will generally have a research octane number above 85 and up to about 102 with the preferred range being from about 90 to 100.

Example 3: The additives of the process have anticorrosion properties as shown by their performance in the National Association of Corrosion Engineers (NACE) Rusting Test. In this test a determination is made of the ability of motor gasolines to inhibit the rusting of ferrous parts when water becomes mixed with gasoline. Briefly stated, the test is carried out by stirring a mixture of 300 ml of the test gasoline and 30 ml of water at 37.8°C with a polished steel specimen completely immersed therein for a test period of 3½ hours. The percentage of the specimen that has rusted is determined by comparison with photographic standards. Further details of the procedure appear in NACE Standard TM-01-72 and ASTM D6651 1P-135 (Procedure A).

The results of this test for a representative compound of the process, anhydro-2-lauroyl-3,4-(N,N'-methyl lauroyl glycyl)-5-hydroxy-1,3-oxazolonium hydroxide, at different concentrations in pounds per 1,000 barrels (PTB) in an unleaded base fuel show that as little as 2.5 PTB of the additive substantially eliminates rusting.

Polymeric Acid plus Lactono-Imidazoline Reaction Product

M.E. Davis and K.L. Dille; U.S. Patent 4,263,014; April 21, 1981; assigned to Texaco Inc. provide an antirust motor fuel composition comprising a major proportion of a mixture of hydrocarbons boiling in the gasoline boiling range, a minor amount of a polymeric acid comprising a dimer or trimer of a dienoic or trienoic acid containing from 16 to 18 carbon atoms and a minor amount of an aminoalkyl or polyalkyl polyamine imidazoline derivative of a hydrocarbon-substituted gamma or delta lactone reaction product.

The polymer acid component of the additive combination of the process comprises a dimer of a dienoic or trienoic acid containing from about 16 to 18 carbon atoms. Specific olefinic acids which can be employed are linoleic, linolenic, 9,11-octadecadienoic and eleostearic acids. Effective polymeric acids can be prepared from naturally occurring materials, such as linseed fatty acids, soy bean fatty acids and other natural unsaturated fatty acids. The preparation of polymeric acids is disclosed in U.S. Patent 2,632,659. Suitable polymeric acids are available commercially, such as Empol 1022 dimer acid, a dimer of linoleic acid. Dimer acid, such as dilinoleic acid or Empol 1022 is well known as a rust inhibitor for a motor fuel composition and will provide a gasoline qualifying under the NACE Rust Test when employed in a concentration of 6 PTB (pounds of additive per 1,000 barrels of fuel), i.e., about 0.002 weight percent.

The hydrocarbon-substituted lactono-imidazoline reaction product additive component of the process is a detergent additive for motor fuel composition. It is prepared in a two-step process, i.e., via the preparation of an intermediate reaction product followed by the preparation of the additive component as described below.

Example 1: *Polyisobutenylsuccinic anhydride reaction product* – To a solution of 126 g (0.05 mol) of crude polyisobutenylsuccinic acid (prepared from polyisobutene of about 1300 molecular weight and maleic anhydride by thermal alkenylation with about 50% unreacted polyisobutene) in a 50 weight percent mineral oil solution was added 1.25 g (0.0125 mol) of concentrated sulfuric acid.

The mixture contained about 0.0125 mol of sulfuric acid or about 0.025 mol of available protons. This mixture was held at 90°C for 3 hours. Infrared analysis of the product from the foregoing reaction showed a high conversion to five-and-six membered lactones, with the yield estimated to be greater than 85 mol percent.

Example 2: *Polyisobutenyl (1290) lactono-aminoethyl imidazoline product* – To 105.4 pounds of polyisobutenyl (1,290) lactono carboxylic acid in 132 pounds of xylene is added 8.3 pounds of diethylene triamine. The molar ratio of amine to acid is 1.46. The mixture is heated to 293°F and held there for 9 hours. About 3.3 pounds of water is collected. The reaction mixture is allowed to cool to room temperature to permit settling of sludge (amine salt of sulfuric acid). The reaction mixture is filtered, and the xylene removed from the filtrate under reduced pressure.

The additive analyses for this product were as follows: mod. neut. No., 2.2; sap. No., 10.9; TBN, 40.00; furol viscosity, 1,580; S, 0.064%; N, 1.89%; MW, 1900. The infrared absorptions at 5.65 microns (lactono) and 6.25 microns (imidazolino) show that the above additive is formed.

Example 3: The objective of the fuel composition of the process is to provide a motor fuel composition which will pass the National Association of Corrosion Engineers Standard TM-01-72 (NACE) Rust Test with a minimum amount of rust inhibitor additive. A passing rust rating in this test is a rating of "trace" or a trace amount of rust. If the major proportion of the treated motor fuel composition has a test rust rating of trace, the fuel composition is judged as having a passing antirust rating.

The fuel composition may contain any of the additives normally employed in gasoline. Thus, the fuel composition can contain an antiknock compound, anti-icing additives, dyes, upper cylinder lubricating oils and the like. The dimer or trimer acid component of the additive combination of the process is employed in a concentration range of from 0.000003 to 0.0003 weight percent, amounts corresponding to about 0.009 and 0.9 PTB. The preferred concentration of this component is from 0.000003 to 0.0001 corresponding to about 0.09 to 0.3 PTB.

The lactono-imidazoline component of the additive combination in the fuel composition is employed in a concentration range of about 0.0001 to 0.02 weight percent, which corresponds to about 3 and 50 PTB. The preferred concentration range from the lactono-imidazoline component is from 0.002 to 0.01 weight percent. Additive A was a 60% active solution of 2-polyisobutenyl (335 MW) lactono-1-aminoethylimidazoline in unreacted polyisobutylene.

The base fuel employed in the following tests was a lead-free gasoline having a Research Octane Number of about 91. This gasoline consisted of about 10% olefinic hydrocarbon and 60% paraffinic hydrocarbons and boiled in the range from about 90° to 370°F.

The antirust properties of the fuel composition of the process and of a comparison fuel composition in the NACE Rust Test is shown in the following table. The amount of additive added to the base fuel is given in pounds per thousand barrels (PTB).

Run	Dimer Acid* PTB	Additive A PTB	NACE Rust Rating Percent of Fuel With More than a Trace of Rust
1.	0.25	—	100
2.	—	10	78
3.	0.25	10	18

*Approximately 85% dimeric acids and 12% trimeric acids of C_{16-18} dienoic or trienoic acid, i.e., Emery Empol 1022.

The fuel composition of Run 1, wherein dimer acid was employed failed the NACE Rust Test 100% of the time. The fuel composition of Run 2 failed the NACE Rust Test 78% of the time. The fuel composition of Run 3, which is representative of this process was outstandingly effective with a NACE Rust Test pass-fail ratio of 82 to 18%.

Acyl Glycine Oxazolines

The process of *R.L. Sung; U.S. Patent 4,266,944; May 12, 1981; assigned to Texaco Inc.* relates to fuels comprising a mixture of hydrocarbons in the gasoline boiling range and acyl glycine oxazolines acting as detergents therein. The compounds of the process are defined by the following structure:

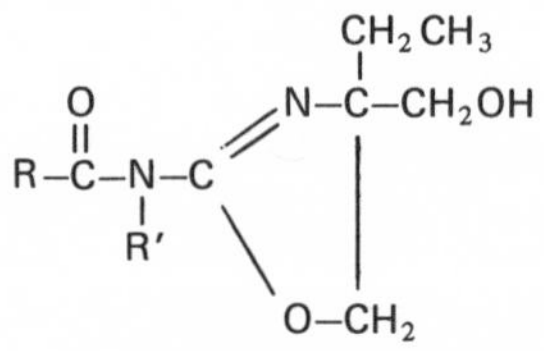

```
                      CH2CH3
                        |
      O              ,N—C—CH2OH
      ||           //     |
   R—C—N—C              |
         |    \           |
         R'    \          |
                 O—CH2
```

where R is lauryl, $C_{11}H_{23}$, oleyl or stearyl; R' is hydrogen or (lower) alkyl. Preferably R and R' taken together contain from 13 to 21 carbon atoms.

The process also provides a motor fuel composition comprising a mixture of hydrocarbons in the gasoline boiling range containing a minor detergent and corrosion inhibiting amount of at least one of the above compounds; preferably, this amount ranges from 20 to 200 parts per thousand barrels of fuel.

The compounds of the process preferably are synthesized by reacting a 2-amino-2-(lower) alkyl-1,3-propanediol with an N-acyl sarcosine in an inert solvent preferably xylene, refluxing the reaction mixture for about 8 hours to azeotrope the xylene and the water of reaction; filtering and stripping the filtrate under vacuum to isolate the product.

N-acyl sarcosines also known as "sarkosyls" previously suggested as corrosion inhibitors for fuels were found to be completely extracted into caustic water bottoms so that the fuels lost their corrosion inhibiting properties. Unexpectedly, it was found in accordance with this process that the oxazolines of these compounds were completely unextractable by acidic or basic water bottoms.

Example 1: *Synthesis of oxazoline of sarkosyl O* – A mixture of 0.7 mol of oleyl sarcosine and 0.7 mol of 2-amino-2-ethyl-1,3-propanediol in 600 parts of xylene was refluxed and water of reaction was azeotroped over. After 8 hours of reflux, the reaction mixture was cooled and filtered, then stripped under vacuum. The residue was analyzed by IR and elemental analysis.

Example 2: A mixture of 0.7 mol of lauroyl sarcosine and 0.7 mol of 2-amino-2-ethyl-1,3 propanediol in 600 ml of xylene was refluxed and water of reaction was azeotroped over. At the end of 8 hours, the reaction was stripped, filtered, and stripped under vacuum. The residue was analyzed by IR and elemental analysis.

Example 3: The additives of the process have anticorrosion properties as shown by their performance in the National Association of Corrosion Engineers (NACE) Rusting Test (See Example 3 of U.S. Patent 4,257,780).

The table below shows the results of this test for a representative compound of the process at different concentrations in pounds per 1,000 barrels (PTB) in and against an unleaded base fuel. The date of the table below show that as little as 5 PTB of the additive substantially eliminates rusting.

Concentration (PTB)	Rust Rating (%)
10	Trace to 1
10	Trace to 1
5	Trace to 1
5	Trace to 1
Unleaded base fuel	50 to 100
Unleaded base fuel	50 to 100

The fuels of the process may contain any additive conventionally employed in gasoline. Tetraalkyl lead, antiknock additives, dyes, corrosion inhibitors, antioxidants and the like can be beneficially employed without materially affecting the additive of the process.

N-Substituted 1,2,3,4-Tetrahydropyrimidines and Hexahydropyrimidines

The process of *B.A.O. Alink; U.S. Patent 4,281,126; July 28, 1981; assigned to Petrolite Corporation* relates to the preparation of N-substituted 1,2,3,4-tetrahydropyrimidines and hexahydropyrimidines. The compositions have a wide variety of uses, including their use as corrosion inhibitors, biocides, scale inhibitors and fuel additives. The tetrahydropyrimidines have the formula shown on the following page.

where R_1 and R_2 are substituted groups and R_3, R_4, R_5 and R_6 are hydrogen, alkyl, alkenyl, aryl, aralkyl, cycloalkyl or heterocyclic. The hexahydropyrimidines have the formula:

where R_1 and R_2 are alkyl, alkenyl, aryl, cycloalkyl, aralkyl or heterocyclic groups or amino substituted derivatives thereof, R_4 and R_3 are hydrogen or substituted groups or may be joined to form a cyclic group and R_5 and R_6 are hydrogen or substituted groups or may be joined to form a cyclic group.

Example 1: *1,3,4,4,6-pentamethyl 1,2,3,4-tetrahydropyrimidine* – A sample of 196 g of mesityl oxide was cooled in an ice bath. Over a 30 minute period a sample of 310 g of 40% aqueous methylamine solution was added at such a rate that the reaction temperature did not exceed 28°C. After the addition was completed, stirring was continued for 10 minutes.

To the cooled reaction mixture was added over a 15 minute period, a sample of 162 g of a 37% formaldehyde solution while maintaining a reaction temperature below 29°C. To the homogeneous solution was added 20 g of sodium hydroxide pellets. The organic layer which separated was removed and the aqueous layer extracted with ether. The ethereal solution was combined with the organic phase and evaporated under diminished pressure.

Distillation yielded a fraction of 218 g, BP (20 mm) 65° to 85°C, which was redistilled to yield a fraction BP (20 mm) of 69° to 75°C; identified as 1,3,4,4,6-pentamethyl-1,2,3,4-tetrahydropyrimidine by the infrared spectrum and NMR spectrum.

Example 2: *1,3,4,4,6-pentamethyl hexahydropyrimidine* – A sample of 55.2 g of 1,3,4,4,6-pentamethyl 1,2,3,4-tetrahydropyrimidine prepared as described in Example 1, was dissolved in 128 g of absolute ethanol. To the stirred mixture was added 8 g of sodium tetrahydridoborate, ($NaBH_4$), in three portions over a 1½ hr period.

The mixture was stirred for 18 more hours and the solvent removed under diminished pressure. The remaining product was treated with water containing a small amount of ammonium chloride and extracted with ether. The ethereal solution was evaporated under diminished pressure and the remaining product distilled under diminished pressure, to yield 36.8 g, BP (20 mm) 73° to 76°C, of a product identified as 1,3,4,4,6-pentamethyl hexahydropyrimidine by its NMR spectrum.

Additives for Gasohol

Fuel compositions typified by gasohol and alcohols which are to be considered for commercial use must possess low corrosion activity; and this may be effected by addition thereto of various corrosion inhibition systems.

The fuel for internal combustion engines which may be treated by the process of *R.L. Sung; U.S. Patent 4,282,007; August 4, 1981; assigned to Texaco Inc.* may contain (1) at least one alcohol selected from the group consisting of ethanol and methanol and (2) gasoline in amount of 0 to 50 v/v of alcohol. The fuel may be an alcohol-type fuel containing little or no hydrocarbon. Typical of such fuels are methanol, ethanol, mixtures of methanol-ethanol, etc. Commercially available mixtures may be employed. Illustrative of one such commercially available mixture may be that having the following typical analysis.

Table 1

Component	Parts
Ethanol	3,157.2
Methyl isobutyl ketone	126.3
Acetic acid	0.256
Methyl alcohol	0.24
Isopropyl alcohol	0.2
n-Propyl alcohol	0.162
Ethyl acetate	0.2

The fuels which may be treated by this process include gasohols which may be formed by mixing 90 to 95 volumes of gasoline with 5 to 10 volumes of ethanol or methanol. A typical gasohol may contain 90 volumes of gasoline and 10 volumes of absolute ethanol.

The fuels to be treated by the process may be substantially anhydrous, i.e., they contain less than about 0.3 vol % water.

In accordance with this process, there may be added to the fuel a minor corrosion inhibiting amount of, as a corrosion inhibiting agent, a reaction product of an aminotriazole and a C_{15-30} hydrocarbyl succinic acid anhydride. The aminotriazoles which may be employed may be 3-amino-1H-1,2,4-triazole, 4-amino-1,2,4-triazole, and 5-amino-1H-1,2,4-triazole.

The corrosive nature of the formulated products may be readily measured by the Iron Strip Corrosion Test (ISCT). In this test, an iron strip (12 x 125 x 1 mm) is prepared by washing in dilute aqueous hydrochloric acid to remove mill scale, then with distilled water to remove the acid, then with acetone, followed by air drying. The strip is then polished with #100 emery cloth.

The polished strip is totally immersed in 110 ml of the test liquid in a 4 ounce bottle for 15 minutes at room temperature of 20°C. 20 ml of the test liquid is poured off and replaced with 20 ml of distilled water. The bottle is shaken as the sample is maintained for 3 hours at 90°F. The percent rust on the strip is determined visually. A second reading is taken after 40 hours. In the following examples all parts are parts by weight unless otherwise specified.

Example 1: In this example which illustrates the mode of practicing the process, there is added to 96 parts of the anhydrous alcohol composition of Table 1, 4 parts of distilled water and 7.68 ppm (2 PTB, i.e., lb additive/1,000 bbl fuel) of the reaction product of equimolar amounts of (1) 3-amino-1H-1,2,4-triazole and (2) polyisobutenyl ($\overline{M}_n$ of 335, corresponding to about C_{24} chain length) succinic acid anhydride, the latter having been prepared by reaction of polyisobutylene and maleic acid anhydride. The reaction mixture is refluxed for 6 hours at 140°C in an excess of xylene solvent; the solvent is stripped off after completion of reaction. The resulting fuel composition was tested in the ISCT; and the Rust and Corrosion Rating determined after 40 hours.

Example 2: *Control* – The procedure of Example 1 was duplicated except that the additive was the reaction product of 3-amino-1H-1,2,4-triazole and polyisopropenyl ($\overline{M}_n$ of 168, corresponding to a C_{12} chain) succinic acid anhydride, the latter having been prepared by reaction of polypropylene and maleic acid anhydride. The fuel composition was tested in the ISCT.

Example 3: *Control* – The procedure of Examples 1 and 2 was duplicated except that the additive was 5-amino-1H-1,2,4-triazole.

Example 4: *Control* – The procedure of Examples 1 through 3 was duplicated except that the additive was 76 PTB of a prior art standard commercial rust and corrosion inhibitor.

Example 5: *Control* – The procedure of Examples 1 through 4 was duplicated except that no additive was present, only 4 parts of distilled water. The results of the Iron Strip Corrosion Test were as follows.

Table 2

Example	40 Hour Rust and Corrosion Rating (%)
1	1*
2	40
3	20
4	50
5	50

*Trace.

From the above table, it will be apparent that the system of Example 1, prepared in accordance with this process, showed less than about 1% rust and corrosion. Control Examples 2 through 5 showed 20 to 50% rust and corrosion which is unsatisfactory.

It is noted that if the additive is made from a higher molecular weight polypropylene than that used in Example 2, e.g., one having a chain length of, e.g., C_{15}, the results will be comparable to those attained with Example 1.

Examples 6 through 8: Results comparable to those of Example 1 may be obtained if the aminotriazole reactant is as follows.

Table 3

Example	Reactant
6	4-amino-1,2,4-triazole

(continued)

Table 3: (continued)

Example	Reactant
7	5-amino-1H-1,2,4-triazole
8	3-amino-1H-1,2,4-triazole*

*But in mol ratio of 2:1 with respect to the polyisobutenyl succinic acid anhydride.

Examples 9 through 11: Results comparable to those of Example 1 may be obtained if the fuel is as follows.

Table 4

Example	Fuel
9	Gasohol*
10	Absolute ethanol
11	Absolute methanol

*Containing 90 vol % gasoline and 10 vol % absolute ethanol.

In this process of *R.L. Sung; U.S. Patent 4,282,008; August 4, 1981; assigned to Texaco Inc.* the fuels treated are identical to those of the preceding patent. In this case there may be added to the fuel a minor corrosion inhibiting amount of, as a corrosion inhibiting agent, a reaction product of (1) an aminotriazole, (2) an isatoic anhydride, and (3) a C_{3-12} polyprimary amine bearing at least one free $-NH_2$ group and at least one $-NHR'$ group wherein R' is a C_{12-18} hydrocarbon group.

The aminotriazoles which may be employed include 5-amino-1,2,3-triazoles, 4-amino-1,2,4-triazoles, 3-amino-1,2,4-triazoles, and 5-amino-1,2,4-triazoles, including those bearing inert substituents, typified by hydrocarbon or alkoxy groups, which do not react in the instant reaction.

The isatoic anhydride which may be employed in the process may be characterized by the following formula.

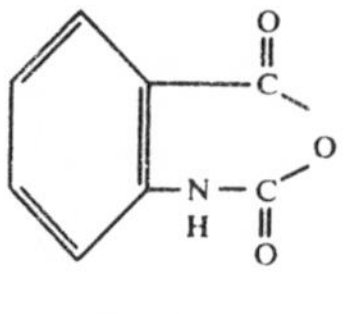

This charge material may bear inert substituents (which do not interfere with the reaction) on the nitrogen atom or on the ring. In the following examples, all parts are parts by weight unless otherwise specified.

Example 1: In this example which illustrates the best mode of practicing the process, there are added to a reaction vessel containing 200 parts of dimethyl formamide solvent, equimolar amounts of isatoic anhydride (33 parts), Duomeen T brand of N-monotallow-1,3-propane diamine (74.8 parts), and 5-amino-1H-1,2,4-triazole (8 parts). The reaction mixture is heated to reflux for 24 hours, filtered hot and then stripped of solvent by distillation.

The additive so-prepared (38.4 ppm corresponding to 10 PTB) is added to 96 parts of the anhydrous alcohol composition of Table 1 (see above patent) and 4 parts of distilled water and the resulting composition was tested in the ISCT to determine the rust and corrosion rating after 40 hours.

Example 2: *Control* – The procedure of Example 1 was duplicated except that the additive was 76 PTB of a commercial rust and corrosion inhibitor and only 3 parts of distilled water is added. The fuel composition was tested in the ISCT.

Example 3: *Control* – The procedure of Examples 1 and 2 was duplicated except that no additive was present, only 4 parts of distilled water. The results of the Iron Strip Corrosion Test were as follows.

Example	40 Hour Rust and Corrosion Rating (%)
1	0
2	25
3	50

Examples 4 through 6: Results comparable to those of Example 1 may be obtained if the fuel is as follows.

Example	Fuel
4	Gasohol*
5	Absolute ethanol
6	Absolute methanol

*Containing 90 vol % gasoline and 10 vol % absolute ethanol.

The fuels treated in this process of *R.L. Sung; U.S. Patent 4,294,585; Oct. 13, 1981; assigned to Texaco Inc.* are the same as those of the two preceding patents. Here the additive used as a corrosion inhibiting agent is a reaction product of (1) an aminotetrazole, (2) an aldehyde or a ketone, and (3) a C_{3-12} polyprimary amine bearing at least one free $-NH_2$ group and at least one $-NHR'$ group wherein R' is a C_{12-18} hydrocarbon group.

The aminotetrazoles which may be employed include 1-aminotetrazoles, 2-aminotetrazole, 3-aminotetrazoles, 4-aminotetrazoles, and 5-aminotetrazoles, including those bearing inert substituents which do not react in the instant reaction typified by hydrocarbon or alkoxy groups. The preferred is 5-amino-1H-tetrazole.

The aldehyde or ketone which may be employed may be one bearing aldehyde and/or ketone groups on a hydrocarbon backbone which later may be derived from alkyl, aryl, alkaryl, aralkyl, cycloalkyl hydrocarbons. Preferred are the C_{1-8} aldehydes; most preferred is formaldehyde which may be employed in 37% aqueous solution or as its trimer paraformaldehyde. In the following examples all parts are parts by weight unless otherwise specified.

Example 1: In this example which illustrates the best mode of practicing the process, there is added to 96 parts of the anhydrous alcohol composition of Table 1 (see U.S. Patent 4,282,007), 4 parts of distilled water and 38.4 parts (corresponding to 10 PTB) of, as additive, the reaction product of equimolar amounts of (1) the Duomeen T brand of N-monotallow-1,3-propane diamine

(74 parts), (2) 5-amino-1,4-tetrazole and (3) 37% formaldehyde (32 parts) which had been refluxed for 105 minutes in 60 parts of absolute methanol and 60 parts of xylene, the product being filtered hot and then stripped of solvent. The resulting final composition was tested in the ISCT, and the rust and corrosion rating determined after 40 hours.

Example 2: The procedure of Example 1 was duplicated except that the additive was 76 PTB of a commercial rust and corrosion inhibitor.

Example 3: The procedure of Example 1 was duplicated except that no additive was present, only 4 parts of distilled water. The results of the Iron Strip Corrosion Test were as follows.

Example	40 Hour Rust and Corrosion Rating (%)
1	0
2	25
3	50

From the above table, it will be apparent that the system of Example 1, prepared in accordance with this process, showed no rust and corrosion. Control Examples 2 and 3 showed 25 to 50% rust and corrosion which is unsatisfactory.

Examples 4 through 6: Results comparable to those of Example 1 may be obtained when the amine reacted is as follows.

Example	Amine
4	Duomeen O brand of N-oleyl-1,3-propane diamine
5	Duomeen S brand of N-stearyl-1,3-propane diamine
6	Duomeen C brand of N-cocoyl-1,3-propane diamine

Examples 7 through 10: Results comparable to those of Example 1 may be obtained when the aldehyde or ketone reactant is as follows.

Example	Reactant
7	Acetaldehyde
8	Propionaldehyde
9	Butyraldehyde
10	Cyclohexylaldehyde

Examples 11 through 14: Results comparable to those of Example 1 may be obtained when the aminotetrazole reactant is as follows.

Example	Aminotetrazole
11	1-aminotetrazole
12	2-aminotetrazole
13	3-aminotetrazole
14	4-aminotetrazole

Examples 15 through 17: Results comparable to those of Example 1 may be obtained if the fuel is as follows.

Example	Fuel
15	Gasohol*
16	Absolute ethanol
17	Absolute methanol

*Containing 90 vol % gasoline and 10 vol % absolute ethanol.

In these three patents, rust and corrosion inhibitors may be added to a fuel in 0.25 to 25 PTB amount, preferably 10 PTB. Alternatively expressed, the inhibitor may be added to a fuel in minor corrosion-inhibiting amount of 0.0001 to 0.01 wt %, preferably 0.004 wt %. Larger amounts may be used but may not be necessary.

The fuels that may be treated in the process of *R.L. Sung; U.S. Patent 4,376,635; March 15, 1983; assigned to Texaco Inc.* are again the same as those of the three preceding patents. In this case, the additive is the reaction product of (1) a benzotriazole, (2) an aldehyde or a ketone, and (3) a C_{3-12} polyprimary amine bearing at least one free $-NH_2$ group and at least one $-NHR'$ group wherein R' is a C_{12-18} hydrocarbon group. The benzotriazole which may be employed include those bearing inert substituents, which do not react in the instant reaction, typified by hydrocarbon or alkoxy groups. In the following examples all parts are parts by weight unless otherwise specified.

Example: In this example which illustrates the best mode known of practicing this process, the additive is prepared by adding 32 parts of 37% formaldehyde over 105 minutes to a refluxing mixture of 74 parts of the Duomeen T brand of N-monotallow-1,3-propane diamine (corresponding to the formula:

$$R'NHCH_2CH_2CH_2NH_2$$

wherein R' is a straight chain C_{18} alkyl group) and 24 parts of benzotriazole in 60 parts of absolute methanol and 60 parts of xylene. The reaction product is filtered hot and then stripped of solvent.

The additive so prepared (7.68 ppm, corresponding to 2 PTB) is added to 96 parts of the anhydrous alcohol composition of Table 1 (U.S. Patent 4,282,007), and 4 parts of distilled water and the resulting composition was tested in the ISCT to determine the rust and corrosion rating after 40 hr; a rating of zero was found.

Comparable results to the above example may be obtained when the amine reacted is Duomeen O brand of N-oleoyl-1,3-propane diamine, Duomeen S brand of N-stearyl-1,3-propane diamine, and Duomeen C brand of N-cocoyl-1,3-propane diamine.

Comparable results to the example may be obtained when the aldehyde or ketone reactant is acetaldehyde, propionaldehyde, butyraldehyde and cyclohexylaldehyde.

Comparable results to the example may be obtained when the benzotriazole reactant is 4-methyl benzotriazole, 5-methyl benzotriazole, 6-methyl benzotriazole, 7-methyl benzotriazole and 4-methoxy benzotriazole.

Comparable results to the example may be obtained if fuel is gasohol containing 90 vol % gasoline and 10 vol % absolute ethanol, absolute ethanol and methanol.

The rust and corrosion inhibitors may be added to a fuel in amounts of 0.25 to 25 PTB, more preferably 1 to 5 PTB.

Hydrocarbyl Alkoxy Amino Alkylene-Substituted Asparagines

S. Herbstman and P. Dorn; U.S. Patent 4,290,778; September 22, 1981; assigned to Texaco Inc. provide a primary hydrocarbyl alkoxy amino alkylene-substituted asparagine represented by the formula:

$$\begin{array}{l} CH_2-\overset{O}{\overset{\|}{C}}-\overset{H}{\overset{|}{N}}-C_3H_6-\overset{H}{\overset{|}{N}}-[C_3H_6O]_x-R \\ \quad | \\ {}^-OOC-CH-\underset{\underset{H}{|}}{\overset{H}{\overset{|}{N^+}}}-C_3H_6-\overset{H}{\overset{|}{N}}-[C_3H_6O]_x-R \end{array}$$

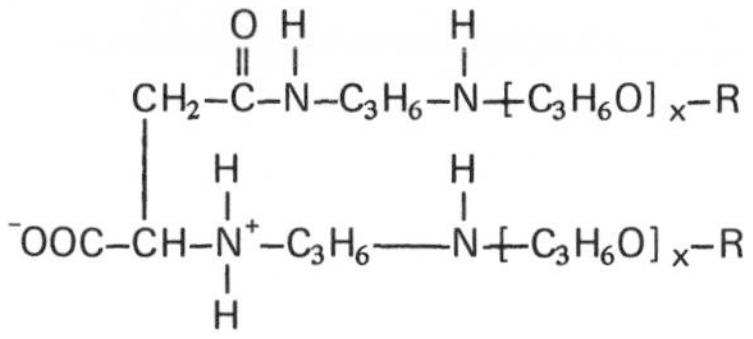

in which R is a primary hydrocarbon radical having from about 8 to 24 carbon atoms and x has a value from 1 to 10.

The compound, which is produced by reacting about 2 mols of a hydrocarbyl alkoxy alkylene diamine with a mol of maleic anhydride to produce a compound characterized by having a plurality of alkoxy and amino groups, exhibits surprising corrosion inhibiting properties as well as essential carburetor detergency properties when employed in gasoline.

The fuel composition of the process prevents or reduces corrosion problems during the transportation, storage and the final use of the product. The gasoline of this process also has highly effective carburetor detergency properties. When a gasoline of the process is employed in a carburetor which already has a substantial buildup of deposits from prior operations, a severe test of the carburetor detergency property of a fuel composition, this motor fuel is effective for removing substantial amounts of the preformed deposits.

The additive is effective in an amount ranging from about 0.0002 to 0.2 wt % based on the total fuel composition. An amount of the neat additive ranging from about 0.001 to 0.01 wt % is preferred, with an amount from 0.001 to 0.007 being particularly preferred, the latter amounts corresponding to about 3 to 20 PTB (pounds of additive per 1,000 barrels of gasoline) respectively.

Example 1: 16 g of maleic anhydride (0.159 mol) are suspended in 115.6 g mineral oil having an SUS at 100°F of 100 and with stirring and nitrogen purge is heated at 100°C for 1 hour. N-1-[n-C_{16-18} alkyl (isopropoxyl)$_{4.1}$]-1,3-propane diamine, 100 g (0.32 mol) is introduced into the oil solution at 100°C over 0.5 hr. The reaction mixture is stirred at 100°C for an additional 2 hr. The reaction product was filtered hot to yield 231 g of a pale yellow liquid. Analysis of the 50% oil solution of the additive was as follows: N, wt % 2.8. The compound produced is represented by the formula:

$$\begin{array}{l} CH_2-\overset{O}{\overset{\|}{C}}-\overset{H}{\overset{|}{N}}-CH_2CH_2CH_2-\overset{H}{\overset{|}{N}}-[CH(CH_3)CH_2O]_{4.1}-CH_2(CH_2)_{14-16}-CH_3 \\ \quad | \\ {}^-OOC-CH-\underset{\underset{H}{|}}{\overset{H}{\overset{|}{N^+}}}-CH_2CH_2CH_2-\overset{H}{\overset{|}{N}}-[CH(CH_3)CH_2O]_{4.1}-CH_2(CH_2)_{14-16}-CH_3 \end{array}$$

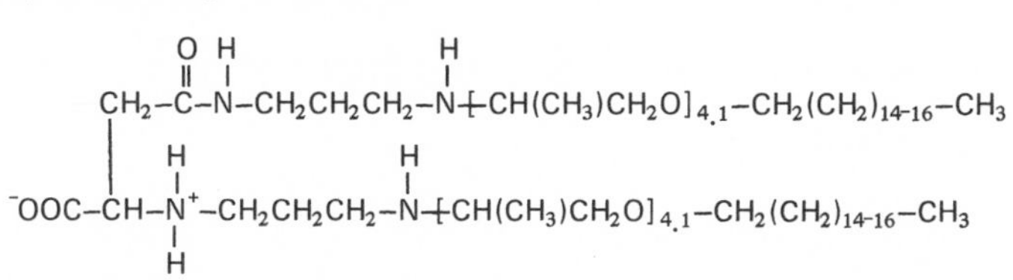

Example 2: 24.5 g (0.25 mol) of maleic anhydride were added to 197.8 g of mineral oil having an SUS at 100°F of 100 and heated to about 100°C under a nitro-

gen atmosphere. 173 g, 0.5 mol of N-1-(n-C_{12-14} alkoxy propylene)-1,3-propane diamine were added over 1 hr at 90° to 100°C. The reaction conditions were maintained for 2 hr at which time the reaction product was cooled and filtered to recover a light yellow liquid. As a 50 wt % solution in mineral oil, the reaction product had the following analysis:

TBN	113
TAN	33.2
% N, wt %	3.5

This product is represented by the formula:

```
        O H             H
        ‖ |             |
     CH2–C–N–CH2CH2CH2–N–CH2CH2CH2–O–(CH2)11-13CH3
     |
     |  H               H
     |  |               |
 ⁻OOC–CH–N⁺–CH2CH2CH2–N–CH2CH2CH2–O–(CH2)11-13CH3
         |
         H
```

The base fuel employed with the additive of this process in the following tests was an unleaded grade gasoline having a Research Octane Number of about 93. This gasoline consisted of about 33% aromatic hydrocarbons, 7% olefinic hydrocarbons and 60% paraffinic hydrocarbons and boiled in the range from 90° to 375°F.

The rust inhibiting properties of fuel compositions of the process was determined in the NACE Test (National Association of Corrosion Engineers) which is a modification of ASTM Rust Test D-665-60 Procedure A. In the NACE Test, a steel spindle is polished with nonwaterproof fine emery cloth. The spindle is immersed in a mixture containing 300 cc fuel and 30 cc distilled water and is rotated at 100°F for 3.5 hr. The spindle is then rated visually to determine the amount of rust formation. A passing result is an average of less than 5% rust. The results of this test are set forth in the table below.

Run	Additive	Concentration, PTB*	Percent Rust
1	Ex. 1	10.0	Trace to 1
2	Ex. 2	10.0	Trace

*PTB is pounds of additive per 1,000 barrels of fuel (unleaded gasoline).

Glycidyl Ether-Polyamine Reaction Products

According to *M.E. Childs; U.S. Patent 4,295,860; October 20, 1981; assigned to UOP Inc.* the reaction products of glycidyl ethers, wherein the alkoxy portion contains from about 6 to 20 carbon atoms, with alkylenediamines, N-alkyl alkylenediamines and N-alkoxyalkyl alkylenediamines are effective carburetor detergents and reduce deposits on various components of internal combustion engines.

Such materials may have additional desirable properties when added to gasoline and used in internal combustion engines. For example, materials of this process may inhibit gum formation, may act as antistalling or carburetor deicing agents, as corrosion inhibitors, and so forth.

The materials may be advantageously employed as a detergent in a broad variety of fuel oil, for example, diesel oil, aviation oil, gasoline, burner oil, etc., although their use in gasoline is particularly advantageous. Such materials when used as additives exhibit detergent properties at concentrations at least as low as 25 ppm, although some may be useful at concentrations as low as 10 ppm. It has been found that gasoline containing from about 15 to 100 ppm of additives of this process has advantages in preventing deposits on working parts of carburetors and on the valves and ports of internal combustion engines.

Example 1: A solution of a glycidyl ether where the alkoxy group was derived from fatty acids containing mainly 12 to 14 carbon atoms (60 g, 0.2 mol) and ethylenediamine (6 g, 0.1 mol) in 70 g mixed xylenes was heated at a temperature in the range from about 125° to 135°C until reaction was complete. The reaction was monitored by following the disappearance of glycidyl ether using gas-liquid partition chromatography (glpc) and generally was complete in 4 to 8 hours. The infrared spectrum of the resulting product showed the presence of a hydroxyl, whereas the nmr spectrum showed the absence of absorption at 3.0 δ characteristic of the glycidyl ether.

Example 2: A suspension of N-tallow-1,3-propylenediamine (33 g, 0.1 mol), the glycidyl ether of Example 1 (30 g, 0.1 mol) and 63 g mixed xylenes were heated at a temperature in the range from about 125° to 135°C until glpc examination showed the absence of glycidyl ether, generally complete in about 6 hr.

Example 3: The experimental procedure was analogous to that described in Example 1, using a glycidyl ether of formula

$$H_{25-29}C_{12-14}OCH_2\overset{O}{\overbrace{CH-CH_2}}$$

(25 g, 0.1 mol), an alkoxypropyl-1,3-propylenediamine where the alkoxy group was comprised of chains containing 12 to 15 carbon atoms (35 g, 0.1 mol) and 60 g xylene. The mixture was heated at a temperature range from about 125° to 135°C until glpc showed the disappearance of glycidyl ether.

Example 4: To show the effect of phenol in reducing reaction times, two experiments were conducted. In both the glycidyl ether was that described in Example 1 (11.6 g, 0.039 mol) and the amine was 1,3-propylenediamine (1.5 g, 0.02 mol) in 13 g mixed xylenes. The reaction temperature was maintained at 100°C, and the disappearance of the glycidyl ether was followed by glpc. To one reaction mixture was added 1 g phenol; to the other reaction mixture no phenol was added. The time for complete disappearance of ether in the absence of phenol was 3 hr; the time for complete disappearance of ether in the presence of phenol was less than 15 minutes.

This experiment shows that phenols can reduce reaction times by a factor of more than 12. Acceleration also was observed upon addition of tert-butyl hydroxyanisole, nonylphenol, and a mixture of various mono- and di-tert-butylphenols.

Trimeric Acid for Alcohol Motor Fuel

Automobiles having conventional internal combustion gasoline engines can be adapted to run on a liquid aliphatic alcohol fuel composition. However, serious corrosion problems have been encountered from the use of fuel compositions containing significant amounts of an alcohol.

A rust-inhibited alcohol fuel composition comprising a major proportion of a lower aliphatic alcohol and a minor rust-inhibiting amount of a trimeric acid produced by the condensation of an unsaturated aliphatic monocarboxylic acid or a hydroxy aliphatic monocarboxylic acid having between about 16 and 18 carbon atoms per molecular is provided in a process by *M.E. Davis and K.L. Dille; U.S. Patent 4,305,730; December 15, 1981; assigned to Texaco Inc.* The alcohol composition of this process prevents or mitigates the problem of carburetor and fuel system corrosion which is critical in this type of fuel composition. 0.0005 to 0.1 wt % of the additive is used.

Many naturally occurring fatty acids such as linseed fatty acids, soya bean fatty acids and the like can be polymerized to produce a trimer acid following the procedure disclosed in U.S. Patent 2,482,761.

This process can also result in the formation of dimeric acids. In general, it has been found that dimeric acids do not impart a high level of corrosion inhibition to a liquid alcohol fuel composition. However, a minor amount of dimeric acid coproduced along with the prescribed trimeric acid, or left remaining mixed therewith following a separation process, that is, an amount of dimeric acid ranging from between 2 to 20%, with the balance being the trimeric acid, which is generally representative of the trimeric acid products available in commerce, can be employed in the alcohol composition of the process. The additional cost of removing the dimeric acid is not justified simply to avoid dilution of the prescribed trimeric acid additive.

Commercially available trimer acid or triethenoid acid is Empol 1040 Trimer Acid. This acid is produced by the polymerization of unsaturated C_{18} fatty acids and is essentially a mixture of about 80% trimer acid and about 20% dimer acid and some residual monobasic acid.

The alcohol base for the composition of this process is a lower aliphatic alcohol having generally from 1 to 4 carbon atoms. The alcohol base for the composition should consist of at least about 90% of alcohol. A preferred concentration of alcohol in the alcohol base is from 95 to 99.8% with the most preferred concentration being from about 98 to 99.5%. The balance of the base composition can consist of water and minor amounts of such impurities which are normally coproduced during the manufacture of the alcohol, namely, acids, formaldehydes and other alcohols. It will be appreciated that the alcohol compositions of this process will be a fuel grade alcohol and will correspond to technical or commercial grades of alcohol.

Example: The corrosion-inhibiting properties of the alcohol fuel composition of the process and of a comparison fuel composition was determined in the Carburetor Metal Corrosion Test described below.

In this test, a clean strip of carburetor zinc metal is placed in a 120 ml tall form bottle. 50 ml of fuel grade ethanol is added to the bottle covering about one-half of the metal strip. 11 ml of distilled water are added to the bottle and the contents mixed by gently swirling for a few seconds. The bottle is stoppered and then stored in the dark at room temperature. Corrosion of the wetted metal surfaces is visually rated after 7 and 14 days storage.

A fuel grade ethyl alcohol consisting of about 92.5% ethanol, about 7.5% water and minor amounts of impurities including a maximum of 5.0 mg/100 ml of residue, a maximum of 3.0 mg/100 ml acetic acid, a maximum of 6.0 mg/100 ml of aldehyde, a maximum of 8.0 mg/100 ml of esters and a maximum of 6.0 mg/100 ml of higher alcohols was employed to evaluate the effectiveness of the fuel composition of the process. The additive employed in the fuel composition of the process was trimer acid (Empol 1040) which consists of about 80% of the trimer of linoleic acid.

The comparison additive employed in the examples below was dimer acid (Empol 1022) which consists of about 75% of the dimer of linoleic acid. The corrosion test results are set forth in the table below.

	. . .% Surface Corrosion. . .	
	7 Days	**14 Days**
Dimer acid (Empol 1022) 2 PTB*	45	100
Trimer acid (Empol 1040) 2 PTB	1	25

*PTB is pounds of additive per 1,000 barrels of fuel.

The foregoing tests demonstrate the effectiveness of trimer acid as a carburetor metal corrosion-inhibiting additive in a fuel grade ethyl alcohol composition.

Aminoalkylimidazoline Derivative of a Sarcosine

The detergent and anticorrosion additive composition of the process of *R.L. Sung and P. Dorn; U.S. Patent 4,305,731; December 15, 1981; assigned to Texaco Inc.* is a 1-aminoethylimidazoline derivative of a sarcosine. It is represented by the following formula:

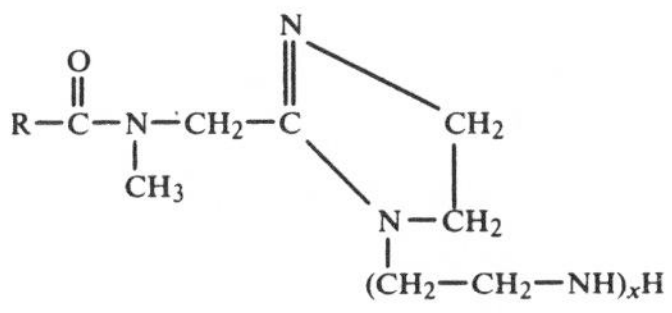

N
O
R—C—N—CH$_2$—C CH$_2$
CH$_3$
N—CH$_2$
(CH$_2$—CH$_2$—NH)$_x$H

in which R is a monovalent hydrocarbon radical having from about 10 to 20 carbon atoms and x has a value from 1 to 3.

The motor fuel composition of this process which has improved detergency and anticorrosion characteristics comprises a mixture of hydrocarbons boiling in the gasoline boiling range and a minor amount of the prescribed imidazoline derivative of a sarcosine.

The additive composition is employed in a concentration ranging from about 0.0002 to about 0.2 wt % based on the weight of the motor fuel composition.

Example 1: *1-Aminoethylimidazoline derivative of N-oleoyl sarcosine* – A mixture of 0.8 mol of N-oleoyl sarcosine, 0.8 mol of diethylene triamine (DETA) and 11 mols of xylene was refluxed and the water of the reaction separated by an azeotrope distillation over a reaction period of 8 hr. The reaction product was then stripped of solvent under vacuum and recovered. The product, 1-aminoethylimidazoline of N-oleoyl sarcosine, was identified by infrared and elemental analysis. The results of these analyses are as follows.

% N	13
Molecular weight	483

Example 2: *1-Aminoethylimidazoline derivative to N-lauroyl sarcosine* – A mixture of 0.8 mol of lauryl sarcosine, 0.8 mol of diethylene triamine (DETA), and 11 mols of xylene is refluxed and the water of the reaction is separated by an azeotrope distillation over the 8 hr reaction period. At the end of that period the reaction product is stripped of solvent under vacuum and recovered. The product, 1-aminoethylimidazoline of N-lauroyl sarcosine, is then analyzed by infrared and elemental analysis. The results of these analyses are as follows.

% N	17
Molecular weight	318

Gasoline blends were prepared from a typical base fuel mixed with specified amounts of the fuel additive of the process. These fuels were then tested to determine the effectiveness of the additive in gasoline. The base fuel employed for demonstrating the effectiveness of the additive composition of the process was an unleaded grade gasoline having a research octane number of about 93. This gasoline consisted of about 30% aromatic hydrocarbon, 8% olefinic hydrocarbon and 62% paraffinic hydrocarbon and boiled in the range from 100° to 380°F.

The rust inhibiting effect of the fuel composition of this process was determined in the National Association of Corrosion Engineers Test (NACE). In this test a mixture of 300 ml of test gasoline and 30 ml distilled water is stirred at a temperature of 37.8°C (100°F) with a steel specimen completely immersed therein for a test period of 3½ hours. The percentage of the specimen that has rust is determined visually and noted. The results of this test are set forth in the following table.

Additive in Unleaded Base Fuel	% Rust*
None	50-100 ck 50-100
5 PTB** of the product of Ex. 1	Tr-1***, Tr-1
10 PTB of the product of Ex. 1	Tr, Tr
76 PTB commercial rust inhibitor	Tr, Tr

*Less than 5% passes test.
**PTB is pounds of additive per 1,000 barrels of fuel.
***Tr is trace.

Tridecyl(Oxypropyl)Amine for Alcohols

In the process of *R.L. Sung and G.J. Sidote; U.S. Patent 4,321,060; March 23, 1982; assigned to Texaco Inc.* alcohols may be inhibited against corrosion by addition thereto of an ether amine.

The alcohol compositions which may be treated by the process may include C_{1-12} alkanols such as water-soluble alkanols including C_{1-4} alcohols. Preferably, the alcohols include methanol, ethanol, propanols, etc. The alcohols may include mixtures of alcohols with each other and/or with other compositions including ketones, esters, hydrocarbons, etc. The alcohol may be in the form of gasohol, a mixture commonly containing 80 to 95 vol %, say 90 vol % gasoline and 5 to 20 vol %, say 10 vol % alcohol. The alcohol may contain water, typically 5 wt %, but preferably it will be anhydrous.

There may be added to the alcohol a minor effective corrosion-inhibiting amount of, as a corrosion-inhibiting additive, an amine having the formula

$$(R-O-R'')_a-NH_{3-a}$$

wherein R contains 1 to 30 carbon atoms and is selected from the group consisting of alkyl, alkenyl, alkaryl, aralkyl, cycloalkyl, and aryl groups and R'' is a divalent hydrocarbon group containing 1 to 30 carbon atoms and is selected from the group consisting of alkyl, alkenyl, alkaryl, aralkyl, cycloalkyl, and aryl groups, and a is an integer 1 to 3.

The preferred compositions may be the primary amines $R-O-R''-NH_2$. The compositions wherein R'' is $(-CH_2-)_3$ may be particularly preferred. A particularly preferred composition may be $C_{13}H_{27}OCH_2CH_2CH_2NH_2$. In the following examples, all parts are parts by weight unless otherwise specified.

Example 1: In this example of the best mode of practicing the process, 7.68 ppm of Armeen EA-13, tridecyl(oxypropyl)amine, (5 PTB) are added as additive to 90 parts of absolute alcohol drawn from a reservoir having the composition given in Table 1 of U.S. Patent 4,282,007 above. The iron strip is observed after six days.

Example 2: In this control example, the test procedure of Example 1 is duplicated except that the additive is 100 PTB of the Ethomid HT/15 brand of

$$R-\overset{\overset{\large O}{\|}}{C}-N(CH_2CH_2OCH_2CH_2OH)_2$$

wherein R is an alkyl group derived from hydrogenated tallow in place of the additive of Example 1.

Example 3: In this control example, no additive is present. The results of the Iron Strip Corrosion Test were as follows.

Example	6 Day Rust and Corrosion Rating (%)
1	Trace
2	25-30
3	30

Example 4 through 6: Results comparable to those of Example 1 may be obtained when the additive or the alcohol is as follows.

Example	Alcohol
4	Gasohol*
5	Absolute ethanol
6	Absolute methanol

*Containing 90 vol % gasoline and 10 vol % absolute ethanol.

Hydrocarbyl Substituted Phenylaspartates of N-Primary-Alkyl-Alkylene Diamines

The process of *S. Herbstman and P. Dorn; U.S. Patent 4,321,062; March 23, 1982; assigned to Texaco Inc.* relates to compounds having utility as carburetor detergents, antirust and antiicing agents in motor fuels.

They are hydrocarbyl-substituted phenylaspartates of N-primary-alkyl-alkylene diamines of the formula:

$$\overset{\oplus}{R}NH_2(CH_2)_3N(H)-C(H)(-C(O)O^{\ominus})-CH_2-C(O)O-C_6H_4-R'$$

and

$$RNH(CH_2)_3N(H)-C(H)(-C(O)O^{\ominus}\,{}^{\oplus}NH_3(CH_2)_3NHR)-CH_2-C(O)O-C_6H_4-R'$$

wherein R is a hydrocarbyl radical having from about 6 to 30 carbon atoms; R' is an alkyl or an alkenyl group having from 6 to 30 carbon atoms.

The base fuel in which the additive of the process is used is a mixture of hydrocarbons boiling in the gasoline boiling range.

The additive of the process is added to the base fuel in a minor amount, i.e., an amount effective to provide corrosion inhibition, deicing properties and carburetor detergency to the fuel composition. The additive is effective in an amount ranging from about 0.0002 to 0.2 wt % based on the total fuel composition. An amount of the neat additive ranging from about 0.001 to 0.01 wt % is preferred.

Example 1: This example illustrates the preparation of a 1/1 mol product which is 50% active in oil. 23.7 g (0.237 mol) of maleic acid was suspended in 252 g of oil and heated to 100°C. To this suspension was added 70 g (0.241 mol) of nonylphenol. The reaction mixture was maintained at 100°C for 3 hr. 150 g of n-oleyl-1,3-propanediamine (Duomeen-OL, 0.24 mol) was added dropwise over 0.5 hr. The reaction mixture was maintained at this temperature for two additional hours. The crude reaction product was then filtered free of any insolubles. The product gave the following analysis: total acid No. (TAN), 19.8; N, 2.1%; total base No. (TBN), 29.7; infrared-ester carbonyl at 1,700 to 1,720 cm^{-1}.

Example 2: This example shows an improved modification of the procedure of Example 1. Nonylphenol (70 g, 0.241 mol) was dissolved in 252 g of oil. 158 g of N-oleyl-1,3-propanediamine (Duomeen-OL, 0.24 mol) was added, and the blend heated to 125°C slowly over 0.5 hr. At 125°C, maleic anhydride (23.7 g, 0.237 mol) was introduced and the reaction heated at 100°C with stirring an additional hour. The reaction product was then filtered.

Example 3: This example shows the preparation of the 2/1 reaction product. 22 g of nonylphenol were introduced to 190 g of oil. At room temperature with stirring, 70 g of N-oleyl-1,3-propanediamine were added. The blend was heated to 125°C and maintained at this temperature for 1 hr. 9.8 g of maleic anhydride were introduced. The reaction mixture was heated at 125°C with stirring an additional 0.5 hr. The reaction was filtered hot. The reaction product contained 34.9% active material. Analytical results: N, 1.6; TAN, 9.85; TBN, 62.9; infrared-ester carbonyl at 1,720 cm^{-1}.

2-Hydroxypropylimidazoles

H.-H. Vogel, R. Strickler, K. Oppenlaender and R. Baur; U.S. Patent 4,323,689; April 6, 1982; assigned to BASF AG, Germany provide 2-hydroxypropylimidazole derivatives of the formula

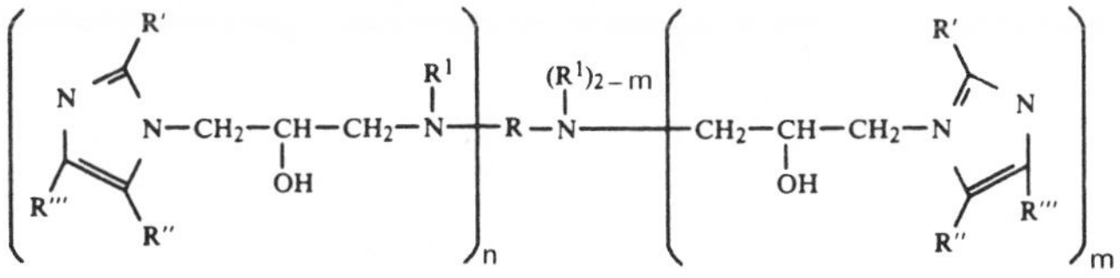

$$\left[\text{Im}(R', R'', R''') - N - CH_2 - \underset{OH}{CH} - CH_2 - \overset{R^1}{N} \right]_n - R - \overset{(R^1)_{2-m}}{N} - \left[CH_2 - \underset{OH}{CH} - CH_2 - N - \text{Im}(R', R'', R''') \right]_m$$

In the formula m is 1 or 2 and n is 0 or 1, R^1 is an aliphatic or cycloaliphatic radical of 6 to 21 carbon atoms, R, if n is zero, is an aliphatic, cycloaliphatic, aromatic or araliphatic radical of 6 to 21 carbon atoms and is preferably identical with R^1, or R, if n is 1, is a divalent aliphatic or aromatic radical of 2 to 15 carbon atoms or is

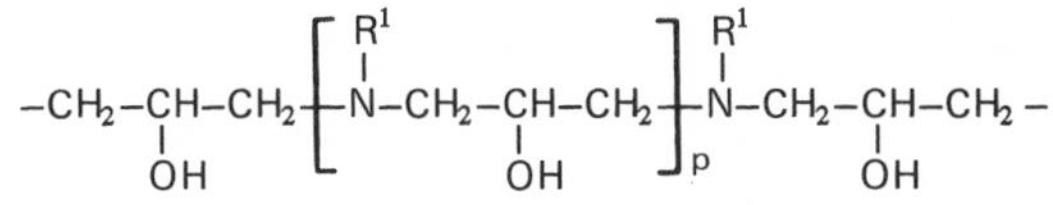

$$-CH_2-\underset{OH}{CH}-CH_2-\left[\overset{R^1}{N}-CH_2-\underset{OH}{CH}-CH_2\right]_p-\overset{R^1}{N}-CH_2-\underset{OH}{CH}-CH_2-$$

p being 0 or 2, and R', R'' and R''' are hydrogen or alkyl of 1 to 4 carbon atoms, and R'' or R''' may also be nitro.

The hydroxypropylimidazoles according to this process are effective as corrosion inhibitors in numerous applications, especially in hydrocarbons.

The applications include industrial heat transfer systems, pumping pipes and other metal equipment for oil-raising equipment, and pipelines, especially fuel pipelines of gasoline engines and diesel engines, as well as all engine components, e.g., carburetors, fuel injection pumps, pistons, etc., and also fuel storage and transportation tanks. Parts which are particularly prone to corrosion are the components of carburetors (which are made from die-cast zinc = 95% zinc, 4% of aluminum and 1% of copper), as well as the lead-lined parts of fuel tanks.

The corrosion inhibitor can be added to the fuel direct, in concentrations of 0.1 to 1,000 ppm, preferably of 1 to 500 ppm. It is also possible to add the inhibitor as part of a commercial fuel additive mixture, consisting of a valve and carburetor cleaner, an oxidation inhibitor, a film-forming anticorrosion additive (to prevent iron corrosion), etc.

The corrosion behavior of water-containing fuels is usually tested on metal coupons in fuel/water mixtures. The action of the product is shown in the examples which follow.

Example 1: *N-Cyclohexyl-N,N-bis-(3-imidazolyl-2-hydroxypropyl)amine* – (a) 68.1 g of imidazole are fused, under nitrogen, in a 500 ml stirred apparatus equipped with a reflux condenser, dropping funnel, stirrer and contact thermometer. 115.0 g of N-cyclohexyl-N,N-bis-2,3-epoxypropylamine (91.7% pure) are added dropwise in the course of one hour, at 100°C. After a further hour, epoxide is no longer detectable in the reaction mixture. The latter is allowed to cool, and 173.0 g of a pasty product are obtained. Basic N: 8.3 meq/g (theory: 8.6 meq/g). Total N: 14.2 meq/g (theory: 14.4 meq/g). OH number: 317.4 (theory: 322.8).

(b) Corrosion test: A 250 ml glass bottle is filled with 100 ml of supergrade gasoline (Erdölraffinerie Mannheim). The product is added in amounts varying from 10 to 1,000 ppm. 4 ml of distilled water are then added. Test coupons of size 50 mm x 20 mm x 2 are sanded down (grade 20 abrasive), degreased with toluene, and weighed. The test bottles are shaken vigorously for one minute so as to disperse the water in the gasoline. The metal coupons are introduced and stored for 14 days at 20° to 25°C. They are then cleaned with 15% strength hydrochloric acid which contains 1% of propargyl alcohol as a cling inhibitor, degreased and dried. The weight loss is determined by weighing (the results being shown in ‰ weight loss). As a rule, multiple determinations are carried out.

Additive		Lead
None		1.84 ‰
Compound from Example 1	10 ppm	1.48 ‰
	50 ppm	1.37 ‰
Additive mixture +	300 ppm	0.45 ‰
	500 ppm	0.48 ‰
Additive mixture plus 1% by weight, based on the mixture, of the compound from Example 1	300 ppm	0.40 ‰
	500 ppm	0.14 ‰

+The additive mixture is a commercial valve and carburetor cleaner, without benzotriazole as corrosion inhibitor.

Example 2: *N-2-Ethylhexyl-N,N-bis-(3-imidazolyl-2-hydroxypropyl)amine* – (a) 68.1 g of imidazole (1 mol) are dissolved in 200 ml of ethyl methyl ketone in a 500 ml stirred apparatus equipped with a dropping funnel, reflux condenser, stirrer and contact thermometer. The reaction mixture is then boiled (80°C) under nitrogen, and 126.3 g of 95.4% pure 2-ethylhexyl-bis-2,3-epoxypropylamine (0.5 mol) are added dropwise in the course of one hour. Finally, the solvent is distilled off; epoxide is no longer detectable in the batch. 194.4 g of pasty N-2-ethylhexyl-N,N-bis-(3-imidazolyl-2-hydroxypropyl)amine are obtained. Basic N: 7.7 meq/g (theory: 8.0 meq/g). Total N: 13.2 meq/g (theory: 13.3 meq/g). OH number: 293.1 (theory: 297.0).

(b) Corrosion test 2: The test was carried out similarly to Example 1 (b).

		Data in ‰ weight loss	
Additive		Lead	Die-cast zinc 95% of Zn, 4% of Al 1% of Cu
None		1.53	3.72
Additive from Example 2	10 ppm	0.80	1.50
	20 ppm	0.33	1.18
	30 ppm	0.34	1.07
	50 ppm	0.48	1.40
Additive mixture without corrosion inhibitor	300 ppm	0.45	0.05
	500 ppm	0.48	0.06
Additive mixture + 1% of the compound from Example 2	300 ppm	0.34	0.04
	500 ppm	0.29	0.04

Alkyl or Alkenyl Succinic Acid-Alkyl Ether Diamine Reaction Products

The process of *C.C. Hendricks, R.L. Godar and K.R. Roux; U.S. Patent 4,326,987; April 27, 1982; assigned to Petrolite Corporation* relates to the reaction products of (1) an alkenyl or alkyl succinic acid or the anhydride thereof (AASA); and (2) an alkyl ether diamine (EDA); and to the use thereof as a corrosion inhibitor.

The ether diamine has the general formula $ROANHA'NH_2$ where R is an alkyl group having about 1 to 18 carbons, but preferably about 8 to 9 carbons. A and A', which may be the same or different alkylene group, having about 2 to 10 carbons, but preferably 3 carbons. The preferred ether diamine is:

$$CH_3(CH_2)_{7-9}O(CH_2)_3NH(CH_2)_3NH_2$$

The reaction products are prepared by mixing the components together at ambient temperature. Since the reaction is exothermic, cooling may be desirable in larger batches.

The molecular weight of tetrapropenyl succinic acid is 284 and

$$CH_3(CH_2)_7CH_2O(CH_2)_3NH(CH_2)_3NH_2$$

is 258. Thus, the stoichiometrical weight ratio of AASA to EDA is about 1.1 to 1. As a corrosion inhibitor the most effective AASA to EDA weight ratio is in excess of 1.1 to 1, with an optimum of about 3 to 1 or greater.

Thus, the AASA to EDA weight ratio can be, for example, from about 10 to 1, such as from about 8 to 1, but preferably from about 6 to 1, with an optimum of about 3 to 1. Stated another way, the reaction product contains an excess of AASA.

The compositions of this process which are soluble or dispersible therein are particularly useful as rust or corrosion inhibitors such as in refined petroleum products such as in gasoline, aviation gasoline, jet fuels, turbine oils, fuel oils, etc. They may be employed in any amount capable of inhibiting rust or corrosion, such as in minor amounts of at least 1 ppm, preferably 25 to 50 ppm. They are particularly effective in inhibition of rust and corrosion in refined petroleum products, such as petroleum distillates in contact with metals such as ferrous or other metal surfaces. In the examples in the table the following compositions were compared:

(A) Tetrapropenyl succinic acid
(B) Ether diamine $CH_3(CH_2)_{7-9}O(CH_2)_3–NH–(CH_2)_3–NH_2$
(A+B) 3 parts (A) and 1 part (B)

Examples 1 through 10: *Static Rust Test –*

(1) Put 2,000 ml furnace oil without additives into 1 gal jug
(2) Add X ml additive (1% solutions)
(3) Shake 15 seconds
(4) Add 20 ml of 0.25% NaCl solution
(5) Shake 30 seconds
(6) Put special Millipore filter cap on jug and invert for 5 min.
(7) Put metal coupons into small canisters (1½ oz seamless tin boxes)

(8) Drain off water phase and some fuel (¼ inch) phase over coupon.
(9) Important: do not use acetone to clean special caps. Rinse over vacuum with IPA followed by hexane.
(10) Pass = no rust after 24 hours

It should be noted that the metal coupons be degreased with benzene then stored in acetone. In the table, $R^1 = CH_3(CH_2)_9$ and $R^2 = CH_3(CH_2)_{13}$.

Static Rust Test (#2 fuel oil)

Ex. No.	Additive	Concentration (ppm)	Rating
1	None	–	Fail
2	$R^1–O–(CH_2)_3–NH–(CH_2)_3–NH_2$	30	Fail
3	$R^2–O–(CH_2)_3–NH–(CH_2)_3–NH_2$	30	Fail
4	B	30	Pass
5	B	20	Pass
6	B	15	Pass
7	A	15	Fail
8	A + B	6.6 + 2.1	Pass
9	A + B	9.8 + 3.2	Pass
10	A + B	13.1 + 4.3	Pass

Perhydrophenanthridines

B.A.O. Alink; U.S. Patent 4,346,223; August 24, 1982; assigned to Petrolite Corp. has found that 2,3,4,5-tetrahydropyrimidines (THP) can be reduced to perhydrophenanthridines by reacting THP with formic acid according to the following equation:

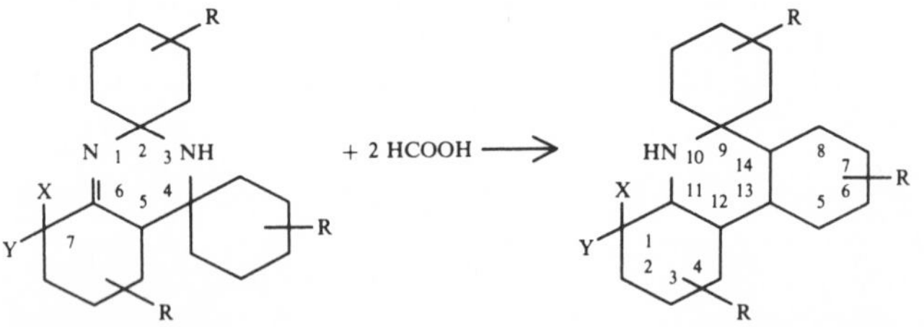

+ 2 HCOOH ⟶

2,2,4,4-Dipentamethylene 5,6-tetramethylene 2,3,4,5-tetrahydropyrimidine

perhydrophenanthridine

where R is hydrogen or a substituted group such as a hydrocarbon group, for example, alkyl and X and Y are hydrogen or a substituted group such as a hydrocarbon group containing a functional group, for example, an alkyl X group where X is a functional group such as nitrilo, carboxyl, etc.

The reaction is carried out by heating the THP with at least 2 equivalents of formic acid but it may be desirable to use more than 2 equivalents such as 2 to 10 equivalents of formic acid. Reaction temperatures are from 60° to 101°C but a temperature of 101°C (reflux) is preferred. Reaction times are from 1 to 24 hours. The products are useful as corrosion inhibitors and fuel stability additives.

Example 1: *2,2,4,4-Dipentamethylene-5,6-tetramethylene-2,3,4,5-tetrahydropyrimidine* – A mixture of 294 g of cyclohexanone and 5 g of ammonium chloride was placed in a pressure reactor. Over a ¾ hour period, 38.8 g of ammonia gas was added. After the addition was complexed, the mixture was stirred for 5 hours at ambient temperature. The product was taken up in toluene and the aqueous phase which separated was discarded. The toluene solution was evaporated under diminished pressure to yield 268 g of 2,2,4,4-dipentamethylene-5,6-tetramethylene-2,3,4,5-tetrahydropyrimidine.

Example 2: *9,9-Pentamethylene perhydrophenanthridine* – A mixture of 21.2 g of 2,2,4,4-dipentamethylene-5,7-tetramethylene-2,3,4,5-tetrahydropyrimidine prepared as described in Example 1 and 35 g of formic acid were refluxed for 18 hours. The excess of formic acid was distilled off under diminished pressure and the resulting product basified with an aqueous sodium hydroxide solution. The product was extracted with ether and the ethereal solution washed with water. Removal of the ether yielded 20 g of a crude product which was analyzed by ms/gc as a mixture of N-formylcyclohexylamine dicyclohexylamine and 9,9-pentamethylene perhydrophenanthridine. Separation by distillation yielded 9.2 g of 9,9-pentamethylene perhydrophenanthridine.

Maleic Anhydride-Alkoxy Propyl Amine Reaction Product for Alcohols

According to *R.L. Sung; U.S. Patent 4,348,210; September 7, 1982; assigned to Texaco Inc.* alcohols may be inhibited against corrosion by addition thereto of a reaction product of maleic anhydride and certain alkoxy propyl amines.

The composition of this process may comprise (1) at least one water-soluble alcohol preferably selected from the group consisting of ethanol and methanol; and (2) an effective corrosion-inhibiting amount of the reaction product of a maleic anhydride and $C_aH_{2a+1}-O-C_bH_{2b}-NH_2$ wherein a is an integer greater than 5, b is an integer greater than 2, and C_bH_{2b} is a straight chain alkylene group.

The alcohol compositions which may be treated by this process may include alkanols such as water-soluble alkanols most commonly including C_{1-4} alcohols. The alcohols may include mixtures of such alcohols with each other and/or with other compositions including ketones, esters, hydrocarbons, etc. The alcohol may be in the form of gasohol, a mixture commonly containing 90 vol % gasoline and 10 vol % alcohol. The alcohol may contain water, typically 5 wt %; but preferably it will be anhydrous. Anhydrous compositions commonly contain 0.004 vol % water. One preferred charge may be 100% anhydrous ethanol.

The rust and corrosion inhibitors may be added to an alkanol in minor corrosion-inhibiting amount of 0.25 to 25, preferably 1 to 20 PTB. The inhibited alcohols of this process, after 40 hours of ISCT, generally show a rust and corrosion rating below about 2 to 3% and frequently as low as trace to 1%. In the following example, all parts are parts by weight unless otherwise specified.

Example: In this example, the preferred reaction product is prepared by adding 147 parts of maleic anhydride to 100 E Pale Oil (720 parts). The mixture is heated at 55°C until the anhydride is dissolved. There may then be added 600 parts of the Armeen EA-13 brand of tridecyloxy propylamine $C_{13}H_{25}O-C_3H_6-NH_2$. Addition of the latter occurs over one hour as the temperature is allowed to rise to not exceeding 90°C. The product is recovered by cooling the reaction mixture and filtering. On analysis it is found that the TBN is 30.1, the TAN is 61.7, and the % N is 2.03.

FOR FUELS AND LUBRICANTS

Aminomethyl Cyclododecanes

The process of *R. Braden and K. Wagner; U.S. Patent 4,251,462; February 17, 1981; assigned to Bayer AG, Germany* is directed to aminomethyl cyclododecanes selected from the group consisting of aminomethyl cyclododecanes, bis-(aminomethyl)cyclododecanes, tris-(aminomethyl)cyclododecanes and mixtures thereof. The process is also directed to a method for producing aminomethyl cyclododecanes, bis-(aminomethyl)cyclododecanes, tris-(aminomethyl)cyclododecanes, and mixtures thereof comprising reacting cyclododeca-1,5,9-triene with carbon monoxide and hydrogen in the presence of a rhodium-containing catalyst at temperatures of from 80° to 180°C and under pressures of from 30 to 900 bars, separating the catalyst off from the hydroformylation product and treating the hydroformylation products with hydrogen at from 50° to 150°C in the presence of ammonia and a hydrogenation catalyst, optionally after separating by distillation into the individual components.

The amines according to this process are valuable corrosion inhibitors which, by virtue of their high hydrocarbon content, show in particular a high level of compatibility with heating oils, lubricants and hydrocarbon-based motor fuels.

Example 1: *Hydroformylation of cyclododecatriene* – The catalyst, 75 mg of tris-(dibenzyl sulfide)tris-chlororhodium and 2.4 g of dicobalt octacarbonyl, and 500 g of toluene are introduced into a fine-steel autoclave. The autoclave is repeatedly purged with a 1:1 gas mixture of carbon monoxide and hydrogen with which a pressure of up to 100 bars is established. The autoclave is heated with stirring to 170°C and the pressure is increased to 200 bars using the same gas mixture, the pressure subsequently being kept constant by the introduction of more CO/H_2 when the pressure falls.

After 1 hour, the temperature is reduced to 110°C and a solution of 500 g of cyclododeca-1,5,9-triene in 1,000 g of toluene is pumped into the autoclave over a period of 3 hours. After another 90 minutes, the reaction mixture is cooled and the autoclave is vented and purged with nitrogen. The reaction solution is filtered. The solvent is distilled off at 1,600 Pa. The reaction product is distilled in a thin layer evaporator at 13 Pa and at a jacket temperature of around 220°C. The composition of the distillate is determined by gas chromatography (column: 1 m Carbowax 6000 on Teflon; heating rate: 15°C/min; 130° to 260°C). The sample is diluted with tetrahydrofuran. The yield was 13.9 mol % monoformyl cyclododecane, 57.7 mol % bisformyl cyclododecane and 26.2 mol % trisformyl cyclododecane.

Example 2: *Aminomethyl cyclododecane* – 235 g of formyl cyclododecane, 2 g of acetic acid, 250 g of tetrahydrofuran and 20 g of Raney cobalt are introduced into a fine-steel stirrer-equipped autoclave. The autoclave is closed and purged with nitrogen. 300 g of liquid ammonia are pumped into the autoclave. The contents of the autoclave are then heated to 110°C under a hydrogen pressure of 80 bars, after which the pressure (up to 120 bars) is kept constant using hydrogen for 45 minutes. Reductive amination is then over. After the ammonia has been evaporated, the catalyst is removed from the cooled reaction mixture by filtration and the solvent distilled off. Distillation at 13 Pa gives a fraction boiling at from 112° to 113°C of which up to 98.5% consists of aminomethyl cyclododecane. n_D^{20} 1.5012, MW: observed 200 (theoretical 197), yield: 86.7%.

Example 3: 50 g of Raney nickel, 1,000 g of methanol and 2 g of phosphoric acid are introduced into an autoclave. After the autoclave has been closed and purged with nitrogen, 700 g of ammonia are introduced. The autoclave is then heated to 95°C under hydrogen pressure so that a pressure of 120 bars is established. At 90° to 100°C/120 bars pressure (the pressure is kept constant by the introduction of more nitrogen), a solution of 500 g of a mixture emanating from the formylation of cyclododeca-1,3,5-triene and containing 7.9%, by weight, of monoformyl cyclododecane, 71.6%, by weight, of bisformyl cyclododecane and 14.7%, by weight, of trisformyl cyclododecane in 1,000 g of methanol is pumped in over a period of 90 minutes. The contents of the autoclave are then stirred for 10 minutes at 105°C/120 bars.

Working up of the reaction mixture in the usual way gave 435 g of an amine mixture which distilled over in a thin-layer evaporator at 25 Pa and at a wall temperature of 170°C. According to its gas chromatogram, this mixture contained 8.5% of mono-, 76.3% of di- and 13.4% of tri-(aminomethyl)cyclododecane.

It is possible by redistilling this mixture to obtain a fraction boiling at 122° to 140°C/1.3 Pa which, in addition to 41% of di-(aminomethyl)cyclododecane, contains approximately 59% of tri-(aminomethyl)cyclododecane and is free from mono-(aminomethyl)cyclododecane.

Hydrocarbyl-Substituted Succinic Anhydride-Aminotriazole Reaction Product

The additive of the process of *R.L. Sung, J.J. Bialy, P. Dorn, W.P. Cullen and J.W. Nebzydoski; U.S. Patent 4,263,015; April 21, 1981; assigned to Texaco Inc.,* which is effective as a rust inhibitor for motor fuels, fuel oils and lubricating oils, comprises the reaction product of a hydrocarbyl-substituted anhydride and an aminotriazole.

The additive is obtained by reacting a hydrocarbyl-substituted succinic anhydride with an aminotriazole at a temperature ranging from about room temperature to about 150°C until the substantial completion of the reaction. This reaction is conducted in the absence of any catalyst but generally in the presence of a solvent to facilitate the reaction. The hydrocarbyl-substituted succinic anhydride reactant is represented by the formula:

```
    H      O
    |     //
R — C — C
    |      \
    |       O
    |      /
   CH2 — C
          \\
           O
```

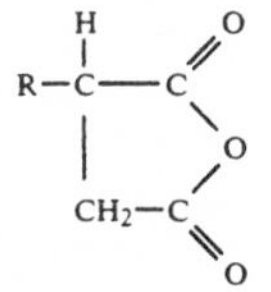

in which R is a monovalent aliphatic hydrocarbon radical having from about 6 to 30 carbon atoms. The hydrocarbon radical can be straight or branched chain hydrocarbon radical and can be saturated or unsaturated. Particularly preferred reactants are the alkenylsuccinic anhydrides in which the alkenyl radical has from about 12 to 24 carbon atoms, preferably 12 carbon atoms.

The aminotriazole reactant is represented by the formula shown on the following page. It will be understood that the hydrogen and the amino radicals are attached at the unsatisfied carbon atom bonds and that they can be interchanged in these positions.

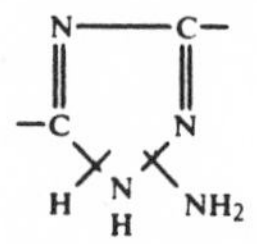

Suitable aminotriazoles include 3-amino-1,2,4-triazole and 5-amino-1,2,4-triazole. The hydrocarbon-substituted succinic anhydride and the aminotriazole are reacted in the proportion of from about 0.75 to 1.25 mols of the aminotriazole per mol of the hydrocarbon-substituted succinic anhydride.

Example: 100 g (1 mol) of tetrapropenylsuccinic anhydride and 100 g (1 mol) of 3-aminotriazole are dissolved in 1,200 ml of xylene. The reaction mixture was refluxed for about 6 hours followed by the removal of the solvent by distillation. The reaction product recovered contained 15.7% nitrogen. Infrared spectroscopy indicated that the product was a mixture of the tetrapropenylsuccinamide of 3-aminotriazole and the alkenylsuccinimide of 3-aminotriazole.

The prescribed reaction product of this process was tested for its corrosion inhibiting properties in gasoline in the Colonial Pipeline Rust Test described below.

A steel spindle, 3 and 3⁄16 inches long and ½ inch wide made from ASTM D-665-60 steel polished with Crystal Bay fine emery paper is used in the Colonial Pipeline Rust Test. The spindle is placed in a 400 cc beaker with 300 cc of fuel sample which is maintained at 100°F for ½ hour. Then 30 cc of distilled water is added. The beaker and contents are kept at 100°F for 3½ hours. The spindle thereafter visually inspected and the percentage of rusted surface area is estimated.

The base fuel employed in the following examples was an unleaded grade gasoline having a research octane number of about 91. This gasoline consisted of about 24% aromatic hydrocarbons, 8% olefinic hydrocarbons and 68% paraffinic hydrocarbons and boiled in a range from about 90° to 375°F. A base blend was prepared from the foregoing base fuel and conventional antioxidant and metal deactivator in the amount of 4.5 PTB (pounds of additive per thousand barrels of the fuel composition). The results are set forth in the following table.

Run		% Rust
1.	Base Blend	50-100
2.	Base Blend + 10 PTB Example	Trace, Trace
3.	Base Blend + 5 PTB Example	Rust Free, Trace
4.	Base Blend + 1 PTB Example	Rust Free, Trace

Modified Aminocarbamate Compositions with Improved Corrosion Properties

It is the object of *R.A. Lewis and L.R. Honnen; U.S. Patent 4,289,634; Sept. 15, 1981; assigned to Chevron Research Company* to provide modified aminocarbamate compositions suitable for use as deposit-control additives and having improved corrosion characteristics towards metal surfaces of internal combustion engines when present in lube oil and fuel compositions. The fuel or lube oil additive comprises the salt of an oxyacid ester of phosphorus and a hydrocarbyl poly(oxyalkylene)aminocarbamate deposit-control compound.

The salts of the process will generally be employed in a hydrocarbon distillate fuel or lube oil. Where used as a fuel additive, the proper concentration of additive necessary in order to achieve the desired detergency, dispersancy and corrosion resistance varies depending upon the type of fuel employed, the presence of other detergents, dispersants and other additives, etc. Generally, however, from 30 to 2,000 weight parts per million, preferably from 50 to 250 ppm of salt per part of base fuel is needed to achieve the best results.

In general, the lubricating oil compositions will contain from about 0.01 to 20 wt % of the salt. More usually, the lubricating oil composition of the process will contain from about 0.5 to 10 wt % of the salt and more usually from about 1 to 8 wt % thereof.

Example 1: *Preparation of alkylphenylpoly(oxybutylene) alcohol* – The experiment was carried out in dry glassware under an inert atmosphere. Potassium (1.17 g, 0.03 mol) was added to 26.34 g (0.1 mol) of a phenol alkylated with propylene tetramer. The mixture was stirred and heated to 50°C for 24 hours until the potassium dissolved. The pot temperature was raised to 80°C and 1,2-epoxybutane (215 ml, 2.5 mols) was added at a rate slow enough to prevent flooding of the condenser. The reaction was stirred and heated at reflux until the pot temperature reached 125°C. The product was extracted into 2 volumes of diethyl ether and washed with two volumes of 0.5 N HCl. Diethyl ether (250 ml) was added to the ethereal layer, and it was washed four times with 250 ml aliquots of water. The solvent was removed and the product was azeotroped with toluene to remove traces of water. A yield of 145 g of a viscous liquid of molecular weight approximately 1,500 was obtained.

Example 2: *Reaction of alkylphenylpoly(oxybutylene) alcohol with phosgene* – Phosgene (14 ml, 0.198 mol) was condensed and transferred to a flask containing 150 ml of toluene. This mixture was cooled and stirred in an ice bath while the poly(oxybutylene) alcohol of Example 1 (140 g, 0.09 mol) was added dropwise. After the addition was complete, the ice bath was removed and the mixture was stirred for about 1 hour. An aliquot was taken, and the infrared spectrum of its nonvolatile residue showed a strong chloroformate absorption at 1,785 cm^{-1}.

Example 3: *Reaction of alkylphenylpoly(oxybutylene) chloroformate with amine* – Ethylenediamine (41 ml, 0.61 mol) was stirred rapidly and cooled in an ice bath. The chloroformate of Example 2 was diluted with four volumes of toluene and added to the ethylenediamine at such a rate that the pot temperature did not exceed 30°C. After the addition was completed, the ice bath was removed and the mixture was stirred for about 1 hour.

The mixture was extracted into 500 ml of hot n-butanol and washed four times with 500 ml aliquots of hot water. The solvent was removed and the product was azeotroped with toluene to remove traces of water, giving 125 g of a viscous amber liquid of molecular weight about 1,600. The product alkylphenylpoly(oxybutylene) ethylenediamine carbamate, i.e., alkylphenylpoly(oxybutylene)-N-(2-aminoethyl) carbamate, contained 1.20% by weight nitrogen and dispersed sludge at 200 to 400 ppm.

Example 4: *Reaction of carbamate with diisooctyl hydrogen phosphate* – Alkylphenylpoly(oxybutylene) ethylenediamine carbamate in about a 50 wt % solution in a heavy aromatic (mainly C_9-aromatics) was heated to about 60°C. To this heated solution was added a second approximately 50 wt % solution of diiso-octyl hydrogen phosphate in xylene solvent in an amount sufficient to provide in the resulting mixture an acid to basic amine equivalent ratio of 0.8 to 1. The resulting mixture was heated and maintained at about 60°C for about one-half hour. A salt concentrate was then produced by stripping the lower boiling fraction of the solvent from the reaction mixture.

Example 5: *ASTM D665 Rust Prevention Test* – This test is a determination of the ability of a fuel or oil containing an additive to prevent rusting of a standard steel specimen under standard conditions, for example, contact with a stirred mixture of water and the fuel or oil tested at a temperature of 38°C for a period of five hours. In this test tap water and 200 ppm wt of the additive in a typical base gasoline were used. The rating was by the NACE scale of A to E as follows where A is no rust, B++ is less than 3 spots, C is approximately 5% of surface rusted, D is greater than 5% rusted and E is 100% rusted.

A comparison in this test of the rust prevention characteristics of the carbamate and the salt was as follows: the carbamate, C and the carbamate salt, B++. These data demonstrate that, relative to the carbamate the salt of this process exhibits markedly superior rust preventing characteristics.

In this process of *R.A. Lewis and L.R. Honnen; U.S. Patent 4,294,714; Oct. 13, 1981; assigned to Chevron Research Co.* for controlling deposition of solids and inhibiting metal corrosion in intake systems and combustion chambers of internal combustion engines the fuel or lube oil additive comprises the salt of a C_{3-30} organic monocarboxylic acid and a hydrocarbyl poly(oxalkylene) aminocarbamate deposit-control compound. The salts of this process are used in fuels and lube oils in the same amounts as are the salts of the above patent.

Example: Preparation of the carbamate is as described in Examples 1 through 3 above.

Reaction of carbamate with organic carboxylic acids – Alkylphenyl poly(oxybutylene) ethylenediamine carbamate in about 50 wt % solution in a heavy aromatic solvent (mainly C_9-aromatic refinery cut) was heated to about 60°C. To the heated solution of the carbamate a second solution containing the organic acid in a suitable solvent, for example, isopropanol, butanol, toluene and the like, was added in an amount sufficient to provide in the resulting mixture an acid to basic amine equivalent ratio of 0.8 to 1. The mixture was heated and maintained at about 60°C for about one-half hour. A salt concentrate was then produced by stripping the major portion of the solvent(s) from the reaction mixture. In the foregoing manner salts were prepared from the carbamate and the following acids.

Formic	Oleic
Lactic	Octanoic
Citric	Stearic
Ethoxyacetic	12 Hydroxystearic
Propionic	$\sim C_{16-18}$ mix*

*Of saturated vs unsaturated acids.

The resulting salts were tested under comparable conditions for rust prevention effects in the ASTM D665 Rust Prevention Test described in the above patent.

Carboxylic Acid Used	ASTM D665 Rust Rating
Formic	D
Lactic	D
Citric	D
Ethoxyacetic	E
Propionic	B+
Oleic	B
Octanoic	B
Stearic	B
12-Hydroxystearic	B+
~C_{16-17}* mix	B++
None (carbamate)	C

*Saturated and unsaturated acids (DuPont DCI-6A).

These data demonstrate that organic acid salts as herein disclosed exhibit, relative to hydrocarbyl poly(alkylene) aminocarbamates per se, superior rust inhibiting action.

Benzoquinone-Amine Reaction Product

An oil-soluble, ashless detergent and corrosion inhibitor is provided by *P.F. Vartanian and J.B. Biasotti; U.S. Patent 4,292,047; September 29, 1981; assigned to Texaco Inc.* It is the reaction product of a benzoquinone, represented by the formula:

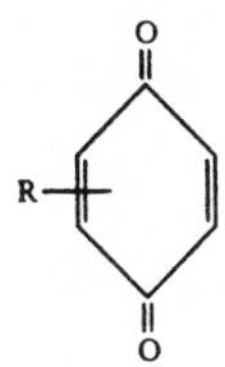

in which R is hydrogen or an alkyl radical having from 1 to 6 carbon atoms, and an amine, represented by a formula selected from the group consisting of RNHR'' and R'_2NR'' in which R is a monovalent hydrocarbon radical having from 8 to 30 carbon atoms, R' is a monovalent hydrocarbon radical having from 6 to 12 carbon atoms, and R'' represents hydrogen or an aminoalkyl radical selected from the group consisting of $-CH_2CH_2CH_2-NH_2$ and $-CH_2CH_2-NH_2$. At least 1.5 mols of the amine per mol of the benzoquinone is used.

Example 1: 81 g (0.75 mol) of benzoquinone, 530 g (1.5 mols) of N-tallow-1,3-diaminopropane and 200 ml of xylene are added to a reaction vessel equipped with a stirrer and a reflux condenser. This mixture is heated and refluxed for about 3 hours, after which it is cooled and filtered to remove unreacted starting materials. The solvent is removed to yield about 560 g of the reaction product.

Example 2: 81 g (0.75 mol) of benzoquinone, 530 g (1.5 mols) of N-soya-1,3-diaminopropane and 200 ml of xylene are reacted as in Example 1. On removal of the solvent by distillation, a substantial yield of the reaction product is designated benzoquinone-N-soya-1,3-diaminopropane is recovered.

Example 3: A mixture of 108 g (1.0 mol) of benzoquinone, 700 g (2.0 mols) of

N-oleyl-1,3-diaminopropane, and 270 g of a paraffinic oil having an SUS viscosity at 100°F of 100 are heated to 200°F for 3 hours, after which it is cooled and filtered to remove unreacted starting material. The reaction yields approximately 1,060 g of material which is 75% active in the desired product and may be used without further modification.

The reaction product of the process is useful as a detergent and/or a corrosion inhibitor for hydrocarbon fuels, mineral oils and lubricating oil compositions. In general, the reaction product of the process is effective as a carburetor detergent and/or corrosion inhibitor in a motor fuel composition comprising a mixture of hydrocarbons in the gasoline boiling range, i.e., from about 90° to 425°F. Broadly effective concentrations as detergent and corrosion inhibitors in fuels range from about 0.005 to 0.20 wt % of the reaction product based on the weight of the gasoline composition. A preferred concentration of the reaction product in gasoline is an amount ranging from about 0.002 to 0.04 wt % which corresponds to about 5 and 100 PTB (pounds of additive per 1,000 barrels of gasoline), respectively. In oil compositions the reaction product can be employed at concentrations ranging from about 0.0005 to 10 wt % based on the total weight of the oil composition with a preferred concentration being from 0.01 to 5 wt %.

Example 4: The ability of the additive to provide corrosion protection was tested in a rust test (similar to the ASTM D-665 procedure). In this test, a polished steel spindle is suspended in 300 ml of additive test fuel at 100°F for one-half hour. At the end of this time, 30 ml of water are added and the stirred mixture is continued at 100°F for 3½ hours. At the end of this time the steel spindle is examined visually for signs of rust with the rating expressed in percent of area covered by rust.

The base fuel employed in this test was a premium grade gasoline having a Research Octane Number of about 99.6 and contained about 2.97 cc of tetraethyl lead per gallon. This gasoline consisted of about 32% aromatic hydrocarbons, 8% olefinic hydrocarbons and 60% paraffinic hydrocarbons and boiled in the range from about 95° to 363°F.

Run	Additive	Dosage, PTB[a]	Rating % Rust
1	None	—	50–100
2	Benzoquinone-N-oleyl-1,3-diaminopropane (Example 3)	25	Trace
3	Benzoquinone-N-oleyl-1,3-diaminopropane (Example 3)	5	1–5
4	Benzoquinone-N-tallow-1,3-diamino-propane (Example 1)	50	1–5

[a]Dosage on a diluent free basis.

Thio-Bis-(Hydrocarbon-Bisoxazolines)

According to *S.J. Brois and A. Gutierrez; U.S. Patent 4,292,184; September 29, 1981; assigned to Exxon Research & Engineering Co.* thio-bis-(hydrocarbon-bisoxazolines) which are the reaction products of thio-bis-(hydrocarbon substituted dicarboxylic acid material), for example, thio-bis-(polyisobutenyl succinic anhydride), with 2,2-disubstituted-2-amino-1-alkanols, such as tris(hydroxymethyl)aminomethane (THAM), and their derivatives are useful additives in oleaginous compositions.

The products have utility in hydrocarbon fuel and lubricating systems as highly stable anticorrosion agents and/or sludge dispersants.

The oxazoline products of this process can be both molybdated with molybdenum to enhance their lubricity activity and borated with boron to enhance the additives' anticorrosion and/or varnish inhibition activities.

"Thio" as the term is generically used herein encompasses sulfur and its congener, i.e., selenium.

The oil-soluble oxazoline reaction products of the process can be incorporated in a wide variety of oleaginous compositions. They can be used in lubricating oil compositions, such as automotive crankcase lubricating oils, automatic transmission fluids, etc., in concentrations generally within the range of about 0.01 to 20 wt %, e.g., 0.1 to 10 wt %, preferably 0.3 to 3.0 wt %, of the total composition. The lubricants to which the oxazoline products can be added include not only hydrocarbon oils derived from petroleum but also include synthetic lubricating oils.

When the products of this process are used as multifunctional additives having detergent and antirust properties in petroleum fuels, such as gasoline, kerosene, diesel fuels, No. 2 fuel oil and other middle distillates, a concentration of the additive in the fuel in the range of 0.001 to 0.5 wt %, based on the weight of the total composition, will usually be employed.

Example 1: *Dithio-bis-[polyisobutenyl-bis-(5,5-bis-methylol-2-oxazoline)]* – 200 g (ca 0.154 mol) of a polyisobutenyl succinic anhydride (prepared via the reaction of polyisobutene and maleic anhydride) having an average molecular weight of 1,300 and a Saponification No. 72, were diluted with 100 ml of methylene chloride and stirred at room temperature under a nitrogen blanket. Then, 10.4 g (ca 0.077 mol) of S_2Cl_2 were added dropwise for a period of one-half hour. The reaction mixture was stirred at room temperature for about 10 hours.

One-half of this product was evaporated and the residue was sparged with nitrogen at 150°C for 4 hours. The resulting dithiobis(polyisobutenyl succinic anhydride) adduct analyzed for 2.08 wt % S and 0.15 wt % Cl.

About 21 g (ca 0.007 mol) of the adduct were diluted with an equal weight of mineral oil (Solvent 150 Neutral) and heated to 120°C. Then 0.1 g of zinc acetate and 3.4 g (0.028 mol) of THAM were added. The reaction mixture was heated to 180°C for 2 hours while sparging with nitrogen and filtered. The oil solution analyzed for 0.9 wt % nitrogen. The infrared analysis confirmed the presence of the aboveidentified tetraoxazoline product.

Example 2: *Dithio-bis-[polyisobutenyl-bis-(5,5-bis-methylol-2-oxazoline)]* – 30 g (ca 0.014 mol) of dithio-bis-(polyisobutenyl succinic anhydride) adduct derived from a polyisobutenyl succinic anhydride with an average molecular weight of 990 and a Saponification No. 107 and S_2Cl_2 were diluted with 32 g of mineral oil (Solvent 150 Neutral) and heated to 120°C. Then 0.1 g of zinc diacetate and 6.9 g (ca 0.057 mol) of THAM were added. The reaction mixture was heated to 180°C for 2 hours with nitrogen sparging and then filtered. The oil solution of the oxazoline product analyzed for 1.07 wt % nitrogen.

Example 3: *Thio-bis-[polyisobutenyl-bis-(5,5-bis-methylol-2-oxazoline)]* – 500 g (ca 0.385 mol) of polyisobutenyl succinic anhydride having a molecular weight of 775 and Saponification No. 84 were dissolved in 60 ml of methylene chloride and cooled to 0°C. While stirring at 0°C under a nitrogen atmosphere, 19.8 g (ca 0.192 mol) of SCl_2 were added dropwise for a period of one-half hour. The reaction mixture was allowed to warm up to room temperature and stirred for about 10 hours.

One-half of this adduct product was dehydrohalogenated by rotoevaporation under high vacuum for 6 hours at about 100°C. The adduct analyzed for 1.34 wt % S and 0.70 wt % Cl. The infrared analysis was consistent with that of a thio-bis-(polyisobutenyl succinic anhydride) adduct.

About 80 g (ca 0.03 mol) of the adduct were diluted with an equal amount of mineral oil (Solvent 150 Neutral) and heated to 130°C. Then 0.1 g of zinc acetate dihydrate was added, followed by the addition of 14.5 g (ca 0.12 mol) of THAM. The reaction mixture was heated slowly to 180°C and kept at this temperature for 2 hours while sparged with nitrogen. The oil solution was filtered and the filtrate analyzed for 1.27 wt % nitrogen. An infrared spectrum of this product confirmed the presence of the aboveidentified tetraoxazoline.

MgO Dispersions

The process of *W.J. Cheng, D.B. Guthrie and D.M. Leiendecker; U.S. Patent 4,293,429; October 6, 1981; assigned to Petrolite Corporation* relates to a stable, fluid magnesium oxide-containing dispersion prepared by the thermal treatment of MgO powder in a nonvolatile fluid in the presence of a stoichiometric or lesser amount of a carboxylic acid as illustrated by acetic acid.

The compositions of this process have a wide variety of uses, such as a combination anticorrosion and acidic neutralization additive for lubricating oils and greases, a combination anticorrosion and acidic neutralization additive during the combustion of fuels such as residual fuel, pulverized sulfur-containing coal, or mixtures thereof, and as corrosion inhibitors, particularly in fuels containing vanadium.

Example 1: This example describes a preparation of an oil dispersion of submicron-sized MgO when the starting magnesium oxide is a reagent grade material. A low stoichiometric amount of acetic acid is used to make magnesium acetate in situ. The amount of water added was calculated to be the amount necessary to make $Mg(OAc)_2 \cdot 4H_2O$; although it is possible that the H_2O added can also react with MgO to form $Mg(OH)_2$, nevertheless the water added is readily removed by the thermal conditions of the method of this process.

To a one-gallon SS pressure reactor having agitator and thermometer was charged a mixture containing 675 g process oil, 227 g naphthenic acid, and 257 g reagent grade magnesium oxide; 33 g H_2O and 74 g glacial acetic acid followed. The contents were heated to 147°C and 12 psig and maintained at that temperature for 3 hours.

A 636.8 g aliquot of the above mixture was heated in a one liter three-necked glass reactor having agitator, thermometer, and Dean/Stark assembly for distillation to remove water with a return of process oil which codistills. The contents of the reactor were heated to 287°C to distill off all the water that would evolve while returning any oil which codistilled. The material was clear and bright. The mass was heated to 315°C with a trace of additional water being removed. The product was calculated to contain 16.2% as Mg. The product upon centrifugation for 2 hours gave virtually no sediment.

Example 2: The importance of low stoichiometric amounts of acetic acid in Example 1 is demonstrated by this example. The procedure of Example 1 was the same except that no acetic acid was used. The resulting product was not clear and bright; upon centrifugation more than 20% separation of solids was indicated.

Example 3: This example describes a ten-gallon preparation according to the procedure of Example 1 except that the entire process was carried out in a single reactor. The starting magnesium oxide was a commercial source of technical grade material.

To a ten-gallon pressure reactor were charged: 33⅛ lb hydrocarbon solvent; 15⅜ lb naphthenic acid; 15⅞ lb magnesium oxide (commercial, tech grade); 4½ lb acetic acid; and 7¼ lb H_2O. The contents of the reactor were heated under pressure at a temperature of 200° to 235°C for 15 hours. The reactor pressure was vented and the contents heated further to effect distillation of any water, etc. Upon reaching 304°C, there was no further water distilling. The product was clear and bright. The magnesium content was calculated at 19.5%, and the sediment obtained was 0.95% after centrifugation for 5 hours; the bulk of the sediment separated quickly.

Stable $Mg(OH)_2$ Suspensions

W.J. Cheng and D.B. Guthrie; U.S. Patent 4,298,482; November 3, 1981; assigned to Petrolite Corporation have found a low-cost, practical and energy efficient method of preparing nonsettling suspensions of $Mg(OH)_2$ which comprises blending $Mg(OH)_2$ powder and acids in a surfactant-containing hydrocarbon and which method makes unnecessary the formation of oil-soluble magnesium compounds for many applications.

In practice the mol-fraction of acid employed is less than 50% of that stoichiometrically necessary for the formation of magnesium salts. It is unexpected that the product of this process is stable in view of the fact that, in the absence of acid, the $Mg(OH)_2$ powder employed in the same process, even with the aid of surfactants, is difficult to suspend without separation.

It is believed that the magnesium salt initially formed in situ is submicron in size and quickly erodes the large particles of $Mg(OH)_2$ powder into smaller particles until equilibration is reached among the species Mg^{2+}, $Mg(OH)_2$ and acid anion, thus resulting in the easy and lasting suspendability of $Mg(OH)_2$ by the surfactant(s).

Since a low stoichiometric amount of acid is employed, the product of this process is in essence a mixture of magnesium salts and magnesium hydroxide of very small particle size.

The compositions of this process have a wide variety of uses, such as a combination anticorrosion and acidic neutralization additive for lubricating oils and greases, as a combination anticorrosion and acidic neutralization additive during the combustion of fuels such as residual fuel, pulverized sulfur-containing coal, or mixtures thereof, and as corrosion inhibitors, particularly in fuels containing vanadium.

Example 1: *Typical suspension procedure* – To a 4-liter beaker are charged 1,890 g kerosene, 94 g (about 0.3 eq) dodecylbenzenesulfonic acid and 64 g (about 1.067 mols) glacial acetic acid. The contents are stirred until homogeneous. With stirrer on, 1,250 g (about 21.44 mols) magnesium hydroxide is added followed by 218 g kerosene and 154 g sorbitan monooleate. A magnesium sulfonate and magnesium acetate are formed in situ. The mass is stirred for 3 hours. The viscosity of the mass is about 5,000 cp. The magnesium content is calculated at about 14.2% for 3,670 g total suspension. There was no separation of layers after 4 months of standing.

Example 2: *Preparation of a magnesium sulfonate dispersant* – To a reactor were charged 94 g (about 0.3 eq) dodecylbenzenesulfonic acid, 1,890 g kerosene and 10 g (about 0.17 mol) of $Mg(OH)_2$. The contents were stirred and heated to 170°C with the reactor closed, pressure rising to 32 psig. The contents were cooled and drained; net weight was 1,985 g. The magnesium content of this product was calculated at 0.21%.

When acetic acid is not used in preparing the suspension, the stability of the suspension, as shown in Example 3, is virtually nonexistent.

Example 3: To a quart jar were added 400 g magnesium dodecylbenzenesulfonate of Example 2 and 30.8 g sorbitan monooleate. The contents were stirred to which was added 248 g $Mg(OH)_2$. Stirring was continued for 1 hour. The mass was uniform in appearance, but, after 1 hour of settling, one-third of the kerosene charged had separated as an upper layer.

When acetic acid is employed in preparing the suspension, the stability of the suspension, as shown in Examples 4 and 5, is astonishingly long-lasting.

Example 4: To a quart jar were added 400 g magnesium dodecylbenzenesulfonate of Example 2, 31 g sorbitan monooleate and 13 g acetic acid. The contents were stirred to which was added 248 g $Mg(OH)_2$. The contents were stirred for 3 hours. The mass was allowed to stand for 3½ months. There was virtually no separation of an upper kerosene layer nor a lower solid layer of $Mg(OH)_2$. The magnesium content was calculated at 15.06%.

The long-term stability of the product of Example 4, even when greatly diluted in kerosene fuel, is shown in Example 5.

Example 5: The product of Example 4 (2.0 g) was dispersed in kerosene (248 ml) in a 250 ml graduated cylinder. The dispersion (calculated at about 1,500 ppm Mg) was then allowed to settle for 3½ months. The upper 10% (25 ml) was removed and analyzed for magnesium content using dry ash/atomic absorption spectroscopy; the Mg content found was 770 ppm or 0.08%. The bottom 3% (7 to 8 ml) was removed and analyzed; the Mg content found was 757 ppm or 0.08%. This data indicates that about 50% of the $Mg(OH)_2$ had remained suspended over a 3½ month period.

Haze-Free Asparagines

It has been found by *B.J. Kaufman; U.S. Patent 4,364,846; December 21, 1982; assigned to Texaco Inc.* that the reaction between maleic anhydride and an N-alkylpropanediamine mixture can be modified so that the resulting primary aliphatic hydrocarbon amino alkylene-substituted asparagine is not hazy and does not form a haze or a precipitate on standing. More specifically, it has been found that the reaction between maleic anhydride and an N-alkylpropanediamine mixture can be surprisingly improved to produce a nonhaze forming or nonprecipitate forming primary aliphatic hydrocarbon amino alkylene-substituted asparagine by effecting the reaction in the presence of a phenolic compound represented by the formula:

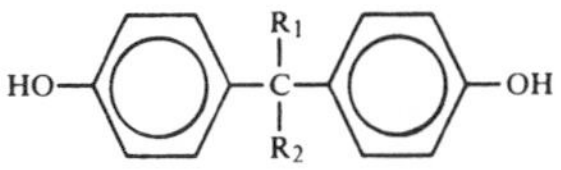

where R_1 and R_2 are independently hydrogen or alkyl groups of from 1 to 9 carbon atoms and R_1 and R_2 contain a total of 0 to 9 carbon atoms. R_1—C—R_2 can also be viewed as a methylene group when both R_1 and R_2 are hydrogen or when at least one of R_1 and R_2 is an alkyl group, R_1—C—R_2 is an alkylidene group of from 2 to 10 carbon atoms.

The alkyl diamine composition which can be employed in this process is represented by the formula: $RNH(CH_2)_3NH_2$ in which R is an alkyl radical having a chain distribution of 0.5% dodecyl; 3.5% tetradecyl; 0.5% pentadecyl; 4.0% hexadecyl; 1.0% heptadecyl; 14.0% octadecyl; 1.5% tetradecyl; 5.0% hexadecenyl; 64.0% octadecenyl; 3.0% octadecadienyl. This material is marketed as Duomeen O and is designated herein as Diamine A. The anhydride which can be employed in this process is maleic anhydride.

The reaction is conducted by reacting approximately 2 to 4 mols of Diamine A with 1 mol of maleic anhydride to produce a primary aliphatic hydrocarbon amino alkylene-substituted asparagine for use as, or for the preparation of, a fuel or lubricating oil additive. A more preferred mol ratio for the reaction is the ratio of about 3 mols of Diamine A with about 1 mol of maleic anhydride.

The phenolic antioxidant is employed at a concentration ranging from about 0.5 to 5 wt % based on the amount of Diamine A employed in the reaction. A preferred phenolic antioxidant concentration is about 1 wt %. The manner of introducing the phenolic antioxidant is not critical. It has been found convenient, however, to mix the antioxidant with the maleic anhydride in the diluent oil before adding the Diamine A.

The following examples illustrate the preparation of a primary aliphatic hydrocarbon amino alkylene substituted asparagine useful as a carburetor detergent and corrosion inhibitor in fuels and lubricants by the practice of this process.

Example 1: A solution containing 261.7 g of diluent oil having an SUS at 100°F of 100 and 31.7 g (0.32 mol) of maleic anhydride is heated to 60°C. 230 g (0.67 mol) of Diamine A are added dropwise to the solution while maintaining the temperature below 80°C. No phenolic antioxidant haze inhibitor is added to this comparison example. After heating the mixture at 100°C for 2 hours, the solution is cooled to 50°C and filtered. The 517 g of reaction product yielded either was hazy or became hazy on standing at room temperature for one month. Often a phase separation occurred in this reaction product mixture.

The analysis of the reaction product by infrared spectroscopy revealed amine carboxylate salt absorption bands at 2100 to 2220 cm^{-1}, and an amide carbonyl stretching band at 1640 cm^{-1}. The infrared analysis also showed an absence of anhydride and succinimide bands.

Example 2: A solution containing 261.7 g of diluent oil having an SUS at 100°C of 100, 3.17 g (0.32 mol) of maleic anhydride and 5.3 g (0.023 mol) of 4,4'-bis-phenol-2,2'-propane (bisphenol A) is heated to 60°C. 230 g (0.67 mol) of Diamine A are added dropwise to the solution while maintaining the temperature below 80°C. After heating the mixture at 100°C for 2 hours, the solution is cooled to 50°C and filtered, yielding 517 g of a clear amber product which does not become hazy on standing under the same conditions as in Example 1.

The above examples illustrate a process wherein the formulation of haze in the reaction product of maleic anhydride and Diamine A is prevented by the addition of a phenolic antioxidant, more specifically 4,4'-bisphenol 2,2'-propane (bisphenol A). The marked reduction or prevention of haze formation by this process enhances the usefulness of the carburetor detergent produced thereby.

FOR HYDRAULIC FLUIDS

Substituted Mono- and Bicyclic Oxazolidines for Automatic Transmission Fluids

In prime movers utilizing a functional fluid for power transmission, including hydraulic fluids and automatic transmission fluids, it is generally necessary to remove heat generated during the operation of the functional fluid.

One approach involves passing the fluid through a heat exchanger utilizing copper as a structural part or in a brazing mixture joining structural parts, e.g., the automatic transmission fluid of a car is frequently controlled by a heat exchanger located in the car radiator and immersed in the radiator coolant. Operational corrosion of the copper results in mechanically catastrophic intermixing of the functional fluid and radiator coolant (ethylene glycol) and/or loss of the fluid. It is necessary to reduce the copper corrosiveness of the fluid circulating in contact with copper so as to extend the operational lifetime of the prime mover or other mechanical device employing the fluid. One approach is to incorporate a compatible anticopper corrosion additive into the fluid.

It has been found by *S.J. Brois, J. Ryer and E.D. Winans; U.S. Patent 4,277,354; July 7, 1981; assigned to Exxon Research & Engineering Co.* that oil-soluble hydrocarbyl substituted analogues of 1-aza-3,7-dioxabicyclo[3.3.0] oct-5-yl methyl alcohols preferably in both the 2 and 8 positions, impart excellent anti-copper-corrosion activity to mineral oils and are particularly stable when added in at least a copper-corrosion reducing amount to a functional fluid, preferably a mineral oil system useful as an automatic transmission fluid (ATF) for prime movers.

The oil-soluble additives of the process can be characterized by the formulas:

```
         O                                   O
       /   \                               /   \
    CH2      \                          CH2      \
     |        CHR                        |        CHR
HOCH2—C—N                  or      HOCH2—C—N         R'
     |        CHR                        |         C
    CH2      /                          CH2      /   R''
       \   /                               \   /
         O                                   O
```

where R represents hydrogen and C_1 to C_{30} hydrocarbyl substituent and R' and R'' may be the same or different and are each C_1 to C_7 hydrocarbyl groups, e.g., methyl, ethyl, tert-butyl, phenyl, etc. The additives of the process are obtained from the reaction of 1 molar proportion of tris(hydroxymethyl)aminomethane (THAM), with at least 2 molar proportions of a C_1 to C_{30} substituted aldehyde or, with the combination of 1 molar proportion of the aldehyde and 1 molar portion of a ketone containing from 3 to 15 carbons.

The oil-soluble additives of this process can be incorporated into a wide variety of functional fluids, at concentrations generally within the range of about 0.01 to 1%, preferably 0.05 to 0.5, weight percent, of the total composition.

When the oil-soluble additives of this process are used as anticopper-corrosion additives for automatic transmission fluids, it has been found that these additives do not deteriorate the frictional properties of the ATF, i.e., these additives are compatible in ATF. The ATF lubricants contain many other additives.

Example 1: *1-Aza-3,7-dioxabicyclo[3.3.0]oct-5-yl methyl alcohol* – 0.1 mol (12.1 g) of THAM was dissolved in an equal weight of water. To the resulting solution in a 250 ml Erlenmeyer flask equipped with magnetic stirrer was added 0.2 mol (6.0 g) of paraformaldehyde. The stirred mixture was heated to 70°C to effect dissolution of the paraformaldehyde and continued for 15 minutes at 70°C to produce the 1-aza-3,7-dioxabicyclo[3.3.0]oct-5-yl methyl alcohol (hereinafter referred to as DOBO) in quantitative yields. The product, after evaporation of water and recrystallization from benzene, melted at 60° to 61°C and analyzed for 49.12% carbon, 7.52% hydrogen and 9.59% nitrogen. This product was not oil-soluble.

Example 2: *1-Aza-3,7-dioxa-2,8-dipropylbicyclo[3.3.0]oct-5-yl methyl alcohol* – 1.5 mols (181.5 g) of THAM and 3.0 mols (216 g) of n-butyraldehyde were added to 200 ml of benzene in a 1 liter flask provided with a Dean Stark trap to collect evolved water. The reactants were heated for 5 hours at from 78° to 102°C with the collection of 54 cc of water. The benzene was then distilled off and the resulting clear-yellow viscous residue was vacuum distilled at 99° to 105°C and 0.08 to 0.1 mm pressure. The product analyzed for 62.7% carbon, 10.1% hydrogen and 6.1% nitrogen.

Example 3: *1-Aza-3,7-dioxa-2,8-diisopropylbicyclo[3.3.0]oct-5-yl methyl alcohol* – The procedure of Example 2 was generally followed except 2 mols (242 g) of THAM and 4.1 mols (296 g) of isobutyraldehyde was admixed with 400 ml of benzene. 65 cc of water was collected after distillation. The product, a colorless oil, analyzed for 62.6% carbon, 9.6% hydrogen and 6.1% nitrogen.

Example 4: The following data is illustrative of the copper corrosion inhibition improvement of ATF lubricants afforded according to this process.

Two commercial ATF lubricants 1 and 2 were examined in the following copper corrosion test in both modified and unmodified form. The copper corrosion test is carried out as follows. A copper specimen 3" x ½" x 1/16" is polished until clean and uniform, washed in hexane, dried and weighed to the tenth of a milligram. 40 cc of the test fluid is placed in a test tube into which the copper bar is immersed, and the test tube thereafter corked with a cork with two 1/8" holes in it. The tube is placed in a 300°F aluminum block for 65 hours. At the end of the time, the specimen is removed, washed in hexane, rubbed vigorously with a paper towel to remove any loose deposits, rewashed and reweighed. The results of the test are shown in the following table.

ATF Lubricant	ATF 1	ATF 2
Unmodified	15.21	14
Modified by addition of 0.2 wt % of product of Example 2	14	7
Modified by addition of 0.3 wt % of product of Example 2	2	–

The additive product of Example 3 was incorporated into an ATF formulation at a 0.09 wt % concentration (based on the entire weight of the ATF formulation) as an anticopper-corrosion inhibitor. The resulting ATF formulation passed the L-2 Friction Test required by the Buick Division of General Motors Corporation and conducted on SAE No. 2 friction apparatus which showed the additive of the process had no adverse effect on the friction characteristics of the ATF; a deemulsibility test; and passed the difficult Turbo Hydromatic Transmission Cycling Test–which is a copper braze corrosion test published in Dexron II Automatic Transmission Fluid Specification by General Motors Co., Detroit, Michigan (see Pub. No. 6137-M 2nd Ed. July 1978, Appendix Page 35).

In this case *H.E. Deen, R.O'Halloran, E.D. Winans, J. Ryer and S.J. Brois; U.S. Patent 4,277,353; July 7, 1981; assigned to Exxon Research & Engineering Co.* provide oil-soluble 4-alkyl substituted mono- and 5-alkyl substituted bicyclic oxazolidines, e.g., 1-aza-3,7-dioxa-5-ethyl[3.3.0] octane that impart excellent anticopper-corrosion activity to mineral oils. They are particularly useful when added in at least a copper-corrosion reducing amount to a functional fluid, preferably a mineral oil system useful as an ATF for prime movers.

The oil-soluble additives of this process can be characterized by the formulas:

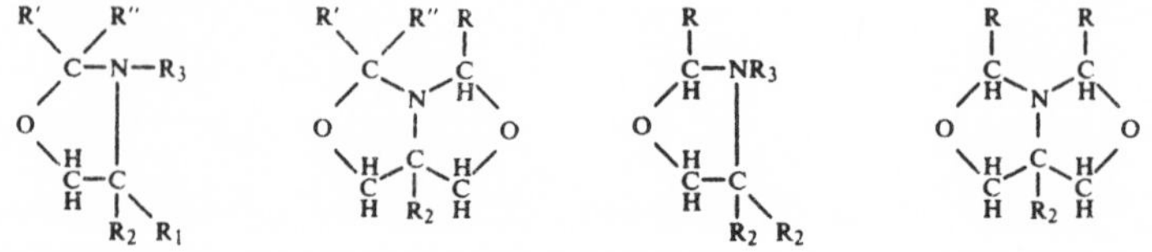

where R is hydrogen or a C_1 to C_{30} hydrocarbyl substituent group; R' and R" may be the same or different and each is a C_1 to C_7 hydrocarbyl substituent group; R_1 is hydrogen, methylol, methyl or ethyl; R_2 which can be the same as R_1 or different is hydrogen, methyl or ethyl; and R_3 is hydrogen or methyl.

The additives are obtained from the reaction of 2 mols of a C_1 to C_{30} hydrocarbyl substituted aldehyde or 1 mol of a ketone having 3 to 15 carbons or 1 mol of the aldehyde and 1 mol of the ketone per mol of an aminoalkanol (includes both the monool and diol) having from 4 to 7 carbons and preferably according to the formula:

$$R_3HN-\underset{\underset{R_2}{|}}{\overset{\overset{CH_2OH}{|}}{C}}-R_1$$

where R_1, R_2 and R_3 are the same as earlier defined.

The oil-soluble additives of this process can be incorporated into a wide variety of functional fluids. They are preferably used in lubricating oil compositions, such as automotive crankcase lubricating oils, automatic transmission fluids, etc., and at concentrations generally within the range of about 0.01 to 1%, preferably 0.05 to 0.5, weight percent, of the total composition.

Example 1: *1-Aza-3,7-dioxa-5-methyl-bicyclo[3.3.0] octane* – 1.0 mol (105 g) of 2-amino-2-methyl-1,3-propane diol was heated 4 hours with 2.0 mols (60 g) of paraformaldehyde in 200 ml of benzene. 37 ml of water were collected in 4 hours in a Dean Stark trap.

The reaction mix was rotoevaporated to remove the benzene and the product was vacuum distilled at 42°C (0.25 mm pressure) to give 126 g of distilled product, i.e., 1-aza-3,7-dioxa-5-methyl-bicyclo[3.3.0] octane, which analyzed for 55.78% carbon, 7.9% hydrogen and 10.65% nitrogen.

Example 2: *1-Aza-3,7-dioxa-2,8-diisopropyl-5-ethyl-bicyclo[3.3.0] octane* – 1.5 mols (178.8 g) of 2-amino-2-ethyl-1,3-propane diol and 3.0 mols (216 g) of isobutylaldehyde were added to 200 ml of xylene in a 1 liter flask provided with a Dean Stark trap to collect evolved water. The reactants were heated for 5 hours at from 110° to 155°C. After rotoevaporation to remove xylene, on vacuum distillation, the product was obtained as a slightly viscous liquid.

Example 3: The following data is illustrative of the copper corrosion inhibition improvement of ATF lubricants afforded according to this process.

The additive product of Example 2 was incorporated into an ATF formulation at a 0.13 wt % concentration (based on the entire weight of the ATF formulation) as an anticopper-corrosion inhibitor. The resulting ATF formulation passed the L-2 Friction Test required by the Buick Division of General Motors Corporation and conducted on SAE No. 2 friction apparatus which showed the additive of the process had no adverse effect on the friction characteristics of the ATF; passed a deemulsibility test; and passed the Turbo Hydromatic Transmission Cycling Test–which is a copper braze corrosion test published in Dexron II Automatic Transmission Fluid Specification by General Motors Co., Detroit, Michigan (see Pub. No. 6137-M 2nd Ed. July 1978, Appendix Page 35).

Poly(Oxyalkylated) Hydrazines

H.F. Lederle and F.J. Milnes; U.S. Patent 4,317,741; March 2, 1982; assigned to Olin Corporation provide selected poly(oxyalkylated) hydrazines of the formula shown on the following page

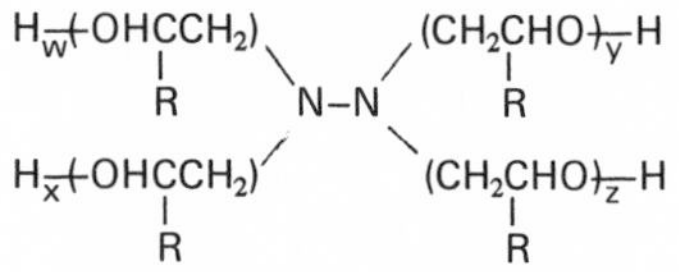

$$\begin{array}{ccc} H_{\overline{w}}\!\!+\!OH\underset{\displaystyle R}{\underset{|}{C}}CH_2) & & (CH_2\underset{\displaystyle R}{\underset{|}{C}}HO\!+_{\overline{y}}H \\ & N{-}N & \\ H_{\overline{x}}\!\!+\!OH\underset{\displaystyle R}{\underset{|}{C}}CH_2) & & (CH_2\underset{\displaystyle R}{\underset{|}{C}}HO\!+_{\overline{z}}H \end{array}$$

where R is selected from hydrogen, lower alkyl groups having from 1 to 4 carbon atoms, phenyl and mixtures thereof; and the sum of w, x, y and z is from about 4 to 20.

One preferred use of the corrosion inhibitors of the process is in hydraulic fluids which are in contact with metal surfaces. Such hydraulic fluid compositions contemplated by the process include hydraulic brake fluids, hydraulic steering fluids, fluids used in hydraulic lifts and jacks. Also included are hydraulic fluids used in hydraulic systems such as employed in heavy equipment and transportation vehicles including highways and construction equipment, railways, planes and aquatic vehicles.

In the following examples all parts and proportions are by weight unless otherwise explicitly indicated.

Example 1: A one-liter, three neck flask was fitted with a magnetic stirrer, thermometer, condenser with a dry ice condenser in tandem, nitrogen inlet and a pressure equalized dropping funnel. The apparatus was swept with nitrogen and then charged with hydrazine hydrate [100 g (2 mols) (64% aqueous solution of N_2H_4)]. While stirring and cooling, the addition of propylene oxide (PO) was begun [464 g (8 mols)]. The addition rate was 1 to 2 drops per second and the reaction temperature was maintained at 30°±10°C by use of an ice bath and by adjusting the dropping rate.

After adding ⅔ of the propylene oxide, the ice bath was removed and the remainder added slowly while heating the reaction flask to 90°±10°C. When the addition was complete, the reaction mixture was heated 2½ hours at 90°C. After cooling under nitrogen, the product was stripped at 80° to 90°C under 1 to 2 mm vacuum to remove water. The elemental analysis of this adduct was as follows: Calculated for $N_2H_4{\cdot}4C_3H_6O$: C, 54.54%; H, 10.61%; N, 10.61%. Found: C, 52.70%; H, 9.98%; N, 10.50%.

Example 2: Hydrazine hydrate [50 g (1 mol) (64% aqueous solution of N_2H_4)] was charged to a N_2 flushed flask used in Example 1 and propylene oxide [172 g (3 mols)] was added over an 8 hour period, maintaining the temperature below 43°C with an ice bath. Additional propylene oxide [290 g (5 mols)] was slowly added over 16 hours while heating between 85° and 100°C to give a viscous, water clear liquid. The product was cooled to room temperature and placed under 0.3 torr vacuum and heated to about 60°C. Considerable amount of volatiles came over. Obtained was 396 g of product which corresponds to the addition of 6 mols of propylene oxide. The elemental analysis for this adduct was as follows: Calculated for $N_2H_4{\cdot}6C_3H_6O$: C, 56.84%; H, 10.53%; N, 7.37%. Found: C, 54.88%; H, 9.62%; N, 6.92%.

Example 3: The compound prepared in Examples 1 and 2 were tested as corrossion inhibitor in a polyglycol-based hydraulic fluid according to the test method set forth in SAE-J1703F. This polyglycol-based fluid with the inhibitor had the following formula:

	Percent by Weight
Triethylene glycol monomethyl ether*	76.1
Polypropylene glycol (MW 1,000)**	20.0
Polyethylene glycol (MW 300)***	3.0
Borax†	0.2
Boric acid	0.2
Compound of Example 1 or 2	0.5

*Poly-Solv (Olin Corp.)
**Poly-G 20-112 (Olin Corp.)
***Poly-G 300 (Olin Corp.)
†$Na_2B_4O_7 \cdot 10H_2O$

After this fluid formulation was made, a bundle of six different metal coupons (previously weighed) was placed in a test jar containing 380 ml of this fluid and 20 ml of water. All of the coupons were completely covered by the solution. After running the test at 100°C for 5 days, the coupons were removed, washed, dried and reweighed. The weight change per square centimeter was then determined for each coupon. The results of this corrosion test in the fluid containing the compounds of Examples 1 and 2 are given in the table below. As can be seen, the fluids containing the inhibitors had a smaller weight change for some metals than did a fluid containing only the borate buffer alone and, thus, offered protection against corrosion. No antioxidant was used in this hydraulic fluid.

	Copper	Brass	Cast Iron	Aluminum	Steel	Tinned Iron
Maximum permissible (SAE J1703F)	0.4	0.4	0.2	0.1	0.2	0.2
Borate buffer	0.67	0.69	0.11	0.01	0.35	0.01
$N_2H_4 \cdot 4PO$	0.03	0.03	0.15	0.05	0.04	0.05
$N_2H_4 \cdot 6PO$	0.03	0.01	0.08	0.02	0.03	0.03

Phosphate Ester Fluids Containing Piperazines

Hydraulic fluids based on phosphoric acid esters have been used for some time. They have characteristics suitable therefor; particularly, they are not easily inflammable, so that they are used above all in systems comprising closed hydraulic circuits such as hydraulic systems of turbines and of pressure casting, continuous casting and press plants.

In accordance with the process of *J. Huebner; U.S. Patent 4,318,817; March 9, 1982; assigned to Mobil Oil Corporation* there is provided a fire resistant fluid comprising a phosphate ester and a corrosion or hydrolysis inhibiting amount of piperazine or a substituted piperazine. Piperazine may be substituted with an alkyl, alkaryl, hydroxy, hydroxyalkyl, alkoxy, alkoxyalkyl, alkoxyaryl, amino, aminoalkyl, alkylamino, alkylaminoalkyl, a dialkylamino or a dialkylaminoalkyl group. The proportion of piperazine or piperazine derivative in fluids according to the process is about 0.005 to 5% by wt, and preferably 0.01 to 0.5% by wt.

The phosphate esters used as the base for fire resistant fluids are mainly aryl esters of phosphoric acid. The triaryl esters are particularly preferred because they are clearly superior in regard to fire resistance but diaryl alkyl phosphate esters may also be employed.

–6–

Natural Gas and Oil Industry Applications

PRODUCTION OPERATIONS

Aqueous Systems Containing Ammonium Carboxylates

W.A. Higgins; U.S. Patent 4,250,042; February 10, 1981; assigned to The Lubrizol Corporation has found a method of inhibiting metal corrosion in earth drilling operations. The method comprises contacting at least some of the metal during its use in the operations with at least one water-soluble, ammonium carboxylate salt made from at least one polycarboxylic acid corresponding to the formula: $R(COOH)_{2-3}$, where R is an alkylene or monohydroxy alkylene group of about 4 to 15 carbons and from at least one amino compound corresponding to the formula: $(R')_3N$, where each R' is independently hydrogen, C_{1-20} hydrocarbyl or a C_{2-20} hydroxyl hydrocarbyl group.

Mixtures of two or more salts can, of course, be used. Generally a corrosion inhibiting amount is at least about 0.01 wt % of the system and as much as up to the saturation point of the inhibitor salt(s) in the aqueous system. Typically, an inhibiting amount is about 0.5 to 5% by wt of the aqueous system.

In the following examples all parts and percentages are by weight unless otherwise specified.

Example 1: A mixture of 20.2 parts sebacic acid, 12.2 parts water and 14 parts ethanolamine is heated with intermittent agitation for 30 min at 80°C. A clear yellow liquid containing 70% of the desired salt is obtained.

Example 2: A mixture of 29 parts suberic acid, 20.4 parts ethanolamine and 21.3 parts water is heated at 80°C for 1 hr with intermittent agitation. A clear yellow liquid containing 70% of the desired salt is obtained.

Example 3: A mixture of 18.8 parts azelaic acid, 13.2 parts water and 12.2 parts ethanolamine is treated in the same fashion as Example 1 to yield a 70% solution of the desired salt.

Example 4: A mixture of 27 parts pimelic acid, 20.4 parts ethanolamine and 20.3 parts water is treated in the same manner as Example 2 to yield a 70% solution of the desired salt.

Example 5: A mixture of 149 parts water and 79 parts di(ethanol)amine is heated to 60° to 70°C. Then 70.5 parts azelaic acid is slowly added. The resulting mixture is heated for 0.5 hr to produce a 50% solution of the desired salt. When tested in the filter paper rust test at 1% in water only 1 rust spot was observed.

Example 6: To a stirred mixture of 376 parts azelaic acid (4 eq) and 208 parts water is slowly added 255 parts of reagent grade ammonium hydroxide (4 eq). The addition causes the mixture's temperature to rise to 60°C. The stirred mixture is held at this temperature for 1 hr to yield a 53% solution of the desired salt in water.

Effectiveness of the inhibitor salt made in Example 6 can be demonstrated in an autoclave corrosion test. Test specimens are prepared from 3½" drill pipe made of J55 steel. Each run consists of exposing three specimens in 100 ml beakers immersed in tap water and a 1% concentration of the inhibitor salt. Each specimen is immersed to 80% in the water, the autoclave is pressured to 2350 psi air or oxygen at 185°F for 20 hr. The specimen surfaces are sandblasted prior to use. Sample weights before and after the test give weight loss data; from this data the corrosion rate and average corrosion rate are calculated. Results in the presence of air and 100% oxygen are shown in the following table. From this data it can be seen that the inhibitor salt substantially prevents corrosion of drill pipe in the presence of air and oxygen.

Inhibitor	Atmosphere	Specimen Weight Loss (g)	Specimen Weight Loss (%)	Corrosion Rate (mpy)	Average Corrosion Rate (mpy)
None	Air	0.3088	0.43	165.2	168
		0.2974	0.42	159.1	
		0.3332	0.47	178.2	
1% Example 6	Air	0.0	0.0	0.0	0.033
		0.002	0.0003	0.1	
		(+0.0004)	(+0.0003)	0.0	
None	100% oxygen	0.0052	0.008	2.78	2.74
		0.0027	0.004	1.44	
		0.0075	0.01	4.01	
1% Example 6	100% oxygen	(+0.0001)	(+0.0001)	0.0	0.0
		(+0.0001)	(+0.0001)	0.0	
		(+0.0003)	(+0.0004)	0.0	

Scavenging Hydrogen Sulfide with Organic Zinc Chelate

The process of *L.L. Carney; U.S. Patent 4,252,655; February 24, 1981; assigned to Halliburton Company* relates to the removal or inactivation of hydrogen sulfide or soluble sulfide ion contamination which is frequently encountered in wells which penetrate subterranean formations such as oil wells, gas wells and the like. Fluids in sewage systems, fluids produced from wells and make-up fluids also frequently contain hydrogen sulfide. By this process, these fluids can be pretreated prior to any hydrogen sulfide encounter or treated to remove hydrogen sulfide contamination after it has occurred. Removal or inactivation of this sulfide ion is necessary to prevent poisoning of surrounding personnel, contamination of the area and excessive corrosion of steel pipe and tools used in the well.

In the practice of this process, the organic zinc chelate is merely added to the drilling fluid which is preferably an aqueous dispersed or nondispersed fluid. The drilling fluid can also be an oil base or emulsion fluid. The oil can be any normally liquid hydrocarbon such as aliphatic hydrocarbon, an aromatic hydrocarbon or mixtures thereof.

The fluid is typically circulated in the well during drilling and other operations so that the concentration of organic zinc chelate should be monitored to maintain a certain concentration, preferably from about several parts per million to several percent, depending on the possibility of encountering hydrogen sulfide. Normally, a concentration of up to about 5 pounds per barrel (ppb) of fluid will be sufficient.

The fluid should also be monitored to indicate the presence of any hydrogen sulfide or sulfide ions which would indicate the needed addition of organic zinc chelate or a need to increase the level of concentration of the organic zinc chelate in the fluid. As a safeguard where hydrogen sulfide is not likely to be a severe problem, chelate concentrations of about 0.25 to 0.5 ppb should be used to scavenge out trace amounts of sulfides that may not be detected by tests on the surface.

The organic zinc chelates not only remove the hydrogen sulfide but do so without adversely affecting rheological properties of the fluid such as thickening or gelling the mud or increasing fluid loss significantly. This is significant in drilling, completing or servicing a well because it is essential to maintain circulation and control of the well when hazardous conditions, such as hydrogen sulfide, are encountered.

In the particular organic zinc chelates of this process the zinc is combined with the organic chelant so that the combination has an ionization or stability constant which prevents formation of insoluble zinc hydroxide which would prevent reaction of the zinc ion with sulfide or make the zinc ion unavailable for reaction with the hydrogen sulfide. Furthermore, the stability constant is such that formation of highly soluble salts which would adversely affect the rheology of the well fluid and thus control of the well are prevented.

The preferred chelating agents are a relatively simple, low molecular weight hydrocarbon base material containing acetic or nitrogen functional groups with a stability constant in the range of about 10 to 16 as described by Chaber and Martell in *Organic Sequestering Agents*. The preferred chelates are relatively simple aliphatic amine acids or salts having at least one tertiary amino group and more than two carboxyl groups or salts. In addition, the chelates should contain at least about 10% zinc by weight. The preferred chelates contain about 15 to 25% zinc and can be blended as concentrates and added directly to the drilling fluid or premixed with water and then added to the drilling mud.

The following are examples of organic zinc chelates, with the stability constant in parenthesis which can be used: dithiotartaric acid (15.82); triethylenetetramine (12.1); ethylene-bis-α,α'-(2-aminomethyl)pyridine (11.5); β,β',β''-triaminotriethylamine (14.65); tetrakis(2-aminoethyl)ethylenediamine (16.24); α,β-diamino-propionic acid (11.5); β-mercaptoethyliminodiacetic acid (15.92); ethylenediamine-N,N'-diacetic acid (11.1); ethylenediamine-N,N-diacetic acid (11.93); ethylenebis-N,N'-(2-aminomethyl)pyridine-N,N'-diacetic acid (15.2); N-hydroxyethylenediaminotriacetic acid (HEDTA) (14.5); ethylenediamine-N,N'-dipropionic-N,N'-diacetic acid (14.5); hydroxyacetic acid; and nitrilotriacetic acid (NTA) (10.45). The preferred organo chelating agent is the NTA.

Compositions for High Temperature, High Pressure Gas Wells

In view of the severe corrosion problems encountered in gas wells producing from very deep high pressure and high temperature horizons wherein no petroleum condensate phase exists at bottomhole conditions but an aqueous or brine phase is present, *S.P. Sharp and L. Yarborough; U.S. Patent 4,295,979; Oct. 20, 1981; assigned to Standard Oil Company (Indiana)* have developed a method for inhibiting corrosion in such wells. The method involves the steps of:

(a) adding up to about 27 pbw of an alkyl amine/100 pbw of an dialkyl disulfide oil producing an amine activated dialkyl disulfide solvent;

(b) adding to the activated dialkyl disulfide solvent at least 40 pbw elemental sulfur/100 pbw dialkyl disulfide oil producing a corrosion inhibitor carrier capable of existing in a liquid phase at bottomhole conditions at the high temperature, high pressure gas well; and

(c) injecting the corrosion inhibitor carrier into the gas well to inhibit corrosion.

The term alkyl disulfide herein should be considered equivalent to the term dialkyl disulfide and they are used interchangeably. The known alkyl sulfide and disulfide sulfur solvents used in preparing the corrosion inhibitor carrier can be found, e.g., in U.S. Patents 3,531,160 and 3,846,311. The disulfide is the preferred form. At ambient conditions, they are liquid dialkyl disulfides and include such compounds as dimethyl disulfide, diethyl disulfide, dioctyl disulfide, ditert-tetradecyl disulfide, and the like.

One particularly useful starting material is a mixture of aliphatic disulfides in which the aliphatic group therein contains from about 2 to 11 carbon atoms, e.g., $(C_2H_5S)_2$, $(C_{11}H_{23}S)_2$, etc., typically those disulfide mixtures produced as a product stream of the Merox process described in *The Oil and Gas Journal,* Vol 57, pp. 73-78, Oct. 26, 1959. Briefly, such mixtures of disulfides are produced by first contacting a refinery hydrocarbon stream containing aliphatic mercaptans with a caustic solution to produce corresponding sodium salt of the mercaptans. The latter are then converted to dialkyl disulfides by air oxidation, simultaneously regenerating the caustic.

In order to enhance the formation of the desired alkyl polysulfide oil, a small effective amount of an amine catalyst or solvent activator is added to the liquid dialkyl disulfide. This addition of the amine in order to enhance the sulfur solvency properties of the dialkyl disulfide can be accomplished by any of the well-known methods found in the art, including the method described in U.S. Patent 3,846,311, except the necessity of the aging step is viewed as being optional. Also, as indicated in U.S. Patent 3,846,311, the aliphatic unsubstituted amines are believed to be uniquely suitable for activating the dialkyl disulfide. The normally liquid lower aliphatic amines of about 4 to 12 carbon atoms were found to be preferred when employed at a concentration of up to about 10 wt % based on the weight of the dialkyl disulfide.

It has been found that achieving full stoichiometric incorporation of amine nitrogen is preferred and consequently, values in excess of 10 pbw amine/100 pbw dialkyl disulfide are useful with values frequently being as high as 20 to about 27 pbw amine. In situations in which very high sulfur solubility is desired, even higher concentrations of amine can be employed. In such cases, the additional amine not only enhances the amounts of sulfur incorporated into the polysulfide oil, but appears to enhance the corrosion inhibition properties of the resulting polysulfide oil. In preparing the activated dialkyl disulfide sulfur solvent, the amine catalyst can be added directly to the dialkyl disulfide and the mixture used immediately to dissolve sulfur, thus forming the desired alkyl polysulfide oil.

Alternately, the amine/dialkyl disulfide mixture can be aged according to the procedure found in U.S. Patent 3,846,311, thus enhancing the solubility of elemental sulfur. In situations where very high sulfur content is desired (e.g., in very high temperature, very high pressure gas wells wherein high molecular weight and/or very low volatility is required), the aged amine activated dialkyl disulfide sulfur solvent is preferred. However, both the aged and the unaged are useful in the process.

Having activated the dialkyl disulfide oil by the addition of an amine, the desired alkyl polysulfide oil is then prepared by dissolving elemental sulfur in the amine activated disulfide oil. The solubility and the rate of dissolution of the sulfur in the dialkyl disulfide/amine mixtures varies somewhat with the source of the sulfur, its particular physical state and degree of pulverization. However, any elemental free sulfur is acceptable for purposes of this process.

The dissolution processes can be performed by conventional techniques with mild heating and agitation being advantageously employed. The specific amount of sulfur to be dissolved will vary according to the needs of the particular gas well to be treated. In principle, the more severe the bottomhole conditions, the greater the sulfur content. It has been found that significant liquid phase at bottomhole conditions can be achieved with as little as about 40 to 60 pbw sulfur/100 pbw dialkyl disulfide with 80 pbw sulfur/100 pbw dialkyl disulfide being particularly useful in that the resulting solution is easily handled, stored, and transported at ambient conditions without appreciable sulfur precipitation or excessive viscosities and solidification.

The process of *S.P. Sharp and L. Yarborough; U.S. Patent 4,350,600; Sept. 21, 1982; assigned to Standard Oil Company (Indiana)* also is concerned with corrosion inhibition under high temperature, high pressure bottomhole conditions in the absence of petroleum condensate phase.

In this case the composition comprises a dialkyl disulfide to which has been added a high molecular weight fatty amine in quantities in excess of about 30 pbw fatty amine/100 pbw dialkyl disulfide.

The high molecular weight amines or fatty amines useful in the process are in principle long chain alkyl amines usually synthesized from naturally occurring fatty acids wherein the alkyl group involved contains 12 or more carbon atoms. The commercial available fatty amines will contain mixtures of alkyl chain lengths since they are derived from fatty acids occurring in nature. Frequently this will also result in an abundance of the even carbon numbered species and the presence of unsaturation such as found in the oleic, palmitic, and the like structures. However, any long chain predominantly aliphatic amine, whether it be a single species with either even or odd numbered carbon atoms or mixtures of these species, is viewed as an acceptable high molecular weight amine for purposes of this process. These fatty amines are preferably waxy solids or semisolids which are easily melted at temperatures characteristic of the gas wells of interest.

The preferred amines will involve carbon chain links of 16 through 30 carbon atoms. This preferred range is consistent with the view that increasing the molecular weight in order to decrease volatility is of paramount importance in achieving the desired liquid phase at bottomhole high temperatures and high pressures. In cases where the fatty amine is a liquid at room temperature (i.e., fatty amines having alkyl chain links predominantly at the lower end of the acceptable range, e.g., approaching C_{12}), the addition of elemental sulfur to the amine activated dialkyl disulfide may be necessary to achieve the desired high molecular weight heavy oil which is capable of existing as a liquid film forming phase at the severe conditions of interest. A subclass of fatty amines which have been found to be particularly useful in the process is the N-alkyl-1,3-propane diamines.

In practicing this process, although it is believed that achieving the full stoichiometric incorporation of amine nitrogen is desirable, significant liquid phase has been observed at high temperatures and high pressures with as little as about 50% stoichiometric quantities of fatty amines. Using the Merox mixture, 50% stoichiometry corresponds to about 30 to 45 pbw high molecular weight amine/100 pbw dialkyl disulfide, depending on the particular molecular weight of the amine. Furthermore, the desired liquid phase has been observed at bottomhole conditions at amine concentrations ten-fold of the lower values (300 pbw amine/100 pbw dialkyl disulfide), certainly in excess of the believed stoichiometry.

Phosphorylated Oxyalkylated Polyols for Highly Oxygenated Systems

The process of *T.J. Bellos; U.S. Patent 4,311,662; January 19, 1982; assigned to Petrolite Corporation* relates to the use of phosphorylated oxyalkylated polyols as corrosion inhibitors in highly oxygenated systems such as experienced in air drilling. These corrosion inhibitors not only inhibit general or overall corrosion but also inhibit localized corrosion of the pitting type.

The oxyalkylated polyols which are phosphorylated according to this process are ideally represented by the formula: $R[O(AO)_nH]_x$ where R is an organic and preferably hydrocarbon moiety of the polyol, OA is the oxyalkylene moiety derived from an alkylene oxide, e.g., ethylene oxide, propylene oxide, butylene oxides, etc., and mixtures or block units thereof, n is the number of oxyalkylene units and x represents the total number of units containing OH groups.

Preferred polyols include glycerol, polyglycerol, trimethanolethane, pentaerythritol, dipentaerithrytol, etc., mannitol, 1, 2, 3-hexanetriol, etc.

A number of processes are known in the art for preparing the phosphorylated polyols. A preferred process is to react polyphosphoric acid with a polyol. The polyphosphoric acid should have a P_2O_5 (i.e., phosphorus pentoxide) content of at least about 72%, preferably about 82 to 84%. A residue of orthophosphoric acid and polyphosphoric acid remains on completion of the reaction. This residue may be as high as about 25 to 40% of the total weight of the phosphorylated polyol. It may either be removed or left in admixture with the phosphorylated polyol. Preferably, the phosphorylated polyols produced by this process are prepared employing amounts of a polyphosphoric acid having about 0.5 to 1 molar equivalents of P_2O_5 for each equivalent of the polyol used. Larger amounts of polyphosphoric acid can be used if desired. By equivalent of the polyol is meant the hydroxyl equivalents of the polyol. The phosphorylated polyols (acid esters) can be partially or completely converted to their corresponding alkali metal salts or ammonium salts by reacting with appropriate basic material.

The compositions are polyfunctional acid phosphate esters of polyhydric alcohols, the esters having the formula $R{+}OPO_3H_2)_x$ wherein R is the hydrocarbyl group of a polyhydric alcohol (i.e., R is any remaining organic residue of a polyhydric alcohol used as the starting material) and x is a number from 2 to 6, the esters often being referred to in the art as phosphorylated polyols.

Also included within the definition of polyol are amine-containing polyols such as of the general formula above where R is amino-containing, e.g., tris(hydroxymethyl)aminomethane, 2-amino-2-ethyl-1,3-propanediol, triethanolamine, etc. The following examples illustrate the oxyalkylated polyols of the process.

Example	
1	1 mol Pentaerythritol plus 4 mols EtO
2	1 mol Pentaerythritol plus 5 mols EtO
3	1 mol Pentaerythritol plus 6 mols EtO
4	1 mol Pentaerythritol plus 7 mols EtO
5	1 mol Pentaerythritol plus 8 mols EtO
6	1 mol Glycerin plus 3 mols EtO
7	1 mol Glycerin plus 4 mols EtO
8	1 mol Glycerin plus 5 mols EtO
9	1 mol Trimethyolpropane plus 3 mols EtO
10	1 mol Trimethyolpropane plus 4 mols EtO
11	1 mol Trimethyolpropane plus 5 mols EtO

The phosphorylated oxyalkylated polyols are employed in the oxygenated systems at a pH of from about 5.0 to 8.5, such as from 6.0 to 8.0, but preferably from about 6.5 to 6.7.

The phosphorylated oxyalkylated polyols are employed in concentrations of at least about 1,500 ppm, but preferably from about 2,000 to 3,000 ppm, these amounts being effective to inhibit corrosion.

Quaternary Polyaminoamides for Deep Well Systems

The process of *D. Redmore and B.T. Outlaw; U.S. Patent 4,315,087; February 9, 1982; assigned to Petrolite Corporation* relates to the quaternized derivatives of the amino-amido polymers of U.S. Patent 3,445,441 and to uses thereof in the processes described in U.S. Patent 3,445,441. They are more effective as corrosion inhibitors, particularly in high temperature and highly corrosive systems than the parent nonquaternized polymers. They are particularly useful in high temperature systems where high concentrations of CO_2 and/or H_2S are present, most particularly in deep gas wells.

The compositions are quaternaries of polymers of the general unit formula:

$$-\overset{R'}{\overset{|}{N}}-\left[A\overset{R'}{\overset{|}{N}}\right]_n-CH_2\overset{R}{\overset{|}{C}}H-\overset{O}{\overset{\|}{C}}-$$

where R is hydrogen or methyl, R' is hydrogen, A is a $-CH_2CH_2-$ moiety of a hydrocarbon polyamine which links amino groups and n is an integer; and cross-linked derivatives thereof, where the polyamine is of the general formula:

$$NH_2\left[CH_2CH_2-\overset{H}{\overset{|}{N}}\right]_{1\text{-}5}H$$

where the quaternizing agent is selected from the group consisting of benzyl chloride, ethylene dichloride and p-dodecylbenzyl chloride.

Example 1: *Polyaminoamide from methyl acrylate and ethylenediamine* – To ethylenediamine (120 g; 2 mols) in a flask fitted with a stirrer, reflux condenser and addition funnel was added methyl acrylate (86 g; 1 mol) during 1 hr at 45° to 55°C. The mixture was heated under reflux for 1 hr (pot temperature 100° to 115°C) and then the flask was fitted with a still head. Methanol (32.4 g) was removed by distillation at atmospheric pressure. The mixture was then heated gradually to 175°C under reduced pressure, 95 to 100 mm, resulting in distillation of excess ethylenediamine (60.1 g). The viscous residue was dissolved in water (113 g). Analysis of product (before dilution with water): total nitrogen 22.2%; basic nitrogen 13%.

Example 2: *Polyaminoamide from methyl methacrylate and ethylenediamine* – To ethylene diamine (60 g; 1 mol) was added methyl methacrylate (100 g; 1 mol) during 20 min with efficient stirring. The reaction mixture was gradually heated to 110° to 115°C during which methanol began to form. The methanol was allowed to distill from the reaction flask as the temperature was increased to 165°C during 2½ hr (32 g methanol was collected). The viscous polymer resulting was dissolved in water (130 g). Analysis (without water): total nitrogen 23.7%; basic nitrogen 13.4%.

Example 3: *Polyaminoamide from diethylenetriamine and methyl methacrylate* – Methyl methacrylate (100 g; 1 mol) was added dropwise in 20 min to diethylenetriamine (103 g; 1 mol) with good stirring. The mixture was gradually heated to 160°C with gradual removal of methanol (35 g) by distillation. The resulting polymer was dissolved in water (170 g).

Example 4: Following the procedure of Example 3 amine AL-1 (polyamine mixture from Jefferson Chemical Co.) (100 g) was reacted with methyl methacrylate (100 g). The resulting gum was dissolved in water (100 g). Analysis (before dissolution in water): total nitrogen 16.59%; basic nitrogen 9.3%.

The following examples illustrate the preparation of alkylated polyaminoamide structures.

Example 5: The aminoamide polymer of Example 1 as a 50% aqueous solution (80 g; 0.37 eq basic nitrogen) was heated with benzyl chloride (43.3 g; 0.34 mol) under reflux for 2 hr (mol ratio alkylating reagent to basic nitrogen of 0.92). The product was diluted to 50% active using isopropanol (43 g). Analysis: Ionic chloride, 14.7%. Calculated: 15.1%.

Example 6: The aminoamide polymer of Example 4 as a 62% aqueous solution (80 g; 0.53 eq basic nitrogen) was heated under reflux with benzyl chloride (46.7 g; 0.37 mol) for 2 hr. After cooling the product was diluted with isopropanol. The molar ratio of alkylating reagent to basic nitrogen was 0.70.

Examples 7 through 10: The following table summarizes additional quaternization reactions carried out by the procedure of Example 5.

Example No.	Aminoamide Polymer	Quaternization Reagents	Equivalency Ratio
7	Example 2	p-dodecylbenzyl chloride	0.70
8	Example 2	ethylene dichloride	0.80
9	Example 4	p-dodecylbenzyl chloride	0.90
10	Example 3	benzyl chloride	1.00

Since the reaction goes substantially (95 to 100%) to completion, at an equivalency ratio of 0.7, 70% basic nitrogen is alkylated; at a mol ratio of 0.9, 90% is alkylated, etc.

The effectiveness of the inhibitors of the process is illustrated by the following tests which are designed to simulate the extreme conditions of deep wells.

CO_2 Procedure: 400 ml of deionized water are charged to a stainless steel test vessel fitted with mild steel coupons, and the liquid is sparged to 1.5 hr with CO_2 to remove the O_2. The vessel is then put on a magnetic stirring hot plate and brought up to temperature (250°F). At this time 100 psi CO_2 was added and the electrodes allowed to freely corrode for 1.5 hr. The corrosion rate is measured with a PAIR meter (blank, R_O). The inhibitor is then added with an overpressure of 100 psi additional CO_2 and the corrosion rates measured (R_1). The percent protection = $(R_O - R_1)/R_O \times 100\%$.

Compound	Concentration (ppm)	Corrosion Rates (mpy) Blank R_O	Corrosion Rates (mpy) Inhibited After 20 Hours R_1	Protection (%)
Example 7	500	450	168	63
Example 8	500	640	227	65
Example 5	500	450	370	18

H_2S/CO_2 Procedure: 400 ml of deionized water are charged into a stainless test bomb. The water is sparged for 1½ hr with nitrogen to remove the oxygen content, followed by addition of 100 psig H_2S and 50 psi of CO_2. The bomb is brought up to temperature (250°F) and the electrodes are allowed to corrode for 1½ hr. The corrosion rate is measured at this time (blank R_O). The inhibitor to be evaluated is charged into the bomb with an additional 50 psi of CO_2 and the corrosion rates measured by PAIR meter. Percent protection = $(R_O - R_1)/R_O \times 100\%$.

Compound	Concentration (ppm)	Corrosion Rates (mpy) Blank R_o	Inhibited After 20 Hours R_1	Protection (%)
Example 6	500	250	24	90
Example 5	500	250	35	86
Example 7	500	300	76	75
Example 8	500	240	46	81

In the above tests, the corresponding nonquaternary polymers gave little, if any, protection.

Prevention of Hydrogen Embrittlement

One major problem encountered in offshore oil and gas pipelines is their susceptibility to corrosion, the problem being especially serious in the riser pipes which carry the hot oil or gas from the ocean bed to the underwater pipes which transport the fuel to the shore. The riser pipes are subjected to considerable stress caused by the oil or gas pressure as well as the pounding action of the waves.

The process of *A.C.C. Tseung, A.I. Onuchukwu and H.C. Chan; U.S. Patent 4,335,754; June 22, 1982* is based on the observation that a major factor contributing to the deterioration of metal substrates, such as steel pipes, in corrosive environments, especially when under mechanical stress in the presence of seawater and H_2S, is embrittlement caused by adsorption of the hydrogen which is cathodically evolved during other corrosion processes.

There is accordingly provided a method of protecting a metal substrate in a corrosive environment against hydrogen embrittlement which comprises providing, in the vicinity of the substrate and in electrical contact therewith, an active substance which has a low over-potential for the cathodic evolution of hydrogen.

The active substance, on which the hydrogen is preferentially cathodically evolved, may be applied in a variety of forms. For example, it may be applied in the form of or as a coating on, a separate member, e.g., an electrode which is in electrical contact with the metal substrate. Alternatively, it may be applied as a coating on, or incorporated in a coating composition which is coated on the metal substrate. In yet other embodiments, the active substance may be formed in situ on the metal substrate or an electrically connected member.

Especially suitable active substances for use in the process are the catalytic compounds described in British Patent 1,556,452. These compounds comprise sulfur, optionally together with oxygen, and at least two metals selected from cobalt, nickel, iron and manganese. Preferred compounds have the formula:

$$A_x B_{4-2x} S_{3.6-4} O_{0.4-0}$$

where x is 0.05 to 1.95 and A and B are any different two metals selected from Co, Ni, Fe and Mn, e.g., Co and Ni, $NiCo_2S_4$ being especially preferred.

These compounds, which may be prepared by treating the corresponding mixed oxides with a sulfur-containing compound, such as H_2S or CS_2, may be provided as porous particles formed by incorporating them with a suitable chemically inert binder, especially polytetrafluoroethylene (PTFE). As described in British Patent

1,556,452 the porous particles, which typically comprise from 1 to 10, preferably from 2 to 6, pbw of binder for each 10 parts of the active compound, may be assembled to form an electrode, e.g., by painting a dispersion of them in the binder on a suitable support, such as a nickel screen, and then air drying and curing. One particularly preferred active substance comprises 3 parts PTFE and 10 parts $NiCo_2S_4$.

For use in the protection of pipelines it is preferred to coat the active substance directly onto the surface to be protected. This may be conveniently effected by coating the surface of the substrate with a solution of the mixed oxide (or a precursor thereof, such as a mixture of metal nitrates, which will decompose on heating), with or without a chemically inert binder, and then converting the mixed oxide by treatment with H_2S or another sulfur-containing compound, e.g., CS_2, which may be preferable in certain applications since it will reduce the possibility of hydrogen permeation into the substrate during the coating process.

In other applications, the mixed sulfide may be applied to the substrate as an ingredient in a paint or similar coating composition (care being taken that it does not become electrically insulated within the composition so that electrical continuity is lost).

Alternatively, the active substance may advantageously be applied as a coating to the metal substrate by electrophoretic deposition; such a technique is especially suitable for the internal coating of pipes on an industrial scale. For example, $NiCo_2S_4$/PTFE dispersion may be electropheretically deposited onto steel substrates followed by optional air drying and curing, at a temperature of from 300° to 350°C.

It was found that iron sulfide exhibits a surprisingly low over-potential for the cathodic evolution of hydrogen and, accordingly, is especially suitable as an active substance for use, in accordance with the process, with steel substrates because it can readily be formed as a coating in situ. For instance, a coating of rust which has naturally or deliberately been formed on the steel surface can readily be converted to FeS by treatment with a suitable sulfur-containing compound such as H_2S or CS_2. Alternatively, a FeS coating may, e.g., be applied in situ on an unoxidized steel surface by anodization followed by treatment with H_2S under alkaline conditions, or by treatment with a mixture of iron chlorides ($FeCl_2 + FeCl_3$) and H_2S in the presence of an alkali such as sodium hydroxide.

The iron sulfide should have a relatively high surface area, e.g., greater than 2 m^2/g, preferably greater than 4 m^2/g and the coating preferably has a surface area which is at least 5 times, and preferably at least 10 times, that of the geometric area of metal substrate.

In some applications, the sulfidizing ingredient may simply be added to, or may already be present in, the fluid flowing through the steel pipe and the process accordingly contemplates the addition of such an ingredient as a method of forming the sulfide (or any other active substance) in situ during the normal operation of the apparatus comprising the metal substrate which is to be protected even though, in general, it will be preferred to apply the active substance to the substrate before it is subjected to the corrosive and embrittling environment.

Compositions for Limited Oxygen Systems

R.L. Martin and E.W. Purdy; U.S. Patent 4,339,349; July 13, 1982; assigned to Petrolite Corporation have found that certain compounds inhibit ferrous metal corrosion in oil field fluids such as found in systems for producing wells and in handling of secondary recovery fluids. These compounds find particular advantage when these fluids contact limited amounts of air, such as about 5 parts per million of oxygen in the corrosion fluids or below. Concentrations of oxygen in the fluids resulting from this contact may be so low that it is not detectable, such as about a few parts per billion, e.g., 10 parts per billion or less; yet the presence of this oxygen can have a strong influence on both corrosivity and the effectiveness of inhibitors. The inhibitors have greatly increased effectiveness in these conditions.

The corrosion inhibitors of this process are characterized in two categories: one category is best for use in sour (H_2S-bearing) fluids that contact oxygen; the second category is best for sweet fluids (CO_2-bearing, with no significant H_2S).

Inhibitors in the first category include fatty quaternary ammonium compounds in conjunction with other components; these other components include one or more of the following: (1) sulfur-oxygen phosphates, (2) polyphosphate esters, and (3) cyclic amidines such as imidazolines.

Inhibitors in the second category include (1) polyphosphate esters in conjunction with one or more of cyclic amidine salts such as phosphate salts of imidazolines of the type described in U.S. Patent 3,846,071, (2) sulfur-oxygen phosphates, and (3) trithiones.

For aqueous fluids the optimum inhibitors in the first category are blends of sulfur-oxygen phosphates with polyphosphate esters and fatty quaternary ammonium compounds; for aqueous/hydrocarbon systems the optimum inhibitors in the first category are blends of cyclic amidines such as imidazolines with sulfur-oxygen phosphate compounds and fatty quaternary ammonium compounds.

For aqueous fluids of the second category, the optimum inhibitors are sulfur-oxygen phosphates and polyphosphate esters; for aqueous/hydrocarbon fluids the optimum blend contains phosphate salts of cyclic amidines such as imidazolines, polyphosphate esters and trithiones.

The patent includes examples of compositions in each category, preparations of compounds and test results.

Polyether Polyamines

The process of *H. Diery, W. Wagemann, J. Weide, R. Deubel and M. Hille; U.S. Patent 4,341,716; July 27, 1982; assigned to Hoechst AG, Germany* relates to polyether polyamines of the formula:

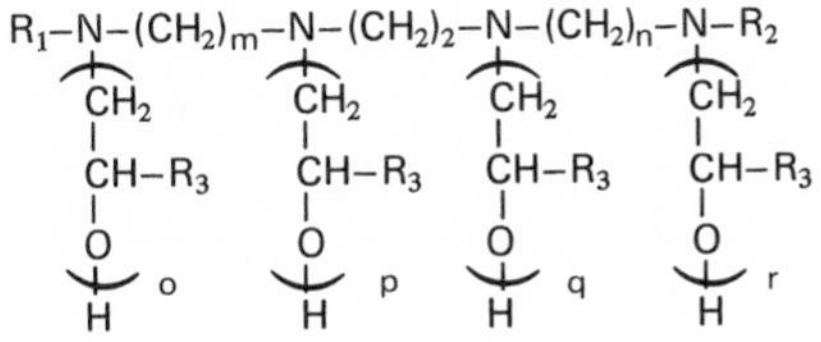

$$R_1\text{-}\underset{\left[CH_2\text{-}CH(R_3)\text{-}O\right]_o H}{N}\text{-}(CH_2)_m\text{-}\underset{\left[CH_2\text{-}CH(R_3)\text{-}O\right]_p H}{N}\text{-}(CH_2)_2\text{-}\underset{\left[CH_2\text{-}CH(R_3)\text{-}O\right]_q H}{N}\text{-}(CH_2)_n\text{-}\underset{\left[CH_2\text{-}CH(R_3)\text{-}O\right]_r H}{N}\text{-}R_2$$

In the formula R_1 and R_2 are identical or different and denote $C_{8\text{-}30}$ alkyl, $C_{8\text{-}30}$ 2-hydroxyalkyl or $C_{8\text{-}30}$ alkenyl or $C_{8\text{-}24}$ alkoxypropyl, m and n are 2 or 3, R_3 is hydrogen or methyl and the sum of o + p + q + r is an integer from 4 to 600.

The polyether amines according to the process, preferably those in which the sum of o + p + q + r is from 4 to 150, and the salts thereof can be used as corrosion inhibitors, demulsifiers and acid retarders in the exploitation and refining of mineral oil and natural gas.

Example 1: A 1 ℓ stainless steel autoclave is charged with 205 g (0.25 mol) of N,N'-ditallow alkylamine propylethylenediamine and the substance is melted. After addition of 0.3 g of freshly pulverized sodium hydroxide, the autoclave is closed, scavenged twice with nitrogen and heated while stirring. Propylene is forced in at an interval temperature of 170° to 175°C. The internal pressure should not exceed 5 to 6 bars, if possible. When the pressure drops further propylene oxide is added, a total amount of 290 g (5 mols). When the addition is terminated, the autoclave is stirred for 2 hr at 170°C. After cooling the autoclave is repeatedly scavenged with nitrogen to remove propylene residues. About 490 g of a reddish-brown liquid product are obtained.

A 2 ℓ stainless steel autoclave is charged with 330 g (0.17 mol) of the product obtained. After addition of 0.2 g of freshly pulverized sodium hydroxide, closing of the autoclave and scavenging with nitrogen, the autoclave is heated to 130° to 140°C while stirring and ethylene oxide is forced in. The ethylene pressure should not exceed 4.5 bars. A total amount of 1,237 g (28.1 mols) of ethylene oxide is added. The content of the autoclave is stirred for another 2 hr at 130°C, cooled to 60°C and scavenged with nitrogen in order to remove the ethylene oxide residues. About 1,560 g of pasty brown final product of the above formula in which R_1 and R_2 are tallow alkyl, R_3 denotes methyl and hydrogen, m = n = 3 and o + p + q + r is 185 are obtained. The product has a molecular weight of about 9,400.

Example 2: 82 g (0.1 mol) of N,N'-ditallow alkyl aminopropylethylenediamine are melted in a 1 ℓ stainless steel autoclave. After addition of 0.15 g of freshly pulverized sodium hydroxide, the autoclave is closed, scavenged twice with nitrogen and heated. At a temperature of 130° to 140°C a total amount of 420 g (9.5 mols) of ethylene oxide is added so that the internal pressure does not exceed 4.5 bars, if possible. When the addition is terminated, stirring is continued for 2 hr at 130°C whereupon the mixture is cooled to 60°C. At that temperature the autoclave is scavenged with nitrogen in order to remove residues of ethylene oxide. About 500 g of brown, waxy product of the above formula are obtained in which R_1 and R_2 denote tallow alkyl, R_3 is hydrogen, m = n = 3 and o + p + q + r is 95. Its molecular weight is about 5,000.

Example 3: This example is intended to demonstrate the efficiency of the compounds according to the process as corrosion inhibitors. The following products were tested:

(1) compound of the above formula with R_1 and R_2 = tallow alkyl, R_3 = H, m = n = 3, sum of o, p, q, r = 10;

(2) compound of the above formula with R_1 and R_2 = tallow alkyl, R_3 = H, m = n = 3, sum of o, p, q, r = 50;

(3) compound of the above formula with R_1 and R_2 = tallow

alkyl, R_3 = H, m = n = 3, sum of o, p, q, r = 20; neutralized with 4 mols of dodecylbenzenesulfonic acid;

(4) compound of the above formula with R_1 and R_2 = tallow alkyl, R_3 = H, m = n = 3, sum of o, p, q, r = 20; neutralized with 2 mols of dodecylbenzenesulfonic acid; and

(5) compound of the above formula with R_1 and R_2 = tallow alkyl, R_3 = H, m = n = 3, sum of o, p, q, r = 20; neutralized with 4 eq of mono- and diphosphoric acid esters of nonylphenoltetraglycol ether.

Test Method: Specimens of carbon steel are immersed for 7 hr at 60°C into saline water (200 g/ℓ) which is continuously stirred and through which carbon dioxide is bubbled and thereafter the loss in weight is determined. Under the above conditions the water contains at most about 0.5 ppm of oxygen. The value of protection in percent indicates the reduction of the metal corrosion with respect to the blank value.

	. . . Protection in Percent . . . . with the Use of	
Product No.	5 ppm	25 ppm
1	77	86
2	80	86
3	73	82
4	81	87
5	54	85

REFINING OPERATIONS

Chelating Agents for Iron Sulfide Removal

Iron sulfide (FeS), also known as troilite, by itself or in combination with organic materials such as asphalt or tar is often deposited on metal surfaces in process and refinery equipment such as heat exchangers, powerformers, etc., particularly where sour fluids, i.e., fluids containing hydrogen sulfide are processed. While troilite deposits are readily dissolved in common acidic solvents such as hydrochloric acid and sulfuric acid, as the troilite dissolves in such solvents hydrogen sulfide is evolved creating a severe occupational and environmental hazard. Further, when the deposits containing iron sulfide also contain organic materials such as asphalt or tar, penetration and rapid dissolution of the deposits is difficult.

By the process of *L.D. Martin; U.S. Patent 4,276,185; June 30, 1981; assigned to Halliburton Company* improved methods and compositions for dissolving deposits containing iron sulfide, including those also containing organic materials, are provided which bring about the efficient removal of the deposits with little or no evolution of hydrogen sulfide.

A composition of the process for removing deposits containing iron sulfide from surfaces with minimal hydrogen sulfide evolution is comprised of a basic aqueous solution of a chelating agent having a pH in the range of from about 8 to 10, and most preferably, a pH of 9 or above.

The chelating agents which are useful in accordance with the process are those chelating acids which in an aqueous solution are capable of dissolving iron sulfide and tightly binding iron ions in the presence of sulfide ions. Particularly suitable chelating acids for use in accordance with the process are citric acid, oxalic acid, nitrilotriacetic acid, alkylenepolyamine polyacetic acids and mixtures of such acids. Examples of the alkylenepolyamine polyacetic acids are ethylenediamine-tetraacetic acid (EDTA), diethylenetriaminepentaacetic acid, N-2-hydroxyethyl-enediaminetriacetic acid, propylene-1,2-diaminetetraacetic acid, propylene-1,2-diaminetetraacetic acid, propylene-1,3-diaminetetraacetic acid, the isometric butylenediaminetetraacetic acids, etc., and mixtures of such acids. Of these, ethyl-enediaminetetraacetic acid is preferred. The most preferred chelating acids for use in accordance with the process are nitrilotriacetic acid and ethylenediaminetetra-acetic acid.

The chelating acid or acids used are preferably present in the composition in an amount in the range of from about 2 to 10% by wt of the composition. When nitrilotriacetic acid or ethylenediaminetetraacetic acid or mixtures of such acids are used, they are preferably present in the composition in a total amount in the range of from about 4 to 8% by wt of the composition, with 4% being the most preferred.

When the deposits being removed are contained on metal surfaces, a corrosion inhibitor such as a mixture of an alkylpyridine and dibutylthiourea is preferably included in the composition. Preferably, such corrosion inhibitor is present in the composition in an amount in the range of from about 0.05 to 0.6% by volume of the composition, most preferably in the range of from about 0.1 to 0.2% by volume.

Another composition of the process for removing deposits containing organic materials and iron sulfide is an emulsified composition comprised of a basic aqueous solution having a pH of from about 8 to 10 of the chelating agent, emulsified with a liquid hydrocarbon solvent for the organic materials in the deposits. While a variety of liquid hydrocarbon solvents can be utilized, ortho-dichlorobenzene, xylene and heavy aromatic naphtha are preferred with heavy aromatic naphtha having a flash point above about 200°F being the most preferred.

Various emulsifiers can be utilized in the composition to provide a rapidly formed stable emulsion. Examples of preferred such emulsifiers are oxyalkylated phenols and the carboxylated salts thereof.

When the deposits being removed are contained on metal surfaces, a corrosion inhibitor such as a mixture of an alkylpyridine and dibutylthiourea is preferably included in the composition. Preferably, such corrosion inhibitor is present in the composition in an amount in the range of from about 0.05 to 0.6% by volume of the composition, most preferably in the range of from about 0.1 to 0.2% by volume.

In order to increase the rate of dissolution of organic materials by the liquid hydrocarbon solvent used, a coupling agent, preferably benzotriazole is also included in the composition in an amount in the range of from about 0.1 to 1.0% by wt of the composition.

Example: An emulsified composition of the process is prepared containing the following components:

Component	Percent
Water Aqueous solution of ethylenediaminetetra-acetic acid (40% by wt EDTA)	58.9% by volume
Ethoxylated castor oil (40 mols ethylene oxide)	10.0% by volume
Dibutylthiourea and an alkylpyridine mixture	0.1% by volume
Heavy aromatic naphtha	30.0% by volume
Benzotriazole	0.3% by weight

The composition is prepared by combining the ethylenediaminetetraacetic acid and corrosion inhibitor with water, combining the emulsifier with the heavy aromatic naphtha and then combining the aqueous phase with the hydrocarbon phase and agitating the mixture to form an emulsion. After the emulsion is formed the benzotriazole is combined with the emulsion.

A 2" specimen of an actual tube from a refinery powerformer having a deposit containing organic material and iron sulfide thereon, i.e., the outside surface of the tube is coated with a layer of deposit 3⁄16" thick and the inside surface is coated with a layer of deposit ⅛" thick, is placed into 200 ml of the abovedescribed emulsified composition. The composition and specimen are heated to 175°F and stirred. After 20 hr, the specimen is removed from the composition and visually inspected. The inspection shows that 80% of the deposit is removed from the specimen.

Glyoxylic Acid-Nonoxidizing Acid Solution for Sulfide Scale

Many sources of crude oil and natural gas contain high amounts of hydrogen sulfide. Refineries processing such crude oil or natural gas commonly end up with substantial amounts of sulfide-containing scale on the metal surfaces in contact with the crude oil or gas. This scale is detrimental to the efficient operation of heat exchangers, cooling towers, reaction vessels, transmission pipelines, furnaces, etc. Removal of this sulfide-containing scale has been a substantial problem because conventional acid-cleaning solutions react with the scale and produce gaseous hydrogen sulfide.

Hydrogen sulfide and acid-cleaning solutions containing hydrogen sulfide can cause severe corrosion problems on ferrous metals. The corrosion can be due to attack by acid and/or ferric ion corrosion.

An aqueous cleaning composition which is described by *G.R. Buske; U.S. Patent 4,289,639; September 15, 1981; assigned to The Dow Chemical Company* contains glyoxylic acid dissolved in an aqueous nonoxidizing acid. This cleaning solution is effective in removing acid-soluble, sulfide-containing scale from metal surfaces (e.g., refinery equipment) without the evolution of gaseous hydrogen sulfide.

Example: A solution of glyoxylic acid (7.5 g) and water (102 ml) was charged to a reaction vessel equipped with a gas scrubber containing 25% aqueous sodium hydroxide. The temperature of the glyoxylic acid/water solution was raised to

150°F in a water bath and iron sulfide (7.5 g) was then added. After the temperature of this mixture reached 150°F, 35 ml of concentrated (36%) hydrochloric acid was introduced and the vessel was quickly sealed. When the acid was first added, there was a brief initial smell of hydrogen sulfide but no detectable amount of hydrogen sulfide after that. Analysis of the sodium hydroxide scrubbing system using an Orion S= electrode gave a zero reading for sulfide. The cleaning solution dissolved all of the iron sulfide and the spent cleaning solution was a clear liquid without any noticeable amounts of solid precipitate. No evolution of hydrogen sulfide gas was observed during the 3 hr test.

In Situ Generation of Aldehyde

W.W. Frenier; U.S. Patent 4,310,435; January 12, 1982; assigned to The Dow Chemical Company has found that the method of chemically cleaning acid-soluble, sulfide-containing scale from a metal surface using an aqueous cleaning composition comprising an aqueous nonoxidizing acid having one or more aldehydes dissolved or dispersed therein is improved by generating the aldehyde in situ. An aqueous acid-cleaning composition comprising hydrochloric or sulfuric acid and hexamethylenetetramine is a preferred composition. The method is of use in refineries processing crude oil or natural gas containing high amounts of hydrogen sulfide.

In cleaning an item or vessel that will require circulation of the aqueous acid-cleaning solution (e.g., a contacting tower, heat exchanger or furnace), it is preferred to ascertain the circulation path within the system using water. The aldehyde generating chemical can then be added to the water after which concentrated acid can be added until the desired cleaning concentration is achieved. This represents a preferred embodiment for cleaning many, if not most, systems since the aldehyde is always present in excess and effective elimination of gaseous hydrogen sulfide is assured.

Example 1: A solution of hexamethylenetetramine (HMTA; 5.5 g) dissolved in 150 ml of 14% by wt aqueous hydrochloric acid was charged to a reaction vessel having a mechanical stirrer and a gas collecting means. Powdered iron sulfide (9.6 g) was added to the stirred aqueous acid solution. The mixture was then heated to 150°F and stirred for 3.5 hr. During this time, approximately 65% of the iron sulfide dissolved and 32 ml of gas collected. This gas was tested for hydrogen sulfide using lead acetate paper. The results were negative.

Example 2: A pipe specimen (1" x 4" x ¼") from a petroleum refinery furnace was obtained which was heavily fouled with a deposit containing iron sulfide. The specimen was placed in a 2 ℓ jacketed flask attached to a gas collecting means. The specimen was submerged in 600 ml of a solution containing 14% by wt aqueous hydrochloric acid and 20 g HMTA and the flask sealed. The contents of the vessel were then heated to 150°F and maintained at this temperature for 6 hr. No measurable amount of gas was observed or collected. The surface of the pipe specimen was essentially free of scale.

The "spent solvent" was analyzed in each of Examples 1 and 2 and found to contain 2.5 and 1.5 wt % dissolved iron, respectively.

Bis-Amides

J. Levy; U.S. Patent 4,344,861; August 17, 1982; assigned to UOP, Inc. provides a method of inhibiting the corrosion of metals in contact with petroleum and petroleum fractions other than gasoline that comprises contacting metal surfaces with a corrosion inhibiting amount of bis-amides resulting from reaction of 1 mol of amine with 1 eq of a dicarboxylic acid.

In protecting materials of construction in oil wells, concentrations of bis-amides from about 1 to 1,000 ppm, based on well liquids, may be beneficial. Preferably, a preliminary period at higher concentrations, from about 5 to 50 times the required steady state concentration, should be employed for several days at the beginning of the treatment.

In protecting refinery equipment, e.g., condensers and fractionating columns, the bis-amides may be used in concentrations generally from about 5 to 50 ppm. To prevent corrosion of condensers, the bis-amides may be injected as a solution in a suitable solvent at a point above that where condensation occurs. The non-volatile inhibitor then will run down the condenser, thereby protecting the metal surfaces from the action of corrosive agents. A preliminary treatment from about 5 to 50 times the steady state concentration for several days is desirable. Protection of metal surfaces in storage tanks may be obtained by applying a solution of bis-amides to such surfaces. Application may be by brushing, by spraying, or simply by introduction as a fog into the vapor space. The solution in kerosene may contain from 10 to 50% or more bis-amides, and may be applied in an amount equal to about 1 gallon solution per 500 to 5,000 square feet of metal surface to be protected.

Example 1: A 500 ml flask was fitted with a magnetic stirring bar and a reflux condenser attached to a Dean-Stark trap so that the pure xylene condensate would automatically be returned to the reaction flask. The flask was charged with 33.6 g N-oleyl-1,3-propylenediamine (0.1 mol), 28.5 g dimer acid (0.05 mol), obtained from Emery Industries as Empol 1018 (83% dimer, 17% trimer, and 2% monomer of 18 carbon atoms), and 60 g xylene. The mixture was heated to reflux with stirring until 2.1 ml of water was collected in the Dean-Stark trap. Total reaction time was about 7 hr. The reaction mixture was a clear, light yellow solution, and was stored and tested for activity without further treatment. Its total weight was 120 g and therefore consisted of a 50% solution of bis-amide.

Example 2: In a manner similar to that described in Example 1, an unsymmetrical bis-amide may be made by heating 24.5 g phenothiazine (0.25 mol), 42 g N-oleyl-1,3-propylenediamine (0.25 mol), 71.2 g dimer acid, Empol 1018 (0.25 mol), and 0.5 g p-toluenesulfonic acid, and 138 ml xylene. The total reaction time necessary to collect a theoretical amount of water was about 5 hr.

Example 3: Materials were tested for anticorrosion properties by NACE standard tests TM-01-72. This test, prepared by the National Association of Corrosion Engineers, is a modification of ASTM D-665. The method involves stirring a mixture of a petroleum product containing dissolved inhibitor with distilled water at 38°C for a definite time with a cylindrical steel specimen immersed in the solution. The test surface is then examined for rust. Experience has shown, according to NACE, that enough inhibitor present to produce B+ or B++ results by this test will control corrosion. The corrosion inhibiting properties of some of the products of this process are summarized in the following table. In all examples listed therein inhibition was tested at a concentration equivalent to 8 pounds inhibitor per 1,000 barrels petroleum, or about 33 ppm. Isooctane was used as the petroleum product.

Corrosion Inhibiting Properties of Bis-Amides

Amine	Acid	Rating
$C_{13}H_{25}O(CH_2)_3NH(CH_2)_3NH_2$	Empol 1018	B+
N-oleyl-1,3-propylenediamine	Empol 1018	B+
N-tallow-1,3-propylenediamine	Empol 1018	B++
N-oleyl-1,3-propylenediamine	Dimer*	B+

*Contains about 75% dimer and 25% trimer acids from monocarboxylic acid of 18 carbon atoms.

Vanadium-Containing Systems

It has been a long standing commercial practice to use aqueous alkanolamine solutions (e.g., a monoethanolamine solution) to absorb acidic acids such as CO_2, H_2S, COS and HCN to condition naturally occurring and synthetic gases. These treated gases may include feed synthesis gases, natural gas and flue gas. Frequently, the conditioning process is practiced by passing a 5 to 30% alkanolamine solution countercurrent to a gas stream in an absorption column to remove the acid gases. The absorbed acid gases may be later forced out of the conditioning solution at higher temperatures and the alkanolamine solution recycled for more absorbing.

Aqueous alkanolamine solutions are not themselves very corrosive toward ferrous metal equipment, however, they become highly corrosive when acid gases are dissolved therein, particularly when the solution is hot. It has been found that both general and local corrosive attack can occur. This is a particular problem in reboilers and heat exchangers where the steel is exposed to a hot, protonated alkanolamine solution. A heat transferring metal surface appears to be especially vunerable.

E.C.Y. Nieh; U.S. Patent 4,371,450; February 1, 1983; assigned to Texaco Inc. provides a corrosion inhibited composition consisting essentially of an aqueous alkanolamine solution employed in acid gas removal service and an inhibiting amount of a combination of an anion containing vanadium in the plus 4 or 5 valence state and an anion containing cobalt in the plus 2 valence state.

Example 1: In this example the equipment involved a set of copper strip corrosion test bombs that met ASTM D-130 specifications. The covers were modified with valves and dip tubes to allow sampling of the liquid phase when the vessel was pressurized due to autogenous pressures. A Teflon coupon mount was attached to the dip tube and a polypropylene liner was fitted to the vessel in a manner so that the test solution was not in direct contact with the body of the vessel.

In a typical experiment, 90 ml of a 50 wt % aqueous monoethanolamine was premixed with carbon dioxide, ammonium metavanadate and certain transition metal salts. The solution was placed in the liner of the vessel. A piece of 1020 mild steel coupon (1.48" x 0.41" x 0.12") with a 0.25" diameter hole for mounting was freshly polished with fine Emery cloth (#JB5R, Red-I-Cut carborundum), followed by rinsing with water and acetone.

The dried clean coupon was then weighed and attached to the Teflon mounting in a manner such that when the vessel was closed the coupon would be totally immersed in the test solution. The vessel was sealed and placed in a 115°±1°C shaker bath for a period of 96 hr. Then the coupon was recovered and cleaned by scrubbing with a bristle brush. When needed, a mild abrasive, Pumace FFF (Central Texas Chemical Co.), was employed for post-test cleaning. After the coupon was clean and dried, weight loss was determined. A series of such experiments provided the results listed in the following table for the corrosion inhibitor screening tests which showed that of the transition metals tested, cobalt noticeably reduced corrosion of the mild steel coupons. In the table MEA is monoethanolamine, low iron grade, <10 ppm Fe (Texas Chemical Co.); CO_2 is mol CO_2 per mol of MEA; Inhibitor A is nickel introduced as nickel nitrate, copper introduced as cupric nitrate, cobalt introduced as cobalt and zinc introduced as zinc nitrate; vanadium is vanadium introduced as ammonium metavanadate; and Corrosion Rate is a measurement of linear penetration in thousandths of an inch per year.

MEA, %	CO_2, mol/mol	Inhibitor A, ppm	Vanadium, ppm	Corrosion Rate, mpy
50.0	0.39	A = Ni, 100	0	170
50.0	0.39	A = Ni, 100	100	26
50.0	0.39	A = Cu, 100	0	39
50.0	0.39	A = Cu, 100	100	47
50.0	0.39	A = Co, 100	0	7
50.0	0.39	A = Co, 100	100	11
50.0	0.39	A = Zn, 100	0	65
50.0	0.39	A = Zn, 100	100	27
50.0	0.39	–	–	24

Example 2: The effectiveness of the cobalt-vanadium inhibitor system was further tested in a 50% aqueous monoethanolamine loaded with 0.39 mol of carbon dioxide per mol of amine reagent. To further increase the corrosiveness of the test, the bath temperature was increased to 120°C. Results of these tests indicated the combination of cobalt and vanadium provided protection to mild steel coupon while either cobalt or vanadium alone was not effective. The results are as follows.

	CO_2	Additives[a]			Post-test Analysis[b]			Cor-
MEA, %	mole/mole	CO, ppm	V, ppm	Fe, ppm	Co, ppm	V, ppm	Fe, ppm	rosion Rate
50.0	0.39	—	100	—	—	116	1134	57
50.0	0.39	—	200	—	—	211	578	22
50.0	0.39	—	300	—	—	296	506	18
50.0	0.39	100	—	—	82	—	1061	39
50.0	0.39	200	—	—	214	—	813	51
50.0	0.39	300	—	—	226	—	570	26
50.0	0.39	100	100	—	82	103	393	19
50.0	0.39	200	100	—	130	88	92	12
50.0	0.39	300	100	—	223	100	100	6
50.0	0.39	100	200	—	66	197	198	12
50.0	0.39	200	200	—	139	199	275	12
50.0	0.39	300	200	—	245	200	7	<1
50.0	0.39	300	200	40	261	163	44	<1
50.0	0.39	300	200	80	267	170	72	<1
50.0	0.39	300	200	120	265	173	109	<1
50.0	0.39	300	300	40	266	278	42	<1
50.0	0.39	300	300	80	266	269	77	<1
50.0	0.39	300	300	120	264	274	108	<1
50.0	0.39	300	400	40	269	388	44	<1
50.0	0.39	300	400	80	268	387	72	<1
50.0	0.39	300	400	120	272	391	108	<1

[a]Cobalt was introduced as cobalt nitrate, vanadium was introduced as ammonium metavanadate, and iron was introduced as aqueous solution of ferrous ammonium sulfate.

[b]By atomic absorption.

In this process of *E.C.Y. Nieh; U.S. Patent 4,372,873; February 8, 1983; assigned to Texaco Inc.* the corrosion inhibited composition consists essentially of an aqueous alkanolamine solution employed in acid gas removal service and an inhibiting amount of a combination of an anion containing vanadium in the plus 4 or 5 valence state and an amine selected from the group consisting of N-aminoethylethanolamine, ethylenediamine, propylenediamine, piperazine, N-aminoethylpiperazine, methyliminobispropylamine and lower alkyl and N-hydroxyalkyl substituted derivatives thereof.

Example 1: Tests were conducted as described in Example 1 of the previous patent. In a typical experiment, 90 ml of a 50 wt % aqueous monoethanolamine was premixed with carbon dioxide, ammonium metavanadate and N-aminoethylethanolamine. A series of such experiments provided the results listed in the following table.

Monoethanolamine[a], wt. %	CO_2[b] m/m	AEEA[c], wt. %	V[d], ppm	Corrosion Rate Mils per year[e]
50.0	0.30	0	0	45
50.0	0.30	1.0	0	22
50.0	0.30	0	100	60
50.0	0.30	0	200	<1
50.0	0.30	1.0	100	<1
50.0	0.30	1.0	100	<1
50.0	0.39	0	0	24
50.0	0.39	0.87	0	30
50.0	0.39	0.87	0	22
50.0	0.39	0	200	30
50.0	0.39	0	300	11
50.0	0.39	0.87	100	<1
50.0	0.39	0.87	200	<1
50.0	0.39	0.87	300	<1

[a]Monoethanolamine, low iron grade, <10 ppm Fe (Texaco Chemical Co.).
[b]Mol CO_2 per mol of MEA.
[c]N-aminoethylethanolamine (Aldrich Chemical Co.).
[d]Introduced as ammonium metavanadate.
[e]The corrosion rate is a measurement of linear penetration in thousandths of an inch per year.

The surprising fact about Example 1 is that the amine coinhibitor, AEEA, has been found by other investigators to increase the corrosion rate of ferrous metal surfaces under heat transfer conditions. For instance, see the first column of U.S. Patents 3,808,140, 3,896,044 and 3,959,170 and *Gas Purification* by Fred C. Riesenfeld and Arthur L. Kohl, Houston: Gulf Publishing Co., 1974, p. 85.

Example 2: Combinations of ethylenediamine (EDA) and vanadium inhibitors were tested in the manner described in Example 1. The results are shown in the following table.

Monoethanolamine, wt. %	CO_2, m/m	EDA[a], wt. %	V, ppm	Corrosion Rate Mils per Year
50.0	0.39	0	0	24
50.0	0.39	0	200	30
50.0	0.39	0	200	21
50.0	0.39	0	300	13
50.0	0.39	0.87	0	22
50.0	0.39	0.87	100	15
50.0	0.39	0.87	200	<1
50.0	0.39	0.87	300	<1

[a]Ethylenediamine (Texaco Chemical Co.).

–7–

Metal Treating Baths

SURFACE CLEANING

Acid-Wool Fat Emulsion

B.E. Svenson; U.S. Patent 4,289,638; September 15, 1981 describes a rust treatment and inhibitor preparation which includes a strong acid, such as concentrated phosphoric acid within an emulsion including wool fat or grease and derivatives. Solvents, stabilizers and absorbing agents are provided in the emulsion. The emulsion enables the acid to remove rust and protect the treated surface from further corrosion without affecting the surface finish.

One particularly preferred preparation is composed of the following components: 33⅓% v/v concentrated phosphoric acid, 25% w/v commercial anhydrous wool grease, 10% w/v light or colloidal kaolin, and methylated or industrial spirits to 100% vol.

The procedure for the production of the rust treatment prepared according to the process may comprise the steps of adding concentrated phosphoric acid and methylated or industrial spirits to wool fat, wool grease and/or derivatives thereof, the resultant heat of reaction being sufficient to melt the wool fat, wool grease and/or derivatives thereof, and adding kaolin with constant stirring to the reaction mixture. The mixture may be added to a homogenizer, preferably while it is still warm, so as to further stabilize the emulsion. Finally the mixture is allowed to cool.

To facilitate the melting of the wool fat, wool grease and/or derivatives thereof, heat may be applied (the heat from a hot water bath being quite adequate). However, as the addition of strong acid to methylated spirits creates enough heat to melt the wool fat, wool grease and/or derivatives thereof, this step is generally unnecessary.

Alternatively the wool fat (et al.) is placed in a reaction vessel and softened to a pastelike consistency by applying heat (e.g., obtained from a hot water bath). The concentrated phosphoric acid is added and stirred into the wool fat. Kaolin in the form of a suspension in the methylated or industrial spirits is added with constant stirring to the resultant mixture. The subsequent heat of reaction enables any previously undissolved wool fat to completely dissolve. Detergent diluted in a relatively small volume of hot water is added with continued stirring. The reaction mixture is diluted to volume with hot water and cooled.

Poly(Oxyalkylated) 1,3,4-Thiadiazoles

The process of *E.F. Rothgery; U.S. Patent 4,306,988; December 22, 1981; assigned to Olin Corporation* is directed to the use of poly(oxyalkylated) 2,5-dimercapto-1,3,4-thiadiazoles of the formula:

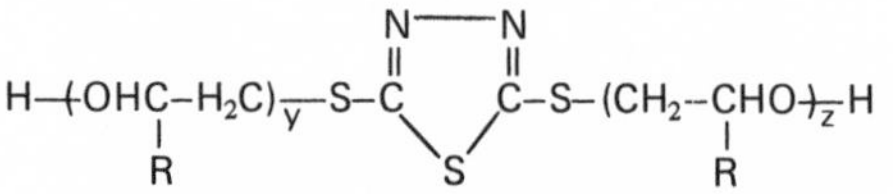

```
                         N———N
                         ||   ||
H—(OHC–H2C)y—S–C     C–S–(CH2–CHO)z—H
      |                  \   /               |
      R                   S                  R
```

where each R is individually selected from H and CH_3; and the sum of y and z is from 2 to about 30, as corrosion inhibitors, particularly in acid metal-treating baths.

Such acidic solutions include mineral acid solutions such as sulfuric acid, hydrochloric acid, or the like. These acidic solutions may be used for acid-pickling baths for the surface cleaning of metals or in similar processes. The preferred amount of this corrosion inhibitor in such acid solutions is preferably at least 0.005% by wt of the solution; more preferably, from about 0.01 to 0.5% by wt of the solution or bath.

In the following examples all parts and percentages are by weight unless otherwise indicated.

Example 1: *2,5-Dimercapto-1,3,4-thiadiazole·6 ethylene oxide* – 2,5-Dimercapto-1,3,4-thiadiazole (15 g; 0.1 mol) was placed in a flask fitted with a dry-ice condenser, thermometer and a dropping funnel with 0.5 g of its disodium salt as catalyst and heated to 90°C in an oil bath. Liquid ethylene oxide was dropped in causing an exothermic reaction and forming a yellow liquid product. The temperature was held at 120° to 130°C while 66 g (1.5 mols) of EO was added. After heating 4 hr, the unreacted EO was swept out in a N_2 stream and 35.5 g of semisolid product collected with an elemental analysis as follows. Calculated for DMTD·6EO: C, 40.63%; H, 6.32%; N, 6.77%; S, 23.24%. Found: C, 39.98%; H, 6.01%; N, 7.15%; S, 23.92%.

Example 2: *2,5-Dimercapto-1,3,4-thiadiazole·12 ethylene oxide* – Dimercapto-1,3,4-thiadiazole (37.5 g; 0.25 mol) was dissolved in 200 ml of dimethylformamide (DMF) in a 3-neck flask as above. KOH (1.2 g) was added and the mixture heated to 116°C while dropping in 44 g of EO over 3 hr to give a yellow solution. The entire mixture was transferred to a pressure vessel. An additional 100 ml of DMF and 2.5 g of KOH was added. After purging with N_2, the vessel was heated to 90° to 95°C while adding 176 g (4 mols) of EO over 1.5 hr. The mixture was post-reacted 1 hr at 95°C. The mixture was transferred to a flask and the solvent and unreacted EO was removed under vacuum to give 254 g of product, a viscous brown liquid with an elemental analysis as follows. Calculated for DMTD·12EO:

C, 46.06%; H, 7.14%; N, 4.13%; S, 14.19%. Found: C, 45.64%; H, 8.19%; N, 3.66%; S, 8.77%.

Example 3: *2,5-Dimercapto-1,3,4-thiadiazole·4 propylene oxide* – A pressure vessel was charged with 75 g (0.5 mol) of 2,5-dimercapto-1,3,4-thiadiazole, 4.5 g of potassium hydroxide and 300 ml of dioxane. The vessel was purged with nitrogen gas, heated to 108°C and 295 g (5 mols) of propylene oxide fed in over a 2 hr period. The temperature was held at 115°±5°C while the pressure varied from 5 to 40 psi. The mixture was post-reacted 1 hr to give a turbid orange solution. Removal of the solvent and unreacted PO under vacuum left 194 g of product with an elemental analysis as follows. Calculated for DMTD·4PO: C, 43.98%; H, 6.81%; N, 7.33%; S, 25.13%. Found: C, 43.81%; H, 6.72%; N, 7.17%; S, 24.76%.

Example 4: *2,5-Dimercapto-1,3,4-thiadiazole·6 propylene oxide* – Dimethylformamide (200 ml) was placed in a 3-neck flask with dimercapto-1,3,4-thiadiazole (37.5 g; 0.25 mol). Potassium hydroxide (1.4 g) was added and the mixture heated to 110°C. Propylene oxide (58 g; 1 mol) was added over 2 hr. This reaction mixture was poured into a pressure vessel with an additional 100 ml of DMF and 3.5 g KOH. The vessel was sealed, purged with N_2 and heated to 90°C while adding 232 g (4 mols) of PO over 1.5 hr. The mixture was post-reacted 1 hr at 90°C. The mixture was transferred to a flask and the solvent and unreacted PO were removed under vacuum leaving 126 g of product, a brown liquid with an elemental analysis as follows. Calculated for DMTD·6PO: C, 47.78%; H, 7.62%; N, 5.57%; S, 19.13%. Found: C, 46.63%; H, 7.78%; N, 7.80%; S, 15.75%.

Example 5: *2,5-Dimercapto-1,3,4-thiadiazole·5 propylene oxide·6 ethylene oxide* –2,5-Dimercapto-1,3,4-thiadiazole (75 g; 0.5 mol) was placed in a pressure vessel with 300 ml of dimethylformamide and 2.5 g of sodium methylate. After purging with N_2 174 g (3 mols) of propylene oxide were added over 1 hr at a temperature of 88°C while the pressure varied from 20 to 60 psi. The mixture was post-reacted 2.5 hr at 80° to 85°C and 35 psi. Ethylene oxide (132 g; 3 mols) was then added over 30 min at 80°C with a pressure of 15 psi and post-reacted at 90°C and a pressure of 15 to 60 psi.

The reaction mixture was neutralized with acetic acid, filtered and the solvent and unreacted PO and EO were removed under vacuum to give 241 g of orange-brown liquid with an elemental analysis as follows. Calculated for DMTD·5PO·6EO: C, 49.13%; H, 8.81%; N, 3.95%; S, 13.37%. Found: C, 46.30%; H, 7.60%; N, 5.33%; S, 14.21%.

Example 6: Stock solutions of the inhibitors were made by weighing 10 g of the test compound into a 100 ml volumetric flask and filling with either ethanol or dimethylformamide to give a 10% w/v concentrate. This concentrate was pipetted into weighed 500 ml flasks and acid added to give 500 g of solution of required inhibitor concentration.

Coupons measuring about 76 x 18 x 1.5 mm made from 1010 steel were sanded with 240 grit paper, degreased in acetone and weighed after drying. After the tests, the coupons were scrubbed with pumice, rinsed with water, acetone and reweighed after drying. All runs were made in duplicate and average values reported. The tests were run in glass flasks and the coupons were suspended from glass rods, two per flask.

The tests at ambient temperature (75°±3°F) were run for 24 hr in 10% hydrochloric acid and in 20% sulfuric acid. The high temperature tests, 190°±1°F, were run for 3 hr in 10% hydrochloric acid and in 5% sulfuric acid.

In Tables 1 through 4 below, the corrosion rate (in mils per year - MPY) and the percent inhibition (% I) are given. The symbols used in the tables are as follows: DMTD = 2,5-dimercapto-1,3,4-thiadiazole; EO = ethylene oxide; and PO = propylene oxide.

TABLE 1

	CORROSION RATE IN MPY IN 10% HCl AT 75° F.					
Compound	1000 ppm	(% I)	500 ppm	(% I)	250 ppm	(% I)
Blank	463		463		463	
DMTD . 6EO	9.5	(97.9)	9.4	(98.0)		
DMTD . 12 EO	27.3	(94.1)	27.9	(94.0)	--	
DMTD . 4PO	18.4	(96.6)	20.8	(95.5)	—	
DMTD . 6PO	11.3	(97.6)	10.2	(97.68)	11.6	(97.5)
DMTD. 5PO . 6EO	12.1	(97.4)	11.2	(97.6)		

TABLE 2

	CORROSION RATE IN MPY IN 20% H_2SO_4 AT 75° F.			
Compound	500 ppm	(% I)	500 ppm	(% I)
Blank	241.5		241.5	
DMTD . 6EO	5.1	(97.9)	5.0	(97.9)
DMTD . 12 EO	12.3	(94.9)	11.8	(95.1)
DMTD . 4PO	5.4	(97.8)	5.1	(97.9)
DMTD . 6PO	4.6	(98.1)	4.9	(98.0)
DMTD. 5PO . 6EO	5.9	(97.6)	5.0	(97.9)

TABLE 3

	CORROSION RATE IN MPY IN 10% HCl AT 190° F.	
Compound	1000 ppm	(% I)
Blank	57,982	
DMTD . 6EO	1,499	(97.4)
DMTD . 12EO	2,637	(95.5)
DMTD . 4PO	21,740	(62.5)
DMTD . 6PO	5,788	(90.0)
DMTD . 5PO . 6EO	3,445	(94.1)

TABLE 4

	CORROSION RATE IN MPY IN 5% H_2SO_4 AT 190° F.	
Compound	1000 ppm	(% I)
Blank	14,079	
DMTD . 6EO	214	(98.5)
DMTD . 12EO	611	(95.7)
DMTD . 4PO	1,041	(92.6)
DMTD . 6PO	570	(96.0)
DMTD . 5PO . 6EO	328	(97.7)

In this process of *E.F. Rothgery; U.S. Patents 4,329,475; May 11, 1982; and 4,349,458; September 14, 1982; both assigned to Olin Corporation* the corrosion inhibitors for acid metal treating baths are poly(oxyalkylated) 2-amino-5-mercapto-1,3,4-thiadiazoles. The compounds have the formula:

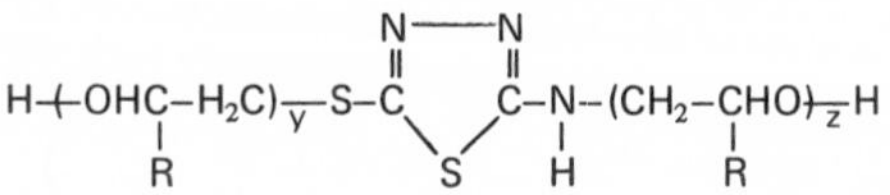

$$H\!\left(OCH(R)\!-\!H_2C\right)_y\!-\!S\!-\!C\overset{N\!-\!N}{\underset{S}{\diagdown\,\diagup}}C\!-\!N(H)\!-\!\left(CH_2\!-\!CH(R)O\right)_z\!H$$

where each R is individually selected from H and CH_3; and the sum of y and z is from 2 to about 30.

Example 1: *2-Amino-5-mercapto-1,3,4-thiadiazole·3 ethylene oxide* – 2-Amino-5-mercapto-1,3,4-thiadiazole (13.3 g; 0.1 mol) was placed in 100 ml of DMF with 0.2 g of NaOH and heated to 80° to 100°C. Ethylene oxide (69.6 g; 1.6 mols) was slowly dropped in over 3 hr. The mixture was cooled and the solvent and unreacted EO removed under vacuum to give 30 g of orange liquid with an elemental analysis as follows. Calculated for AMTD·3EO: C, 36.23%; H, 5.70%; N, 15.84%; S, 24.23%. Found: C, 36.90%; H, 5.91%; N, 15.81%; S, 19.43%.

Example 2: *2-Amino-5-mercapto-1,3,4-thiadiazole·8 ethylene oxide* – 2-Amino-5-mercapto-1,3,4-thiadiazole (53.2 g; 0.4 mol), KOH (3.4 g) and DMF (300 ml) were charged in a pressure vessel. After purging with N_2, it was heated at 95° to 100°C while adding 176 g (4 mols) of EO over 1 hr. The mixture was post-reacted 3 hr at 100°C. Removal of the solvent and unreacted EO in vacuo left 194.6 g of viscous brown liquid with an elemental analysis as follows. Calculated for AMTD·8EO: C, 44.67%; H, 7.09%; N, 8.68%; S, 13.22%. Found: C, 42.85%; H, 7.16%; N, 8.57%; S, 10.41%.

Stock solutions were prepared and tests run as described in Example 6 of the above patent. Results are shown in the following tables. In the tables AMTD = 2-amino-5-mercapto-1,3,4-thiadiazole and EO = ethylene oxide.

TABLE 1

CORROSION RATE IN MPY IN 10% HCl AT 75° F.

Compound	1000 ppm	(% I)	500 ppm	(% I)
Blank	463		463	
AMTD . 3EO	29.4	(93.7)	28.2	(93.9)
AMTD . 8EO	28.6	(93.8)	26.0	(94.4)

TABLE 2

CORROSION RATE IN MPY IN 20% H_2SO_4 AT 75° F.

Compound	500 ppm	(% I)	250 ppm	(% I)
Blank	241.5			
AMTD . 3EO	15.7	(93.5)	13.8	(94.3)
AMTD . 8EO	12.9	(94.7)	11.8	(95.1)

TABLE 3

CORROSION RATE IN MPY IN 10% HCl AT 190° F.

Compound	1000 ppm	(% I)
Blank	57,982	
AMTD . 3EO	3,637	(93.7)
AMTD . 8EO	5,310	(90.8)

TABLE 4

CORROSION RATE IN MPY IN 5% H_2SO_4 AT 190° F.

Compound	1000 ppm	(% I)
Blank	14,079	
AMTD . 3EO	450	(96.8)
AMTD . 8EO	737	(94.8)

Synergistic Mixture for Aluminum Protection

It has been found by *B.D. Oakes; U.S. Patent 4,370,256; January 25, 1983; assigned to The Dow Chemical Company* that a synergistic reduction in the corrosion of aluminum in acid media is obtained when various known nitrogen containing or dextrin corrosion inhibitors are employed in combination with certain anionic surfactants. The corrosion inhibiting composition consists of (1) a suitable corrosion inhibitor selected from the group consisting of:

(A) Heterocyclic, nitrogen-containing aryl, alkyl, and alkylaryl monocyclic and polycyclic compounds, with from 5 to about 14 carbon atoms which may contain one or more alkanolamine moieties;

(B) Alkanolamines with from 2 to about 14 carbon atoms;

(C) Thiourea and thioureas substituted with no more than about 12 carbon atoms;

(D) Dextrin; and

(E) A mixture of the foregoing A through D; and

(2) in admixture with an anionic sulfate or sulfonate surfactant selected from the group consisting of alkyl, aryl, and alkylaryl sulfates and sulfonates, and their alkali metal salts.

Thus the corrosion inhibiting composition of the process comprises two parts, an anionic surfactant, and a known corrosion inhibitor. While the precise amounts of each will vary with the particular components involved, the surfactant generally comprises from about 90 to 25% of the inhibitor composition, and usually from about 25 to 75%. The known corrosion inhibitor makes up the rest of the composition.

Of the amines it is preferred to use acridine, benzotriazole, ethanolamine, hexamethylenetetraamine, morpholine, pyrazine, pyrrole, and forms of these which are substituted with aliphatic moieties so long as the molecule contains a total of no more than about 14 carbon atoms. It is also preferred to use a complex amine made according to the method of U.S. Patent 3,077,454, which comprises allowing an amine to react with formaldehyde, acetophenone, an organic acid, and hydrochloric acid. Dextrin and forms of dextrin substituted with no more than 6 to 8 carbon atoms, may be used. Of the thioureas, particularly thiourea and para-tolyl thiourea may be used. Particularly effective surfactants are the sodium alkyaryl sulfonate such as sodium dodecylbenzene sulfonate, sodium salt of dodecylated sulfonated phenyl ether, and the sodium hydrocarbon sulfonate.

An acid cleaning composition may be made containing water, acid, a composition to inhibit corrosion of metals other than aluminum, and the abovedescribed corrosion inhibitor composition. In commercial practice it is preferred to make up the composition to contain about 20 wt % acid, and then dilute with water to obtain the desired cleaning concentration.

A particular preferred embodiment comprises 20% hydrochloric acid, 1% acridine and 2% of the sodium salt of dodecylated sulfonated phenyl ether. When diluted to contain 1% hydrochloric acid, it will have 0.05% acridine and 0.1% surfactant. Another embodiment would contain 20% hydrochloric acid, 6% of a corrosion inhibitor formulated according to U.S. Patent 3,077,454, and 2% sodium hydrocarbon sulfonate.

METALWORKING

Amine Salts of Monoaminoalkylene Dicarboxylic Acids

Rust-inhibiting compounds, especially for aqueous systems such as tool-lubricating emulsions for machine tools and which consist of amine salts of a number of monoaminoalkylene dicarboxylic salts are described by *E. Brandolese; U.S. Patent 4,273,664; June 16, 1981; assigned to Snamprogetti SpA, Italy.* These rust-inhibitors are used in combination with water and an alkanolamine.

The salts according to the process can be defined as the amine salts of the secondary monoamides and tertiary monoamides of the homolog series of the dicarboxylic acids having from 3 to 10 carbon atoms, wherein the two substituents bound to the amide nitrogen are selected from an alkyl radical having from 1 to 18 carbon atoms, a phenyl radical, a phenyl radical substituted by F, Cl, Br, NO_2 or an alcoholic radical having from 2 to 5 carbon atoms and the other substituent is selected from hydrogen, an alkyl radical having from 1 to 4 carbon atoms, an alcoholic radical having from 2 to 5 carbon atoms, an oxyethylene radical, an oxypropylene radical, or a mixed oxyethyleneoxypropylene radical having a degree of oxyethylation or oxypropylation, respectively, of from 1 to 20.

The salts are used in aqueous solution at a concentration of from 0.5 to 50% by wt, individually or in admixtures. The salts according to the process, especially when the acidic function of the amide is salified by mono-, di- or triethanolamine, are advantageously admixed, so as to improve the lubricating power of the composition, preferably with a water-soluble, oxyethylated-oxypropylated polyglycol at a concentration of from 0.1 to 20% by wt relative to the salt.

For the preparation of the salts, the acids and the amines can be used in the stoichiometric ratios, or, more particularly, an excess of amine can be employed (ratio of the amine to the acid between 1.5 and 3).

In practice, for reasons of cost and lower toxicity, there are used, preferentially, salts of mono-, di- and, above all, triethanolamine, it being generally preferred to operate with an excess of the amine base (usually 1 mol of acid for 2 mols of base).

Example 1: This is a composition containing 30% of the triethanolamine, 30% of water and 40% of rust-preventing agent composed by the triethanolamine salt of the N-ethanolanilido maleic acid having the formula:

$$C_6H_5\text{–}N(CH_2\text{–}CH_2OH)\text{–}\overset{\overset{\displaystyle O}{\|}}{C}\text{–}CH{=}CH\text{–}COOH$$

in which the ratio acid-to-base is 1 to 2.

The rust-inhibiting power of this composition, assessed with the IP 125 test, is such as to give no corrosion on iron at a dilution in water of 1 to 60 (IP 125 test positive to 1.5%). For IP 125 test it is intended herein to refer to the standard corrosion test prescribed in the specifications published by the Institute of Petroleum as No. 125.

Example 2: This is a composition containing 30% of triethanolamine, 30% of water and 40% of the rust-preventing triethanolamine salt of the N-(2-hydroxypropyl)anilido adipic acid having the formula:

$$C_6H_5{-}N(CH_2{-}CHOH{-}CH_3){-}\underset{\underset{O}{\|}}{C}{-}(CH_2)_4{-}COOH$$

in which the acid-to-base ratio is 1 to 2.

The rust-inhibiting power of this composition, evaluated with the IP 125 test, is such as not to give corrosion on iron at a dilution in water of 1 to 200 (IP 125 test positive to 0.5%).

Example 3: This is a composition containing 30% of triethanolamine, 30% of water and 40% of the rust-preventing triethanolamine salt of the N-(2-hydroxypropyl)anilido succinic acid having the formula:

$$C_6H_5{-}N(CH_2CHOHCH_3){-}\underset{\underset{O}{\|}}{C}{-}(CH_2)_2{-}COOH$$

in which the acid-to-base ratio is 1:2. This composition has an IP 125 test which is positive to 2%.

Organophosphonate Layer for Lubricated Cupreous Sheet

The mass production of small parts and articles of copper and copper alloys by individual or successive working and/or forming operations has long been known to require, among other items, the selection of the proper lubricant in order to prolong tool life and also to attain the desired unmarred surface appearance of the finished article. However, specialized lubricant compositions have at times been required in order to facilitate the working and forming operations, by effecting a sizeable reduction in friction. Quite often, these compositions must be removed from the workpiece promptly after the operation, in order to avoid undesired tarnishing or harmful corrosion effects.

Thus, need has arisen for simplification of the manufacturing procedures by elimination of the need for specialized lubricant compositions and the requirement for prompt and frequent cleaning operations.

According to *E.J. Caule; U.S. Patent 4,312,922; January 26, 1982; assigned to Olin Corporation* copper or copper alloy sheet or foil displaying improved forming and working properties is prepared by the provision over its surface of a thin coating containing a copper salt of an organophosphonic acid, and subsequently, an outer film of lubricant strongly retained by the coating is applied. To form the coating, the sheet or foil is immersed for a short time in an aqueous solution containing a phosphonic acid, rinsed and dried, the treatment being combined with or preceded by oxidation of the sheet surface, and the lubricant film is applied prior to forming operations, preferably as the final step in the production of the cupreous sheet.

The phosphonic acid, structurally an organic substitution product of phosphoric acid, is preferably one in which one of the three hydroxyl groups of phosphoric acid has been replaced by a monovalent hydrocarbon radical, which may be substituted or unsubstituted and which may be saturated or unsaturated, as by including ethylenic or carbonyl bonds. Generally, such organophosphonic acids or salts thereof are characterized as having a hydroxyl group of phosphoric acid replaced by a hydrocarbon radical, a carbon atom of which is linked directly to the phosphorus atom of the acid. Such radicals may include additional substituents which may display C to N or C to O linkages.

The treatment may be effected with an aqueous solution containing a low to moderate concentration of the phosphonic acid component or components, preferably ranging from about 0.1 to 30 vol % for liquid acids or corresponding weight percent limits for solid phosphonic acids, preferably in the range of about 0.1 to 40% by wt.

The treating solution also preferably includes a low to moderate concentration, such as about 0.1 to 15.0% by wt, preferably 0.2 to 5.0% by wt of oxidizing agent, such as sodium or other alkali chromate or dichromate, or nitric acid (100%) at a concentration of about 0.05 to 10.0 vol %, preferably about 0.05 to 2.0% by vol HNO_3. Other known oxidizing agents of similar activity may be used at a comparable dilute or moderate concentration effective for the purpose, but generally with avoidance of such vigorous oxidizing conditions as might cause substantial decomposition of the phosphonic acid.

Furthermore, it may at times be convenient to apply an oxidation step separately, prior to the treatment with the phosphonic acid component or components. Such procedure may be advisable, e.g., in instances where the treating solution shows signs of some instability, as by change in color, when stored in solution in the presence of oxidizing agent.

The outer lubricant film is applied by immersion in or spraying with the lubricant at room temperature or at a higher temperature, up to about 100°C, and removing excess lubricant by draining or wiping.

Thus-treated sheets are closely similar in appearance to initial samples before treatment. However, the presence of a substantially pore-free coating, transparent and invisible to the naked eye, was established by improvements effected thereby in a number of properties, even after prolonged storage in laboratory cabinets. In particular, treated samples remained untarnished after being subjected in tests to hydrogen sulfide vapor or to laboratory atmospheres containing other pollutants.

Molybdate Compound

M.S. Vukasovich and D.R. Robitaille; U.S. Patent 4,313,837; February 2, 1982; assigned to AMAX, Inc. provide a water-based metalworking fluid which inhibits corrosion of metal surfaces which the fluid contacts during a metalworking operation. The fluid contains dissolved therein a small but effective amount to inhibit corrosion of the metal surfaces by the fluid of a corrosion inhibitor consisting essentially of a molybdate compound and one or more compounds selected from the group consisting of nitrites, borates, alkanolamines, amine borates, amine salts of unsaturated fatty acids, alkanolamine sarcosinates, alkanolamine phosphates, and alkali, morpholine, and alkanolamine salts of arylsulfonamido carboxylic acids.

The process is applicable to metalworking fluids which are aqueous solutions, and is also applicable to metalworking fluids which are oil-water emulsions such as emulsions in water of about 1 to 45 wt % of an oil such as paraffinic or naphthenic mineral oil, sperm oil, lard oil, vegetable fats or oils, and fatty acid esters of animal or vegetable fats or oils, together with optional emulsifying agents such as petroleum sulfonates, amine soaps, rosin soaps, naphthenic acids, and optional lubricating agents such as sulfur, chlorine or phosphorus products available for this purpose.

The fluids are useful in inhibiting corrosion in metalworking operations of ferrous metals and nonferrous metals such as copper or cupreous metals, aluminum and zinc.

Particularly advantageous molybdate compounds are sodium molybdate or sodium molybdate dihydrate, which are available commercially and are readily soluble in water.

The water-based metalworking fluid can contain MoO_4^{-2} anion at a concentration from about 0.03 to 1.3 wt % or sodium molybdate dihydrate at from about 0.05 to 2.0 wt %.

The pH of the metalworking fluid of the process is generally in the range of about 8 to 10.

Example: The corrosivity of a given fluid formulation for iron was determined by a cast iron chips corrosion test which was specifically designed to evaluate the corrosion protection afforded by metalworking fluids. This test uses gray cast iron chips, produced by the dry shaper machining of a gray cast iron ingot, which were cleaned in acetone and then dried.

About 8 g of dry chips are placed in a 50 ml glass beaker and covered with 30 ml of the test fluid for 15 min. The fluid is then decanted and the wet chips are spread uniformly on a 9 cm Whatman No. 1 filter paper on a glass plate. The chips, paper and plate assembly are transferred to a closed cabinet maintained at a relative humidity of 80 to 90% at room temperature. After 4 hr, the wet chips are discarded and the filter paper is dried under ambient conditions.

The extent of rust spotting on the paper is used to measure the corrosivity of the fluid. A corrosivity rating of 1 is given for no rust spotting, a rating of 2 is given for slight spotting, and a rating of 3 is given for gross rusting of the chips. Tests are performed in triplicate for each fluid and results are averaged.

The fluids were made up by adding given amounts of one or more compounds to given amounts of synthetic low hardness water of high chloride and sulfate-concentrations, simulating a relatively highly corrosive water supply.

The following table correlates the amount(s) of corrosion inhibiting components with the corrosivity rating obtained in the cast iron chips corrosion test described above.

Compound	Concentration, Percent by Weight																				
Sodium molybdate dihydrate	0.5	–	0.3	–	–	0.2	–	0.2	–	0.5	–	0.5	–	0.5	–	–	0.5	–	–	0.3	0.5
Sodium nitrite	–	0.8	0.4	–	0.8	0.4	–	–	–	–	–	–	–	–	–	–	–	–	–	–	–
Triethanolamine	–	–	–	0.4	0.4	0.2	–	–	–	–	–	–	–	–	–	1.0	0.5	–	1.0	0.3	0.3
Sodium tetraborate decahydrate	–	–	–	–	–	–	1.0	1.0	–	–	–	–	–	–	–	–	–	–	–	–	–
Monoethanolamine borate	–	–	–	–	–	–	–	–	1.0	0.5	–	–	–	–	–	–	–	–	–	–	–
Triethanolamine oleate	–	–	–	–	–	–	–	–	–	–	1.0	0.5	–	–	–	–	–	–	–	–	–
Triethanolamine-N-lauroyl sarcosinate	–	–	–	–	–	–	–	–	–	–	–	–	1.0	0.5	–	–	–	–	–	–	–
Triethanolamine phosphate	–	–	–	–	–	–	–	–	–	–	–	–	–	–	1.0	1.0	0.5	–	–	–	–
Sodium benzenesulfonamido acetate	–	–	–	–	–	–	–	–	–	–	–	–	–	–	–	–	–	1.0	1.0	0.5	–
Corrosivity Rating	3	2	1	3	2	1	3	2	3	1	3	2	3	2	3	3	1	3	3	1	2

Reducing Cobalt Leaching from Tools

A large percentage of industrial cutting tools used to drill, cut, grind, and mill metals are made of tungsten carbide particles held together by a cobalt bonding agent. In a few instances the bonding agent may be nickel or platinum. The drilling, cutting, milling, or other metalworking step requires the application of a liquid coolant or lubricant at the area of contact between the metal surface being machined and the drilling, cutting or milling tool.

Although water or mineral oil can be used alone as a coolant or lubricant the practice has been to add compounds which increase the lubricity and cooling ability of the liquid and which delay its deterioration. These added compounds often, however, contain sequestering (chelating) agents and moieties.

It has been observed that when cutting, drilling or milling tools which are made up of tungsten carbide particles bonded with cobalt or nickel metal are exposed to these cutting fluids containing chelating agents, the cobalt or nickel is leached away. Leaching of the cobalt or nickel matrix from the tool leaves a residue of carbide particles and results in premature failure of the tool.

The process of *J.M. McChesney and P.E. Landers; U.S. Patent 4,315,889; Feb. 16, 1982; assigned to Ashland Oil, Inc.* comprises contacting with a solution of a triazole or thiadiazole compound a cobalt or nickel surface exposed to leaching action by a liquid agent. Ordinarily the triazole or thiadiazole compound will be contained in the liquid leaching agent.

The triazole compounds utilized in this process are benzene triazole and tolyl triazole. The thiadiazole compounds are 2,1,3-benzothiadiazole, 2,5-dimercapto-1,3,4-thiadiazole, the disodium salt thereof, and di(triethanolammonium)dimercaptothiadiazole.

Example 1: *Oil-base coolant and lubricant* – An oil-base coolant and lubricant was prepared by mixing the following compositions:

	Parts by Weight
Mineral oil	81
2,1,3-Benzothiadiazole	3
Extreme pressure additive*	10
Rust preventative **	3
Mist suppressant	3

*Mixture of chlorinated paraffins and sulfurized fats
**Tectyl 477; Ashland Oil, Inc.

Example 2: *Oil-in-water-emulsion concentrate and coolant and lubricant composition* – A concentrate was prepared by mixing the following components:

	Parts by Weight
Sodium sulfonate	18
Oleic acid	1
Triethanolamine	1
Bactericide*	1
Antifoaming agent**	1
Extreme pressure agent***	10
Tolyltriazole	3
Mineral oil	65

*IMC Chemical Group, Inc.
**Napco NDW; Diamond Shamrock, Inc.
***Chlorinated paraffins

This concentrate was then used to prepare a metalworking coolant and lubricant by diluting it in a ratio of 3 parts by weight of concentrate to 97 parts of water.

Example 3: *Water-base concentrate and coolant and lubricant composition* – A concentrate was prepared by mixing the following components:

	Parts by Weight
Boric acid	6
Diethanolamine	12
Arylsulfonamidocarboxylic acid*	4
Bactericide**	1
Antifoaming agent***	1
Sodium gluconate	2
Tolyltriazole	5
Polyglycol (lubricating additive)†	64

*Hostacar H Liquid; American Hoechst Corporation, having a specific gravity of 1.17±0.05, solidification point of -16°C and acid number of 161±10
**IMC Chemicals Group, Inc.
***Napco NDW; Diamond Shamrock, Inc.
†This component is optional but its use increases lubricity

This concentrate was diluted in a ratio of 3 parts of concentrate to 97 parts of water to form a water-base metalworking coolant and lubricant.

Tests of metalworking coolant and lubricant compositions in each of the preceding examples were conducted. The resistance of specimens of tungsten carbide particles bonded together with cobalt to leaching by the liquid coolants and lubricants was observed to be substantially reduced.

BURNISHING AND POLISHING

Benzotriazole in Aluminum Polishing Composition

T.R. Rooney; U.S. Patent 4,251,384; February 17, 1981; assigned to Albright & Wilson Ltd., England has found that certain aromatic organic compounds have a beneficial effect in reducing the occurrence of transfer etch in aluminum polishing solutions. The presence of such etch inhibitors therefore permits the proportion of sulfuric acid in an aluminum polishing solution to be substantially increased.

The process therefore provides an aluminum polishing solution comprising phosphoric acid, nitric acid, sulfuric acid and dissolved copper, which additionally comprises as an etch inhibitor, an organic compound comprising an aromatic ring having at least two hetero atoms conjugated therewith.

Preferably the etch inhibitor has a benzene or benzo ring fused to a heterocyclic ring, e.g., a five-membered heterocyclic ring, e.g., benztriazole is particularly effective. Substituted benztriazoles in which the benzene nucleus is substituted with, e.g., hydroxyl, alkoxy, amino, nitro or alkyl groups are also operative as are halo-substituted benztriazoles.

The etch inhibitor is preferably present in a proportion of from 0.05% by wt up to 0.7% or higher.

Example: A chemical polishing solution was prepared containing 45% w/w H_3PO_4 (1.75 SG), 50% w/w H_2SO_4 (1.84 SG), 1.5% w/w diammonium phosphate, 0.25% w/w copper sulfate, 2% nitric acid (1.50 SG), the rest being water. The bath was aged to 30 g/ℓ Al by dissolving aluminum and the nitric acid content readjusted to 2% w/w. Components of HE9 alloy and BA211 bright trim alloy were polished in this bath for 3 min at 100°C and subjected to various drainage times before rinsing in hot water. It was found that at drainage times greater than 10 sec a grey "transfer etch" appeared on the upper surface of components and could not be removed in 50% nitric acid desmutting solution.

To the above polishing solution 3 g/ℓ benztriazole was added and the tests carried out again. Transfer etch appeared only after a drainage time of 25 to 30 sec, in contrast to the above solution without benztriazole. The solution was used for polishing until the aluminum content rose to 35 g/ℓ and a further 2 g/ℓ benztriazole was added. This solution continued to give good results and no loss of benztriazole could be detected. The solution was maintained in the usual way by adding fresh polishing solution and nitric acid as required. The replenishing solution contained 5 g/ℓ benztriazole.

Rust and Discoloration Retarding Burnishing Composition

The essential component, hereinafter sometimes referred to as Component X, of the compositions of this process of *B.E. Unzens; U.S. Patent 4,367,092; Jan. 4, 1983; assigned to Roto-Finish Company, Inc.* is a combination of the compounds 6-[methyl(phenylsulfonyl)amino] hexanoic acid and 2,2',2"-nitrilotris(ethanol), preferably but not necessarily in proportions of approximately one to one having the approximate empirical formula $C_{13}H_{19}NO_4S{\cdot}C_6H_{15}NO_3$. This Component X constitutes the essential agent for retarding discoloration and inhibiting rust formation in any finishing operation involving either the employment of steel burnishing balls or the finishing of ferrous or nonferrous metal parts or workpieces.

The process is particularly advantageous when employed for that particular area of finishing known as burnishing. According to this aspect of the finishing field, case-hardened or other steel burnishing balls are tumbled, vibrated, or rotated together with the parts to be finished for purposes of imparting a bright luster and smooth surface finish to the parts involved.

Preferred acidic burnishing compounds comprise the following essential ingredients and optional ingredients: essential ingredients—water, organic acid, amphoteric surfactant, Component X; optional ingredients, one or more of which may be included—detergent, wetting agent, chelating agent; and preferred optional ingredient—organic acid salt, preferably an organic acid alkali metal salt, such as the sodium or potassium salt.

Following are a list of terms used in the examples:

Mafo 13 Mod I; Supplier: Mazer Chemical; Ingredients: potassium salt complex N-stearyl amino acid; Type of component: amphoteric surfactant.

Surfynol 104H; Supplier: Air Products Co.; Ingredients: 75% solution of tetramethyl decynediol in ethylene glycol; Type of component: wetting agent.

Armohib 31; Supplier: Armak; Ingredients: aliphatic nitrogen-containing compound; Type of component: acid inhibitor.

LCI 815; Supplier: Lisle Chemical; Ingredients: approximately 1:1 mixture of 6-[methyl(phenylsulfonyl)amino] hexanoic acid and 2,2',2"-nitrilotris(ethanol); Type of component: sold as replacement for sodium nitrite inhibitor, normally used and sold for use in alkaline solutions. This material is designated "Component X."

Example 1: *XL-440 formula –*

	Percent by Weight
Water	70
Citric acid	12
Sodium citrate	12
Mafo 13 Mod I	5
LCI 815 (Component X)	1

The foregoing formula, because of the addition of Component X, has been found to reduce the attack of the organic acid on case-hardened steel. If the Component X is used at a 1 to 2% concentration by weight, it prevents the steel media from rusting for a period of up to one week, depending upon ambient conditions. This, of course, includes not only the steel burnishing balls which are employed during the burnishing process, but also the material being burnished. It has also been observed that this formula imparts a brighter finish to nonferrous metals than is possible using a nitrogen based acid inhibitor, of the type commonly employed in acidic burnishing compounds, such as Armohib 31. In actual tests, the burnishing compound set forth in the foregoing formulation has been found to impart a bright, lustrous finish to both nonferrous and ferrous metals, employing case-hardened steel burnishing balls as the burnishing medium.

Example 2: *Comparative formulation employing usual acid inhibitor in acid burnishing compound–L-543 formula* – The comparative burnishing compound had the following formula, and included an aliphatic nitrogen type of acid inhibitor as is usual in such methods. The formulation was as follows:

	Percent by Weight
Water	69.9
Citric acid	12.0
Sodium citrate	12.0
Mafo 13 Mod I	5.0
Surfynol 104H	1.0
Armohib 31	0.1

Example 3: *Comparative test results and data* – For purposes of measuring corrosion-inhibition properties, solutions of 1% and 2% of the acidic burnishing compound of Example 1 were compared with solutions of the acidic burnishing compound of Example 2 at the same concentrations. To each concentration, samples of mild steel and case-hardened steel media were subjected for a period of 24 hr by immersion. They were thereafter measured for a weight loss. The results are as follows:

	Burnishing Compound of			
Results	Example 2 at 1%	Example 2 at 2%	Example 1 at 1%	Example 1 at 2%
Steel media gram weight loss	0.0167	0.0354	–	0.0032
Steel media percent weight loss	0.012	0.024	–	0.0022
Mild steel coupon gram weight loss	0.0104	0.0094	0.0060	0.0060
Percent weight loss per hour	0.0394	0.0336	0.0216	0.0216

The steel samples and the case-hardened steel burnishing balls remained clean and bright with very little, if any, discoloration, during the foregoing reported tests employing the burnishing compound of Example 1.

In contrast, the steel samples and the case-hardened steel media became black during the experiments conducted employing the acidic burnishing compound of Example 2.

In the latter case, it was observed that the burnishing operation carried out employing a 2% acidic burnishing compound of Example 2 were blacker and darker than when the 1% solution was employed.

STRIPPING RESINS AND ALLOYS

Removing Phenolic Resins from Aluminum

E.P. Cornwall and W. Batiuk; U.S. Patent 4,290,819; September 22, 1981; assigned to The Boeing Company provide a chemical composition and method of using the same which effectively removes a phenolic resin-asbestos composite liner from aluminum and its alloys with a minimum of etching or corrosion of the alloy surface exposed to the solvent. In particular, the cleaning composition of the process is capable of breaking the strongly crosslinked bonds of the phenyl six-carbon ring forming phenolic resins, while substantially preserving the surface integrity of the metallic alloy substrate free from excessive intergranular attack and end grain pitting.

The composition used consists essentially of nitric acid, a hexavalent chromium chosen from the group of chromium trioxide and sodium dichromate and water in the proportions of 0.01 to 0.12 mol of the hexavalent chromium and 1 to 1.7 mols of H_2O per mol of HNO_3.

This composition is heated and maintained in the range of 110° to 180°F, preferably at about 160°F. The aluminum alloy suitably masked or covered where no coating of the phenolic resin composite liner is present, is immersed for about one hour. After immersion, the remaining phenolic resin composite liner material can usually be rinsed from the aluminum substrate with pressurized water. If necessary, the immersion process may be repeated to remove any remaining coating. If the phenolic resin composite liner has been sufficiently removed, the aluminum alloy can be given a final water cleaning and then dried.

Example 1: A scrapped rocket booster case was obtained. The engine nozzles and ancillary holes were sealed with an over-cured silicone compound (STB 5-82 Dow Corning Silastic E). The exterior was covered with four coats, approximately 0.012 to 0.020 inch of a chemical milling maskant such as Adcoat 828, a solvent-based vinyl composition used for masking in chemical milling operations.

The solvent solution used was made up in 10-gallon quantities by heating 7 quarts of water to 150° to 160°F. Eight pounds, 5 ounces of chromium trioxide (CrO_3) were slowly added and stirred until completely dissolved. Eight gallons of technical grade 70% nitric acid were then added with constant, vigorous stirring. The solvent area was well ventilated and protective equipment was used to prevent injuries. The solution thus prepared was heated to 160°F and poured into the booster case, suspended inside a small stainless steel tank for safety and handling reasons. Metal and phenolic test specimens were immersed in the solvent solution.

Exothermic reaction caused solution temperature to increase to 168°F; however, after 35 minutes the temperature started down. No external heat was supplied. After one hour, the temperature was 155°F, the solution was poured off and the booster case rinsed inside and out.

Only a few specks of phenolic asbestos liner compound remained plus the part of the insulating cap in the booster rocket nozzle sockets which had been covered by the silicone sealer. Five more minutes immersion would have removed all exposed liner material.

The following results were obtained on the test specimens immersed in the solvent solution during the foregoing test: phenolic etch rate, 2.2 mils/side/hour; aluminum etch rate, 0.18 mil/side/hour; and intergranular attack and pitting, none.

After ultrasonic determination of wall thickness the booster case was sawed in two lengthwise. A two-inch square test piece was removed and immersed in the working solution for one more hour.

The etch rates on the phenolic composite resin and aluminum were such that 0.005 inch of phenolic and up to 0.0004 inch of aluminum were removed.

Example 2: Two hundred grams of chromium trioxide were slowly added to one liter of 70% by volume of nitric acid. This solvent solution was stirred slowly until all the chromium trioxide was dissolved. The solvent solution was maintained at approximately 160°F during the entire process by both exothermic reaction and by the application of heat and/or cooling of the tank.

Small segments of 2014-T6 aluminum approximately 0.120 x 2 x 2 inches along with coupons of a phenolic resin composite approximately 0.016 x 1 x 2 inches were immersed in the solvent solution for one hour. Specimen weight loss was used to calculate etch rates. It was found that the solvent solution removed 3.7 mils of phenolic composite resin per side, per hour and removed 0.16 mil of aluminum per side, per hour. The ratio of phenolic to aluminum alloy removal was found to be 23 to 1 for this solvent solution.

Selective Stripping Solutions

Improved solutions are provided by *F.A. Brindisi, Jr., T.W. Bleeks and T.E. Sullivan; U.S. Patent 4,302,246; November 24, 1981; assigned to Enthone, Incorporated* for selectively stripping alloys containing nickel with gold, phosphorus or chromium from substrates formed from alloys containing iron with chromium and in some instances nickel, or nickel rich, chromium bearing alloys, some of which also contain iron. These solutions comprise concentrated nitric acid, at least one chloride salt, an organic corrosion inhibitor and, optionally, an organic surface active agent. An improved method utilizing the solution of the process is also provided and facilitates stripping of these alloys in substantially reduced time, without degradation of the underlying substrate.

The solution and method of the process have particular utility for selectively stripping gold/nickel brazing alloys used in jet engines, where minimal time and avoidance of degradation of the underlying substrate is of utmost importance and criticality.

The organic corrosion inhibitors used in the stripping solutions of the process are acetylenic alcohols and, preferably, 2-butyne 1,4-diol, exthoxylated butyne, 1,4-diol or propargyl alcohol. The organic corrosion inhibitor must be effective to inhibit degradation of the substrate by the stripping solutions of the process, in view of the fact that they contain strong oxidizing agents. Preferably, the organic corrosion inhibitor is provided in a concentration ranging from about

1 to 30 g/ℓ, based on the amount of nitric acid, and, most preferably, between about 1 to 3 g/ℓ. The corrosion inhibitors are believed to contribute not only to the operability of the stripping solution in substantially reduced times of exposure, without degradation or adverse effect upon the substrate, but they also allow for operation over a wider range of temperature and concentration of chloride ion.

Example 1: A working solution was prepared by admixture of the following: 1 ℓ concentrated nitric acid (70%, 42° Be); 19 g potassium chloride; 19 g ferric chloride; and 1.65 g 2-butyne 1,4 diol. This solution contained 21 g/ℓ of chloride ion. A specimen, comprising two strips of 410 Stainless Steel, each 1" x 1" x 0.0625", joined to form a "T" with a braze of gold/nickel alloy, containing 80% gold and 20% nickel, was immersed in the solution for 30 min. An ultrasonic transducer was used to administer ultrasonic agitation. The solution temperature ranged from 70° to 78°F (21.1° to 25.6°C), with an average temperature of 74°F (23.3°C). Inspection of the specimen after 30 min revealed 100% removal of the gold/nickel braze without attack or degradation of the specimen substrate.

Example 2: A working solution was prepared by admixture of the following: 1 ℓ concentrated nitric acid (70%, 42° Be); 19 g potassium chloride; 19 g ferric chloride; and 3.16 g ethoxylated butyne 1,4 diol. The ethoxylated butyne 1,4 diol was used comprised of butyne 1,4 diol ethoxylated with ethylene oxide in a ratio of ethylene oxide to butyne diol of 1.8:1.0. This working solution contained 21 g/ℓ of chloride ion. A specimen, comprising two strips of 410 Stainless Steel, each 1" x 1" x 0.0625", joined to form a "T" and brazed with a gold/nickel alloy containing 80% gold and 20% nickel, was immersed in the solution for 30 min. An ultrasonic transducer was used to administer ultrasonic agitation. The temperature ranged from 66° to 76°F (18.9° to 24.4°C), with an average temperature of 71°F (21.7°C). Inspection of the specimen after 30 min revealed 100% removal of the gold/nickel braze, without attack or degradation of the specimen substrate.

Example 3: A working solution was prepared by admixture of the following: 1 ℓ concentrated nitric acid (70%, 42° Be); 19 g potassium chloride; 19 g ferric chloride; and 1.07 g propargyl alcohol. This working solution also contained 21 g/ℓ of chloride ion. A specimen, similar to those used in Examples 1 and 2, was immersed in the solution for 30 min. Ultrasonic agitation was again administered and the temperature ranged from 70° to 74°F (21.1° to 23.3°C), with an average temperature of 74°F (23.3°C). Inspection of the specimen after 30 min revealed 98% removal of the gold/nickel braze, with no attack or degradation of the specimen substrate.

Example 4: For purposes of comparison, the following solution of nitric acid and hydrochloric acid was prepared to demonstrate the unsatisfactory performance of such solutions, due to their attack upon and degradation of the substrate alloy. 56 g of hydrochloric acid (37%, AR grade) was admixed with 1 ℓ of concentrated nitric acid (70°, 42° Be), yielding a chloride ion concentration of 20 g/ℓ. A specimen similar to those used in Examples 1 through 3 was immersed, with ultrasonic agitation, in this solution at a temperature ranging from 66° to 81°F (18.9° to 27.2°C). After 15 min, severe attack was observed on the specimen substrate. After a total of 45 min, while effectively all of the gold/nickel brazing alloy was removed, the specimen substrate was severely attacked and degraded.

Company Index

The company names listed below are given exactly as they appear in the patents, despite name changes, mergers and acquisitions which have, at times, resulted in the revision of a company name.

Inventor Index

U.S. Patent Number Index

NOTICE

Nothing contained in this Review shall be construed to constitute a permission or recommendation to practice any invention covered by any patent without a license from the patent owners. Further, neither the author nor the publisher assumes any liability with respect to the use of, or for damages resulting from the use of, any information, apparatus, method or process described in this Review.